RIDDLE OF THE FEATHERED DRAGONS

Yale

UNIVERSITY PRESS

New Haven & London

Alan Feduccia

RIDDLE OF
THE FEATHERED
DRAGONS

....

Hidden Birds of China

Published with assistance from the foundation established in
memory of Philip Hamilton McMillan of the Class of 1894, Yale
College.

Yale University Press books may be purchased in quantity for educational, business, or promotional use. For information, please
e-mail sales.press@yale.edu (US office) or sales@yaleup.co.uk
(UK office).

Designed by Lindsey Voskowsky. Set in Fournier MT type by
Newgen North America.
Printed in the United States of America.

Library of Congress Cataloging-in-Publication Data

Feduccia, Alan.
   Riddle of the feathered dragons : hidden birds of China / Alan
Feduccia.
      p. cm.
   Includes bibliographical references and index.
   ISBN 978-0-300-16435-0 (hardback)
   1. Birds, Fossil—China.   2. Birds—Evolution.   3. Evo-
lutionary paleobiology.   I. Title.
QE871.F43   2012
568—dc23   2011015862

A catalogue record for this book is available from the British
Library.

This paper meets the requirements of ANSI/NISO Z39.48–1992
(Permanence of Paper).

10 9 8 7 6 5 4 3 2 1

To Sir Gavin Rylands de Beer (1899–1972),
who cast a flood of light on the field of
avian evolution

The truth is rarely pure and never simple.

Oscar Wilde, The Importance of Being Earnest

CONTENTS

....

ACKNOWLEDGMENTS

Ideas on bird origins have been rapidly evolving since the great bonanza of Chinese Lower Cretaceous fossils began to be unearthed beginning in the early 1990s, so likewise the views expressed in this book metamorphosed in the hopes of conforming to the best explanations of the latest evidence; and therefore this book owes a debt to many people. For valuable discussion and input I thank David Burnham, Stephen Czerkas, Frietson Galis, Richard Hinchliffe, Lian-hai Hou, Frances James, Theagarten Lingham-Soliar, Paul Maderson, Larry Martin, Gerd Müller, Storrs Olson, John Pourtless, John Ruben, and Zhonghe Zhou. The entire manuscript was reviewed by Dominique Homberger, Lian-hai Hou, Frances James, and an anonymous paleontologist; and I particularly thank John Pourtless for suggestions and adroitly editing and making suggestions for all the chapters at an early stage. Additionally, chapters and sections were critically read by Walter Bock, David Burnham, Tom Kaye, Larry Martin, Ann Matthysse, Pavel Pevzner, David Pfennig, John Ruben, Steven Salzberg, and Zhonghe Zhou. Illustration contributions are individually acknowledged, but I would particularly like to thank Stephen Czerkas, who kindly permitted the use of many of his excellent drawings. I also owe a huge debt of gratitude to Susan Whitfield, artist-illustrator in the Department of Biology at the University of North Carolina, Chapel Hill, who formatted the figures and skillfully rendered many of the illustrations. The book was enthusiastically received by editor Jean Thomson Black, expertly formatted by Joyce Ippolito, and superbly edited to completion by Laura Jones Dooley.

THEME

How the euphoria of paleontologists at having solved
the century-and-a-half-old mystery of
bird origins and the origin of flight
was premature, generated by unfounded
confidence in a rigid methodology that excludes geologic time
and biological phenomena, some extravagant in nature;
and how we have yet to discover the true ancestry of birds

INTRODUCTION

The news media's feeding frenzy on the recent dinosaur bonanzas seems never ending, with publications reporting almost weekly "astounding" breakthroughs on the origin of birds ranging from deciphering the color of supposed dinosaur "protofeathers" to unearthing birds' actual ancestors. Under "*T. rex* Discoveries," *USA Today* reported on 29 December 2009 that "for a critter dead 65 million years, *T. rex* had a pretty good year as relatives turned up in study results almost weekly." Accompanying the article was an illustration of the small "ancestor" clothed in a pelt of presumed protofeathers. Just a year earlier, a "rex-chicken" nexus based on a dubious report in *Science* of preserved protein from the sixty-eight-million-year-old dino-icon was covered in the Associated Press: "It looks like chickens deserve more respect. Scientists are fleshing out the proof that today's broiler-fryer is descended from the mighty *Tyrannosaurus rex*"—not from a mere *Allosaurus* or *Alioramus* but from the 8-ton, mighty *T. rex!* Bolstered by publications in such prestigious journals as *Science* and *Nature*, the bandwagon on the direct descent of birds from earthbound "feathered" dinosaurs (presumably hot-blooded) has become an unchallengeable orthodoxy, such that those who oppose the consensus view find themselves unable even to offer contrary evidence or opinions, which are no longer considered respectable science. "The debate is over," proclaims the paleontology editor of *Nature*. "The sheer number of shared characters between maniraptorans (raptor theropods) and early birds is compelling," writes one book reviewer: "The results have convinced all but the tiniest band of ornithologists."[1]

Amid the haze of this hyperbolic rhetoric is an all-too-common but dangerous scientific development: the ascendency of a consensus view that has quickly evolved into a dogma. As the great twentieth-century philosopher of science Karl Popper noted presciently in 1994, "There is an even greater danger: a theory, even a scientific theory, may become an intellectual fashion, a substitute for religion, an entrenched ideology."[2]

The popular press has distorted the arguments on bird origins out of proportion. Simply put: Are birds derived from dinosaurs? Both sides agree that birds are closely allied with theropod dinosaurs; the only question is at what level. If direct descent from earthbound theropods—the current orthodoxy—is correct, then its corollary—that all avian flight adaptations and sophisticated anatomical aerodynamic architecture evolved in a context other than flight—must also be part of that doctrine. For such a theory to be true stretches biological credulity and must therefore involve an extreme in special pleading. Until the 1970s it was considered the consensus view that birds and dinosaurs shared common "thecodont," or basal archosaur, ancestry. This view requires neither that flight evolved from the ground up from the worst possible anatomical plan for the origin of flight (bipedal with foreshortened forelimbs) nor a nonflight origin of aerodynamic architecture.

Much of the discussion of this dino-bird nexus is considered a direct outcome of what has come to be known as the "cladistic revolution" in systematic biology, a transformative methodological upheaval during the second half of the twentieth century initiated by German biologist Willi Hennig in the 1950s and stated more formally in his book *Phylogenetic Systematics* in 1966. Hennig's phylogenetic systematics provided the basis of a formal algorithmic system for divining the family tree of life by comparing shared, derived characters (synapomorphies), entirely skeletal in the case of dinosaurs, which are coded by the investigator and dealt with by computer algorithms that align ancestor-descendant relationships by matching taxa with increasing numbers of shared, derived characters. Largely because of its simplicity and its ability to "solve," albeit in naive fashion, long-term, intractable problems of the field, cladistics soon became a dominating force, blending ideology with science. A number of scientific advancements have derived from cladistic thinking, but many scientists practicing evolutionary biology in recent decades without a pledged commitment to this phylogenetic catechism have been severely criticized, and many, including myself, have given talks at which they were the target of emotionally charged, derisive comments from audience members. In a blustering review of my book *The Origin and Evolution of Birds*, Kevin Padian, referring to cladistic methodology, forcefully noted that such methods are "now generally preferred by the systematic community, including reviewers for the . . . National Science Foundation." Commenting on this review, Peter Dodson detected a tone implying "the ultimate threat in science, the withholding of funding, for dissenters."[3]

This tendency toward indoctrinating disbelievers and infidels led geneticist John Avise to write a review entitled "Cladists in Wonderland," in which he describes cladists' views using words uttered by fictitious characters from Lewis Carroll's well-known tales *Alice's Adventures in Wonderland* and *Through the Looking-Glass*. The Red Queen's command "Off with his head!" was applied to Harvard's eminent evolutionary biologist Ernst Mayr for his enduring support of the Biological Species Concept. Before his death at the age of one hundred, Mayr and I carried on a lively correspondence for a number of years, and in one handwritten letter dated 16 April 1999, he quipped that, "Of course, cladistics is a religion and once you are a 'born again' cladist you can ignore anything in conflict with the dogma." Some years ago, in 1996, the distinguished Russian paleontologist Evgeny Kurochkin reflected a similar disdain for the certitude of the more ardent practitioners of the field, noting in a letter to me that "it looks like . . . biology in the 1940s and 1950s in former Soviet Union, which have not fundamentals of Lysenko biology." For those unfamiliar with this name, under Stalin, Trofim Lysenko (1898–1976) initiated a campaign of defamation, firing, arrest, and even death of many scientists who did not adhere to his pseudoscientific views. Lysenkoism set back the advancement of Soviet biology and genetics by decades.[4]

The overarching problem in cladistic approaches has been practitioners' tendency to ignore or dismiss contradictory evidence. History is filled with examples of overwhelmingly supported theories that were subsequently refuted, including everything from eugenics in the 1940s and global cooling in the 1970s to the volcanic origin of moon and earth craters until the 1960s. In the current approach, hypotheses are not truly tested; rather, evidence is marshaled to "prove" and "bolster" what is already "known" from cladograms. Even some ardent cladists have recognized the problem with this methodology and have urged the adoption of a more falsificationist approach.[5]

Paleontology is a tale of two disciplines. One is led by outstanding adventurers and professional scientists who are firmly grounded in both geology and biology and have made dramatic discoveries. Mixed in with these pillars of the field are an array of practitioners, ranging from amateurs to professionals, who blend a strange amalgam of rocks and biology with iconoclastic, often preconceived beliefs. Among the amateurs are outstanding scholars and scientists who deserve to be heard equally with professionals, but others lack any meaningful scientific background and clutter the field with misinformation that is all too frequently popularized in the press. Controversial since its inception, perhaps because of its inherent uncertainty of interpretation, the study of fossils tends to attract people who stake their claims and take intractable views that become inviolate dogma, impervious to evidence. Among the more amusing websites available for this group is the message board "Dinosaur Mailing List" (sponsored by the University of Southern California), whose mission is "to foster dialogue on science pertaining to dinosaurs." Although there is much of interest on the site, sifting the grain from the chaff can be a daunting task, and numerous amateurs use the site to vent their anger at dissenters, standing like Swiss Guards at the Vatican over all matters pertaining to dinosaurology, including entries in *Wikipedia*. After I was interviewed as *Discover*'s "scientist of the month" in 2003 (the article was titled "Plucking apart the Dino-Birds"), a new thread appeared on the Dinosaur Mailing List site. "'Tis Time to Get Medieval on Alan Feduccia" stated: "Let Feduccia make his bold claims. . . . Let the General go into battle with his army. . . . But!!!! On the other side of the table will be sitting our knights." A follow-up entry even called for "a scientific crucifixion." It indeed seems that the weaker the science, the more contentious and vitriolic its adherents.[6]

Following the dramatic discovery of the Late Jurassic early bird *Archaeopteryx* in the early 1860s, ornithology slowly progressed toward a more complete knowledge of the evolution of birds. However, since the development, especially during the last three decades, of phylogenetic systematics (cladistics), claims are now made that most of the age-old problems of avian evolution have finally been resolved, and the fact that "birds are living dinosaurs" is considered the major triumph of the field. In fact, as we shall see, this idea has its origin in a single fossil discovery of the late 1960s.

These claims are made even though there has been no smoking gun, no equivalent of DNA, no outside check—just a new method of analysis. The methodology of cladistics has become a system in which large numbers of skeletal characters are emptied into the phylogenetic hopper, with no distinction between primitive (plesiomorphic) characters, which carry no phylogenetic significance, and phylogenetically critical derived (apomorphic) characters, to be dealt with by the computer. Often, characters are cherry-picked, critical key characters are swept into an ocean of trivial ones, and scant attention is paid to the co-correlation of innumerable skeletal characters that are inseparably melded in animals with particular lifestyles (such as bipedalism in dinosaurs or foot-propelled diving in birds) in which large arrays of skeletal features form tightly linked character complexes. For example, loons and grebes, which can be shown not to be closely related by innumerable studies from DNA to embryology, almost invariable slot cladistically as a clade (a monophyletic group, derived from common ancestry).

A recent study of avian phylogeny employing more than 2,954 characters seemed to do little to advance the field and, like many other recent cladistic analyses, is exemplary of a trend back toward the so-called phenetic approach of the 1960s and 1970s, the goal of which was to elucidate overall similarity, often blending convergence with true relatedness. One reviewer noted that this massive study produced "a low ratio of phylogenetic signal to 'noise' in the data."[7]

Today's cladistic approaches tend to group animals as ecological equivalents (ecomorphs) without any necessary regard to actual relatedness. In reality, it has been new fossil finds and other discoveries that continue to provide grist for the phylogenetic mill, resolving seemingly insoluble problems in avian evolution. As we shall see in this book, more frequently than not, cladistic analyses of living groups of animals are overturned by DNA or other molecular comparisons, a luxury not available as a check for the purely skeletal comparisons of paleontology. We can only ponder how many cladistic analyses of groups represented solely by fossils would be disputed or overturned if they could be scrutinized by molecular comparisons, and we eagerly await whole genome comparisons that are just around the corner.[8]

In times past, paleontologists tended toward greater caution in interpreting fossils, but today, fantastic claims are made with abandon by amateur and professional alike, and each new fossil find, regardless of age or completeness, is often promoted as a missing link or a solution to some major paleontological mystery. Perhaps this approach relates in part to Sayre's Law (named for the political scientist Wallace Stanley Sayre), often stated as, "In any dispute the intensity of the feelings is inversely proportional to the stakes at issue." This law has most recently been attributed to former secretary of state and Harvard professor Henry Kissinger, who noted that low stakes were the reason why academic politics are so bitter. Yet while that might have been true in the past, it is certainly no longer the case. Today, academic jobs at research universities are highly prized, with great prestige, high salaries, and more freedom than in almost any other profession. Publishing prizes abound, substantial annual raises are common in the United States, and in China, where salaries are quite low, sizable monetary awards, reported in 1999 to range from six hundred dollars (Nanjing Institute of Geology and Paleontology) to four hundred dollars (Institute of Vertebrate Paleontology and Paleoanthropology) for each paper published in *Science* or *Nature*. In exchange for the monetary "goodies," universities expect great discoveries, and that's exactly what they get.

A quick look at recent paleontological papers in prestigious journals shows a new, related trend of piling on a veritable alphabet soup of authors, with little distinction as to who did what. The winners for 2009 were a controversial description of the theropod *Limusaurus* (which "helps clarify avian digital homologies") in *Nature*, with fifteen authors, and an article in the *Proceedings of the Royal Society* describing hadrosaur skin (a "creature with bird-like skin," according to *Discover*), with seventeen.

Another trend is the dismissal of geological time in interpreting phylogeny. One study of a seventy-million-year-old, presumably ornithurine

bird presumes to offer "insight into the assembly of the modern flight apparatus," and a description in *Science* of a dromaeosaurid of similar age claims to show that "miniaturization preceded the avialan node and the origin of flight." Both specimens are from the Campanian stage of the Late Cretaceous period, probably dating to well over a hundred million years after the origin of avian flight. Let us also recall that in 1998, it was claimed that in the Chinese Early Cretaceous, at twenty to thirty million years after the earliest known bird, *Archaeopteryx*, there were present all stages for not only the evolution of birds and avian flight but all stages in feather evolution. Harvard's late dean of vertebrate paleontology Alfred Sherwood Romer, never one to be impressed by academic hype, would wittily assert, when asked about the term "missing link," that all such a discovery accomplishes is to create two more missing links.[9]

This book had its inception some years ago as a normal academic work about the origin of birds and the genesis of avian flight. It was intended to complement the two previous efforts I had made in books published by Harvard University Press (*The Age of Birds*, 1980) and Yale University Press (*The Origin and Evolution of Birds*, 1996; 2nd edition, 1999). As I dug deeper into the happenings of the past decade, however, my work became a sleuthing journey. I uncovered so many difficulties in current theories about the origin and evolution of birds that I could no longer consider the prevailing approach to be valid science. Small wonder that the insidious Creationists have jumped into the fray; one Creation Science Movement article in 2005 was entitled "Feathered Dinosaurs and the Disneyfication of Palaeontology."[10]

Strong-willed people who unrelentingly cling to their favored theories, regardless of contrary evidence, seem to dominate the field, and constant appeal is made to a consensus of opinion. Perhaps this is true of many fields. Even the journal *Science* is replete with commentaries asserting that a "consensus has been reached." What does that have to do with science? The word "consensus" has no place in science and is never a validation of any hypothesis, yet one frequently sees trust in "consensus" for validation of important scientific concepts. In one recent book paleontologist Donald Prothero indicated that 99 percent of paleontologists agree that birds are derived from dinosaurs while not bothering to define what a dinosaur actually was or what the debate is all about.[11] As the late novelist Michael Crichton, MD, said: "I regard consensus science as an extremely pernicious development that ought to be stopped cold in its tracks. Historically, the claim of consensus has been the first refuge of scoundrels; it is a way to avoid debate by claiming that the matter is already settled. Whenever you hear the consensus of scientists agrees on something or other, reach for your wallet, because you're being had. . . . There is no such thing as consensus science. If it's consensus, it isn't science. If it's science, it isn't consensus. Period."[12]

In the 1950s, with the discovery of DNA, the expectations for biology were sky-high, and for the most part they have exceeded our wildest hopes and dreams, especially in genetics and molecular biology. Alas, the same is not true in other fields, and paleontology has been especially disappointing. Despite spectacular new fossil finds from the Early Cretaceous of China in the past decade and the brilliant work of many paleontologists, the field has been marred by constant unfounded speculation, often exacerbated by vivid imaginations of artists who are also amateur paleontologists. Every fantastic idea seems to add momentum to the *Fantasia* that paleontology has increasingly become. The field has evolved from one recognizing the importance of "falsification" as a basic principle of the scientific method to one in which

workers attempt only to verify what is deemed to be already known; in other words, a "verificationist approach" has in essence replaced the normal practices of science. An unjustified expectation is that fossil bones will invariably give up clues to their genealogy, masked by millions of years of adaptations that permit the organism to exist in a particular geologic time, in a specialized adaptive zone. As Al Romer would quip, "An animal cannot make a living as a generalized ancestor."

This book is my view of the current status of the field, and I hope it will contribute to a better understanding of the controversy, providing what I feel is the best fit for the current data, a step toward potential consilience of the conflicting views. The goal of this book is not to resolve complicated problems of avian origins that we are just beginning to understand but to encourage the new generation of students not to be bound by a faith-based reliance on computer-based slanted lines that confer an aura of precision and truth to what is in reality speculation. It is my hope that this book will provide, through a historical approach, the beginning of new approaches to possible solutions of seemingly intractable problems and provide an island of refuge for all views in the turbulent waters of the controversy. In recent years, the debate on bird origins has been favorably compared with equivalently heated debate on human origins as well as with the Middle East conflict. I am not a scientific evangelist, so I have no interest in the "conversion" of anyone to the views expressed here. I have an abhorrence of scientific proselytizing. Ultimately, science will speak for itself.

1

ROMANCING THE DINOSAURS

····································

Blame to Go Around

Dinosaurs are God's gift to television and the newspapers, just as science fiction is the lifeblood of the supermarket tabloids.

Keith Thomson, "Dinosaurs, the Media and Andy Warhol," 2002

"Romancing" refers to an ardent emotional attachment, and nowhere in paleontology does the term apply more aptly than to dinosaurs, which have long been an intense focus of almost every child's fascination. But enchantment with the terrible lizards reached a crescendo with the 1993 Michael Crichton blockbuster movie *Jurassic Park*, featuring an amusement park of cloned dinosaurs, and the view that birds are living dinosaurs ascended from theory to an unchallengeable orthodoxy. Propelled to the forefront over two decades earlier by John Ostrom's discovery of the famous raptor *Deinonychus* (considered "a bizarre killing machine"), in the late 1960s, the entire world of dinosaur paleontology would never be the same. By the mid-1980s Jacques Gauthier, now, like the late Ostrom, also of Yale University, would set the hypothesis that birds are derived from theropod dinosaurs in a formal algorithmic, phylogenetic (cladistic) context.[1] Gauthier's analysis is widely considered to be definitive and is still cited as the seminal paper in support of a theropod origin of birds.

In 1996, I published *The Origin and Evolution of Birds*, which questioned much of the currently accepted view of avian origins, supporting instead a common ancestry of birds and dinosaurs from early archosaurs (or more derived dinosauro-

morphs) and flight origin from the trees-down (arboreal), as opposed to the dinosaurian, ground-up (cursorial) genesis.[2] The book received high praise from such notables as Ernst Mayr and garnered the 1996 Award for Excellence in the Biological Sciences from the American Association of University Publishers.[3] A number of paleontologists were unwilling to consider alternatives to their orthodoxy, however, and in a review of the book that ran in the now-defunct journal *Lingua Franca* in that fall, Gauthier asserted, "We basically try and ignore [Feduccia]. For dinosaur specialists, it's a done deal," and "The bird people trust him, and so he's poisoning his own discipline."[4] Many dinosaur paleontologists revolted at criticisms of their arguments and were dismissive of any evidence that would undermine not only the image of dinosaurs as highly intelligent endotherms (hot-blooded creatures), many with feathers for an insulatory pelt, but also avian flight having evolved from the ground up from highly derived, earthbound theropod dinosaurs. Two principal advocates of a theropod origin of birds and its corollaries, Mark Norell of the American Museum of Natural History and Luis Chiappe, then of the same institution, wrote an immoderate review published in *Nature* entitled "Flight from Reason."[5] This review was sufficiently polemical that paleontologist

Peter Dodson of the University of Pennsylvania commented, "Such a sulfurous heading poisons the well of scientific discourse and seems unworthy of an otherwise respected and responsible journal."[6] Unfortunately, Norell and Chiappe's strident review is illustrative of a growing trend in the scientific and popular literature on the origin of birds.

Two years later, in 1998, an editor of the preeminent science journal *Nature*, Henry Gee, triumphantly announced, "Birds are dinosaurs: the debate is over."[7] Many paleontologists have supported this science by consensus, and their mantra is aptly summarized by Richard Prum, also of Yale University: "It is time to abandon debate on the theropod origin of birds."[8] Not only is the origin of birds a subject upon which no further debate is permissible, an unchallengeable orthodoxy, but, as paleontologist Christopher Brochu proclaimed in 2001, the origin of birds from theropod dinosaurs is "no longer the subject of scholarly dispute."[9] Brochu's position is clear: if the origin of birds from theropod dinosaurs is no longer the subject of scholarly dispute, then those who dispute it are not scholars; it is unscientific to dispute the origin of birds from theropod dinosaurs. Prum agrees, stating that "current critics of the theropod origin of birds are not doing science."[10]

Examination of the literature fueling the consensus view of the origin of birds directly from already highly derived theropod dinosaurs reveals considerable deficiencies.[11] Birds may well be living theropod dinosaurs, but the popularized consensus view in favor of a theropod origin of birds derives in large measure from the uncritical editing of poorly argued and documented manuscripts combined with media sensationalism. The rest stems from an almost canonical adherence to the results of the current methodology of the field of systematics, which deals with the reconstruction of life's history. Known as cladistics or phylogenetic systematics, in paleontology this field is based entirely on the computer analysis and coding of large numbers of skeletal features to seek most-parsimonious trees of life. Once consensus on the theropod origin of birds was reached and codified by cladistic analyses, the field took on a verificationist approach, attempting to prove preconceived ideas about avian origins, flight and feather genesis, and endothermy in dinosaurs.

For cladistics, the only characters of interest are apomorphic, or "derived," character states that, when matched between two taxa as synapomorphies, tell of common descent. While theoretically sound, such an approach depends on the characters being homologous (similar because of descent from a common ancestor), but the catch is that animals possess a morphology that is adaptive to their way of life and their environment. Thus, animals that do similar things tend to look similar, and many characters represent a homoplasy, a character shared by a cause *other* than common ancestry, such as convergence or parallelism, a very common occurrence in vertebrate history, where animals come to look alike because of adaptations for a particular lifestyle. A primary concern of this methodology, especially in paleontology, is that the only test of validity is another competing cladistic analysis, and in the case of living birds, for example, the vast majority of phylogenies produced by cladistics have been refuted by molecular comparisons. As Philip Gingerich of the University of Michigan has aptly noted, "The problem is that we expect too much of morphology in asking it to tell us the genealogy of organisms as well as what they look like."[12] Most recently, Ronald Jenner has emphasized that great care must be taken in analyzing comparative morphology to minimize subjectivity and bias.[13] Rather than being viewed as ultimate solutions to problems of phylogeny, phylogenetic hypotheses should be treated as an exploratory method. Such methods are useful for comparing and evaluating hypotheses, but they must be handled with extreme caution. With this in mind, this book takes another critical look at the problem of bird origins.

The Chinese Gold Rush

In recent years, the news from paleontology has been dominated by discoveries from the Chinese gold rush for the beautifully preserved fossils of the Early Cretaceous Jehol Biota in the northeastern Liaoning Province, some 300 kilometers (180 mi.) northeast of Beijing. Historically, the farmers of Songzhangzi had largely ignored the rocky outcrops around their village, but in the late twentieth century they struck fossil gold when they realized that people would pay large sums of money for fossil skeletons. Impoverished locals began excavating the fossils, and their peaceful, rural life turned topsy-turvy. Soon "the hillsides were honeycombed with grave-sized shafts and the fabulous specimens began to emerge," often with paleontologists working alongside local farmers.[14] As could be expected, fossils were sold to the highest bidder, often for a fraction of what they would eventually bring through illicit trade, and Chinese fossils ended up for sale everywhere, especially in Europe and North America. With hordes of foreign scientists clamoring to study the fossils, and a mini-economic boom for local farmers who could earn up to $1,200 (two years' income) for bird skeletons, fossils ended up all over the world.[15] Chinese dinosaur paleontologist Xing Xu reported in 2000 that there were "assembly line factories" where workers pieced together fossils for the black market, and chimaeric fossils were often forged to increase their value.[16] The problem continues today. "The fake fossil problem has become very, very serious," with more than 80 percent of marine reptile specimens on display in Chinese museums estimated to have been "altered or artificially combined to varying degrees." In 2011 many local museums were reported to have collections chock-full of fakes.[17]

One of these fossils, named *Archaeoraptor* and hailed as a missing link between dinosaurs and birds, is the most infamous example of this unfortunate situation. Unveiled at a National Geographic press conference in October 1999, the fossil was featured the next month in *National Geographic* magazine in an article by art editor Christopher Sloan in which he discussed newly discovered feathered dinosaurs from China and the origin of birds, including speculation on feathers in *T. rex*: "*Tyrannosaurus rex* may have had them . . . at an early stage. Hatchlings would have shed their downy feathers as they grew."[18] In addition to making unfounded, wild speculations on the fossils, Sloan, unfamiliar with the rules of biological nomenclature, called the forged fossil "*Ar-

chaeoraptor liaoningensis*" and stated that it would later be formally named as such. The response was immediate. Writing in the museum's newsletter, *Backbone*, Storrs L. Olson of the National Museum of Natural History denounced the publication of a scientific name in a popular magazine without peer review. On 1 November 1999, Olson wrote an open letter to Peter Raven, then chairman of the Committee on Research and Exploration for National Geographic, stating: "This is the worst nightmare of many zoologists—that their chance to name a new organism will be inadvertently scooped by some witless journalist."[19]

The demise of *Archaeoraptor* was rapid. That December, Xing Xu, dinosaur expert at Beijing's Institute of Vertebrate Paleontology and Paleoanthropology (IVPP), notified National Geographic of a counterslab of a small, supposed theropod known as a dromaeosaur that contained *Archaeoraptor*'s tail. The issue was finally resolved in August 2002 by the careful analysis of Stephen Czerkas of the Dinosaur Museum in Blanding, Utah, who had initially bought and studied the specimen, and Xing Xu; and by Zhonghe Zhou from the IVPP and colleagues, who in November 2002 published a note in *Nature* entitled "*Archaeoraptor*'s Better Half."[20] In the end, science corrected itself; the tail and hind limbs of the "*Archaeoraptor*" composite were in fact those of a small, birdlike dromaeosaur named *Microraptor zhaoianus*, and the foreparts were those of a fish-eating bird, *Yanornis martini* (=*Archeovolans repatriatus*), named for paleontologist Larry D. Martin of the University of Kansas. Indeed, Czerkas has gone on to make major contributions to the study of bird origins. Unfortunately, in the meantime, largely because of sensational coverage in the popular press, Creationists had seized on the story in an attempt to discredit paleontology and evolution.

In his open letter to Peter Raven, Storrs Olson asserted that National Geographic had "reached an all-time low for engaging in sensationalistic, unsubstantiated, tabloid journalism," and "The idea of feathered dinosaurs . . . is being actively promulgated by a cadre of zealous scientists acting in concert with certain editors at *Nature* and *National Geographic* who themselves have become outspoken and highly biased proselytizers of the faith."[21] Although the scandal was resolved through the self-corrective process of science, it is worth noting that it would not have occurred had a more critical attitude toward dinosaurs and the origin of birds prevailed in the scientific and popular literature. In illustrating the degeneration of scientific discourse with respect to this issue, Olson's letter clearly illustrated that the highly respected magazine *National Geographic* and a major scientific journal, *Nature*, were incapable or unwilling to consider critically the question of the origin of birds.

Unveiling in Florida

Shortly before returning the *Archaeoraptor* fossil to China in May 2000, Czerkas, an artist and paleontologist of impeccable integrity, presented an analysis of the specimen in a presentation entitled "A New Toothed Bird from China" at a special meeting held on 7 April at Fort Lauderdale's Graves Museum of Archaeology and Natural History, now defunct. The meeting, as science writer Constance Holden described it, "was one of the most unusual coming-out parties Florida had ever seen. By day, some of North America's top dinosaur experts debated fine points of the evolution of birds and dinosaurs. After dark, they mingled with the cream of Fort Lauderdale society, supping, dancing, and drinking cocktails around a live alligator, assorted skeletons, and a dinosaur carved in ice."[22]

The posh conference was held in honor of a small, newly unveiled, birdlike dromaeosaur from the Late Cretaceous of Montana, *Bambiraptor feinbergi*. The generic name of the specimen refers to its diminutive size, while its specific epithet honors the museum supporter Michael Feinberg of Hollywood, Florida, who purchased it from the discoverers, the Linster family, for an undisclosed sum, and supported its reconstruction by David Burnham at the University of Kansas, who would later write his doctoral dissertation on the newly discovered fossil and other aspects of its paleobiology.[23] Little *Bambiraptor* was hailed as the most birdlike dinosaur discovered to date, although it postdates the earliest known bird, *Archaeopteryx*, by 80 million years.

Bambi was to be the crown jewel of the newly created Florida Institute of Paleontology at the Graves Museum. The symposium, which I attended, drew some 150 scientists, a medley of dealers, collectors, and graduate students, who presented papers on aspects of paleontology, mostly related to the presumed relationship of birds and dinosaurs. As at any conference, the quality of presentations varied; excellent papers were given by archosaur anatomist Lawrence (Larry) Witmer of Ohio University, developmental biologist Roger Sawyer of the University of South Carolina, and others. Among the more speculative presentations was a paper by William Garstka on the DNA of *Triceratops*. He and his team of molecular biologists claimed to have extracted DNA from fossil bones of the Late Cretaceous dinosaur. Shortly after the symposium Constance Holden reported the finding in *Science*: "They said it couldn't be done. But a team at the University of Alabama just may have succeeded in extracting some DNA from a dinosaur. And guess what it resembles: a turkey. If the work pans out, the scientists say, it will be the 'first direct genetic evidence to indicate that birds represent the closest living relatives of the dinosaurs.'"[24]

At the meeting, a polite gentleman came up to me and introduced himself as Jonathan Wells. I did not know at the time that Wells is a well-known Creationist, having received a PhD in religious studies from Yale and, in 1989, a second doctorate from Berkeley in molecular and cell biology. We had a pleasant discussion, and he asked me about the *Triceratops* DNA, which I thought was fantastical, as did most of my colleagues at the meeting. During the year that followed, Wells published a Creationist work entitled *Icons of Evolution*, and in it he recalled the story of the *Triceratops* DNA:

The DNA Garstka and his colleagues found was 100 percent identical to the DNA of living turkeys. Not 99 percent, not 99.9 percent, but 100 percent. Not even the DNA obtained from other birds is 100 percent identical to turkey DNA (the next closest match in their study was 94.5 percent, with another species of bird). In other words, the DNA that had supposedly been extracted from the Triceratops bone was not just similar to turkey DNA—it was turkey DNA. Garstka said he and his colleagues considered the possibility that someone had been eating a turkey sandwich nearby, but they were unable to confirm that. At first, when Garstka presented his findings I thought it was an April Fool's joke—but it was already April 8. Then I looked around to see whether anyone was laughing—but no one was.[25]

Wells singled out one other presentation for ridicule, that of Kevin Padian, who preached to the audience about what was, and what wasn't, science:

> Just before Garstka spoke, Berkeley paleontologist Kevin Padian had blasted critics of the bird-dino theory for being unscientific. Padian explained that, as President of the National Center for Science Education, he spends a lot of time telling people what science is and what it isn't. . . . Padian called critics of the dino-bird hypothesis unscientific because (he claimed) they offer no empirically testable alternative hypotheses. The evidence the critics cite is based on . . . isolated observations, rather than on a method (cladistics) that is "fully accepted by the scientific community." And . . . although "science is not a vote," the cladistic method is endorsed by the National Science Foundation, major peer-reviewed scientific journals, and "the majority of experts." Therefore, criticisms of the bird-dino hypothesis "ceased to be science more than a decade ago," and the "controversy is dead."[26]

It is not in the least bit amusing that a Creationist could come away from the Graves symposium and write, "This isn't science. This isn't even myth. This is comic relief."

Padian, whom Storrs Olson dubbed in a review "the Elmer Gantry of the theropod crusade," has formerly headed the National Center for Science Education, whose mission includes fostering the teaching of evolution in the schools, but his overall approach appears more likely to stoke the Creationists' flames than to encourage engage-

ment.[27] Small wonder the Creationists are thriving and thoroughly enjoying much of the junk science introduced into the current dino-bird debate.

Dinosaur *Fantasia*
A Heart of Stone for Hot-Blooded Dinosaurs

Scandals are sensational, but uncritical editing that permits careless and flawed articles to appear in esteemed journals is more troubling. Examples are particularly common among articles that address dinosaur physiology. Consider the case of the 4-meter-long (13 ft.) Late Cretaceous hypsilophodontid ornithischian *Thescelosaurus*, nicknamed Willo. In 1993 a specimen was discovered with what was said to be an intact heart exhibiting the detail of internal chambers. In a paper coauthored by Paul Fisher of the Biomedical Imaging Facility of North Carolina State University, dinosaur paleontologist Dale Russell of the North Carolina Museum of Natural History, and colleagues and published in the prestigious journal *Science* in April 2000, the authors suggested that *Thescelosaurus* had a four-chambered heart with a single aortic arch (as in mammals and birds).[28] They of course concluded that dinosaurs were endothermic (in support of entrenched dogma), even though ectothermic crocodilians also have a four-chambered heart (albeit with both systemic [aortic] arches). In an interview with science writer Virginia Morell, the team asserted: "a heart that beats like a bird's suggests that this dinosaur's metabolic rate would have been more like that of an endotherm (a warm-blooded animal) than an ectotherm (a cold-blooded animal), providing yet another feature shared by dinosaurs and their putative feathered relatives."[29] Continuing the sensationalism, the *Guardian* published an article on 21 April titled "Dinosaur Heart a Shock to Evolution" in which Russell was quoted as saying, "Not only does this specimen have a heart, but computer-enhanced images of its chest strongly suggest it is a four-chambered, double-pump heart with a single systemic aorta, more like the heart of a mammal or a bird than a reptile."[30] Of course, the physiology of a Late Cretaceous ornithischian is not relevant to the origin of birds because they are thought to have nothing to do with avian origins, splitting off from theropod dinosaurs more than 220 million years ago, shortly after the origin of the Dinosauria, yet enthusiasm for a connection between birds and dinosaurs was somehow destined to creep into the discussion.

Fisher and colleagues' paper was sufficiently speculative that, unusually, it attracted widespread criticism. It was later revealed that *Science* had accelerated the manuscript review process to coincide with the opening of a new seventy-one-million-dollar building for the North Carolina Museum of Natural History in Raleigh, so that the exotic fossil discovery, which the museum had purchased, could be the focus of the gala event. Dale Russell told Tim Friend of *USA Today*, "The director of the museum asked *Science* to push ahead to coincide with the opening. The turnaround time for the paper was short, and it was close, but they made it."[31]

Papers submitted to *Science* are customarily reviewed for several months before acceptance or rejection. In this case, the paper was received for review on 9 March and accepted for publication on 27 March. Numerous paleontologists criticized Fisher and colleagues' findings, including Tim Rowe of the University of Texas and Paul Sereno of the University of Chicago.[32] In a paper with another colleague, Rowe and Sereno concluded that the grapefruit-sized structure in question was a lump of minerals, a literal heart of stone. As they noted, such ironstone concretions, precipitated by bacteria, are common in the Montana's Hell Creek Formation, where the fossil was found: "Ironstone concretions are notorious for producing suggestive and misleading shapes. . . . The object studied . . . fails both geological and anatomical tests of its unprecedented identification."[33]

Moreover, even if a four-chambered heart were present in *Thescelosaurus*, a four-chambered heart is also present in crocodilians; the presence of a four-chambered heart in dinosaurs would therefore not remotely constitute conclusive evidence in assessing their physiology. Indeed, it would be more surprising if dinosaurs did not have a circulatory system more advanced than most living reptiles such as turtles and lizards, perhaps more crocodile-like in structure. Last, the specimen was from the Late Cretaceous, postdating the origin of birds by at least 100 million years or more. Clearly, assertions that the specimen offers any insight about the physiology of dinosaurs, let alone the origin of birds, should be regarded with extreme skepticism. As one scientist told *USA Today*, "I think this whole thing happened because dinosaurs are a good sell for *Science*." Commented another, "I have a very cynical view. . . . I think this paper was pushed through because it was going to make a splash."[34] Incredibly, as late as December 2010, a website maintained by the Center for the Exploration of the Dinosaurian World, a collaboration between North Carolina State University and the North Carolina Museum of Natural Sciences, insists: "A team of scientists in North Carolina and Oregon used medical technology to probe an iron-stained concretion inside the dinosaur's chest. With the aid of imaging equipment and software, they were able to reconstruct 3-dimensional structures through the interior of the concretion. The images reveal a heart that was more like that of a bird or a mammal than those of reptiles, adding substantially to evidence suggesting that at least some dinosaurs had high metabolic rates."[35] The site features images of the famous Willo, with arrows pointing to the imaginary left and right ventricles and interventricular septum!

Most recently, a study by Timothy Cleland and colleagues at the same institution examined the *Thescelosaurus* heart using a higher-resolution computed tomography and other techniques and concluded that "microstructural examination of a fragment taken from the "heart" was consistent with cemented sand grains, and no chemical signal consistent with a biological origin was detected.[36] Perhaps poor Willo should be laid to rest.

Hot-Blooded Dinos

One speculative, poorly edited paper could be excused, but there are other examples; indeed, before the *Thescelosaurus* debacle there was the article "Thermophysiology of *Tyrannosaurus rex*: Evidence from Oxygen Isotopes," published in *Science* in 1994.[37] Its authors, paleontologist Reese Barrick and geochemist William Showers, then both of North Carolina State University, attempted to measure body temperature in extinct reptiles by measuring ratios of certain isotopes of oxygen in fossil bones. Before obtaining any meaningful baseline with known organisms of established provenance, they selected the great *Tyrannosaurus rex* for study. Barrick and Showers reasoned that because oxygen-16 is taken up in higher proportions at higher body temperature, it would follow that different parts of a body should register differences if the body temperature varied along a gradient, as in modern ectothermic organisms like reptiles. If this were so, then if an organism had an even body temperature throughout—that is, if that organism were endothermic—then there should be little evidence of a temperature gradient. They concluded that the apparent lack of intrabone and interbone variation in a variety of dinosaurs signaled the presence of homeothermy and high metabolism.

Geochemist Yehosua Kolodny of the Institute of Earth Sciences of the Hebrew University of Jerusalem and colleagues, who pioneered such isotope studies, questioned whether fossil bone actually preserves the original "signal" from the living organism, pointing out that signals can be altered during burial, when bone is mineralized, or by groundwater. They noted that the reconstruction of the thermophysiology of fossil organisms through analysis of stable isotopes of oxygen requires the assumption that apatitic fossils are isotopically unaltered parts of their original skeletons, and they presented evidence to question that assumption.[38] Kolodny and colleagues suggested that the distribution of stable oxygen isotopes in fossils might provide information about the burial environment but that it could mislead attempts to interpret an organism's physiology; indeed, there are similar values for bones of coexisting fossil fish, dinosaurs, and other reptiles at differing latitudes. The analysis of Barrick and Showers was clearly speculative and premised on a facile understanding of the degree to which isotopic studies can elucidate reconstructions of the physiology of extinct organisms. Nevertheless, Barrick and Showers's paper appeared in *Science*, typically the trigger for articles in the *New York Times* and other major news media, and it is still widely cited.

Yet another important theoretical question, apparently considered by neither the authors nor *Science* reviewers, concerns how one would distinguish large inertial homeotherms (animals that maintain a fairly constant daily body temperature) from genuinely endothermic organisms (those that produce endogenous body heat). It is not implausible that because of their large size, many dinosaurs in the exceptionally warm Cretaceous may have been inertial homeotherms, as has been suggested many times, rather than genuine endotherms, and therefore were quite active. Even if it is granted that Barrick and Showers's methodology and its underlying assumptions are sound, homeothermy might not show any isotope gradient; their analysis would therefore not be able to distinguish homeothermy from genuine endothermy. Consequently, even granting the soundness of Barrick and Showers's scientific methods and assumptions, their study was from the start incapable of providing definitive evidence as to the metabolic rate and physiological energetics of dinosaurs. The basic underlying assumptions of their study were not in accord with either the original work on oxygen isotopes or the physiological literature. As comparative physiologist John Ruben of Oregon State University has stated, "Extremity vs core body temperatures in extant birds and mammals are often as variable as those of ectotherms."[39]

It has often been noted that among the problems faced by endotherms is that great iconic flesh-eaters of the Cretaceous such as *T. rex* would have had to eat about ten times as much food as it would have as an ectotherm of the equivalent size. This is probably an exaggeration, since the tenfold figure refers to human-sized endotherms compared to ectotherms, but the difference would still be huge, and a threefold increase in food requirements for endothermy would not be an excessive estimate. In the exceptionally warm Cretaceous it would have been extremely disadvantageous for large animals to have been endotherms, and we see even today in the tropics the evolution of large size in ectotherms, not endotherms. The success of ectothermy in today's tropical regions is reason enough to argue that *T. rex* would likely not have evolved the burden of endothermy in such a setting.

Fortunately, science marches on, and new, carefully conducted studies by French scientists and colleagues comparing oxygen isotope compositions of tooth phosphate of extinct marine reptiles (ichthyosaurs, plesiosaurs, and mosasaurs) to those of coexisting fish have produced beautiful results, showing that some of these reptiles may have been able to sustain constant body temperatures (that is, homeothermy). Given the spectrum of temperature regulation regimes in modern vertebrates—from homeothermic leatherback turtles, which are gigantothermic, to "regional endotherms" such as tunas and lamnid sharks, to brooding pythons that produce endogenous heat to brood eggs—it is not surprising that a multitude of regulatory mechanisms evolved during the Mesozoic, including particularly inertial homeothermy and gigantothermy among its behemoths. These studies will require additional verification, but they represent a good start, and their approach is not aimed at verifying any preconceived views.[40]

As for dinosaurs' thermal biology, James Farlow, a paleontologist at Indiana University, provides an apt précis of the field:

> Unfortunately, the strongest impression gained from reading the literature of the dinosaur physiology controversy is that some of the participants have behaved more like politicians or attorneys than scientists, passionately coming to dogmatic conclusions via arguments based on questionable assumptions and/or data subject to interpretations. Many of the arguments have been published only in popular or at best semi-technical works, accompanied by rather disdainful comments about the stodgy "orthodoxy" of those holding contrary views; what began as a fresh way of considering paleontological problems has degenerated into an exercise in name-calling. All of this has made the whole field of dinosaur studies suspect in the mind of scientists.[41]

Can't Get Enough of Tyrannosaurus rex!

Between 2005 and 2008, *Science* and other eminent journals published a series of papers from a team led by Mary Schweitzer of North Carolina State University and the North Carolina Museum of Natural History with colleagues including spectrometry expert John Asara of Harvard Medical School.[42] Led by the famed dinosaur tracker Jack Horner, a team of paleontologists recovered a specimen of *T. rex* in Montana in 2000, and Horner presented parts of its fossilized femur to Schweitzer, his former student, in 2003. As the years progressed, work in her laboratory claimed to show not only soft tissue represented by elastic, translucent blood vessels containing red blood cells but the possibility of recovering intact proteins. In *Science* in April 2007, Asara, Schweitzer, and colleagues announced that through mass spectrometry they had recovered seven preserved fragments of protein, five of which matched sequences of collagen, the major component of bone, from birds, specifically chicken. As the bevy of papers by Schweitzer and colleagues emerged, the authors' claims of recovering fragments of the original protein as well as well-preserved soft tissue rose from the level of speculation to high probability. The authors were quick to observe that "dinosaur protein sequence, including collagen, should be most similar to that of birds among extant taxa, according to other phylogenetic information."[43] One might question, however, just how similarity in protein sequences of dinosaurs and birds might differ if birds actually derived from theropod dinosaurs or shared a common ancestry. There might be little discernable difference between the two competing hypotheses, but none of the normally critical considerations seemed to be of concern. The media printed sensationalistic headlines, including "Study: *Tyrannosaurus Rex* Basically a Big Chicken," and the *New York Times* reported that their work "opens the door for the first time to the exploration of molecular-level relationships of ancient, extinct animals." One reporter from the *Guardian* even heralded "the tantalizing prospect that scientists may one day be able to emulate Jurassic Park by cloning a dinosaur."[44]

These findings on *T. rex* were soon subject to widespread skepticism and criticism, and within less than two years, in addition to numerous negative comments in various news media, three major rebuttals appeared, two in *Science*. It was particularly difficult not to express skepticism following a 2006 *Discover* magazine interview with Barry

Yeoman entitled "Schweitzer's Dangerous Discovery" in which Schweitzer apparently shared her belief that she could smell the distinctive odor of death in dinosaur bones:

> Once, when she was working with a *T. rex* skeleton harvested from Hell Creek, she noticed that the fossil exuded a distinctly organic odor. "It smelled just like one of the cadavers we had in the lab who had been treated with chemotherapy before he died," she says. Given the conventional wisdom that such fossils were made up entirely of minerals, Schweitzer was anxious when mentioning this to Horner. "But he said, 'Oh, yeah, all Hell Creek bones smell,'" she says. To most old-line paleontologists, the smell of death didn't even register. To Schweitzer, it meant that traces of life might still cling to those bones.[45]

One can only ask how such a phenomenon could occur, since the smell associated with the decomposition of dead corpses implies biological tissue and involves two processes, autolysis, or nonbacterial (aseptic) breakdown of the body's tissues and cells due to enzymatic "self-digestion," and bacterially mediated putrefaction, from environmental bacteria as well as endogenous bacteria within the corpse's digestive system. The decay produces cadaverine and putrecine, foul-smelling diamines from protein hydrolysis that remain near the ground while intestinal microorganisms respire, releasing hydrogen sulfide and methane, which causes bloating. Generally the corpse begins to smell within a day, and the odor continues until decomposition is completed. How, then, could Schweitzer smell death on the bones after 68 million years? Was the autolytic process still occurring on the biological tissue, or was there still bacterially mediated putrefaction? If either was the case, then many proteins, including those of myriad bacteria, would have been detectable by methods much simpler than mass spectrometry.

Interviewed in the same *Discover* piece, Jeffrey Bada, an organic geochemist at the Scripps Institution of Oceanography, stated that he could not imagine sixty-eight-million-year-old soft tissue being preserved in *T. rex*. Bada suggested that

the cellular material Schweitzer and her colleagues found represented contamination from outside sources. Even if the *Tyrannosaurus* specimen had died in a colder, drier climate than Hell Creek, environmental radiation would have degraded its body: "Bones absorb uranium and thorium like crazy. You've got an internal dose that will wipe out biomolecules."[46]

Tom Kaye of the Burke Museum of Natural History of the University of Washington and his colleagues used scanning electron microscopy and other techniques to examine similar bones, and their reinterpretation showed extensive mineralized and nonmineralized coatings in porous bone trabeculae of both dinosaur and mammalian species across time. Their findings indicated that these coatings represent bacterial biofilms, commonly found throughout nature. Bacterial biofilms often form endocasts, and when these endocasts are dissolved out of the bone, they can mimic real blood vessels and osteocytes.[47] In an interview with *New Science*, Jeff Hecht reported Kaye as stating: "We cracked open a lot of bones and spent hundreds of hours on an electron microscope examining them."[48] His conclusion: the soft material was most likely not from dinosaurs but from bacterial films that grew on cavities inside the bone long after death. Although biofilms are more familiarly thin, sticky layers, the biofilms Kaye found produced branching hollow filaments when they coated the inside of blood vessel cavities in bone. He and his colleagues then used carbon dating to show that the biofilms are not ancient but of relatively modern origin. A comparison of infrared spectra to the material revealed further that these modern biofilms "share a closer molecular make-up than modern collagen to the coatings from fossil bones."[49]

Kaye and his colleagues also noted that "blood cell size iron-oxygen spheres found in the vessels were identified as an oxidized form of formerly pyritic framboids." The geologic term "framboid" describes a microscopic feature common to certain sedimentary minerals, particularly pyrite (FeS_2), and is derived from the French *framboise*, or raspberry, in reference to the reddish (iron) pigment apparent under magnification. Interestingly, framboids were once thought to be fossilized bacterial colonies or microorganisms. Meanwhile, the inscription at the exhibit at the North Carolina Museum of Natural History reads, "A tiny thing that could change the world of paleontology: a red blood cell from a *Tyrannosaurus rex*, floating in a blood vessel. Both should have rotted away millions of years ago."

Recall that the bones of *Tyrannosaurus rex* from which Schweitzer and her colleagues had supposedly recovered soft tissue and extracted protein sequences are from approximately 68-million-year-old rocks. The oldest well-substantiated vertebrate DNA is probably not older than 500,000 years, from mammoths, cave bears, and other biota taken from frozen tissue in Siberia and Greenland; corroborated soft tissue dates back to about 300,000 to 400,000 years at most; and other DNA from ice cores in Greenland dates to 450,000 to 800,000 years.[50] It thus strains credulity to suppose that soft tissue, particularly blood vessels, and protein sequences have survived over 68 million years; at a minimum, acceptance of the claim requires unequivocal evidence. In the absence of such evidence, it is more reasonable to conclude that the alleged soft tissue and protein represent contamination by substances of modern origin. Media sensationalism knows no caution, however, and thus the popular press offered such whimsical headlines as "*T. rex* Tasted Like Chicken" and "*T. rex* Closer to Gizzards than Lizards."[51] During the same period, in conjunction with a dinosaur exhibit, the American Museum of Natural History café was serving dinosaur nuggets, grilled dinosaur sandwiches, and dinosaur Caesar salads. By contrast, the media scarcely noticed the work of Tom Kaye and his colleagues.

There have been counterarguments to the rebuttals, but whether in the final analysis the interpretation of these structures turns out to be correct, the explanation by Kaye and his colleagues is certainly a more conservative and plausible one than that published in the high-profile articles in *Science*. It is dismaying that the considerations of Kaye and his colleagues were apparently never pondered by the Schweitzer team in their attempts to show that protein was present in dinosaur bone, thus providing further proof that birds were derived from theropod dinosaurs. Such extraordinary claims should have prompted *Science* to insist on painstaking evaluation before publication. However, it is difficult to dismiss the possibility that the thrill of media sensationalism and the lure of lucrative headlines may have overridden rational considerations.

In addition to the efforts of Kaye and his colleagues, University of California San Diego computational biologist Pavel Pevzner and colleagues, Mike Buckley and colleagues, and others have questioned Schweitzer's data analysis and conclusions, arguing in separate papers that the protein fragments are most likely modern contaminants or statistical artifacts and not original *T. rex* proteins.[52] Their analysis was hampered by Asara and Schweitzer's refusal to release their data, which is extraordinary in the field of molecular biology. This refusal was even more disturbing because there were just seven preserved fragments among tens of thousands of "junk spectra"—strings of peptide letters that Asara's spectrometer could not match up with anything in the database. Pevzner compared the recovery of a handful of protein fragments from tens of thousands of "junk spectra" to a monkey banging away on a typewriter and randomly producing seven words. Could one then legitimately conclude that the monkey could spell? Compounding the skepticism, Asara and colleagues continued to refuse to release the data from the junk spectra.

Eventually, in the fall of 2008, following much criticism, Asara posted all 48,216 spectra without restrictions in an online database. Shortly thereafter, however, computational biologist Matthew Fitzgibbon and proteomics expert Martin McIntosh turned up an unexpected result. They showed that a peptide that hadn't appeared in the original papers produced a match, not to collagen, but to a hemoglobin peptide found in ostriches.[53] This discovery led to speculation that Asara could have been dealing with a contamination problem, since he also worked with ostrich proteins in his laboratory. Contamination could perhaps explain the matchup of the peptides with chicken.

Schweitzer, Asara, and colleagues' final article appeared in the spring of 2008, after *Science* had received Pavel Pevzner's rebuttal article and after the journal's editors had received and published the criticism by Buckley and colleagues.[54] *Science* had earlier summarily rejected the paper by Tom Kaye and his colleagues. Fortunately, *PLoS ONE* published Kaye and colleagues' paper in the late summer of 2008, making *Science*'s rejection of the work at least appear inappropriate.

In his highly regarded blog *Genomics, Evolution, and Pseudoscience*, Steven Salzberg, director of the Center for Bioinformatics and Computational Biology at the University of Maryland, aptly summarized the situation:

> What I find most reprehensible on their [the editors of *Science*] part is that they published both the Buckley et al., and the Pevzner et al. critiques as "Technical Comments"—which means they appear online *only*, not in the print edition. Both the Asara and Schweitzer articles, by contrast, appeared in the print edition, which means they will be read more widely. If *Science* truly cared about getting this story right, they would publish the critiques just as prominently as the original article. It seems that *Science* is eager to get publicity for a "discovery," but not so eager for publicity when it turns out the discovery is false.[55]

In an attempt to satisfy the skeptics, Schweitzer and fifteen additional authors, including Asara, published yet another paper in *Science* in May 2009 on protein sequences of a Cretaceous hadrosaur, *Brachylophosaurus canadensis*, an ornithischian dinosaur about eighty million years old. This time they attempted to answer some of the contamination questions by handling the fossils with sterile instruments from the beginning, like a CSI investigation.[56] Once again they identified fragments from collagen, the protein that forms the major component of bone, and claimed that the fragments represented original dinosaur protein, and as before, they discovered that the dinosaur, this time an ornithischian, most closely resembles modern birds, particularly the ostrich. Although the team took extra precautions in collection to avoid additional contamination, they continued to ignore the possibility of preservational contami-

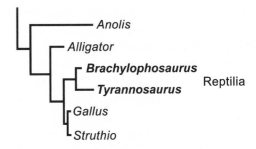

Fig. 1.1. Consensus cladogram from Schweitzer et al.'s 2009 study of supposed protein sequences of the Late Cretaceous hadrosaur, including mastodon *Mammut* and *T. rex* from previous studies. According to this cladogram, theropod dinosaurs (represented by *Tyrannosaurus*) come out closer to ornithiscian dinosaurs (*Brachylophosaurus*) than to birds (*Gallus* and *Struthio*), a finding totally at odds with the current hypothesis for a theropod origin of birds. By this scheme, a basal archosaur origin of birds hypothesis is equally likely. From M. H. Schweitzer et al., "Biomolecular Characterization and Protein Sequences of the Campanian Hadrosaur *B. canadensis*," *Science* 324 (2009): 626–31. Reprinted with permission from AAAS.

nation. Taphonomic problems aside, however, the study has met with much less skepticism from the scientific community. In the meantime, a paper published in July 2009 in the *Journal of Proteome Research* reported that three of the peptides from the original *T. rex* paper were a strong match to what appears to be an ancient form of collagen. The authors conclude, "In summary, we find nothing obviously wrong with the *T. rex* mass spectra: the identified peptides seem consistent with a sample containing old, quite possibly very ancient, bird-like bone, contaminated with only fairly explicable proteins."[57]

Stephen Salzberg further noted in his blog of 24 June 2009 that Martin McIntosh of the Fred Hutchinson Cancer Research Center discovered that the original *T. rex* sample contained hemoglobin. Since no one would expect to find hemoglobin in bone tissue, especially from a sixty-eight-million-year-old fossil, it caught the attention of those who were skeptical about the original findings. More important, McIntosh explained that a chemical modification of the particular hemoglobin made it more likely to be contamination. He submitted a paper on his results to *Science*, but it was rejected, presumably because it questioned findings that *Science* had endorsed. It was subsequently learned that Asara had indeed used his mass spectrometer equipment also to examine samples of ostrich bone. Obviously, if the *T. rex* sample was contaminated, then all the results come into question, including those from the hadrosaur analysis.

Aside from the acrimony involved in these publications, the rebuttals and counterrebuttals, and the dizzying effects of trying to decipher the validity of the results, one could legitimately ask, If one were truly interested in the question of ancient preservation of soft tissue, DNA, and protein, why would one initially jump back to sixty and eighty million years, when no known such preservation before has exceeded about a million years, and even then mainly comes from samples from the frozen Arctic? Whatever happened to "controls" in science? Why not work back to well-preserved mammal or reptile fossils of five, ten, or twenty million years to see when the drop-off occurs for reasonable preservation? The other factor that must be addressed is the possibility of contamination during taphonomy, or preservation, when an entire ecosystem immediately develops following death to dismantle the corpse. As Carter and colleagues have noted, "Each CDI [cadaver decomposition island] is an ephemeral natural disturbance that . . . acts as a hub by receiving . . . materials in the form of dead insects, exuvia and puparia, faecal matter (from scavengers, grazers and predators) and feathers (from avian scavengers and predators). . . . As such, CDIs are a specialized habitat for a number of flies, beetles and pioneer vegetation, which enhances biodiversity in terrestrial ecosystems."[58] No amount of "CSI-style" sterile investigation technique could possibly deal with this level of contamination from the time of death to the time of discovery.

These matters are given scant attention because people are interested in solving sensational problems, in this case proving that dinosaurs were hot-blooded creatures like living birds and mammals. In a strange twist, the conclusions from a recent paper on the Campanian hadrosaur *Brachylophosaurus* are to be seen in the authors' phylogenetic analysis, based on collagen protein data, that shows the theropod *T. rex* more closely related to the ornithischian hadrosaur than either is to birds (chicken and ostrich). One could certainly argue from this cladogram for support of a basal archosaur origin of birds, a competing theory to a direct theropod dinosaur origin.

A well-substantiated finding of ancient dinosaurian soft tissue, protein, or DNA would be savored by all, but mistakes can lead to an avalanche of subsequent misinterpretations. Whether the claims of Schweitzer and her colleagues are ultimately corroborated, the primary casualty here is science itself. In the future, perhaps, more measured studies will resolve whether there is verifiable protein in eighty-million-year-old fossil bones and whether *Tyrannosaurus rex* is a "big chicken," as so many seem to hope.

In 2006 Schweitzer was interviewed by Barry Yeoman of *Discover* and revealed her openly evangelical religious convictions: "Ironically, the insides of Cretaceous-era dinosaur bones have only deepened Schweitzer's faith. 'My God has gotten so much bigger since I've been a scientist,' she says. 'He doesn't stay in my boxes.'" According to the interview, Schweitzer adorned her office wall with a verse from the book of Jeremiah: "For I know the plans I have for you, declares the Lord, plans to prosper you and not to harm you, plans to give you hope and a future."[59] Although any scientist's beliefs are irrelevant to the veracity of their data, for many the blending of science and religion erodes confidence in Schweitzer's conclusions on *T. rex* tissue and protein. In October 2005, *Discover* writer Jack Hitt interviewed me concerning her original soft tissue findings. He quoted Schweitzer as stating of the tissue within the *T. rex* bone, "We have a little movie of our tweezers stretching the material. You can see the blood vessels flexing inside." I said, "Give me a break." I stand by that statement.[60]

Déjà Vu All Over Again

Having a paper published in *Science* is arguably a big event in the life of any scientist, and such an event can bear importantly on promotion and even tenure, so when papers are published with results that don't inspire confidence in a large percentage of the scientific community, one must question editorial practices. *Science* published another fantastic paper on dinosaurs back in 1994 by Scott Woodward, a microbiologist at Brigham Young University.[61] Woodward claimed to have extracted DNA from eighty-million-year-old dinosaur bone fragments from a coal mine in eastern Utah. The team supposedly used the DNA as the template in a polymerase chain reaction that amplified and sequenced a portion of the gene encoding mitochondrial cytochrome b. Although the announcement appeared prominently in the *New York Times* and *New Scientist*, scientists would later discover that the bones themselves could not be identified as even dinosaurian, and many, including Schweitzer, debunked the results as laboratory contamination. All Woodward could say was, "It is likely that the bone fragments belong to a Cretaceous Period dinosaur or dinosaurs." According to the *New York Times*, "Dr. Woodward said other laboratories would be given samples for independent testing. But much of the material was destroyed in the tests already done, and a collapsed tunnel has closed the coal mine, preventing the search for more bone fragments."[62]

Fast-forward fifteen years, and a group of eight scientists, headed by Pavel Pevzner and colleagues and including myself, wrote a letter dated 14 August 2009 to Jack Horner (copied to Schweitzer and Asara), requesting access to the specimen of *Tyrannosaurus rex* used in the studies published in *Science* for independent verification of the results. Horner asked that Schweitzer reply, and in her response we were informed that the remaining bone from the specimen was now on display and was thus not appropriate for molecular analyses. Further, the remaining untreated bone designated for molecular studies resides at North Carolina State University and is under study by her and her graduate students and therefore unavailable for study. She further suggested that we obtain our own specimens. Horner later told our working group that he would send a sample of his bone on display for analysis. However, John Asara informed us that the curated sample was hardened by being treated with the deep-penetrating chemical Vinac and that the bone

would have been exposed to degrading elements at room temperature and thus would not be suitable for examination. One can only ask why at the time of the initial studies material was not set aside for reexamination by colleagues. For all practical purposes, the discovery of putative *T. rex* proteins is not reproducible. The stated policy for research published in *Science* is clear: "All data necessary to understand, assess, and extend the conclusions of the manuscript must be available to any reader of *Science*. Fossils or other rare specimens must be deposited in a public museum or repository and available for research." Although it is true that the *T. rex* fossil is available for research, any bridge to the past has been destroyed, if it ever existed.

In the December 2010 issue of *Scientific American*, under the heading "World Changing Ideas 2010," Schweitzer describes her discoveries of blood cells, blood vessels, and proteins from various Late Cretaceous dinosaurs. Brushing aside criticism, she describes how material from "Big Mike," "Brex," and "Brachy" built her story of ancient proteins and predicts that such studies "might eventually sort out how extinct species are related, . . . how members of a lineage changed at the molecular level," and "how dinosaurs and other species responded to major environmental changes."[63] In the meantime, Jack Horner and science writer James Gorman explore in *How to Build a Dinosaur: Extinction Doesn't Have to Be Forever* the hypothesis expressed by Robert Bakker in a 1993 *NOVA* program, "The Real Jurassic Park," how it might be possible to reverse engineer a dinosaur by turning back on genetic "switches" in birds for such long-lost dinosaurian traits as teeth, long tails, and other reptilian novelties to produce a "chickenosaurus"—which might look something like *Velociraptor*.[64] In reality, a number of living birds, such as the flightless ratites, do have long tails, especially in the embryo—kiwi embryos, for example, have a reptilian tail of ten or so vertebrae— but the tail is definitively not dinosaurian, simply reptilian. Chickens have been induced to grow teeth in experiments where the epidermis responds to tooth-inducing clues from both mouse and lizard oral embryonic tissue, as well as in the talpid2 chicken mutant. Again, the teeth are not in any way dinosaurian; rather, they are archosaurian, closely resembling the conical teeth of crocodilians and in turn the early bird *Archaeopteryx*.[65] When a French team announced their tooth resurrection experiments, the press popularized the finding with the expression "quand les poules auront des dents" (English equivalent: "pigs might fly"), so one can only imagine the English headlines when Horner releases his results of a veritable living *Velociraptor* and opens his "Jurassic Barnyard."[66]

In the meantime, Pevzner and colleagues continue to work on the ancient protein problem, and perhaps by the time of this book's publication, some resolution on the proposed existence of dinosaur proteins will have been published. The reader should note, however, that through all this turmoil and popular press hyperbole, millions of taxpayer dollars have been spent on possibly nonexistent dinosaur proteins.

The late astronomer Carl Sagan popularized the well-worn saying, "Extraordinary claims require extraordinary evidence." To date, none of the extraordinary claims published on dinosaur soft tissue or protein has lived up to these lofty expectations. Perhaps more important, some of the major scientific journals, which should be exercising the highest possible editorial standards, appear to have questionable judgment when considering potentially sensational findings. Just as *Science* rushed publication in 2000 on the dinosaur "heart of stone," years ago the National Science Foundation awarded a grant to Jack Horner to study *T. rex* blood cells, timing the announcement to coincide with the theatrical release of *Jurassic Park*.[67] Funding from major governmental agencies and the subsequent papers published in top journals are used as guideposts for public policy, bearing on everything from the disbursement of research money to initiating formidable governmental programs, so in a very real sense the public trust is at stake. Regrettably, the greatest visibility is often awarded to discoveries that are later shown to be no more than junk science.

Writing on "science made glamorous," Keith Thomson, former director of the Oxford Museum of Natural History, notes:

Nowhere have scientists pursued an often too transient fame further than in paleontology, and particularly with respect to dinosaurs. Dinosaurs are God's gift to television and the newspapers, just as science fiction is the lifeblood of the supermarket tabloids. Tyrannosaurs and little green men—sure winners, both. . . . For today's hungry media, a cornucopia of new dinosaur science and new-style paleontologists, some marketing themselves as a cross between Indiana Jones and the mountain men of the Old West—with beards, boots, silly hats, unwashed shirts and unedited opinions—have been a gift from heaven.[68]

In the final analysis, if the science of dinosaur paleontology does not embrace a more self-critical ethos, it will degenerate into endless speculation and tabloid sensationalism.

Why Get Involved?

A decade has passed since Henry Gee pronounced in *Nature*: "The debate is over." Since then, the consensus view that birds are derived from theropod dinosaurs has grown even more strident. Given this environment, why would anyone write a new book reopening the question of bird origins? In part, as Lawrence Witmer warned, "Dogma is a scary thing." In addition, many workers scarcely concur with the conclusions of such recent high-profile publications as Luis Chiappe's *Glorified Dinosaurs* (2007), a major review on the Mesozoic radiation of birds by Chiappe and Gareth Dyke published in the *Annual Review of Ecology and Systematics* (2002), and Mark Norell and Xing Xu's 2005 paper on feathered dinosaurs in the *Annual Review of Earth and Planetary Sciences*. Through the work of biased editors, a love affair with dinosaurs by the popular press, and the suppression of other valid views through selective citation and rejection of manuscripts, those who find fault with the current orthodoxy for bird origins have effectively been silenced.[69]

Since the so-called dinosaur renaissance in the early 1970s, most skeptics have tried to remain open-minded to new concepts, and many remain skeptical of the speculation that has characterized the dinosaur paleontology. My early (first modern) synthesis of the field of avian evolution, *The Age of Birds* (1980), firmly questioned many aspects of the dinosaurian-bird nexus, particularly the cursorial origin of flight, which was then considered a corollary of, and inextricably linked to, the theropod origin of birds.[70] In writing the book, I had reached no concrete conclusions on avian origins, although the ground-up hypothesis for the origin of flight had always appeared terribly flawed. The furor that arose after its publication was a red flag, a signal that the science was weak. If the evidence were so overwhelming, why would anyone have been foolish enough to dispute it?

Conference in Eichstätt

In 1984, when interest in avian and flight origins had reached a peak, there was an international conference on *Archaeopteryx* and the origin of birds in the southern Bavarian village of Eichstätt. The meeting featured the famous Jura Museum, known for housing the world's widest array of Late Jurassic Solnhofen fossils, including beautiful fishes, ramphorhynchoid pterosaurs, a huge collection of dragonflies, and the then newly discovered Eichstätt *Archaeopteryx* specimen. The conference featured a display of three of the *Archaeopteryx* specimens (as well as the counterpart of the isolated feather) alongside the small Solnhofen theropod *Compsognathus*, and it included field trips to all the major Solnhofen localities. Some sixty scientists, including major figures in the field, attended. Most, like myself, were eager to see the various Solnhofen localities and talk with colleagues.

Numerous papers canvassing all perspectives on the origin of birds and flight were presented. At the end of the conference many were surprised when the ornithologist Kenneth Parkes of the Carnegie Museum (an outspoken advocate of feathers having evolved in an aerodynamic context) took the podium and requested a vote on the two paramount questions of bird origins: a direct dinosaurian or basal archosaurian ancestry of birds and dinosaurs, and a ground-up or trees-down (cursorial versus arboreal) origin of flight. The vote was clear: overwhelming *yeas* for a dinosau-

rian origin of birds, and overwhelming *yeas* for an arboreal origin of flight. While many of us were surprised at putting scientific questions to a vote, we were not surprised by the results. Yet the paradox that apparently escaped most attendees was: How could one have both an origin of birds from theropods and an arboreal origin of flight? All the known early theropods or theropod-like dinosaurs known from the Triassic, including *Herrerasaurus*, *Coelophysis*, and *Syntarsus*, have similar body plans: they are obligate bipeds with the forelimbs reduced to half the length of the highly developed hind limbs and a vertical pubis (or propubic condition) ending in a pubic boot. It was not possible (or at least highly improbable) that flight could have evolved in an arboreal context within a lineage characterized by such a rigid anatomical plan designed for a ground-dwelling lifestyle. As a consequence, many of the papers and discussions favoring a dinosaurian origin of birds were oriented toward the derivation of palatable models for how flight could have evolved from the ground up in birds (which would be exceptional among all vertebrates). These included a hypothetical protoavis using protowings for balance while running (proposed first in 1879 by Samuel Wendell Williston of the University of Kansas and later advocated by the Hungarian Baron von Nopcsa in 1907) and possibly taking off downhill.[71] A problem immediately arose with this explanation: the maximum speed for an animal of the size under consideration was about a third of that necessary to attain lift. All the models seemed to stretch feasible biomechanical limits, while the arboreal model was biologically facile, and a vast array of other vertebrates have attained flight via the trees, taking advantage of the cheap energy provided by gravity. One notable review commented that among the best and worst contributions were those concerned with the "lifestyle" of *Archaeopteryx*, but most concluded that the urvogel (or "first bird") was more birdlike than previously thought and was certainly an active flier. The most bizarre explanations for the origin of flight included one proposal that the protobird lived in mountainous terrain, jumping from steep slopes to gain speed for flight and another that viewed *Archaeopteryx* as an aquatic animal that evolved flight by bounding from one wave crest to another. Reviewer Storrs Olson appropriately termed these outlandish proposals, respectively, the "ground-down" and the "all wet" theories.[72]

The Eichstätt conference received considerable international attention, and the published volume of the proceedings is still widely cited.[73] The paleontological world was beginning to favor a theropod-bird connection and the cladistic movement was gaining momentum. Cladist advocate Kevin Padian admonished attendees to abandon other theories because his research found seventy-two indisputable synapomorphies (derived characters) linking dinosaurs and birds. Yet many of these features were primitive, highly adaptive, or co-correlated. Michael Howgate noted perspicaciously that if it could be demonstrated that any one of the key characters (characters that I have subsequently termed "trump characters"—that is, character complexes with interactive developmental pathways) did not fit, the entire phylogeny would be easily overturned.[74] Many of the characters were problematic; noted Howgate, "Although the consensus was in favor of a theropod ancestor for *Archaeopteryx*, the possibility that one key synapomorphy, if falsified, would reduce all the other synapomorphies to the status of parallelism and hence irrelevant to the debate, should be a lesson to the more dogmatic cladists."[75]

Selective Citation and Biased Publication

There has been an alarming trend in the field of paleontology to dismiss opposing views and to provide biologically bizarre explanations for bird origins and flight that fit current views of the dinosaur-bird nexus. Selective citation has also increasingly become an effective and pernicious form of scientific censorship.[76] Although most authors must necessarily be selective in their citation because of the voluminous literature, as I certainly am in this book, many works on paleontology simply omit mention of publications that don't conform to their views. A notable example is the 2007 *Glorified Dinosaurs* by paleontologist Luis Chiappe of the Los Angeles County Museum of Natural History. Chiappe promotes current theories that dinosaurs evolved protofeathers but ignores evidence that the so-called dino-fuzz may be no more than supportive skin collagen fibers in *Sinosauropteryx* (see chapter 5), a view supported by a series of papers by Theagarten Lingham-Soliar.[77] Regardless of the eventual outcome from studies of these features, all sides of the debate should be acknowledged. Although Chiappe's book mentions that John Ruben of Oregon State and colleagues viewed the protofeathers as frayed collagen fibers, it does not cite papers by functional morphologist Walter Bock of Columbia University, who was among the first to put forth an arboreal theory for the origin of avian flight.[78] Bock's insight and attention to detail has earned him international distinction in both evolutionary biology and functional anatomy. Further, there is no mention or citation of papers by perhaps the greatest figure to study vertebrate flight in recent decades, Ulla M. Lindhe Norberg of Sweden's University of Gothenberg, whose 1990 book *Vertebrate Flight* brought the analysis of flight to a new quantitative level of sophistication.[79] Her work strongly favors an arboreal origin of flight for birds, pterosaurs, and bats (first proposed by Charles Darwin).

More important, there is no mention in *Glorified Dinosaurs* of two well-known papers by distinguished Polish, Chinese, Mongolian, and Japanese scientists providing ample evidence that oviraptorosaurids, described in a cover article in *Nature* as feathered dinosaurs (see chapter 5), are in fact secondarily flightless birds, or "Mesozoic kiwis."[80] As one of the articles asserts, "Oviraptorosaurs were birds rather than dinosaurs and they were more closely related to birds than to velociraptorines."[81] Should this be the case, much of the thesis of *Glorified Dinosaurs*, that both birds and avian flight arose from earthbound theropods, is invalid. Also ignored is Stephen Czerkas's substantial case that dromaeosaurs are avian derivatives and not ancestors.[82] Paleontologists Gregory Paul and George Olshevsky support differing versions of this scenario, as do many other authors.[83]

The work of ornithologist Richard Prum of Yale, which fits the view of the current orthodoxy expressed in the book, is praised, but the work of developmental biologists Roger Sawyer, Loren Knapp, Lorenzo Alibardi, Paul Maderson, and others viewing the derivation of feathers as embryological homologues of reptilian scales or some variation thereof is ignored.[84] Last, there is selective portrayal of the evidence on claw geometry, including a misleading illustration by John Ostrom showing only the bony unguals without their outer, epidermal rhamphathecal sheaths, which provides a false impression of claw geometry.[85] There is little mention of the work of Derek Yalden of the University of Manchester, who wrote the most important paper on the topic using extensive analyses of claws to show conclusively that those of *Archaeopteryx* closely matched those of trunk-climbing mammals and birds.[86] There is no mention of my work on avian claw geometry published in *Science* in 1993.[87] There is a citation to, and a figure from, James Hopson's work attempting to show on the basis of the relative lengths of phalanges in the third toe that *Archaeopteryx* and the first beaked bird, *Confuciusornis*, were consistent with birds that are neither fully arboreal nor fully terrestrial.[88] Of course, there are hardly any birds that fit a fully arboreal or fully terrestrial description. However, there is no mention of the paper in the same volume as Hopson's by Zhonghe Zhou and James Farlow, in which they used a different metric of phalangeal statistics to show that "both *Archaeopteryx* and *Confuciusornis* were probably not much different in life style from the protobirds that pioneered life in trees."[89] These egregious cases of selective citation further damage what is already considered among the weaker fields of biology and undermine the credibility of the views expressed in the book.

Birds Are Dinosaurs: What Is the Question?

The question commonly posed by the popular press of whether birds are derived from dinosaurs is an overly simplistic portrayal of the problem, an overstatement of the conflict designed to elicit a controversy. All involved in the debate agree that bird ancestry is within the large group of Mesozoic reptiles called the Archosauria (archosaurs), to which both groups of dinosaurs, pterosaurs, crocodilians, and birds belong, so birds have a close affinity with dinosaurs regardless of their precise alignment. Yet to the casual observer of the debate on bird origins, the matter in question is simple: Are birds derived from dinosaurs or some other group of Mesozoic reptiles? This formulation of the question is imprecise because dinosaurs include a vast array of extinct reptiles, ranging from hadrosaurs and ankylosaurs to the behemoth sauropods, ostrich mimics, and of course *Tyrannosaurus rex*. Adding to the confusion are such questions and answers as those posed by Kevin Padian in the *Encyclopedia of Dinosaurs*: "What, finally are dinosaurs, and can this taxon have any stability? . . . To be a dinosaur, . . . a given animal must be a member of a group descended from the most recent common ancestor of birds and *Triceratops*."[90] Yet only through a cladistic analysis, which may in fact be faulty, does one ascertain the most recent common ancestor of birds and *Triceratops*. For our purposes, we are interested in the physical characterization of the immediate putative ancestors of birds, the large group Theropoda and a smaller subset, the maniraptoran theropods, which includes Ostrom's *Deinonychus*.

The debate centers on the unchallengeable orthodoxy that, simply stated: birds are the sister group, derived relatively late in time either directly from, or sharing a common ancestor with, a highly specialized group of Cretaceous theropod dinosaurs known as maniraptorans (often more explicitly dromaeosaurs), characterized by Chinese Early Cretaceous microraptors, and later forms such as *Deinonychus* and *Velociraptor*, which are thought to have originated from coelurosaurs, a group of small, early theropods.

Varying versions of this statement abound, however. In *The Rise of Birds*, Sankar Chatterjee writes: "The evolution of birds from dromaeosaur-like ancestors is now widely accepted among . . . paleontologists."[91] Confusion soon arises when he notes that "both groups [dromaeosaurs and birds] show skeletal differences that must have developed after they diverged from a common stock during the Middle Triassic." The oldest dromaeosaurs are known from the Early Cretaceous and possibly the Late Jurassic. In *Vertebrate Paleontology*, Michael Benton of the University of Bristol follows closely with "birds are derived theropod dinosaurs, related closely to the dromaeosaurs or the deinonychosaurs."[92] And Luis Chiappe closely links the origin of birds from theropods to the ground-up origin of flight: "Birds are descendants from theropod dinosaurs. . . . Nonavian theropods such as *Velociraptor*, *Compsognathus*, and *Tyrannosaurus* were clearly terrestrial cursors. Thus, the ancestral mode of life of birds was that of a cursorial biped. Inferences about the habits of *Archaeopteryx* should be made within this framework and not the inverse."[93] Similarly, Kevin Padian and Chiappe state that "some of the closest relatives of birds include the sickle-clawed maniraptoran *Deinonychus*. . . . The cursorial hypothesis is strengthened by the fact that the intermediate theropod ancestors of birds were terrestrial."[94]

In the prevailing view, then, birds are derived from animals similar to the dromaeosaurs *Deinonychus*, *Velociraptor*, and their allies, so the ancestor of birds likely shared similar morphological traits such as obligate bipedalism. In *Glorified Dinosaurs*, Chiappe opts for a common ancestor of maniraptorans: "Animals such as *Velociraptor* and *Deinonychus* shared a common ancestor with *Archaeopteryx* as opposed to being the direct ancestors of birds."[95] But the debate is just this: whether birds are living theropod dinosaurs and whether flight evolved from the ground up rather than from trees down. As has often been noted in the past two decades, the consensus picture is replete with problems.

Let us begin with the basics. Dinosaurs and their immediate progenitors are grouped within a more inclusive taxon, the Dinosauromorpha, and they are in turn nested within the Archosauria, which include not only dinosaurs but pterosaurs, crocodilians, and birds. It has been almost 170 years since Richard Owen coined the term "dinosaur" in 1842, and although at one time Dinosauria was considered polyphyletic, most current workers consider the group to form a fairly well

defined clade.[96] As I noted in *The Origin and Evolution of Birds*: "Are birds derived from dinosaurs? This would depend entirely on what one defines as a dinosaur, or Dinosauromorpha. . . . The real discussion over the past several decades, however, is whether birds derive from advanced theropods, such as the dromaeosaurs."[97] Of course, another possibility, entertained most recently by myself, Larry Martin, Stephen Czerkas, Frances James, John Pourtless, and others, is that birds might be derived even earlier in time from more primitive archosaurs.[98] In a sense, then, the controversy over avian origins has been overstated. Protagonists on all sides of the debate concur that birds are nested within the Archosauria, closely allied with dinosaurs. The two views contrast an earlier origin of birds from a common basal archosaur or dinosauromorph stem-ancestor of theropods and birds (a view not dramatically different from that of Gerard Heilmann in 1926) versus the more popular hypothesis that birds are the sister group of derived theropods, relatively late in time (the view aptly promoted by Yale's John Ostrom beginning in the early 1970s).[99] The latter view, however, means that there exists a hypothetical ancestor of birds and dromaeosaurs, yet to be discovered, a ghost lineage, because none of the early theropods from the Late Triassic shows any deviation from the typical morphology of a highly specialized obligate bipedalism combined with tiny foreshortened forelimbs. The consensus view thus suffers from the same problem as that of a common ancestry of birds and dinosaurs from a basal archosaurian stem-ancestor in the Late Triassic or Early Jurassic—that is, the existence of so-called ghost lineages.

Nevertheless, the question of a dinosaurian origin of birds as currently formulated invariably encompasses the following postulates: (1) birds are either the sister group of and are derived from dromaeosaurid-like (or closely related) theropods or share an immediate common ancestor therewith; (2) birds did not originate before the mid-Jurassic; (3) flight originated from the ground up (this traditional position is now modified by the discovery of the volant, arboreal *Microraptor*);

(4) *Archaeopteryx* is a derived maniraptoran or dromaeosaurid and is not primitive and ancestral to birds; (5) there was a digital frame shift, unique in amniotes and without any apparent selective context, resulting in birds shifting from a hand with digits II, III, and IV to a hand with digits I, II, and III (as in theropods) despite their pentadactyl (five-fingered) hand ground plan; and (6) all the sophisticated flight architecture of birds, including remiges, evolved in nonflight context in endothermic theropods as exaptations (preadaptations). These issues are discussed in the ensuing chapters.

Current Hypotheses

Two essential hypotheses are supported by substantial evidence, and both are considered in this book: first, the consensus maniraptoran theropod hypothesis and, second, the early archosaur hypothesis, here combined with the dinosauromorph hypothesis, both of which acknowledge an avian relationship to dinosaurs, albeit one that is more distant, but which differ mainly in the timing of the branching of Aves.

The current so-called consensus hypothesis, as noted, is usually called the "theropod hypothesis" and, put succinctly, states that the sister group (clade) of Aves is within the maniraptoran theropod dinosaurs. Maniraptorans ("hand snatchers") are a clade that includes the birds and theropods characterized by elongate arms with a tridactyl hand and a semilunate ("half-moon-shaped") bone in the carpus, or wrist, that is also found in modern birds as the semilunate carpal, which permits rotation of the hand at the wrist, enabling flight. Maniraptorans also exhibit a furcula (fused clavicles) and ossified sternal plates. In some of the most birdlike forms, such as the dromaeosaurs and the basal troodontid *Sinovenator*, there is a birdlike, elongate, backward-pointing pubic bone known as the opisthopubic condition. Advanced birdlike troodontids and oviraptorosaurs have a propubic condition in which the pubis is vertical or oriented slightly forward, but the presence of an opisthopubic condition in a basal troodontid suggests that the propubic condition is not the primitive condition for the group. In addition, modern

pennaceous feathers and flight remiges are known from most of the maniraptoran groups (except the therizinosaurs). No one as yet has a good grasp on the relationships of the rather weird therizinosaurs, which are a group of generally large "ground sloth–like" herbivorous theropods with elongate hands and coarsely serrated, lanceolate teeth, but their affinity with theropods is questionable, and they are not dealt with here. There has also been considerable debate on the systematic position of the alvarezsaurids, a group of small, enigmatic birdlike animals with highly attenuated forelimbs adapted in Late Cretaceous forms such as *Mononykus* for digging and with a tubular snout suggestive of feeding on colonial insects. They are variously placed as the sister group of the Ornithomimosauria or as primitive members of the Maniraptora. The recently discovered Chinese, early Late Jurassic *Haplocheirus* supports maniraptoran status (although the ornithomimosaur is nearly as well supported) and extends the range some sixty-three million years.[100] The greatly foreshortened forelimbs of *Haplocheirus* still retained a grasping function, but it curiously lacked the suite of birdlike features, such as fused wrist bones and backward-facing pubis, which may indicate a convergence toward the avian condition by the Late Cretaceous. Given their enigmatic status, alvarezsaurids will not be considered further here.

Most current workers, including myself, agree that the most birdlike of the maniraptorans, the Troodontidae and the Dromaeosauridae, which are united as a Deinonychosauria, are the sister group of Aves, which appears to be deeply nested within the maniraptorans and, more specifically, most likely within the Deinonychosauria. Therefore it is reasonable to conclude, according to current commonly used nomenclature and as a working hypothesis, that "birds are maniraptoran theropods," and if birds are "avian maniraptorans" then the remaining groups, the Oviraptorosauria, Troodontidae, Dromaeosauridae, Alvarezsauridae, and possibly Therizinosauridae, are nonavian maniraptorans. Frances James and John Pourtless have called the hypothesis that birds are nested within the Maniraptora the "BMT" hy-

A   **B**

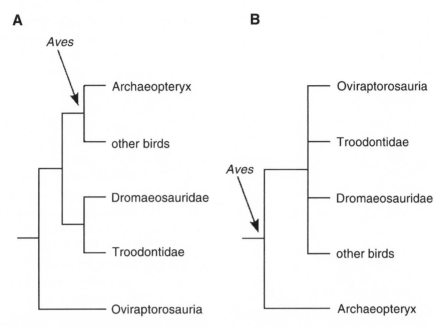

Fig. 1.2. Alternative relationships of birds and selected maniraptorans. (A) The sister group of Aves is within the maniraptoran theropod dinosaurs as in fig. 1.3 (the BMT, birds-are-maniraptoran-theropod-dinosaurs topology). (B) Selected maniraptorans are included within Aves. Reproduced with permission from F. C. James and J. A. Pourtless IV, "Cladistics and the Origin of Birds: A Review and Two New Analyses," *Ornithological Monographs*, 66 (2009): 1–78; permission conveyed through Copyright Clearance Center.

pothesis ("birds are maniraptoran theropod dinosaurs"), and they consider the oviraptorosaurs, dromaeosaurs, and troodontids to be "core maniraptorans," a usage that will be followed here.[101] Various core maniraptorans have suites of derived avian features, including pennaceous contour feathers, flight remiges on both hands and hind limbs; nonserrated, waisted (constricted at the base) avian-like teeth; uncinate processes of the ribs; double-condyle quadrates; laterally facing glenoid sockets; retroverted (opisthopubic) pubes; distal caudal vertebrae fused into a pygostyle; an avian hand with a bowed carpometacarpus; highly recurved and laterally compressed manual and pedal claws; and a reversed second toe, or hallux.

Jacques Gauthier in 1986 named the Maniraptora in 1986 for a branch-based clade (monophyletic assemblage) containing all dinosaurs closer to modern birds than to the ornithomimids, the "bird-mimic theropods," primarily of the Late Cretaceous.[102] Although the name refers to their

"seizing hands," many apparently used their hands in other ways. The most primitive dromaeosaurs, for example, the Early Cretaceous microraptors, were four-winged and used their hands both in flight and to climb trunks. It is unfortunate that the word "raptor" was used as a vernacular for *Velociraptor* and then popularized in the movie *Jurassic Park*, as it has historically been used to designate raptorial birds such as hawks, eagles, and owls and, as we shall see, may be inappropriate for most of the Maniraptora. We have already seen that an "apomorphy-based" (derived character–based) definition included their possession of elongate, winglike arms and birdlike fingers, characters by which they can be easily recognized.

The BMT hypothesis originated in the work of Yale University's John Ostrom, discussed in chapter 6.[103] His discovery of the famous, man-sized dromaeosaurid Cretaceous *Deinonychus* led him to extensive comparisons of the skeletons of that raptor and the earliest bird, the

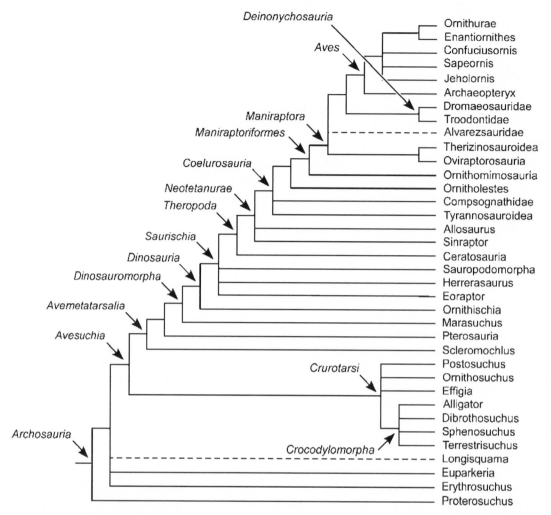

Fig. 1.3. Generally accepted phylogeny of the Archosauria, including the topology of the BMT hypothesis. Reproduced with permission from F. C. James and J. A. Pourtless IV, "Cladistics and the Origin of Birds: A Review and Two New Analyses," *Ornithological Monographs* 66 (2009): 1–78. Permission conveyed through Copyright Clearance Center.

Late Jurassic *Archaeopteryx. Deinonychus*, with its sickle-like claw on the second digit of the foot and its birdlike semilunate carpal, was considered a bipedal predator whose forelimbs were preadapted for flight, and it has therefore been assumed that obligate bipedalism preceded flight and, necessarily, that flight originated from the ground up. Ostrom considered *Archaeopteryx* to be a cursorial predator, incapable of powered flight. In fact, in his early work, Ostrom considered *Archaeopteryx* incapable of all flight and viewed it as largely a terrestrial predator. Later Jacques Gauthier and his

doctoral professor Kevin Padian, using cladistic analyses, posited that Aves had a sister group relationship with either the Deinonychosauria or more exclusively with the Dromaeosauridae.[104] In many texts the avian ancestor is pictured as a nonflying, terrestrial dromaeosaur, not unlike the Late Cretaceous, turkey-sized *Velociraptor* ("swift seizer").

The BMT hypothesis has been widely hailed as one of the most significant outcomes of the advent of modern phylogenetic systematics and a major discovery of twentieth-century biology, although in reality it was accomplished by John Ostrom's discovery and study of *Deinonychus* in the 1960s, and that revelation occurred in the absence of phylogenetic systematics or cladistics, to which he did not adhere. Yet the theory has been so pervasive in paleontology that many, as noted earlier, have actually called for ending debate on bird origins. But others in the field believe that the situation is far more complex. As we shall see, Gregory Paul has been largely responsible for reopening the inquiry by proposing that maniraptorans with suites of birdlike characters are in fact best explained as being secondarily flightless birds, more derived than the earliest known basal bird, the Late Jurassic *Archaeopteryx*, but still ultimately derived within the coelurosaurian theropods.[105]

Among the major concerns with the original BMT hypothesis are its assumptions of homology for contentious characters, particularly the carpal, manual, and tarsal elements. Another problem that has persisted for decades is that a traditional dinosaurian origin of birds, as envisioned by Ostrom, Padian, and most paleontologists, involves the biophysically difficult ground-up, or cursorial, origin of flight, as opposed to the extremely facile trees-down, or arboreal, theory, by which flight has originated in all other volant groups of vertebrates. The recent discovery of small basal dromaeosaurids such as the four-winged *Microraptor*, however, make it entirely plausible that an arboreal origin of flight could be incorporated into a modern BMT hypothesis. Despite these discoveries, many of the long-term and hard-core advocates of the BMT hypothesis have been unwilling to consider alternatives to flight origin from the ground up, not withstanding its improbability.

The Archosauria

Although BMT is the current orthodoxy, the consensus hypothesis, there are alternatives, some of which must be seriously considered. To begin, virtually all paleontologists find the evidence compelling that the origin of both birds and dinosaurs lies within the clade Archosauria, a Triassic group of reptiles characterized by their diapsid skulls (with two temporal openings), an antorbital fenestra, a lateral mandibular fenestra, and serrated teeth implanted in sockets. Basal archosaurs

include such well-known genera as *Proterosuchus*, *Erythrosuchus*, and the classic basal archosaur (or pseudosuchian thecodont) *Euparkeria*, from the Early Triassic of South Africa. As one can see from the diagram on page 26, the crown group of Archosauria is called the Avesuchia, a term employed by British paleontologist Michael Benton, and it includes two daughter clades, the Crurotarsi ("cross-ankles," the crocodile line), a name that has largely replaced the older term Pseudosuchia, and its sister group the Avemetatarsalia or Ornithodira ("bird-ankles," the bird line), which includes all forms closer to birds than to crocodilomorphs.[106] The crurotarsans are characterized by such groups as the convergently crocodile-like phytosaurs, the armored, herbivorous aetosaurs, the large, predatory poposaurs, the crocodilians (alligators, crocodiles, and gavials), and the erect-limbed rauisuchians. Crurotarsans appeared in the Early Triassic and were the dominant terrestrial carnivores by the Middle to Late Triassic, the latter their heyday. Except in ankle structure the Late Triassic rauisuchians shared so many synapomorphies with theropod dinosaurs that Sankar Chatterjee mistakenly thought them directly related to tyrannosaurids. Although they form a separate lineage, many of their characters developed convergently with theropod dinosaurs.[107]

The sister clade of the crurotarsans is the Avemetatarsalia, characterized by their hind-limb modifications for obligate bipedal posture, with elongate tibiae and compact, elongate metatarsi with reduced fifth metatarsals. This clade was established by Benton in 1999 for avesuchians closer to dinosaurs than to crocodilians.[108] The avemetatarsalians include the Pterosauria, Dinosauria, and Aves, and this is the clade that largely concerns us in this book. The basal members of this clade were small, arboreal quadrupeds or facultatively bipedal forms, and they branched into the pterosaurs and the small, ground-dwelling dinosauromorphs, early forms of which were less than a meter (3 ft.) long and carnivorous. The dichotomous split of dinosaurs into Ornithischia ("predentates") and Saurischia occurred shortly after their origin in the Late Triassic. Reversion to a four-limbed posture, known as quadrupedalism, evolved several times

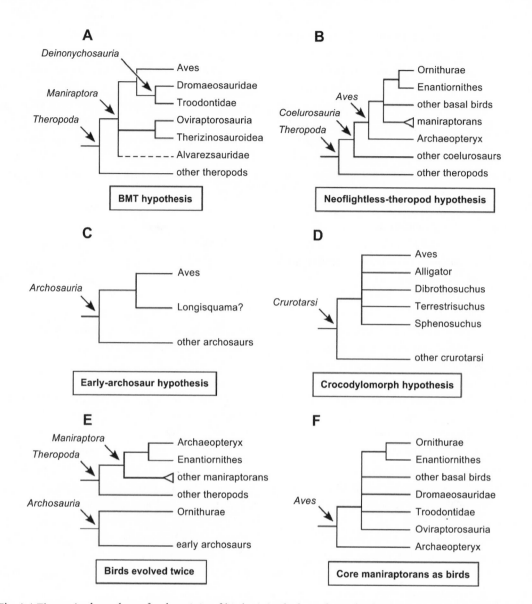

Fig. 1.4. Five major hypotheses for the origin of birds. (A) The hypothesis that birds are maniraptoran theropod dinosaurs (BMT) and that the sister group of birds is the Deinonychosauria (or the Dromaeosauridae). (B) The neoflightless theropod hypothesis, that the sister group of birds lies among the coelurosaurian theropod dinosaurs but not among the Maniraptora, which are viewed as flightless birds. (C) The early archosaur hypothesis, that the sister group of birds is an early arboreal archosaur. (D) The crocodylomorph hypothesis, for which various topologies have been proposed: a sister group relationship between birds and crocodylomorphs (represented by *Alligator, Dibrothosuchus, Terrestrisuchus,* and *Sphenosuchus*) and Aves as nested within Crocodylomorpha but branching off before Crocodylia, which is represented by *Alligator,* or with Aves as the sister clade of Crocodylia. The topology shown is a polytomy that could be resolved in favor of any of these topologies. (E) The hypothesis that "birds" evolved twice: one lineage is *Archaeopteryx,* Enantiornithes, and the Maniraptora, and the other is Ornithurae. Hypotheses B, C, and D include or could include the topology shown in (F), that at least three clades of maniraptorans (Dromaeosauridae, Troodontidae, and Oviraptorosauria) were radiations within Aves, whose members were at varying stages of flight loss or flight. Reproduced with permission from F. C. James and J. A. Pourtless IV, "Cladistics and the Origin of Birds: A Review and Two New Analyses," *Ornithological Monographs* 66 (2009): 1–78. Permission conveyed through Copyright Clearance Center.

independently among the ornithischian dinosaurs and among the large, herbivorous saurischian sauropods, presumably following increase in body size.

Alternatives and Modifications to the BMT Hypothesis

Given the diversity of the maniraptorans, our discussions here will be concentrated largely on the "core maniraptorans": oviraptorosaurids, troodontids, and dromaeosaurs. Frances James and John Pourtless of Florida State University re-analyzed the origin of birds from a cladistic point of view in order to evaluate whether the BMT hypothesis is well supported, as is generally claimed, and to evaluate alternative hypotheses for bird origins within a cladistic, phylogenetic framework. Their overall conclusion is that "because of circularity in the construction of matrices, inadequate taxon sampling, insufficiently rigorous application of cladistic methods, and a verificationist approach, the BMT hypothesis has not been subjected to sufficiently rigorous attempts at refutation." They point to the viable alternatives of an early archosaur hypothesis and a variant of the crododylomorph hypothesis as worthy of continued investigation, along with the consensus theropod hypothesis.[109]

James and Pourtless have nicely categorized five hypotheses for bird origins through time, as shown in the figure on page 28. Hypothesis A, the BMT hypothesis, that birds are maniraptoran theropod dinosaurs. Hypothesis B, The neoflightless-theropod hypothesis, promoted by Gregory Paul, proposes that the sister group of birds lies among the coelurosaurian theropod dinosaurs but not among the Maniraptora, which are flightless birds. The early archosaur hypothesis (C) is that the sister group of birds is an early arboreal archosaur, resembling *Longisquama*. This hypothesis is somewhat confused in that others have included Triassic Dinosauromorpha as part of this hypothesis, with forms such as *Marasuchus* (=*Lagosuchus*) and *Lagerpeton* from the Middle Triassic of Argentina, as potential ancestors. Jacques Gauthier and Kevin Queiroz in 2001 gave

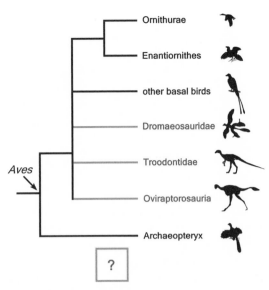

Fig. 1.5. If core maniraptoran theropods (Dromaeosauridae, Troodontidae, and Oviraptorosauria) were flightless and flying birds, more derived toward modern birds than *Archaeopteryx*, then the hypothesis that birds are maniraptoran theropods would lose most of its current support, and the origin of birds would have to be evaluated in the light of at least four other hypotheses. Reproduced with permission from F. C. James and J. A. Pourtless IV, "Cladistics and the Origin of Birds: A Review and Two New Analyses," *Ornithological Monographs* 66 (2009): 1–78. Permission conveyed through Copyright Clearance Center. Icons adapted from N. Longrich from various sources, "Structure and Function of Hindlimb Feathers in *Achaeopteryx lithographica*," *Paleobiology* 32 (2006): 417–31, and following L. Chiappe, "Basal Bird Phylogeny: Problems and Solutions," in L. M. Chiappe and L. M. Witmer, eds., *Mesozoic Birds: Above the Heads of Dinosaurs* (Berkeley: University of California Press, 2002), 448–67; and X. Xu and F. Zhang, "A New Maniraptoran Dinosaur from China with Long Feathers on the Metatarsus," *Nature* 92 (2005): 173–77.

the name Panaves ("all birds") to the Avemetatarsalia but included all Aves, Dinosauria, Pterosauria, and a variety of Triassic archosaurs, such as the dinosauromorphs *Marasuchus* and *Scleromochlus*.[110] Given the taxonomic uncertainty of many of these poorly known groups, it would appear preferable not to restrict hypothesis C to forms such as *Longisquama* but to include Triassic dinosauromorphs such as marasuchids, and that will be the scheme followed in this book. Hypothesis D, the crocody-lomorph hypothesis, with birds as a sister group to the Triassic crocodylomorphs, the sphenosuchids, gained support during the 1970s and has some interesting supporting evidence, but is not generally considered a major competing hypothesis today. However, it may at some time be reopened if sufficient evidence emerges. Hypothesis E, the product of Evgeny Kurochkin, posits that "birds evolved twice," producing one clade of *Archaeopteryx*, en-antiornithines, and maniraptorans, derived within Theropoda; and a separate clade of Ornithurines, derived from early archosaurs.[111] Last, in hypothesis F, James and Pourtless characterize a scenario that could incorporate hypothesis B, C, or D, in which at least three clades within Aves (Dromaeo-sauridae, Troodontidae, and Oviraptorosauria) are later radiations within Aves, at various stages of flight and flightlessness, a theory advocated prominently by Gregory Paul, with a coelurosaur dinosaurian ancestry.[112] However, in more recent papers by myself and Stephen Czerkas, the ancestor is assumed by the current evidence to be an early "basal" archosaur or dinosauromoph.[113] Yet the absence of satisfactory character-based definitions for any of the groups under discussion makes it extremely difficult to get a handle on the problem.

As an interesting aside, although subjected to thorough review and published in the prestigious journal *Ornithological Monographs*, James and Pourtless's paper is almost universally ignored by paleontologists and is not cited by papers dealing with theropods and bird origins. It is a virtual certainty that the James and Pourtless monograph will encounter rebuttals, and there will be counter-rebuttals and rejoinders, but the important point is that all these cladistic phylogenies often hang by a spider's thread and their entire geometry can change rapidly.

The complexity of animal phylogenies is truly astounding, yet current cladograms tend to portray a simplistic view, often giving the impression of a straight-line, or orthogenic, progression and not accommodating the complexities of evolution. For example, mosaic evolution, the phenomenon by which major evolutionary changes take place in stages, is common, so that the rate of evolution in one functional system may vary from that of other systems. Arrested development and massive convergent evolution are other common phenomena inadequately portrayed by cladistics. One need only compare ancient attempts to deal with phylogeny, as exemplified by those of Ernst Haeckel, to the certitude on display in modern cladograms,

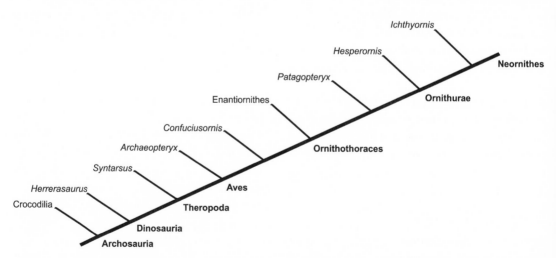

Fig. 1.6. Characteristic cladistic representation of dinosaur-bird relationships, showing the linear, simplistic view of evolution, without the complex bushiness reflecting the reality of the evolutionary process. By this diagram, one gets the impression of an orthogenic or straight-line evolution, in a unilinear manner, without the complexities inherent in the evolutionary process, including mosaicism and other known phenomena that do not lend themselves to such a presentation. Modified from A. Chinsamy-Turan, *The Microstructure of Dinosaur Bone* (Baltimore: Johns Hopkins University Press, 2005).

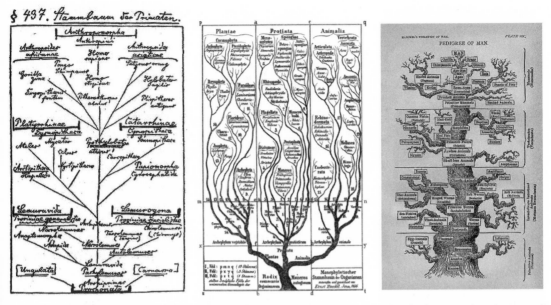

Fig. 1.7. Three "Trees of Life" constructed by Ernst Haeckel, the great German biologist, embryologist, and popularizer of Darwin's theory, showing the appearance of early attempts at phylogenetic trees. Despite its faults, Haeckel's tree, unlike the simplistic cladistic representations, shows the inherent complexities of such an undertaking. Darwin thought that the phylogeny of organisms was expressible metaphorically as a *Tree of Life*. Left, a 1960s Tree of Life showing a tree leading to *Homo sapiens*; middle, from *Generelle Morphologie der Organismen* (1866), with three kingdoms of plants and animals; and right, a "Pedigree of Man" from *The Evolution of Man* (1879).

to illustrate the problem. Current cladograms convey an aura of certainty to what is in fact speculation. This is particularly alarming when one considers that substantial numbers of such character-based phylogenetic analyses of living groups are summarily overturned by DNA or genomic comparisons. We need only remember that in 1900 the president of the Royal Society proclaimed that everything of importance had already been discovered by science. If we have learned anything through the years, it is that certainty is a failed historical enterprise.[114]

Defining Theropoda

Although the clade Dinosauria can be defined by character-based (apomorphy-based) criteria, relying on the nature of the unique postaxial (posterior) reduction of manual digits (in which digits I, II, and III dominate), in combination with a symmetrical reduction of pedal digits and some other features, a character-based definition of Theropoda is less easily achieved, partly because the group as currently defined cladistically is so diverse and partly because many of the Triassic archosaurs, including crurotarsans, attained large suites of theropod-like apomorphies. As noted, the Late Triassic rauisuchian *Postosuchus*, for example, were it not for the ankle region, would most likely qualify cladistically as a theropod (possibly a ceratosaur), and it is not too surprising that Chatterjee concluded erroneously that it might be basal to tyrannosaurids.[115] *Postosuchus* has suites of theropod-like characters, however, including serrated, theropod-style teeth with interdental plates, an intramandibular joint, and a theropod-like skull and pubis, with a pubic boot.

As unsatisfying as it may be, the only way to create a clade Theropoda is via cladistic analyses through which the group emerges as a monophyletic assemblage, with many theropod-like taxa excluded as outgroups because of their tarsus or other key characters. At present there does not appear to be any simple character-based definition of

Theropoda. This book explores the possibility that at least some maniraptorans belong within Aves, an exciting avenue of investigation opened up with the massive discovery of beautifully preserved Lower Cretaceous fossils from China. Are the so-called feathered dinosaurs in fact members of the early radiation of birds, at various stages of flight and flightlessness, and, like the living ratites, did they come to superficially resemble theropods through their acquisition of the flightless condition?

The Debate Continues

As the debate continues, it is increasingly clear that most workers on both sides of the debate were at least partly wrong, but in many cases for the right reasons. Early theropods are anatomically constrained, with large hind limbs designed for a cursorial, obligately bipedal life style, with a strict parasagittal gait (upright and beneath the body—in such a gait, limbs move parallel to the backbone [in parasagittal plane] and held vertically, rather than sprawling sideways), and forelimbs invariably reduced to half or less the length of the hind limbs. Among the major objections to an earlier posited common ancestry of birds and dinosaurs (the so-called thecodont or basal archosaur hypothesis) was that small dinosaurs, like the Late Jurassic *Compsognathus*, lacked clavicles (in modern birds fused to form a wishbone, or furcula), and Gerhard Heilmann in his *Origin of Birds* in 1926 noted that reinstatement of the clavicles in the birds would represent a violation of Dollo's Law, the proposition that evolution does not normally reverse itself. What about the current theory? A far more complex reversal would be exactly what the current orthodoxy calls for: to reelongate already foreshortened forelimbs.

Scientific success today consists largely in the constant production of major discoveries; the currency is new ideas. Our culture prefers to declare clear winners and, more important, losers with respect to any major scientific question. This battleground scenario is ginned up and driven by the popular press, which constantly seeks to portray high-stakes battles within the sciences; the more bitter the controversy, the better. Yet science generally progresses through the efforts of honest,

hardworking, and dedicated scientists who often make breakthroughs through accumulated knowledge gained over decades, their careers spotted with both disappointments and exciting periods of discovery. Going through the chapters, we shall see how Thomas Huxley incorrectly interpreted many features of the lancelet amphioxus, but he was correct in assigning it a position near the ancestry of vertebrates; he was also incorrect in arguing that ratites were derived from earthbound dinosaurs through a nonflying stage. Richard Owen erred in denying that *Archaeopteryx* was a reptile-bird intermediate, but he was correct when he argued for arrested development as the mechanism by which ratites evolved from flying ancestors. German embryologist Ernst Haeckel was incorrect in the formulation of his Biogenetic Law, but he was correct in the assertion of a general link between ontogeny and phylogeny. Much later, John Ostrom correctly noted the similarity of *Deinonychus* to birds, but his conclusion that *Deinonychus* was therefore close to the ancestry of birds was incorrect, and his insistence that avian flight must consequently have evolved from the ground up was equally incorrect. Various colleagues and I were incorrect in believing that *Deinonychus* (and other dromaeosaurs) had nothing to do with the origin of birds because of its late geological occurrence and its possible convergence on the morphology of birds, but we were almost certainly correct in insisting that flight originated from the trees down. Gregory Paul was likely incorrect in viewing dinosaurs as endothermic, but his hypothesis that some birdlike dinosaurs were actually derived from birds and not the reverse is most likely correct.

The goal of this volume is not a complete renovation of the field; rather, it is written only with the hope that people interested in the origin of birds will begin to see that all that glitters in the field of dinosaur paleontology is not gold. It is my wish that readers will consider the alternative view, elaborated upon in the pages that follow, that in fact a vast array of "hidden birds" once dwelled among China's feathered dragons.

As Thomas Huxley quipped in a letter to Charles Darwin, "The revolution that is going on is not to be made with rose-water."[116]

2

WHAT DID
EVOLUTION'S HIGH
PRIEST SAY?

My good and kind agent for the propagation
of the Gospel—i.e. the devil's gospel.

Charles Darwin to T. H. Huxley,
8 August 1860

In December 1874, Thomas Henry Huxley, passionate and eloquent defender of Darwin's theory and commonly dubbed "Darwin's bulldog," wrote to Ernst Haeckel, the famed German embryologist and an early champion of Darwinian ideas. The subject was the anatomical and embryological similarity of *Amphioxus* (a small, primitive, fishlike chordate whose Latin name means "sharp at both ends") to the true vertebrates: "My Dear Haeckel . . . it is demonstrable that *Amphioxus* is nearer than could have been hoped to the condition of the primitive vertebrate—a far more regular and respectable sort of ancestor than even you suspected."[1] Amphioxus (now a collective term for some thirty species of lancelets, superseded by two generic names, most included in *Branchiostoma*) was originally described as a slug by P. S. Pallas in 1774. Almost a century later, Alexander Kowalevsky showed that amphioxus, together with the closely allied tunicates (sea squirts), were of chordate affinity, perhaps ancestral to vertebrates. Following Kowalevsky's work, the classification of the lancelets fluctuated considerably. Lancelets are today placed in the subphylum Cephalochordata, or "head chordates," so named because the notochord extends into the anterior cephalic region as an adaptation for burrowing in the marine substratum. They were variously considered, as by

Huxley, as the direct ancestors of vertebrates, as primitive offshoots of the early vertebrate lineage, and even as degenerate fishlike creatures with little connection to the ancestry of vertebrates; the last view was popular when I took comparative anatomy as an undergraduate in 1963.

Amphioxus Illustrates the Difficulty of Phylogenetic Reconstruction

In the 1980s Charles Jenner, an invertebrate biologist in the Department of Biology at the University of North Carolina, rekindled my interest in the lancelets. Jenner had been a graduate teaching assistant under eminent anatomist and vertebrate paleontologist Alfred Sherwood Romer at Harvard. He was an avid and accomplished collector, and on one occasion he brought in live amphioxus specimens, collected along the coast, and an early stage ammocoetes larva of a lamprey (a undisputed jawless vertebrate) of similar size, collected in a mountain stream, where lampreys spawn. (It takes up to seven years for the ammocoetes larva to undergo metamorphosis, marine species returning to their ancestral ocean home.) When placed side by side, the amphioxus in a saltwater aquarium, and the larval lamprey in fresh water, astonishing anatomical and behavioral similarities

Fig. 2.1. Left, Thomas Henry Huxley at twenty-one, in 1846; right, Huxley in about 1880; from Leonard Huxley, *Life and Letters of Thomas Henry Huxley*, 2 vols. (London: Macmillan, 1900).

were inescapable. Although lancelets split from vertebrates more than 520 million years ago, they provide clues as to how early vertebrates may have evolved.

In 1911, Charles Walcott described the Middle Cambrian *Pikaia gracilens* from the famous Burgess Shale as a polychaete worm with regular segmentation. Decades later, in 1979, Simon Conway Morris, invertebrate paleontologist at Cambridge University, fellow of the Royal Society (elected at the age of thirty-nine), and author of *The Crucible of Creation*, reexamined the Burgess Shale fossils and concluded that the 5-centimeter-long (1.5 in.) *Pikaia* was not a worm but a cephalochordate.[2] Morris placed *Pikaia* with the living lancelets and considered the group to be the prime candidates for vertebrate ancestry. The existence of a cephalochordate 520 million years ago strongly suggested a phylogenetic position close to the ancestry of vertebrates. (Another fossil cephalo-

chordate, *Cathaymyrus* from the Early Cambrian of China, is now known.) *Pikaia* and other cephalochordates display unusual anatomical specializations like the forward extension of the turgid axial skeletal rod (the notochord), an adaptation for burrowing and filter feeding in the substratum. In overall structure, however, with Z-shaped muscular segments (myotomes), finlike metapleural folds along the lower edges of the body, a hollow dorsal spinal cord, and slits in the pharynx, the lancelets seem an excellent candidate for vertebrate ancestry.

Biologists have increasingly taken an interest in the living lancelets, for obvious reasons, not the least of which is that they occur in great masses in Tampa Bay, an additional attraction for geneticists seeking biomass for experimental work. Among the noteworthy researchers are Linda and Nicholas Holland of the Scripps Institution of Oceanography. In 2001, following years of intensive

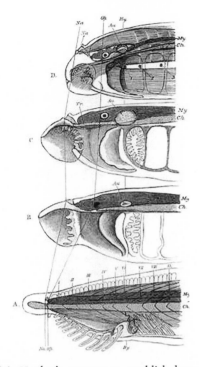

Fig. 2.2. Huxley's attempt to establish homologies between amphioxus and the jawless fishes, using the larval lamprey, or ammocoete. Head region of amphioxus (A), followed above by three stages (B, C, D) in the development of a lamprey, *Petromyzon*. He was convinced that the amphioxus had equivalents of the cranium and brain: "It possesses very well marked and relatively large divisions of the cerebro-spinal nervous axis and of the spinal column, which answer to the encephalon and the cranium of higher Vertebrata," an overstatement perhaps derived from euphoria over such impressive anatomical similarity. From T. H. Huxley, "Preliminary Note upon the Brain and Skull of Amphioxus Lanceolatus," *Proceedings of the Royal Society of London* 23 (1874–75): 127–32.

genetic studies, the Hollands wrote an article on amphioxus as a model for the ancestral vertebrate.[3] In 2007, in a cover article in *Nature*, the genome sequence of the Florida lancelet *Branchiostoma floridae* was unveiled, reinvigorating the study of the origin of vertebrates.[4] The sequencing of the amphioxus genome is a triumph for the methodology of the biological sciences. Persistent advances in the study of protochordates have tracked increasing complexity in technology throughout the past 236 years since the original description of am-

phioxus, resolving first the confusion regarding its true identity, then its phylogenetic position and its homologies with vertebrates. Finally, painstaking efforts have culminated in the complete sequencing of the genome, showing that our early evolution was most likely characterized by genomewide duplications and subsequent reorganizations in the vertebrate lineage. These genome duplications provided the additional genetic variation exploited by the incipient vertebrates in producing features such as the neural crests and complexities of the nervous system and head that we identify as salient characteristics of vertebrates. Surprisingly, however, more recent comparison of the genomes of humans, sea squirts, and lancelets shows that the lancelet lineage diverged long before the vertebrates and that the lowly sea squirts may be closest to ancestry of vertebrates.[5] Most recent authors have regarded amphioxus as the closest relatives of the true vertebrates on the basis of phylogenetic analyses of some ten to fifteen characters not seen in the sea squirts, but genomic comparison argues against this conclusion. Although systematics today seeks a system in which molecular and cladistic morphological phylogenies converge to the same result, there is little reason to believe that the two data sets will ever yield identical phylogenies. As is often the case in the reconstruction of the evolutionary history of a group of organisms, the morphological and genetic evidence conflict, the latter often revealing long-concealed cases of convergence.

Although appealing, this story of the triumph of biological methodology belies the actual history of the situation. It is true that Huxley recognized the affinities of amphioxus and vertebrates, but in his letter to Haeckel, noted above, he made some egregious errors, noting that he had found the skull and brain of *"Amphioxus,"* "both of which are very large (like a vertebrate embryo's)." Of course, there is no cranium in amphioxus, and there is no anterior expansion of the brain, with its three distinctive divisions, as in vertebrates. The best approach today, employing Hox genes, has been to establish definitive homology between the cells of the amphioxus cerebral vesicle and the vertebrate diencephalic forebrain, as well as between

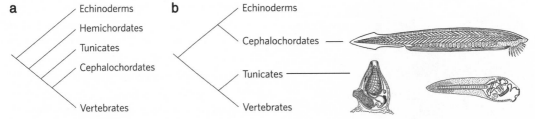

Fig. 2.3. Morphological versus molecular phylogenies for the origin of vertebrates, to illustrate the still intractable problem of phylogenetic reconstruction based on character-based cladograms. (a) The classic textbook view, still popular today, based on morphology, implies a small increase in complexity from simple, sedentary ancestors to the motile, fusiform vertebrates. Modified from H. Gee, "Careful with That Amphioxus," *Nature* 439 (2006): 923–24. (b) The phylogeny suggested by recent phylogenomic evidence. Right, adult of amphioxus (upper) and a sea squirt or tunicate, along with its larval form. This is a classic example of the difficulty of character-based phylogenies, which is all we have for paleontology, and which in modern vertebrates are largely in conflict with molecular data. Morphology and molecular data do not always coincide. After Y. Delage and E. Hérouard, "Les Protochordés," in Delage and Hérourd, *Traité de Zoologie Concrète*, vol. 8 (Paris, 1898).

the amphioxus endostyle (mucous-secreting gland used in filter feeding) and the vertebrate thyroid gland.[6] Huxley, however, was convinced that amphioxus was "a far more regular and respectable sort of ancestor than . . . suspected," and he turned out in the long run to be reasonably correct, in the sense that cephalochordates have many features associated with the chordate to vertebrate transition.[7] But he was dead wrong on the anatomy of the head of amphioxus and its true phylogenetic position, and the story vividly illustrates the extreme difficulty of deciphering the evolutionary history of earth's creatures, or establishing a *Tree of Life*, Darwin's metaphor for the ascent of life forms through time, or the modern phylogenetic tree. Huxley was, like other workers in the realm of Victorian science, wrong on any number of other aspects of vertebrate descent.

Most of us who have worked on amphioxus made the same mistake as Huxley regarding phylogeny, but there is a profound lesson here in the intractable problems involved in the extraordinarily difficult problem of phylogenetic reconstruction. Indeed, as we shall see, despite the euphoria of modern phylogenetic systematics and its methodology at having solved vast problems of the field, the majority of phylogenies, particularly in the case of birds, have been firmly refuted by subsequent molecular systematics. This is to say not that molecular systematics is the final solution to the problem but rather that it has cast serious doubt on the uncompromising faith in cladistic methodology alone by its ardent adherents. Resolution of phylogeny by character-based cladistics is a far more complex issue than has been recognized; and it follows of course that problems in reconstruction phylogeny from the fossil record alone, as is the case with avian origins, are compounded because there is no possibility of independent molecular comparisons.

Huxley and Owen

Historical overviews of the debate concerning the origin of birds often cite Huxley as an authority on all things dealing with evolution, and he was the first to propose a comprehensive theory that birds were derived from dinosaurs. Indeed, the current consensus that birds are derived from dinosaurs, initiated by Yale's John Ostrom of Yale in the early 1970s, is often viewed as a rejuvenation of Huxley's arguments in a more modern context. Following in this vein, many modern authors insist that the reptilian bird *Archaeopteryx* is little more than a feathered dinosaur. It is therefore instructive to examine exactly Huxley's views about the urvogel and the origin of birds.

A year after the discovery of a beautifully preserved skeleton of the small Late Jurassic theropod

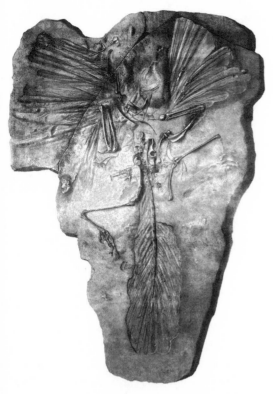

Fig. 2.4. London specimen of *Archaeopteryx*. Courtesy Stephen Czerkas; copyright Stephen A. Czerkas.

Fig. 2.5. The small Solnhofen theropod *Compsognathus*. Courtesy Stephen Czerkas; copyright Stephen A. Czerkas.

dinosaur *Compsognathus* (*kompsos*, "elegant"; *gnathos*, "jaw"), described by Johann A. Wagner in 1859 and 1861, the first Jurassic feather and then the first skeletal specimen of the earliest known bird were found in the same lithographic limestones in Solnhofen in southern Bavaria.[8] I dubbed *Archaeopteryx* (*archaios*, "ancient"; *pteryx*, "wing") an "avian Rosetta Stone," noting that it perfectly fulfills the Darwinian expectation of a fossil intermediate between two major groups of organisms, in this case reptiles and birds.[9] Yet it was controversial from the beginning. One can imagine the excitement such a discovery created, just two years after the publication of Darwin's *On the Origin of Species*, with Victorian science on the march and intense debate on the theory of evolution aroused by the appearance of a bird with reptilian features and teeth.[10] This story has been told many times and need not be repeated in detail here.[11] Suffice it to say that two main protagonists emerged: Sir Richard Owen, leader of British science as well as of the British Museum, and Huxley, eloquent defender of Darwinian evolutionary theory.

Owen and Huxley had squared off in a long, open-ended debate, and the two saw eye to eye on almost nothing. Most notably, Owen, who in 1842 coined the term "Dinosauria" ("terrible lizard"), was no friend of the Darwinian movement, and although not a Creationist by today's standards, he did believe in a "continuous creation" by which new forms of life were created from time to time from basic anatomical blueprints, so-called archetypes.[12] Perhaps the most important early triumph for Huxley was his famous debate with Samuel Wilberforce, the bishop of Oxford, in 1860 in the Oxford University Museum of Natural History, at a meeting of the British Association for the Advancement of Science. Wilberforce had supported the building of the museum as a center not only for Oxford's science departments but also for the study of the wonders of God's creations, including man. In the confrontation, now widely portrayed as the Oxford Evolution Debate, representatives of the church and science debated the validity of evolution, and although the debate is said to have been witnessed by more than a thousand spectators (it is hard to believe so many could have

Fig. 2.6. Caricatures of Wilberforce ("Soapy Sam"), left, and Huxley ("Darwin's bulldog"), right, published in *Vanity Fair* in 1869 and 1871, respectively.

crowded into the room), there are few eyewitness accounts, and many legendary renditions are available.

Wilberforce, known as Soapy Sam (from Benjamin Disraeli's comment that the bishop's manner was "oleaginous, saponaceous"), was among the greatest public speakers of his day, and for his encounter with Huxley he had been coached by none other than Richard Owen, an earlier opponent of Huxley on the question of whether man was closely related to the great apes. Huxley's well-known demolition of Wilberforce in the debate was pivotal in propelling him into a position of public prominence as an advocate of Darwinism. The debate has attained an epic status, based

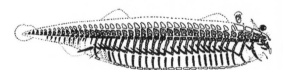

Fig. 2.7. Owen, in taking homology to its conclusion, noted that a common anatomical plan existed for all the vertebrates, and for each class of vertebrates as well. This plan, the *archetype*, illustrated here, is Owen's basic structural plan for vertebrates, but not a hypothetical ancestral vertebrate. His archetype was an "idea" in the divine mind that "foreknew all its modifications." Where Owen saw archetypes, Darwin saw ancestors. From R. Owen, *On the Archetype and Homologies of the Vertebrate Skeleton* (London: J. van Voorst, 1848).

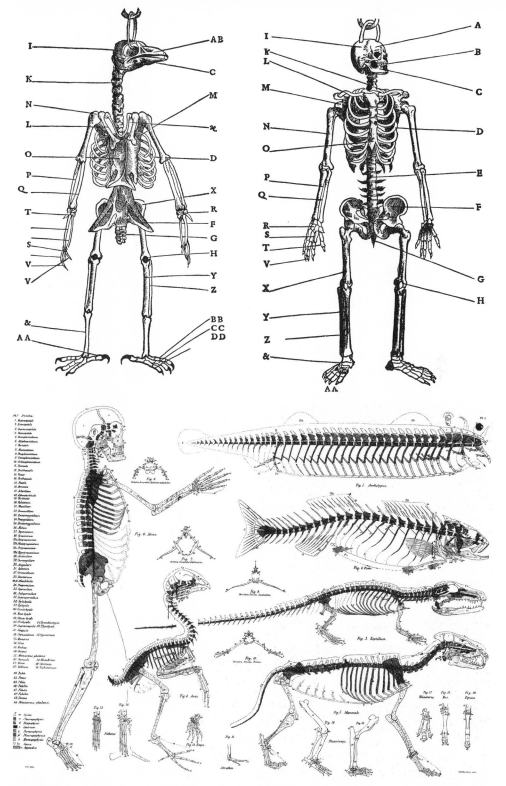

Fig. 2.8. Above, comparison of the skeletons of bird and man in Pierre Belon's 1555 book on birds (*L'Histoire de la nature des oyseaux*), representing the earliest reasonable illustrations of skeletal homology. Below, Richard Owen's vertebrate archetypes (from his book *On the Nature of Limbs*, 1849), showing homologies of the limbs of vertebrates. Owen was quite close to a concept of evolution, but his strict Victorian religious beliefs interfered with his scientific reasoning.

on Wilberforce ridiculing evolution and asking Huxley whether he was descended from an ape on his grandmother's or grandfather's side. As with all legends, the accounts of what happened vary, but one version is recounted in the October 1898 issue of *Macmillan's Magazine*: "On this Mr Huxley slowly and deliberately arose. A slight tall figure stern and pale, very quiet and very grave, he stood before us, and spoke those tremendous words—words which no one seems sure of now, nor I think, could remember just after they were spoken, for their meaning took away our breath, though it left us in no doubt as to what it was. He was not ashamed to have a monkey for his ancestor; but he would be ashamed to be connected with a man who used great gifts to obscure the truth."[13] Huxley's fame was forever secured!

Owen's notorious 1860 review of *On the Origin of Species* only exacerbated the conflict with Huxley. In his review, Owen tried, unsuccessfully, to undercut Darwin's priority in the discovery of a palatable theory of evolution by claiming that evolution proceeded via continuous creation, Owen's "axiom of the continuous operation of the ordained becoming of living things," this coming after Owen's staunch opposition to Darwin's theory of evolution by natural selection.[14] Huxley was infuriated by Owen's attempt to stake claim to an evolutionary theory. Huxley's rage notwithstanding, Owen was in fact close to developing a theory of evolution a decade before *On the Origin of Species*, but he was too constrained by the Victorian sociopolitical climate of the day to venture beyond its norms. In 1849 he revealed similarities in structure from species to species, suggesting some underlying idealized plan, the archetype; thus, where Owen saw an archetype Darwin would see an ancestral condition.[15] Darwin's fame was so pervasive that in his later years Owen alternately professed ignorance on any theory of evolution, denied it, or claimed to have come up with the idea some time before Darwin. Nevertheless, his contribution to the concept of homology was substantial, and one must credit Owen as a pioneer not only in concise anatomical nomenclature but also for being the first to clearly distinguish between the familiar concepts of analogy

and homology.[16] And although Owen's personality, combined with his conflict with Huxley, gained him a soiled reputation, his lasting contributions are numerous. He crafted the first clear definition of homology, as "the same organ in different animals under every variety of form and function," and he was a pioneer in establishing that all vertebrates were constructed on the same basic skeletal plan, or archetype, the simplest common form of all vertebrates.[17] Some of Owen's anatomical and paleontological work, including his treatise on the cranial morphology of crocodiles, is of sufficiently high quality that it can be consulted as important references for research today, and his revelations on flightless birds were revolutionary.

The Urvogel Is Discovered

The London specimen (the first discovered skeletal specimen) of *Archaeopteryx* (German *urvogel*, "original bird") arrived in London in November 1862, having been purchased along with 1,703 other Solnhofen specimens for a sum of seven hundred pounds, but the specimen was under the control of Richard Owen. He believed that *Archaeopteryx* was a bird, indeed the earliest bird, that had been transmuted during the process of continuous creation from a long-tailed pterosaur such as *Ramphorhynchus*, also known from Solnhofen. For Owen, the causes of the evolutionary process remained unknown, but one could discern general sequential change: "Every bone in the bird was antecedently present in the framework of the Pterodactyle."[18] Huxley, too, who published his first paper on the origin of birds in 1868, would focus more on proof of evolution and was inclined to regard *Archaeopteryx* as a true bird.[19] Darwin's approach to *Archaeopteryx* is also intriguing, for although he was clearly interested in it as an intermediate form between reptiles and birds, he never mentioned any possible dinosaurian affinities of the urvogel. Thus, only two years after the discovery of the London specimen, Darwin wrote in January 1863 to his close friend Hugh Falconer, a Scottish paleontologist and botanist who had studied the specimen, asking for more information: "I particularly wish to hear about the wondrous bird; the case has delighted me, because no group is so isolated as Birds. I much wish to hear when we meet which digits are developed; when examining birds two or three years ago, I distinctly remember writing to [Charles] Lyell that some day a fossil bird would be found with the end of the wing cloven, i.e. the bastard-wing and other part, both well developed."[20]

The sixth and final edition of *On the Origin of Species*, published in 1872, well after Huxley's discourse on a possible dinosaurian nexus to birds and more than a decade following the origin edition, also neglects to mention any dinosaurian affinities of *Archaeopteryx*. Darwin's comments in the final edition are characteristically conservative, and no mention is made of the controversy surrounding the putatively transitional position of *Archaeopteryx*: "That strange bird, the Archeopteryx, with a long lizard-like tail, bearing a pair of feathers on each joint, and with its wings furnished with two free claws, has been discovered in the oolitic slates of Solenhofen. Hardly any recent discovery shows more forcibly than this, how little we as yet know of the former inhabitants of the world."[21]

Since it is often asserted that the perceived similarity of *Archaeopteryx* to dinosaurs was among Huxley's principal arguments in favor of the view that birds were derived from dinosaurs, it is intriguing, with his influence on Darwin, that Darwin never explicitly identified *Archaeopteryx* as a transitional form between dinosaurs and birds.

Huxley on Dinosaurs

Birds and reptiles had long interested Huxley, and as early as 1866 he had lectured on the similarity of the chicken to the tortoise, noting that the scaly-legged birds were "an extremely modified and aberrant Reptilian type." The stork and the "snake it swallows" were placed by Huxley in a new taxon, Sauropsida.[22] It was Huxley's study of the Middle Jurassic theropod dinosaur *Megalosaurus* (the first dinosaur to be described in the scientific literature), at Oxford's Museum of Natural History that led him ultimately to conclude that the origin of birds must lie with the dinosaurs, noting particularly the similarity of the pelvic girdle and the hind limbs. The same year he studied the

9-meter (30 ft.) bipedal ornithischian herbivore *Iguanodon* in the British Museum, noting that it was "a sort of cross between a Crocodile & a kangaroo with a considerable touch of a bird about the pelvis & legs." Huxley's "dinosaurian origin of birds" made little distinction between saurischian and ornithischian dinosaurs on account of their overall similarity, despite differences in pelvis and hind limb. Huxley further was struck by the backward position of the pubis (opisthopuby) of the ornithischian *Hypsilophodon*, noting that it "affords unequivocal evidences of a further step towards the bird." Likewise, Louis Dollo, who supervised the excavations of the Belgium specimens of *Iguanodon*, made extensive comparisons with birds and found them to be remarkably similar (often "identical").[23]

Huxley was subsequently introduced to the London specimen of *Archaeopteryx*; John Evans, noted museum paleontologist, had just detected a jaw with teeth on the matrix and later found the skull. Huxley had little to say about *Archaeopteryx*, which Owen considered an unusual bird, only noting, in an attempt to discredit Owen, that his adversary had confused the right foot with the left. It is a strange twist that the *Archaeopteryx* fossil, a nearly perfect intermediate form between reptiles and birds, seemed to spoil Huxley's neat linear descent of birds, flowing smoothly from the dinosaurs to the ratites (ostriches and their allies). Giant birdlike fossil footprints had already been described from the Triassic of North America, and this suggested to Huxley that large moalike birds were living alongside their dinosaurian ancestors. Huxley had doubts about the position of *Archaeopteryx* in the scheme of things; for him, it was "more remote from the boundary-line between birds and reptiles than some living Ratitae are."[24]

In his early years Huxley had been heavily influenced by German science, especially the work of embryologist Karl Ernst von Baer. The progression of Huxley's zoological research and especially his adherence to the new Darwinian revolution was largely parallel to and strongly guided by the zoological school of the University of Jena, headed by Ernst Haeckel and Carl Gegenbaur, both of whom were heirs of von Baer's tradition, and they both strongly supported the application of evolution to studies of animal life.[25] In 1864, shortly after the publication of *On the Origin of Species*, Gegenbaur, a noted embryologist and anatomist, had studied the small, chicken-sized theropod *Compsognathus* from the same Solnhofen deposits that produced *Archaeopteryx*, and he had noted its birdlike leg and ankle. Indeed, Gegenbaur can legitimately be credited as the first to have promoted the view that birds were of dinosaurian ancestry. *Compsognathus* satisfied Huxley's search for an ancestor of birds, and in an 1868 lecture, he put his ideas into words, arguing that one could easily see how the small dinosaurs with their robust hind limbs evolved into the ancient flightless birds, represented today by ostriches, emus, and kiwis, and that these, through forms similar to *Archaeopteryx*, had evolved into modern birds.[26]

In 1867 Huxley had defined the ratites by their palatal structure, considering it primitive among living birds, and he envisioned the "struthious birds" as ancient relics of a once great evolutionary event: "Though comparatively few genera and species of this order now exist, they differ from one another very considerably, and have a wide distribution. . . . Hence, in all probability, the existing Ratitae are but the waifs and strays of what was once a very large and important group."[27] To be brief, Huxley proposed to divide the class Aves into three major orders: the Saururae, represented only by *Archaeopteryx*; the Ratitae, large flightless birds and the small kiwi; and the Carinatae, which would include all other living birds.

Paradoxically, Huxley's ideas on the sequence of avian evolution do not seem to mesh with what Darwin wrote in *On the Origin of Species*. Darwin held the view that living ratites were derived from flying ancestors through the Lamarckian view (the theory of the inheritance of acquired characters) that increased use of the hind limbs and disuse of the wings. Interestingly, Darwin's adversary Richard Owen, whom Huxley had ridiculed, had the most reasonable biological explanation for the evolution of wings of flightless birds, holding that the ratites developed by a process of "arrested development of the wings unfitting them for flight." For Owen, the ostriches were not progeny of dinosaurs but rather a classic case of "Buffon's belief in the origin of species by way of degeneration."[28] Owen's views essentially represent the predominant modern view for the evolutionary origin of most flightlessness in birds, which can be documented to occur through a process of arrested development, or paedomorphosis, a proposal fully developed by embryologist Gavin de Beer, also of the British Museum.[29]

There are heroes and villains in the arguments of the Victorian period as well as today, perhaps invented to make a good storyline. In this case anyone supporting Darwin's view would be seen as heroic (Huxley), and anyone in opposition, villainous (Owen). Huxley had an illustrious, adventurous career of near-heroic proportion. Born near London to a family of few means, Huxley studied medicine at the University of London and in 1845 entered the naval medical service, cruising in the waters of the Indian Ocean and South Pacific, near Australia and New Guinea, for four years (1846–50) aboard HMS *Rattlesnake*. The voyage, which among other things resulted in Huxley meeting his future wife in Sydney, Australia, gave him the opportunity to study the anatomy of sea creatures such as fish, mollusks, and sea worms. He characterized the voyage in his diary: "The wanderings of a man among all varieties of human life and character, from the ball-room among the elegancies and soft nothings of society to the hut of the savage and the grand untrodden forest."[30]

Huxley sent back numerous important papers to the Royal Society, and in 1850, on his return to London, he discovered that he had secured a reputation as a formidable scientist. He was elected a fellow of the Royal Society before the age of twenty-six and was awarded a Royal Medal in 1852, though he was still quite poor. Huxley left the navy in 1854 (he resigned following his refusal to return to ship—for not being paid) and assumed a position as lecturer on natural history at the Royal School of Mines, which propelled him to a position in which he would eventually become the vocal defender of Darwin's theory of evolution by natural selection. He was elected president of the Royal Society in 1883.

Books on the origin of birds invariably cite Huxley as the founder of the theory of a dinosaurian origin of birds. Yet Huxley's views on the evolution of birds are so distant from the current consensus view as to render them almost irrelevant. Indeed, with respect to ratite origins, Owen was correct and Huxley in error. And yet, Owen's mistake on the orientation of the feet of the London *Archaeopteryx* would give him a mark of infamy forever, whereas Huxley's characterization of a skull and brain in amphioxus was to be forgotten as a forgivable oversight. Moreover, Owen's contributions to anatomy and the British Museum were considerable, and Owen was in good company among scientists who did not believe in evolution at the time of Darwin's publication of *On the Origin of Species*. Louis Agassiz of Harvard University, who considered *Archaeopteryx* to be a reptile possessing certain birdlike features, is but one notable figure who opposed Darwin. Although he is held in great admiration today, he was openly a racist and a supporter of apartheid.[31] History has its way of sorting out the good from the bad, often for the wrong reasons.

Even among Darwin's supporters, views on *Archaeopteryx* varied greatly. John Evans claimed that the urvogel linked "the two great classes of birds and reptile" and that "its extreme importance as bearing upon *The Origin of Species* must be evident to all."[32] He does not mention dinosaurs. When Darwin himself mentions the urvogel in *The Descent of Man*, published in 1871, he writes that "the old Dinosaurians are intermediate in many important respects between certain reptiles and certain birds—the latter consisting of the ostrich tribe (itself evidently a widely-diffused remnant of a larger group) and of the Archeopteryx, that strange Secondary bird [referring to its geologic occurrence] having a long tail like that of the lizard."[33]

Huxley's view of the origin of birds was also beset by incongruous geological timing, of little consequence for today's phylogeneticists. Huxley could not accept the idea, implied by the discovery of the *Archaeopteryx* in Late Jurassic rocks, that the transition to birds could have occurred in a mere 150 million years. Huxley wanted the avian ancestors to be at least some two hundred million years old, which would place their origin in the Triassic. He was plagued by Edward Hitchcock's description of gigantic footprints from the Triassic

of Connecticut, predating *Archaeopteryx* by some seventy million years, and although the tracks would turn out to be those of bipedal theropods, Huxley's best guess was that they belonged to giant flightless birds, like the moas of New Zealand that Owen had described.[34] Huxley did not believe that the ultimate bird ancestor was the magpie-sized *Archaeopteryx*; he looked to flightless giants like today's emus, rheas, and ostriches. Disqualification of the urvogel because of its young geologic age would push Huxley toward considering it an offshoot of the main avian lineage, part of a later radiation of long-tailed birds with claws on the fingertips.

Interestingly, we face a similar dilemma today. As we shall see, discoveries from China have placed Mesozoic birds of modern aspect, the ornithurines, with their fully developed flight apparatus and all the pertinent anatomical development, in the Early Cretaceous. There they occur at the same age as the archaic but dominant land birds of the Mesozoic, the sauriurines, a mere twenty to twenty-five million years after *Archaeopteryx*. Other archaic Chinese urvogels are being described at a rapid pace, some perhaps from the Middle to Late Jurassic, possibly predating *Archaeopteryx*, and new birdlike tracks (unlikely to be of theropod origin) have been described from the Triassic of Argentina but may be somewhat younger.[35] Could it be that the true origin of birds is much earlier than the present fossil record indicates and that *Archaeopteryx* is simply an evolutionary dead end, an offshoot of the main lineage leading to modern birds, as Larry Martin suggests?[36]

Huxley's views on the position of *Archaeopteryx* seem at times difficult to assimilate. He noted in 1868, for example, that "in many respects, *Archaeopteryx* is more remote from the boundary-line between birds and reptiles than some living Ratitae are." However, in a series of lectures in New York in 1876 (his *American Addresses*), he argued that *Archaeopteryx* filled a morphological interval between the existing avian groups, demonstrating that "animal organization is more flexible than our knowledge of recent forms might have led us to believe.... But it by no means follows ... that the transition from the reptile to the bird has been affected by such a form as Archaeopteryx."[37]

For Huxley, *Archaeopteryx* was an aberrant bird, an "intercalary" type, as he called it, distinguishing it from the "linear" or transitional type.

Huxley on Ratites and Dinosaurs

Huxley truly felt that ratites were the key to bird origins, the ancient direct derivatives from dinosaurs, yet such a position would be perceived as concurring more or less with the discoveries of Sir Richard Owen, and that could never be allowed. It was Owen who, in a paleontological tour de force in 1839, made his historic announcement, based on a single fossil bone, that "there had existed and perhaps still exists in New Zealand a race of struthious [ostrichlike] birds of larger and more colossal size than the ostrich or any known species."[38] The fragment of femur shaft in question, from a bird larger than an ostrich, was shown to Owen by one Dr. John Rule, who thought it to be from a colossal eagle, recently extinct, but known to the Maori. But Owen, who at first doubted that any bird could have been so large or that such a bird bone could have come from New Zealand, finally became convinced of its avian affinity and gave it the name *Dinornis* (*deinos*, "terrible" or, more aptly, "wondrous"; *ornis*, "bird") following his earlier epitaph for the Dinosauria—in the case of the moa, more wondrous than terrible. Complete skeletons would shortly be recovered, proving that the bone fragment did indeed belong to a giant flightless bird that stood some 3.6 meters (12 ft.) tall and vindicating Owen in his identification.[39] This dynamic anatomical deduction was considered a remarkable achievement in comparative anatomy and propelled Owen to renown. In a real sense, then, Owen was the first to describe what would be Huxley's ultimate avian ancestor, and although Huxley nearly concurred on several fronts with Owen, their personal conflict prevented any such happening.

Huxley's thought was transformed during the production of his monumental work on the classification of living birds and a trip to Oxford's Museum of Natural History in October 1867. He had hitherto avoided Oxford, a well-known center of the anti-Darwinian movement, and John Phil-

Fig. 2.9. Sir Richard Owen was the first person to recognize that a fragment of a bone shown him in 1839 came from a large extinct bird, a moa. With additional fossils, he managed to reconstruct the entire skeleton. In this photo, published in 1879, he stands next to the largest moa, *Dinornis giganteus*, holding the bone fragment he examined forty years earlier. Owen was the driving force behind the establishment of the British Museum of Natural History in 1881. From R. Owen, *Memoirs on the Extinct Wingless Birds of New Zealand*, vol. 2 (London: John van Voorst, 1879), pl. 97.

lips, the curator in charge, was one of Huxley's antagonists. Huxley visited the museum to study the marine reptiles collected from nearby deposits of Jurassic seaways, but Phillips kindly showed Huxley the remains of the theropod *Megalosaurus*, the first dinosaur to be properly described by the eccentric paleontologist William Buckland, curator at the Oxford Museum before Phillips. (Buckland was noted for his ambitious attempt to eat one of every animal species on earth!) Huxley's epiphany came when he noticed a bone labeled as part of the *Megalosaurus* shoulder girdle. He quickly discerned that it was mislabeled and that it closely resembled the ilium of the birds he had been laboring over. For Huxley, the presence of a bone nearly identical in birds and *Megalosaurus* immediately suggested that the two groups might be related. Could the two have shared a common ancestry? The cautious Phillips, and Huxley, with whom he was often at odds on interpretation, could at least agree that Huxley had discovered an as-yet unnoticed and uncanny similarity between birds and dinosaurs. And Huxley, always concerned with the proper stratigraphic alignment of fossils, had found in *Megalosaurus* a fossil that more or less fit the bill. Megalosaurs are primitive theropods known from the Early Jurassic and possibly from the Late Triassic; the actual genus is known from the Lower and Middle Jurassic of England. Megalosaurs (now considered spinosauroids) were obligate bipeds, reached a length of about 6 meters (20 ft:), had a pubis ending in an expanded "boot," as in most theropods (unlike Mesozoic birds), and retained four digits (I–IV), in both the forelimb and hind limb. The most completely known early megalosaur is the Middle Jurassic *Eustreptospondylus* on display at the Oxford Museum, which closely resembles and is quite similar to the Upper Jurassic *Allosaurus*, although more primitive and now considered distinctive from megalosaurids.

To follow up on his remarkable observation, Huxley had little choice but to spend some unpleasant time in the collections of the British Museum, under Owen's control, which housed not only the sole *Archaeopteryx* specimen but most of England's dinosaur remains. Huxley's *Iguanodon*, unlike Owen's quadrupedal reconstruction, would be more birdlike, especially in the pelvis and legs, as he wrote to Phillips, who readily accepted Huxley's view.[40] Today, of course, *Iguanodon*, an ornithischian dinosaur, is not by any analysis relevant to the origin of birds.

Huxley presented his famous talk on the bird-dinosaur nexus on the evening of 7 February 1868, at the Royal Institution, in essence a response to Owen's *Archaeopteryx* talk of 1862. His lecture "On the Animals Which Are Most Nearly Intermediate between Birds and Reptiles" would set the stage for future debate on the origin of birds.[41] The talk was really more on evidence for transitional forms bolstering the concept of evolution, so Huxley began by stating: "We who believe in evolution are often asked to produce solid evidence to prove our theories. If one animal group can evolve into another then why do we have so many gaps in the fossil record? Where, ask our critics, are the missing links?"[42] Huxley thereupon introduced his audience to the two critical fossils, *Archaeopteryx* and the small theropod *Compsognathus*, both from the Solnhofen limestone. After presenting a formidable list of similarities between the two, primarily in the hind limbs, he exclaimed, "Birds are evolved from dinosaurs and the proof is here in these fossils. These are your missing links."[43]

Yet Huxley's presentation and his views on the origin of birds are not wholly consistent, both Adrian Desmond, historian of nineteenth-century British science and Huxley's biographer, and Paul Chambers, author of the book *Bones of Contention* (on the question of *Archaeopteryx* and the origin of birds), have noted. On one hand, Huxley used *Archaeopteryx* and *Compsognathus* to defend the existence of missing links, but on the other, he had never before used *Archaeopteryx* as the missing link between birds and dinosaurs, and certainly Darwin never used it.[44] His remark that "in many respects *Archaeopteryx* is more remote from the boundary-line between birds and reptiles than some living Ratitae are" does not support the view that *Archaeopteryx* is a missing link. When one looks carefully at Huxley's lecture, it is clear that he was presenting a complex view of avian evolution, distinguishing two major types of birds, the flightless ratites, which he considered ancient, and the smaller songbirds, which represented the myriad of more recently evolved birds. In essence, Huxley believed that the songbirds had evolved from the ratites and that the ratites were the key transitional forms, having emerged from the dinosaurs. For Huxley, as noted above, *Archaeopteryx* was an "intercalary" type, an evolutionary dead end, and not a "linear" or truly transitional type.

In fact, it was *Compsognathus* that intrigued Huxley the most, and he saw in the tiny theropod a link to birds: "a still nearer approximation [than *Archaeopteryx*] to the "missing link" between reptiles and birds."[45] The little *Compsognathus* had sprung to light when Carl Gegenbaur had earlier noticed its birdlike leg and ankle, and Huxley actually wondered if the little theropod was itself feathered. But the confusion arose because he had consistently argued that the ancestor of birds must date to the Triassic, at which time he thought that giant ratites had existed. Huxley thus believed that birds were descended from dinosaurs, but certainly not through *Archaeopteryx*.

Enter Seeley

Harry Grovier Seeley, an expert on pterosaurs and dinosaurs, immediately objected that the similarities between birds and dinosaurs that Huxley had noted were indicative only of a shared mode of locomotion, and not necessarily of shared ancestry. Although Huxley rejected Seeley's assertions, in his paper on the classification of the Dinosauria in 1870, he turned the tables on him, proclaiming that the similarities of pterosaurs and birds were indeed convergent, the result of physiological action. (Seeley is remembered today as the first to classify dinosaurs by their pelvic structure, when in 1887 he cleaved Dinosauria into the Saurischia, "lizard-hipped," and the Ornithischia, "bird-hipped.")[46] In *Dragons of the Air* (1901), his most popular work, Seeley argued that birds and pterosaurs were closely allied and that, contrary to Owen's interpretation, pterosaurs were endotherms capable of active flight.[47] In addition, he argued that *Archaeopteryx* was a pterosaur-bird intermediate. Seeley's objection that the similarities in the hind limbs of chickens and dinosaurs were due "to the functions they performed rather than to any actual affinity with birds" (followed by a rebuttal of Huxley's observations bone by bone from Owen) was prescient.[48] Huxley had simply compared the hind limbs of a living, highly derived, running bird to a bipedal, ground-dwelling dinosaur, and the assertion that the similarities between the two were the result of convergence rather than descent has been repeated time and time again (Benjamin Mudge, 1879; Louis Dollo, 1882, 1883; Wilhelm Dames, 1884; W. K. Parker, 1887; Max Fürbringer, 1888; Henry Fairfield Osborn, 1900; Robert Broom, 1913; Gerard Heilmann, 1926;

George Gaylord Simpson, 1946; Gavin de Beer, 1954; Alfred Sherwood Romer, 1966; and many more recent authors).[49] Among the most influential evolutionary biologists of his time, George Gaylord Simpson emphatically asserted: "Almost all the special resemblances of some saurischians to birds, so long noted and so much stressed in the literature, are demonstrably parallelisms and convergences."[50] It would therefore seem that those who do not fully accept the current orthodoxy of the dino-bird nexus are hardly unscientific or heretical. Their arguments have an extensive pedigree.

Paleontologists typically consider Huxley's views of the evolution of birds to be an authoritative statement on the derivation of birds from dinosaurs, but it is still not clear what Huxley actually believed. A variety of conclusions could be reached from Huxley's musings, and as Huxley himself wrote to Darwin on 17 March 1869, concerning the controversies surrounding *On the Origin of Species*, "A good book is comparable to a piece of meat, and fools are as flies who swarm to it, each for the purpose of depositing and hatching his own particular maggot of an idea."[51] Even Darwin was not without fault in many of his evolutionary scenarios. For example, we see Darwin in the sixth edition of *On the Origin of Species* speculating on the evolution of true wings from the penguin: "What special difficulty is there in believing that it might profit the modified descendants of the penguin, first to become enabled to flap along the surface of the sea like the loggerheaded duck, and ultimately to rise from its surface and glide through the air?"[52]

We know today that this is nonsense: penguins represent a highly derived avian morphology and literally fly through the water with their paddle wings. The Sphenisciformes (penguins) was an evolutionary dead end and did not give rise to other avian lineages. Indeed, the evolutionary sequence that produced penguins is the opposite of that envisioned by Darwin, with hypothetical oceanic birds, such as diving petrels, providing a suitable pseudophylogeny. These oceanic birds of the Southern Hemisphere dive into the water, then swim with short wing-driven swims, and then fly out of the water. By this scenario, birds resembling the diving petrels ultimately gave up flight altogether.[53] And although Darwin was the first to propose the arboreal theory for the evolution of flight, specifically in bats, he nevertheless ignored the fact that a dinosaurian origin of birds via the ratites would imply an improbable cursorial origin of flight.

Huxley and Galliformes

Given both the confusion and the extensive citation of Huxley's quotation on the similarity of the hind limb of the chicken and a dinosaur, it is essential to delve deeper into its meaning. Chickens belong to the order Galliformes, a well-defined group of semiterrestrial, ground-foraging birds that also comprises the megapodes, guans and curassows, turkeys, grouse, New World quails, pheasants and partridges, and guinea fowl. The origin of the order is thought by some recent works to predate the Cretaceous-Tertiary boundary by millions of years, but alas, such an assumption is at present speculative, because the fossil record of the order extends only to the Eocene, and Eocene galliforms are quite primitive.[54] If the origin of Galliformes predates the Cretaceous-Tertiary boundary by such a vast time span, it seems implausible that Eocene fossils of the group would display such primitive morphology or that no Cretaceous fossils would have been uncovered.

Their date of origin notwithstanding, Galliformes are characterized by hindquarters highly adapted for a running, cursorial existence like theropod dinosaurs, but the structure of their hindquarters is quite distinctive. Moreover, what some call the ascending process of the astragalus is actually attached in neornithine birds (including chickens) to the calcaneum, not the astragalus, and the first toe is rotated in a somewhat posterior direction, quite unlike that of *Compsognathus*, which is mesial, reduced, and elevated. Contrary to Huxley's assertions on the similarity of the hind limb of the Dorking fowl and *Megalosaurus*, there are major problems with homology, and the similarity is superficial.

As barnyard fowl, galliform birds are easily handled in the laboratory, and their eggs are easily obtained and incubated. As a result, most of what we know of avian embryology stems from studies of the development of chickens. This is unfortunate for avian embryology because galliforms are among the most precociously developed of all the avian orders, and they have a huge flight apparatus that develops early: the major sternal bones develop between the eighth and twelfth days of the twenty-one-day incubation period. Thus, galliforms have an almost fully developed flight apparatus at birth, and some megapodes can fly right out of their eggs! Unlike neotenic ostriches, for example, which are born with a yet-to-be-absorbed yolk sac, the chick yolk sac is completely drawn into the abdomen the day before hatching. Because of the precocious development of the sternal apparatus and associated flight musculature, flight loss is rare among galliforms, the one exception being a recently extinct giant megapode of New Caledonia that lost the ability to fly through evolution of gigantism rather than by the normal paedomorphic process of arrested development.

It is a major popular and pervasive misconception that galliform birds are "terrestrial" and mainly earthbound. They forage largely on the ground, but the vast majority roost in trees, and they are among the strongest short-distance "burst" fliers and can take off from a standing position on the ground. Birds like turkeys invariably leave the ground at night to roost in trees, even with their young, and can often be seen high in big deciduous trees in large assemblages. It is not inaccurate to say that the Galliformes are a somewhat aberrant order, certainly in no way typical of the land bird radiation. Further, as members of the modern lineage of birds, the Neornithes, they have little relation to the great Mesozoic radiation and therefore little bearing on bird origins. Their appeal among paleontologists is that they have well-developed sacral regions and hind limbs for running and thus superficially resemble theropod dinosaurs.

Huxley's reliance on the similarity of the hind limbs of the chicken and dinosaur is thus not just confusing but clearly based on overall similarity rather than careful analysis. First, by "dinosaurs," Huxley presumably meant both the ornithischian *Iguanodon* and the theropods *Megalosaurus* and *Compsognathus*. Second, many of the similarities

Huxley noted might be due to convergence or parallelism, but the ornithischian is not on the lineage thought to have given rise to birds, and thus any similarities therewith are convergent. Third, there is superficial but little similarity between the hind limbs of the chicken and even the early bird *Archaeopteryx*. *Archaeopteryx* specimens present some evidence, especially in the claws and feet, of an arboreally adapted anatomy but little evidence of the detailed structure for a specialized cursorial life, as seen in chickens.

Yet, despite these problems, many works on bird and vertebrate evolution going back to Ostrom's work in the early 1970s and continuing to the present attempt to bolster the current orthodoxy by referring to the absolute authority of Thomas Huxley.[55] Most use his well-worn statement on the similarity of the hind limb of the chicken and "dinosaurian," here blindly quoted by Donald Prothero in his text *Evolution: What the Fossils Say and Why It Matters*: "And if the whole hindquarters, from the ilium to the toes, of a half-hatched chick could be suddenly enlarged, ossified, and fossilized as they are, they would furnish us with the last step of the transition between Birds and Reptiles; then there would be nothing in their characters to prevent us from referring them to the Dinosauria."[56]

Back to Ratites

Huxley was clearly stuck on the large, flightless ratites, the "scanty modern heirs of the great multitude of creatures which once connected Birds with Reptiles" as the secret to the origin of birds: "All birds have a tarso-metatarsus, a pelvis, and feathers, such, in principle, as those possessed by *Archaeopteryx*. No known reptile, recent or fossil, combines these three characters, or presents feathers, or possesses a completely ornithic tarso-metatarsus, or pelvis. *Compsognathus* comes nearest in the tarsal region, *Megalosaurus* and *Iguanodon* in the pelvis." Huxley commented on the ornithischian *Iguanodon* in his 1868 address that it "walked, temporarily or permanently, upon its hind legs." Here, then, were "extinct Reptiles which approached these flightless birds [ratites],

not merely in the weakness of their fore-limbs, but in other and more important characters."[57]

We see further evidence of Huxley's strong belief in this theory in a letter he wrote to Ernst Haeckel only two weeks after his presentation to the Royal Institution in 1868: "I am engaged [in] a revision of the Dinosauria, with an eye to the 'Descendenz Theorie.' The road from Reptiles to Birds is by way of Dinosauria to the Ratitae. The bird 'phylum' was struthious, and wings grew out of rudimentary forelimbs."[58] Even Oxford's John Phillips, who did not concur with Darwinian thought, was more or less convinced of the affinity of birds, through ratites, with dinosaurs, writing to Huxley, "The more I reflect on the monsters the more grows my faith in their struthious [ostrich-like] affinities."[59]

By the late 1860s, then, Huxley, who had ascended from his ancestral home above a butcher's shop to become a middle-class, working-man's hero, had finally made a big play with his new theory of bird descent, but along the way he had treated many complex biological problems oversimplistically. Regarding the origin of both vertebrates and birds, Huxley had offered viable theories, but closer examination of the evidence eventually undermined their validity. Nevertheless, text after text gives the impression that Ostrom rekindled Huxley's theory of a dinosaurian origin of birds, so the dinosaurian origin is thought to be the work of none other than Huxley. As an example, Prothero proclaims that Huxley "could not help but notice that *Archaeopteryx* was a classic 'missing link' between birds and dinosaurs."[60] Like most popular science writers, he does not seem to be aware that the word "dinosaur" is in this context almost meaningless because it is too general, including both ornithischians and saurischians, and that Huxley instead viewed the small theropod *Compsognathus* as the actual missing link between dinosaurs and the struthious birds, the large flightless ratites.

Since its description by Richard Owen, *Archaeopteryx* has been generally regarded as a bird with strange reptilian features, and we see no better illustration of this than in Darwin's *Descent of Man*, in which he writes: "We have seen that the

Ornithorhynchus [duck-billed platypus] graduates towards reptiles; and Prof. Huxley has made the remarkable discovery, confirmed by Mr. Cope and others, that the old Dinosaurians are intermediate in many important respects between certain reptiles and certain birds—the latter consisting of the ostrich-tribe."[61]

In Huxley's view, therefore, the forms intermediate between birds and dinosaurs were the ostrichlike (struthious) birds, and Huxley included both the theropods and the ornithischians (not today considered relevant to the origin of birds) in his comparisons of birds and dinosaurs. Huxley, Darwin, and most of their contemporaries usually considered *Archaeopteryx* to be a strange, albeit perfect, side branch in bird evolution. It is therefore not surprising that Sir John Evans claimed that *Archaeopteryx* features compared favorably with those of a magpie. Even Huxley, as late as 1885, in an article in the December issue of *Nineteenth Century*, referred to "the 'fowl—the Jurassic first bird, *Archaeopteryx*." And Huxley's biographer Adrian Desmond appropriately notes of the urvogel that for Huxley, "It was a cousin, the royal line having already run from the 'bird-leg' dinosaurs to ostrich-like birds."[62] Thus, although most contemporary authors who write about the origin of birds, though they admit that Huxley's views are irrelevant to modern theories of bird origins, nevertheless are quick to cite the authoritative and legendary Huxley as the originator of the concept of a dinosaurian origin of birds, with *Archaeopteryx* as the key player. This was not the case. As the recent scholarship of Desmond and Paul Chambers has shown, it is questionable whether Huxley *ever* endorsed the idea that *Archaeopteryx* was the so-called missing link between birds and dinosaurs. As they note, Huxley's proclamation that "In many respects *Archaeopteryx* is more remote from the boundary-line between birds and reptiles than some living Ratitae are" is not a resounding endorsement of the urvogel as a living dinosaurian, and there is no doubt that Huxley had a more complex view of avian evolution.[63]

Considering all of Huxley's writings on the origin of birds, he may well have felt more comfortable with a theory of avian descent by common an-cestry with the Dinosauria, from basal archosaurs or dinosauromorphs, rather than a direct descent from theropods. Amazingly, as noted in chapter 1, this is the best interpretation of the results of the recent paper by Schweitzer and colleagues, who claim to have recovered collagen from an eighty-million-year-old hadrosaur. According to their cladogram, theropod dinosaurs come out closer to ornithischian dinosaurs than to birds (*Gallus* and *Struthio*), a finding totally at odds with the current hypothesis for a theropod origin of birds, and indicative of a basal archosaur origin.[64] It would therefore seem prudent, given the rancorous nature of the current debate on the origin of birds, to define the issues with care and clarify the two major sides of the debate. As noted, the two major competing theories are: (1) that birds are the sister group or are derived directly from highly derived theropod dinosaurs, the maniraptorans (Ostrom and current consensus); or (2) that they are the sister group or are derived from a common stem, basal archosaur (or slightly later dinosauromorph) ancestor, earlier in time (minority view). As we shall see in the chapter on flightless birds, there were others who followed Huxley and argued that ratites descended directly from dinosaurs and were not secondarily flightless. Today, the evidence against such a scenario is overwhelming.[65]

Robert Broom and Gerard Heilmann

Recall that Huxley found the similarities between *Compsognathus* and the ratites to be striking: "Surely there is nothing very wild or illegitimate in the hypothesis that the phylum of the Class of Aves has its foot in the Dinosaurian Reptiles—that these, passing through a series of such modifications as are exhibited in one of their phases by *Compsognathus*, have given rise to [birds]."[66]

One could thus say that although its formulation was imprecise and focused on the flightless ratites, Huxley was the first to suggest in detail a dinosaurian origin of birds, although in his proposal common descent cannot be excluded.[67] However, Huxley's contemporary Benjamin Mudge, of the University of Kansas, among many others, thought dinosaurs too diverse and too specialized

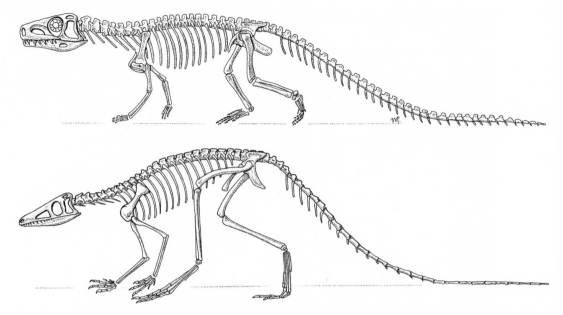

Fig. 2.10. Robert Broom's small "thecodont" *Euparkeria* (top), from the uppermost Lower Triassic of South Africa, was a creature afforded a central position in discussions of the evolution of archosaurs, and considered by Heilmann to be a possible ancestor of both dinosaurs and birds. The lagosuchids (bottom), characterized by *Marasuchus* (*Lagosuchus*) from the Middle Triassic of Argentina, were also small archosaurs, thought to be possibly the sister group of both dinosaurs and pterosaurs, and lagosuchids are popularly classified as dinosauromorphs. From P. Wellnhofer, *The Illustrated Encyclopedia of Pterosaurs* (New York: Crescent Books, 1991). Reproduced with permission; courtesy Peter Wellnhofer.

to have been avian progenitors: "The dinosaurs vary so much from each other that it is difficult to give a single trait that runs through the whole."[68] The situation changed dramatically when Robert Broom, a talented Scottish paleontologist, who had been describing a wealth of vertebrate fossils from the now famous Lower Triassic Karoo deposits of South Africa, revealed in 1913 the "pseudosuchian thecodont" *Euparkeria capensis* (currently thought to be a crurotarsan), which he believed to be a suitable common ancestor of both birds and dinosaurs. *Euparkeria* was a member of a newly discovered group of diapsid reptiles called "thecodonts" (basal archosaurs; thecodont is now considered a paraphyletic grouping) that lived from the Late Permian through the Triassic (approximately 250 to 200 million years ago), a period before dinosaurs or birds.[69] Clearly related at some level to the higher archosaurs (including the pterosaurs, birds, dinosaurs, and crocodilians), they were believed by many to be the immediate common ancestors of these various groups—that is, that "thecodonts"

or basal archosaurs were the common ancestral stock of the higher archosaurs. Based on skeletal similarities, Broom saw in the thecodonts, and specifically in *Euparkeria*, a candidate for an immediate common ancestor of birds and dinosaurs. Moreover, it was a candidate that could overcome the objections posed by a direct descent from dinosaurs. Instead of the orthogenetic progression from dinosaurs to birds via *Compsognathus*, endorsed by both Huxley and O. C. Marsh of Yale, Broom believed that birds could not be descended from dinosaurs; rather, they were related as cousins. Even today there has been a revival of interest in *Euparkeria* and its possible affinities with birds, and Gregory Paul notes: "Few dispute that *Euparkeria* is a suitable ancestral type for birds. But it is suitable only in the same sense that an early Cenozoic, primitive, generalized primate is a suitable ancestral type for humans. . . . *Euparkeria* is important to bird origins only in that it represents the earliest Mesozoic and most primitive archosaur stem stock from which the entire dinosaur-bird

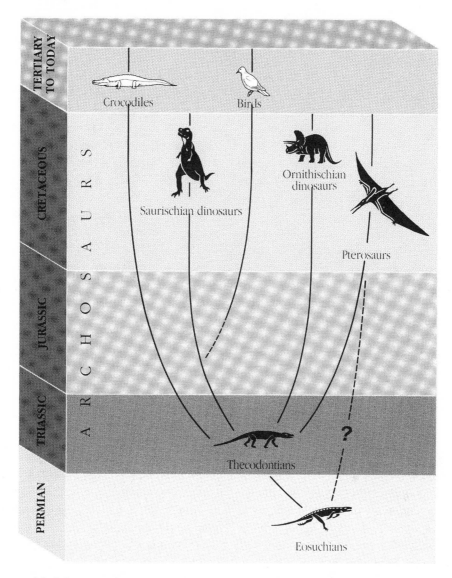

Fig. 2.11. Simplified diagram of probable lines of descent, showing the two classic possible theories for the ancestry of pterosaurs, from Permian eosuchians or from later Triassic thecodonts; and the ancestry of birds, which is either directly from saurischian, theropod dinosaurs (current dogma), as shown, or early from a common thecodont ancestor shared with dinosaurs (Heilmann's view). Pterosaur origins are still much in doubt. From P. Wellnhofer, *The Illustrated Encyclopedia of Pterosaurs* (New York: Crescent Books, 1991). Reproduced with permission; courtesy Peter Wellnhofer.

clade later evolved. For that matter, *Euparkeria* is a good ancestral type of virtually all archosaurs."[70]

Regardless of the lack of centrality of *Euparkeria* to the current debate, within a reasonably short period following the Huxleyan era, the idea emerged that dinosaurs and birds shared an immediate common ancestry through the basal archosaurs or "thecodonts" as well as its corollary, that

flight originated from the trees down, as opposed to the necessity of a ground-up flight origin from Huxley's ratite proposal.

The Danish artist Gerhard Heilmann, who claimed to be a supporter of Darwin by age fourteen, played a vital role in establishing Broom's theory as the dominant view on the origin of birds for most of the twentieth century. Heilmann wrote a remarkable book entitled *The Origin of Birds*, a two-hundred-page monograph published in English in 1926 (the Danish original, *Fuglenes Afstamning*, was published in 1916) that contained four sections filled with his own superb drawings, as well as photographs of fossil birds, including the two specimens of *Archaeopteryx* then known.[71] The overall conclusion from the first three sections was that "from its remains *Archaeopteryx* might be characterized as a reptile in the disguise of a bird." It was in the fourth section that Heilmann diverged from the view that birds were directly descended from dinosaurs. He begins: "We can therefore with absolute certainty maintain that the birds have descended from the reptiles. Of this we cannot, in future, entertain the faintest shadow of a doubt." But for Heilmann, the question had not been satisfactorily answered as to which group? Heilmann quickly dismissed the possibility of a bird origin from pterosaurs, and although his comparisons between theropods and birds (including *Archaeopteryx*) were quite favorable, the presumed absence of a furcula (clavicles) in dinosaurs ruled them out as potential ancestors: "It would seem a rather obvious conclusion that it is amongst the [theropods] that we are to look for the bird-ancestor. And yet, this would be too rash, for the very fact that the clavicles are wanting would in itself be sufficient to prove that these saurians could not possibly be the ancestors of birds."[72]

Heilmann concluded, as did Broom, that both birds and dinosaurs shared an immediate common ancestor among the "thecodonts." Advocates of a theropod origin of birds have argued that "Heilmann's dilemma," the absence of clavicles in dinosaurs, has now been overcome by the discovery of a furcula in at least some theropods, but now we know that furculae are present not only in dinosaurs but also in basal archosaurs, and even prosauropods, so their primitive or plesiomorphic occurrence renders them currently irrelevant to the discussion.[73] Heilmann was aware of Dollo's Law concerning the nonreversibility of evolution, which in general holds up today, especially at the level of major structural transitions, although at the microevolutionary or population level it is hardly viable.

Thus, in Heilmann's view, after the clavicles had supposedly been lost in the coelurosaurian dinosaurs, exemplified by *Compsognathus*, they could not have been regained in birds. Yet today's dinosaur-bird orthodoxy is based on a coelurosaurian (maniraptoran) ancestry of birds and dromaeosaurs, even though all the potential ancestral forms of coelurosaurians, as we currently know them, had massive hind limbs, were obligate bipeds, and had forelimbs reduced to half or less the size of the hind limbs. In almost all modern discourses on the origin of birds, the deficiency of Heilmann's argument is invariably mentioned, but there is no mention of the problem of deriving birds from coelurosaurs, with their hypertrophied hind limbs and foreshortened forelimbs. Even if it were possible to reelongate already shortened forelimbs, what would they look like? If they reelongated proportionally, then the hand should be quite short (since the hand is the first element to shorten in secondarily flightless, neotenic birds), with an elongate radius/ulna, following the general proportions of the coelurosaur limb. No secondarily flightless birds are known to have reelongated formerly shortened forelimbs.

Heilmann's argument was more complex and did not rely solely on the apparent absence of clavicles in dinosaurs. Importantly, he was first to inextricably link the origin of birds with the origin of avian flight, and he offered a convincing argument for an arboreal origin of flight, as Darwin, in *On the Origin of Species*, had for bats. Heilmann's "Proavis" was a hypothetical arboreal "thecodont" (basal archosaur) that lived approximately 250 million years ago, presumably in the Late Permian or Triassic, a time similar to that envisioned by Huxley for bird origins. *Archaeopteryx*, for Heilmann and many others, was simply a 150-million-year-old bird, not a missing link between birds and dinosaurs, but an early bird with many reptilian features; it was a missing link between reptiles and birds only.

Possible Triassic Clues?

Primarily in 1868 and 1870, Huxley thought that the ancestors of birds were likely to be found within the dinosaurs, though the exact group remained elusive and he thought that birds descended by a nonflight stage through large, earthbound ratites. Yet Huxley was very aware of the problem of geologic time, and his view was that the divergence of birds from dinosaurs occurred during the Paleozoic and that bird ancestors were most appropriately searched in the Triassic, where large, ratitelike tracks had been discovered, although they were later determined to be those of theropods.

In 1986 the paleontological world was startled by the announcement from Sankar Chatterjee of Texas Tech University that he had discovered a fossil bird, "*Protoavis*," in the 225-million-year-old west Texas Dockum Formation (Late Triassic). Such a fossil, if verified as avian, would predate *Archaeopteryx* by some seventy-five million years and push the origin of birds back to the dawn of dinosaurs. Chatterjee claimed that the two crow-sized specimens, featuring ulnar quill nodes, strut-like coracoids, and a keeled sternum and furcula, were temporally closer to the origin of birds than the Solnhofen urvogel. Chatterjee's claims were problematic. The National Geographic Society, which sponsored the research, made the dramatic announcement without providing any documentation, and it was not until 1991 that a monograph

on the specimens was published.[74] The fossils, if verified, would have extended the geologic range of Aves by some seventy-five million years, but unfortunately they were fragmentary, without feather remains, and required considerable interpretation, and John Ostrom was quick to point out deficiencies in the fossils.[75] As Walter Bock of Columbia University, who examined the fossils, noted, "If one was being severely critical, few features are conclusively avian."[76] As could be expected, an acrimonious debate ensued, and at this writing there is no additional evidence that would confirm the avian status of the fossils. Surprisingly, in his book *The Rise of Birds*, Chatterjee readily accepts the current view that Aves are the sister group of dromaeosaurs, a group known with certainty only from the Early Cretaceous (perhaps the Late Jurassic). The result is that Chatterjee placed the split between dromaeosaurs and birds early in the Triassic, or early in the Jurassic, timings that appear implausible. In essence, if *"Protoavis"* were avian, it would have serious implications for the current view that birds are derived from theropods.

In an apparent historical replay from Huxley's era, putative bird tracks of presumed late Triassic age have recently been reported from Argentina.[77] However, as one can see from the photograph on

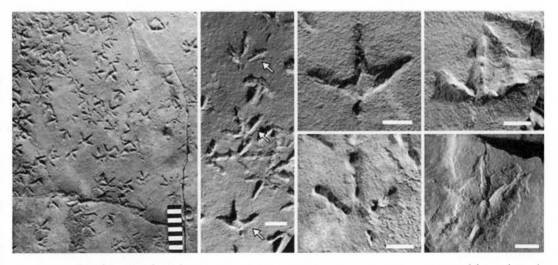

Fig. 2.12. Birdlike footprints from the Late Triassic Santo Domingo Formation. From R. N. Melchor, Silvina de Valais, and J. F. Genise, "Bird-Like Fossil Footprints from the Late Triassic," *Nature* 417 (2002): 936–38. Copyright 2002. Reprinted by permission from Macmillan Publishers Ltd.

page 53, this time the tracks, produced in ephemeral fluvial ecosystems with shallow ponds and mud-flats, would be extraordinarily difficult to assign to a gathering of small theropods. Even if the tracks were of Early Jurassic age, or even younger as some believe, such a find would still be remarkable and would presumably push back both the origin of birds and the dichotomous branching of sauriurine and ornithurine birds. The tracks resemble those of modern ornithurines, sandpipers and plovers, and exhibit a reversed hallux (some elongate). Jorge Genise and colleagues from the Museo Paleontológico Egidio Feruglio of Chubut, Argentina, studied track formation in a modern environmental analogue, a coastal freshwater pond near Chubut, seasonally occupied by the sandpipers *Calidris bairdii* and *C. fuscicollis*.[78] Their comparisons suggested that the modern setting was quite similar to the paleoenvironment, and their field observations distinguished twenty-one behaviors in the modern species that produced distinct traces and four modern footprint profiles related to substrate conditions, including nine related to flight capacity, as well as probing and pecking in shallow water along the shore. Of the behaviors of modern *Calidris* sandpipers, five were identifiable with certainty from the hundreds of tracks and trackways, including: walking (the most common), walking with a zigzag path, making short runs, probing, and landing with legs directed forward. The authors are clearly concerned about the age of tracks that match remarkably with ornithurine shorebirds, and they caution that because of their apparent modern aspect, "the Santo Domingo track site could be younger than supposed." Regardless of the age of the rocks and the final verdict on this important discovery, even if the tracks are of Early Jurassic age or even slightly younger, they would carry many of the same implications.

As we shall see, discovery of primitive ornithurine birds from the Early Cretaceous of China, nearly coeval with ancient sauriurine birds (including enantiornithines and basal birds and *Archaeopteryx*), may indicate that the avian lineage is much older than currently thought, as is the dichotomy of the ancient and modern-type birds. Although we have learned much from these recent discoveries and have clarified aspects of the Cretaceous radiation of birds, much remains to be learned concerning avian ancestry and the origin of flight.

Conclusion

Are the proclamations of evolution's high priest and the subsequent debate on the origin of birds of any lasting interest or utility, even heuristic? The scientific answer is, of course, an emphatic no. Only the true scientific resolution of the problem of the origin of birds will endure. On the other hand, Huxley's views are quoted, as if they have some great authority, in almost every text and treatise on the origin of birds, suggesting that the opinion of Darwin's Bulldog may indeed constitute some form of evidence. The greatest statesman of our time, Sir Winston Churchill, made famous Edmund Burke's statement that "those who do not know history are doomed to repeat it," but despite this great admonition, we have today practically come full circle in the debate on the origin of birds. We have journeyed from Huxley's ratites as the key to the origin of birds from dinosaurs, with an implied cursorial origin of flight, to the current view that birds are living theropods, with Cretaceous dromaeosaurs (maniraptorans) as the key to their evolution. Yet, unlike Huxley's approach, for the modern practice of phylogenetic systematics the geologic record is no longer essential evidence, the "new totem," as Adrian Desmond notes that it had become in Huxley's time. On the contrary, it is today discarded as irrelevant. It is important to consider the establishment of Huxley as an early authority to the current unchallengeable orthodoxy of a "dinosaurian" origin of birds and ground-up flight, because appeal to tradition is often used as a critical defense of the current consensus view of avian origins.

3

THE ICONIC URVOGEL
WAS A BIRD

·······································

In certain particulars, the oldest known bird
does exhibit a closer approximation to the
reptilian structure than any modern bird. . . .
The leg and foot, the pelvis, the shoulder-
girdle, and the feathers . . . are completely
those of existing ordinary birds.

*T. H. Huxley, "On the Animals Which Are
Most Nearly Intermediate Between the Birds
and Reptiles," 1868*

Before the astounding recent discoveries of avian
and dinosaurian fossils from the Mesozoic of
China, the only substantial evidence for the earliest evolution of birds from their reptilian beginnings came from fossil specimens from the approximately 150-million-year-old deposits from
the Late Jurassic of Bavaria, specimens known to
the Germans as the urvogel, or ancient bird, discovered almost 150 years ago. The *Archaeopteryx*,
as it was named, played a significant role through
the years not only in discussions of avian and flight
origins but also in discussions of evolutionary
processes.

In 1861, less than two years after the publication of Darwin's *On the Origin of Species*, German paleontologist Christian Erich Hermann
von Meyer of Frankfurt wrote to H. G. Brown,
publisher of the *New Yearbook of Mineralogy*, an-
nouncing that the feather of a Late Jurassic bird
had been recovered the previous year from the
lithographic limestone deposits at Solnhofen,
a short distance from Munich. The gray-black
feather was approximately 60 millimeters (2.5 in.)
in length, and the outer feather vane was roughly
half the width of the inner. Despite the Jurassic
age of the fossil, it was indistinguishable from the
secondary flight feather (inner set of wing feathers) of a modern bird, such as a magpie or a crow.

It was clearly a flight feather of a Jurassic bird,
and it possessed the airfoil qualities seen in the
feathers of modern flying birds. Within a month,
Meyer further announced that a nearly complete
skeleton of a Jurassic bird had been discovered
from the same Solnhofen limestone, this time from
the Langenaltheim region, a short distance from
the locality at which the feather was discovered;
he named it *Archaeopteryx lithographica* ("ancient
wing" from the lithographic limestone).[1]

From a historical perspective, the timing of
these discoveries seemed no less than miraculous.
With Queen Victoria at the helm of Britain, the
world's most powerful empire, and the Industrial
Revolution bolstering economies in Europe, particularly France and Germany, there was a sense
of optimism and a great interest in more than
mere human survival that spilled over to science
and even social reform, producing a golden age.
Darwin's book was just two years in print and the
public's interest in scientific discoveries was at an
all-time high: a fossil thought by some to be intermediate between two higher groups of vertebrates, birds and reptiles, was not going to be passively catalogued and set aside.

Nowhere were finer fossils to be found than
in the Solnhofen quarries, which were beginning
to provide for the first time a spectacular window

Fig. 3.1. Scenes of Solnhofen limestone quarries near Eichstätt, Bavaria, still active today, including a worker meticulously separating slabs of limestone, as well as a view of one of the quarry yards, with quarried limestone awaiting processing. Photographs by the author, 1984.

on the Jurassic Period. The Solnhofen deposits have been mined since antiquity to provide limestone for construction and even to pave roads to Rome, and the fossils themselves had been prized since the late Stone Age, when they were valued as ornaments. Up to the late eighteenth century Solnhofen limestone retained considerable value for use in buildings, and by the mid-nineteenth century it would be pivotal for the lithographic printing process. Moreover, in the mid-nineteenth century, a lucrative traffic in Solnhofen fossils emerged, and serious collectors began amassing extensive holdings. The best-known collector was an obstetrician in Pappenheim and a Royal Bavarian District medical officer, Dr. Karl Häberlein,

who accepted fossils in lieu of money for medical services. His interesting story has been told many times, but he amassed a huge and beautiful collection of some seventeen hundred Solnhofen fossils, including a newly discovered specimen of a birdlike skeleton, and after three months offered the fossil for sale. Efforts to obtain the specimen were made across Europe, and not far away in Munich, the German court tried to secure the specimen for the Bavarian State Collection of Paleontology and Historical Geology.

J. Andreas Wagner, a staunch opponent of Darwin's theory, vehemently opposed the view that there might exist a transitional animal between birds and reptiles and feared that the new fossil would enhance the Darwinian view. He expressed his view in 1861 in a paper entitled "A New Reptile Supposedly Furnished with Bird Feathers."[2] As a gesture of his opposition, Wagner ignored von Meyer's avian name for the urvogel and provided his own designation: *Griphosaurus* (*gryps*, "mythical beast"; *sauros*, "lizard"). Wagner's anti-Darwinian stance had colored his perception of the urvogel, and though he had never seen the fossil, he remained confident of its classification as a pterosaur: "In conclusion, I must add a few words to ward off Darwinian misinterpretation of our new saurian. At the first glance of the *Griphosaurus* [*Archaeopteryx*] we might certainly form a notion that we have before us an intermediate creature, engaged in the transition from the saurian to the bird. Darwin and his adherents will probably employ the new discovery as an exceedingly welcome occurrence for the justification of their strange views upon the transformation of animals but in this they will be wrong."[3] A similar anti-Darwinian view was also expressed by the original describer, Hermann von Meyer, transcribed from a letter from von Meyer to Sir John Evans in an article by Evans on the newly discovered cranium and jaw, published in 1865:

> It would appear that the jaw really belongs to the *Archaeopteryx* and arming the jaw with teeth would contradict the view of the *Archaeopteryx* being a bird or embryonic form of bird. But after all, I do not believe that God formed His creatures after the systems devised by our philosophical wisdom. Of the classes of birds and reptiles as we define them, the Creator knows nothing, and just as little of a prototype, or of a constant embryonic condition of the bird, which might be recognized in the *Archaeopteryx*. The *Archaeopteryx* is of its kind just as perfect a creature as other creatures, and if we are not able to include this fossil animal in our system, our shortsightedness is alone to blame.[4]

Häberlein realized the significance and therefore the monetary value of the fossil, and he shrewdly allowed no one to draw it, although interested parties were permitted to examine it for authenticity,

and finally a sketch and then an accurate lithograph were produced.

After the failure of German efforts to obtain the fossil, the British made a major effort to obtain the treasure, and anti-Darwinian Sir Richard Owen, superintendent of the Natural History Department of the British Museum, and George Waterhouse, keeper of the Geology Department, recommended that the museum place a bid for the fossil. Häberlein accepted the museum's offer and sold what would become known as the "London specimen," along with 1,703 other fossils, for seven hundred pounds (amounting to two years' acquisition budget for the museum), which was designated as a dowry for one of Häberlein's six daughters. In November 1862 the controversial fossil arrived in London, where it resides today.

Owen believed that new life-forms were created from time to time as modifications of a basic archetypal plan, and he insisted that the *Archaeopteryx* was a bird, albeit the earliest one, that had been transmuted from a long-tailed pterosaur, exemplified by the Solnhofen *Ramphorhynchus*.[5] Owen thus concluded that *Archaeopteryx* supported his theory of archetypes: "Every bone in the bird was antecedently present in the framework of the Pterodactyle. . . . Some pterodactyles had long tails and all had toothed jaws. A bird of the oolitic period [=*Archaeopteryx*] combined a long tail of many vertebrae with true avian wings . . . and . . . we discern . . . a retention of a structure embryonal and transitory in the modern representatives of the class, and a closer adhesion to the general vertebrate type."[6]

The Berlin specimen, most famous of the *Archaeopteryx* specimens, was discovered by Jakob Niemeyer, in 1876 or 1877, from a quarry on the Blumenberg River, just outside Eichstätt, about 30 kilometers (19 mi.) east of the "London specimen" and feather localities. Niemeyer exchanged the fossil gem for a cow with Johann Dörr, and it was eventually acquired by the son of Dr. Häberlein, Ernst Otto Häberlein, who purchased the fossil from the quarry owner, who thought it was merely a pterosaur. The fossil was put up for sale in 1881, and after four years of negotiations, industrial magnate Dr. Werner von Siemens purchased the

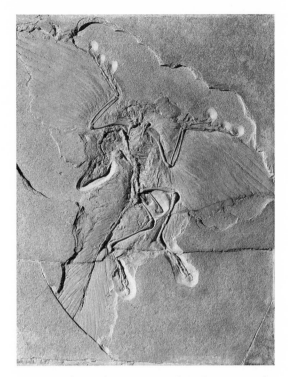

Fig. 3.2. The Berlin specimen of *Archaeopteryx*, the "Avian Rosetta Stone," discovered in 1876 or 1877 in a limestone quarry near the Bavarian town of Eichstätt. Courtesy Daniela Schwarz-Wings, Museum für Naturkunde, Berlin; copyright Christoph A. Hellhake, Munich, Germany.

specimen and then resold it to the Prussian Ministry in 1881 for twenty thousand marks. Both the magnificent fossil and the counterslab were presented to the Museum für Naturkunde in Berlin (formerly East Berlin, where I studied them in 1978), where they reside today.

The beautifully preserved "Berlin specimen" is to this day universally considered among the world's finest fossils, not only in exquisite preservation, but for its significance as an almost perfect Darwinian intermediate linking two classes of vertebrates. I have previously used the term "avian Rosetta Stone" to describe this magnificent fossil bird, which is certainly both historically and scientifically among the most important natural history specimens in existence. Unlike the London specimen, the Berlin *Archaeopteryx* is articulated in a natural pose with outstretched wings exhibiting an elliptical wing-profile characteristic of many

modern birds. Primary and secondary flight feathers are preserved in precise alignment as in modern birds. The long tail, as in the London specimen, exhibits a pair of rectrices, or flight feathers of the tail, attached laterally to each caudal vertebra. Crucially, most of the skull is completely preserved.

Although it was initially assumed that the Berlin specimen was the same species as the London example, German paleontologist Wilhelm Barnim Dames, who first described the specimen, considered it a distinct species and named it *Archaeopteryx siemensii*.[7] In the 1920s, Bronislav Petronievics, overemphasizing disparity in size and other differences between the London and Berlin specimens, placed the Berlin specimen in a separate genus *Archaeornis*, theorizing that his *Archaeornis* was ancestral to all modern birds; and, in a strange twist to Huxley's view of ratites as dinosaur descendants, Petronievics proclaimed that the ratites (ostrich and allies) descended separately from the London *Archaeopteryx!*[8]

As an interesting aside, John Ostrom of Yale University revealed, through a discovery by his secretary Miriam Schwartz in 1983, that O. C. Marsh had attempted to obtain the Berlin specimen for Yale University's Peabody Museum of Natural History in March 1879. F. A. Schwartz (no relation to Miriam Schwartz) offered to sell the Berlin specimen along with other Solnhofen fossils for ten thousand dollars. Within the envelope containing the offer letter were two tracings of the Berlin specimen, which most importantly show contour feathers along the back, feathers on the legs, and a tuft of feathers on the back of the head, all of which have since been prepared away. These important drawings provide valuable evidence that *Archaeopteryx* was fully feathered. In a wonderful story, Ostrom lamented the loss of the "Yale specimen," calling it "the one that flew the coop."[9]

Additional Specimens

Almost a hundred years elapsed before a third *Archaeopteryx* specimen emerged, in 1956. The so-called Maxberg specimen was discovered in a quarry shed by a university student. The new specimen was poorly articulated (presumably as a result of extensive decomposition before fossilization), but it provided evidence that the metatarsals of *Archaeopteryx* were partially fused (presaging the condition in modern birds). Unfortunately, the Maxberg specimen is now lost to science. The fourth specimen of *Archaeopteryx* was uncovered through the efforts of John Ostrom. While studying pterosaurs in the Teyler Museum in Haarlem, Netherlands, in 1970, he noticed a faint impression of feathers on the slab of a Solnhofen specimen identified as a pterosaur and thus happened on what would become known as the Teyler or Haarlem specimen. The specimen had in fact been discovered six years before the discovery of the London specimen and feather, but Hermann von Meyer had mistakenly described it as the pterosaur *Pterodactylus crassipes*. This specimen is notable for the beautiful preservation of a finger claw with its horny sheath.[10]

In 1973, F. X. Mayr of Eichstätt, founder of the famous Jura Museum, announced the recovery of a new *Archaeopteryx* specimen from just north of Eichstätt. It had actually been discovered some twenty years earlier, but it was mistakenly thought to be a juvenile specimen of the small Solnhofen theropod *Compsognathus*. Described in 1974 by Munich paleontologist and *Archaeopteryx* expert Peter Wellnhofer, it is approximately one-third smaller than the London specimen, and this size difference resulted in it later being described as a distinctive genus *Jurapteryx*, though this classification has not been widely accepted.[11] The Eichstätt specimen, on display in the Jura Museum, is well preserved, with a virtually complete skull, but the feather impressions are faint and the furcula is missing (likely lost in the fossilization process).

In 1987, a complete skeleton, some 10 percent larger than the London specimen, was discovered and dubbed the Solnhofen specimen; it was later described by Andrzej Elzanowski as a separate genus and species, *Wellnhoferia grandis*, on the basis of large size and details of pelvic limb morphology.[12] This discovery was followed in 1993 by the recovery of the Aktien-Verein specimen, a nearly complete, somewhat smaller skeleton with longer limb bones and strongly curved, arboreally adapted claws, described by Peter Wellnhofer, and initially thought to have ossified sternal plates, which turned out to be coracoids.[13] Yet the specimen showed arboreal features, including strongly curved claws, and based on the anatomy of the Aktien-Verein specimen, Wellnhofer argued that *Archaeopteryx* "was capable of powered flight and could not have been such a poor flyer as often has been stated."[14] The Aktien-Verein specimen is now known as the Munich specimen and is housed in the Paläontologisches Museum in Munich.

Three more specimens of *Archaeopteryx* have been discovered since 1993, including the eighth, fragmentary Bürgermeister-Müller specimen, discovered in 1997 and currently housed at the Bürgermeister-Müller Museum. Another fragmentary fossil was found in 2004.[15] The most recently discovered *Archaeopteryx*, however, is a complete, well-preserved fossil called the Thermopolis specimen, named for Thermopolis, Wyoming, known for its hot springs, and now as the town in which the Wyoming Dinosaur Center houses the only American specimen of the urvogel. The specimen was long held in a private collection and was described in 2005.[16] The almost complete skeleton most closely resembles the Berlin and Munich specimens in size, and although Elzanowski recognized four species and two genera within the Archaeopterygidae, most authors consider that there is but a single species, and a recent statistical study by pterosaur expert Christopher Bennett of Fort Hays State University in Kansas, concluded that the size variation among the ten specimens, ranging from the largest Solnhofen to the smallest, Eichstätt specimen, at 50 percent the size of the Solnhofen example, is representative of a growth series.[17] Bennett's and other studies have reached various conclusions, and it is doubtful that we will ever be certain of the number of genera and species involved.[18]

The most interesting proposal coming from study of the Thermopolis specimen of *Archaeopteryx* is that it exhibited a number of previously unknown theropod features, most notably a hyperextendible second toe with a sickle claw, otherwise unknown except in dromaeosaurs and troodontids, and the hallmark of such well-known

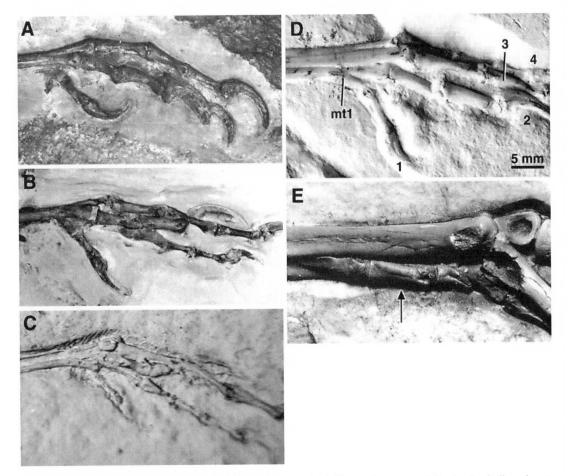

Fig. 3.3. Left feet of *Archaeopteryx* (scaled to the approximate same size): (A) London; (B) Berlin; (C) Eichstätt, right foot reversed; and (D) Thermopolis specimens, showing the reflexed hallux and the large size of the recurved hallucal ungual. The Berlin specimen shows that the articulation of metatarsal 1 is distally located on metatarsal 2, allowing the opposability of the hallux with the other digits (photograph by S. Tarsitano). (E) Pes (photograph rotated for comparison) of *Compsognathus longipes*, a Late Jurassic nondromaeosaurid terrestrial theropod, showing the reduction of the hallux (the arrow points to the hallux and metatarsal 1); the flat hallux claw is in the typical theropod position, parallel and mesial to the tarsus, as in *Velociraptor*. Abbreviations: mt1, first metatarsal; 1–4, digit numbers. From A. Feduccia, L. D. Martin, and S. Tarsitano, "*Archaeopteryx: Quo Vadis?*" *Auk* 124 (2007): 373–80. Photographs by S. Tarsitano and A. Feduccia; Thermopolis foot cropped from photograph of specimen from E. Mayr et al., "A Well-Preserved *Archaeopteryx* Specimen with Theropod Features," *Science* 310 (2005): 1483–86. Reprinted with permission from AAAS.

raptors as *Deinonychus* and *Velociraptor*. Mayr and his colleagues also note that the hallux is not fully reversed, whereas in most other specimens it is. Their conclusions were challenged on the basis that the presence of a hypertrophied and hyperextendible second toe, as in dromaeosaurs, is not conclusive, and as one can see from the photographs, the second digit and claw of the Thermopolis specimen clearly do not resemble those of dromaeosaurs, where the penultimate phalanx is greatly shortened or truncated.[19] Mayr and colleagues' statement that "the absence of a fully reversed first toe indicates that *Archaeopteryx* did not have a perching foot" is misleading because many modern birds with a hallux less than fully reversed or without a hallux perch in trees, so regardless of

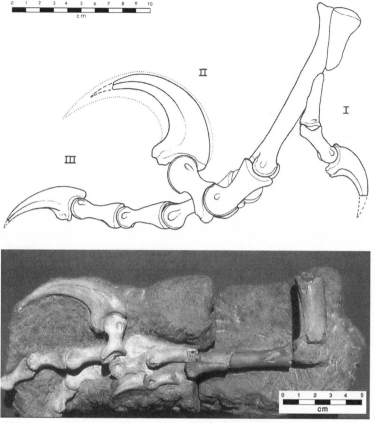

Fig. 3.4. Lower panel: right pes of *Velociraptor mongoliensis* before preparation, with sickle claw of digit 2, along with penultimate phalanx, in retracted position (H. Osmólska, pers. comm.), showing: (1) the reduction of the hallucal ungual; (2) the parallel nature of the hallux in deinonychosaurs; and (3) the more proximal attachment of the metatarsal 1 on metatarsal 2, as compared with metatarsal 1 in *Archaeopteryx*. Note the ventral extension of the joint surface of the penultimate phalanx of digit 2 (not present in *Archaeopteryx*) that allows for the hyperextension (dorsiflexion) of the ungual in deinonychosaurs. Hyperextension is achieved by the extension of the ungual claw of digit 2 along with its penultimate phalanx, which is truncated, unlike that of *Archaeopteryx*. Upper panel: drawing of the pes of *Deinonychus* (rotated for comparison to the photograph), showing nearly identical morphology to that of *Velociraptor*. Note the incorrect position of the hallux in the drawing, presumably to make it look more birdlike. From A. Feduccia, L. D. Martin, and S. Tarsitano, "*Archaeopteryx: Quo Vadis?*" *Auk* 124 (2007): 373–80. Photograph of *Velociraptor* courtesy H. Osmólska; pes of *Deinonychus*, modified and reversed to compare with the right pes of *Velociraptor*. Modified after J. H. Ostrom, "Osteology of *Deinonychus antirrhopus*, an unusual theropod dinosaur from the Lower Cretaceous of Montana," *Bulletin of the Peabody Museum of Natural History* 30 (1969): 1–169).

the degree of reversal of the hallux in *Archaeopteryx*, it was an adaptation for grasping branches.[20]

Looking back on the tortured history of this iconic fossil and the ten skeletal specimens now known, we see that it has variously been described as a pterosaur, a bird, and a small coelurosaurian dinosaur. The London and Berlin specimens were once considered disparate lineages, and the various specimens have been described as a single species, multiple species, and distinctive genera. In addition, the famous urvogel was thought originally by Thomas Huxley to be a bird, albeit primitive, with some reptilian features, notably the teeth and tail; and he termed it an "intercalary" bird, distinguished from the "linear" or truly transitional type.[21] As we shall see, from the past three decades

Fig. 3.5. Gregory Paul's depiction of *Archaeopteryx* as a terrestrial dromaeosaur, shown to be invalid. From G. S. Paul, *Dinosaurs of the Air: The Evolution and Loss of Flight in Dinosaurs and Birds* (Baltimore: Johns Hopkins University Press, 2002), 160, fig. 9.1. © 2002 Johns Hopkins University Press. Reprinted with permission of The Johns Hopkins University Press.

to the present, *Archaeopteryx* has been considered an earthbound dinosaur that could not fly, a side branch of the main lineage leading to modern birds, and, as interpreted by the evidence here, a flying, primarily arboreal bird with diverse behaviors. The problem is, of course, that the behavior of fossil animals cannot always be accurately determined through analysis of anatomy, and behavior invariably precedes phylogeny into new adaptive zones. So, although *Archaeopteryx* shows little in the way of major adaptations for a particular mode of life, it was no doubt capable of a variety of behaviors, and as its claws and reversed hallux show, it was at a minimum an arboreal, volant bird.

The Urvogel Is Focal to the Question of Bird Origins

Fossils require interpretation, and interpretation means that different people, with different backgrounds and perspectives, will inevitably reach disparate conclusions, and that is what differentiates the imprecise science of paleontology from the more experimental fields in which results can be replicated or rejected. Varied interpretations have characterized the literature on the ten skeletal specimens of *Archaeopteryx*. Yet, since the discovery of the London specimen in 1861, one theme

continues: it is a bird, a recognizable avian. In his final comments on *Archaeopteryx* in 1876, Huxley appears somewhat perplexed by the urvogel, and as noted, for him it occupied an "intercalary" but not transitional position between reptiles and birds. Gavin de Beer in his 1954 monograph on *Archaeopteryx* presents an extensive three-page, tabular summary of the various views that have been held regarding the systematic position of the urvogel, included thirty-six detailed positions expressed by various authors, from near the point of its first discovery to the publication of his monograph.[22]

The image of *Archaeopteryx* as a primitive bird, which was firmly established by Gerard Heilmann's *Origin of Birds*, changed suddenly when Robert Bakker, who is credited with the "dinosaur revolution" of the early 1970s, began a major dinosaur revisionism with a paper in *Scientific American* in 1975 entitled "Dinosaur Renaissance."[23] He then viewed *Archaeopteryx* as an earthbound theropod: "In spite of its very birdlike appearance, *Archaeopteryx* was closely related to certain small dinosaurs and could not fly." Bakker was a student of John Ostrom at Yale University, and although Ostrom was the author of the first papers on endothermy in dinosaurs and many of the associated ideas that Bakker later popularized, Bakker's intelligence, charismatic style, and flashy presentations gained him considerable notoriety during the 1970s and beyond, especially after the publication in 1986 of *The Dinosaur Heresies*, which greatly bolstered the movement.[24] His portrayal of dinosaurs as endothermic, energetic, and highly intelligent archosaurs, as well as his flamboyant style—at times a preacher, a teacher, and an artist—gained him near legendary status. As one writer noted: "Until Bakker, brontosaurs languished in swamps, doomed outsized monuments of reptilian isolation and torpor; now they crashed across savannas in organized herds, doting on their young. No scientist has done more than Bakker to advance the view of dinosaurs as energetic, swift, bright-colored, cunning, persevering, supremely well adapted, socially complex—in sum, thrilling—animals."[25] As noted, the major momentum behind Bakker's thought and his "dinosaur revolution" came from the work of John Ostrom of Yale University and his discovery of the early Cretaceous

Fig. 3.6. Above, John Ostrom's attempt at reconstruction of a hypothetical stage in early bird evolution, showing a pre-*Archaeopteryx* stage (top) and *Archaeopteryx* (bottom). The proto-Archaeopteryx illustrates an early stage in the enlargement of feathers on the hands and arms as aids in catching insects, or insect swatters. The enlarged tail feathers are hypothesized as aerodynamic stabilizers that enhanced agility and quick maneuvering while chasing prey. For a terrestrial origin of flight and elongation of flight feathers, some selection force had to be envisioned. Scale bars = 5 cm. From J. H. Ostrom, "Bird Flight: How Did It Begin?" *American Scientist* 67 (1979): 46–56. Reprinted with permission. Below, various versions of Ostrom's model have appeared even though in the literature the "insect swatter" model never gained much momentum outside the paleontological community. Upper left, modified after Ostrom 1979; upper right, modified after D. E. G. Briggs, "Extraordinary Fossils," *American Scientist* 79 (1991): 136; lower left, modified after R. T. Bakker, "Dinosaur Renaissance," *Scientific American* 232, no. 4 (1975): 58–78; lower right, modified after P. C. Sereno, "The Evolution of Dinosaurs," *Science* 284 (1999): 2144. Today the model is widely used to depict various Chinese Lower Cretaceous feathered creatures, including the earliest beaked bird, *Confuciusornis*, usually demonstrably incorrectly.

dromaeosaur *Deinonychus* from Montana in the 1960s, a theropod that exhibited many similarities to birds, particularly in the structure of the hand and wrist, with its semilunate carpal element similar to that of *Archaeopteryx* and modern birds. Ostrom viewed *Archaeopteryx*, as did Bakker, as a more or less terrestrial theropod, insulated with feathers for endothermy, and learning to fly from the ground up; their views presaged the modern paleontological perspective. Although many still view the urvogel as more of an earthbound theropod with some flight capabilities, many recent workers have found it to be much more birdlike than imagined. But regardless of speculation on its anatomy, habits, and relationships, *Archaeopteryx* has traditionally occupied a position of centrality with respect to virtually all discussions on avian and flight origins.[26]

The Avian Wing of *Archaeopteryx*

The most obviously avian features of the urvogel are its wing feathers, which in almost all respects are identical to those of modern birds, even at the microstructural level. In 1978, at a time when the urvogel was being reinterpreted as an earthbound theropod, as program chairman for the annual meeting of the American Ornithologists' Union at Haverford College, I invited Alan Charig of the British Museum and John Ostrom to give plenary presentations on *Archaeopteryx*. Charig talked on the ongoing charges of fraud by British astronomer and cosmologist Sir Fred Hoyle with respect to the *Archaeopteryx* specimens (long since demolished and dismissed), and Ostrom spoke on the actual fossil specimens, asserting that *Archaeopteryx* could not fly and used its wings to catch insects.

Fig. 3.7. A profound student of birds, Harrison "Bud" Tordoff served in the 352nd Squadron, 353rd Fighter Group, US Army Air Force, in World War II, attaining the rank of captain and becoming an ace. He flew eighty-three missions in a P-51 Mustang in support of ground forces. Bud said he knew how peregrine falcons felt, having dived from altitudes of 25,000 feet in his Mustang to shoot down one of the newly developed German fighter jets. Photograph collage courtesy of Carrol Henderson.

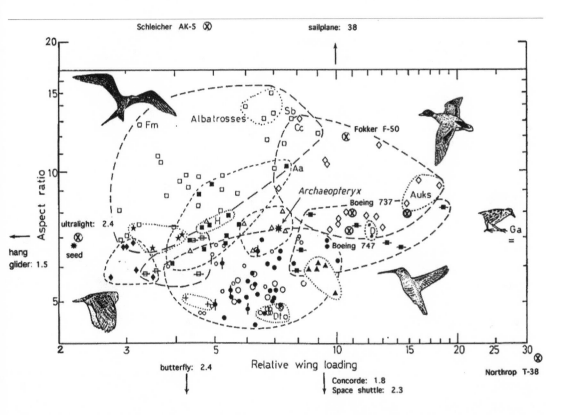

Fig. 3.8. Aspect ratio versus relative wing loading in various birds. The encircled foraging groups are defined solely by the similarity of flight and foraging modes, not by systematic affinity. Abbreviations: Aa, *Apus apus* (common swift); Cc, *Cygnus cygnus* (whooper swan); D, diving petrels; Df, Darwin's finches; Fm, *Fregata magnificens* (magnificent frigatebird); Ga, *Gallirallus australis* (weka, a flightless New Zealand wood rail); H, swallows; Sb, *Sula bassana* (northern gannet). Note particularly the position of the urvogel *Archaeopteryx*. Data from sources from U. M. Norberg and R. Å. Norbergand C. J. Pennycuick, in: U. M. Norberg, *Vertebrate Flight* (Berlin: Springer, 1990); and U. M. L. Norberg, "Bird Flight," *Proceedings of the 23rd International Ornithological Congress, Beijing, Current Zoology* (formerly *Acta Zoologica Sinica*) (2004): 921–35. Reprinted with permission, courtesy Ulla M. Lindhe Norberg.

Ostrom had brought along for display and examination casts of all the known urvogel specimens. At this meeting one of my mentors, Harrison "Bud" Tordoff, and I almost simultaneously noticed on the casts that the wing of *Archaeopteryx* was not only remarkably birdlike but that the individual primary flight feathers were well preserved. As I commented to Tordoff, the feathers were so well preserved that one could actually discern that the vanes were asymmetric.

Tordoff, a flying ace in World War II, was immediately impressed by the same feature, which he said struck "like a bolt of lightning." These were the feathers of a flying bird, for why else would the feathers be designed like individual airfoils?

In other words, the narrow leading edge of the asymmetric feathers gives each feather an airfoil cross-section. In modern birds, as we noted in our paper published the following year in *Science*, the feathers of strong fliers have narrow leading-edge vanes, and in poor fliers, the asymmetry is less obvious, while the feathers of many flightless birds are actually symmetrical.[27] We concluded that "the shape of the feathers seems to show that *Archaeopteryx* had an aerodynamically designed wing and was capable of at least gliding flight." If the urvogel could not fly, one would surely expect that the primary feathers would be symmetric, as in most truly flightless birds. We noted, "Any argument that *Archaeopteryx* was flightless must explain selection for asymmetry in the wing feathers in some context other than flight." Our paper on the asymmetrical flight feathers of *Archaeopteryx* received considerable attention, as it was the first new evidence that the famous urvogel was a flying bird, and it contradicted the views of Ostrom and Bakker.

Contributing to the argument that the feathers of the urvogel were indistinguishable from those of a modern bird, Åke Norberg of the University of Gothenberg showed that the primary flight feathers exhibit curvature as an expression of their complex aerodynamic design, and Siegfried Rietschel showed that the same feathers are uniquely like those of modern birds in the possession of a reinforcing furrow along the ventral shaft.[28] As Ulla Norberg, also of the University of Gothenberg, summarized about birds in general, "The inherent aeroelastic stability of the feather is achieved passively by an appropriate combination of the three main feather characteristics . . . (1) vane asymmetry, (2) the curved shaft, and (3) the great flexural stiffness dorsoventrally . . . achieved partly by a ventral furrow of the feather shaft."[29] All three features are characteristic of the feathers of *Archaeopteryx*. By the Late Jurassic the earliest known bird had clearly developed the firm, bladelike pennaceous feather of modern birds and was, as the late Canadian evolutionary biologist D. B. O. Savile noted, "appreciably advanced aerodynamically."[30]

The elliptical wing shape of *Archaeopteryx* is also recognizably avian, comparing with those of modern woodland birds, which are designed for maneuvering in and out of thick brush and dense vegetation. Such wings are characterized by low aspect ratios, a uniform pressure distribution over the wing surface, and pronounced camber (convex curvature). They are typically found in gallinaceous species, doves, woodcocks, and most passerines. The elliptical wing is less effective than some other designs at high, sustained speeds, but it is suited for twisting flight in dense foliage. Wing loading and aspect ratio are two measures of morphological parameters that functionally describe the shape of a flying animal, such as wing lengths, wing areas, and body and wing masses. Wing loading is the ratio of weight to wing area in a flying animal (or airplane). The aspect ratio

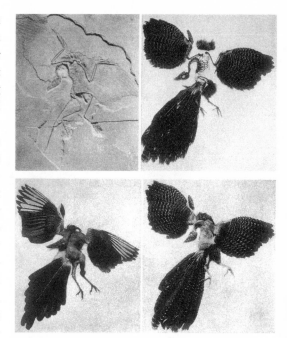

Fig. 3.9. *Archaeopteryx* look-alikes. Upper, the Berlin *Archaeopteryx*, left, compared with its "look-alike," the Australian pheasant coucal (*Centropus phasianinus*), right, prepared brilliantly by Oskar Heinroth in 1923, in the same posture, showing the remarkable superficial similarity in overall form and proportions. However, the coucal is mainly terrestrial and, unlike *Archaeopteryx*, has loosely constructed tail and wing feathers, as well as a less elliptical wing, and its claw arc measurements give a mean of 87.5 (N = 10), typical of ground-dwellers and unlike the highly curved pedal claws of *Archaeopteryx*. Below, similar comparisons of a magpie (*Pica pica*, avg. length 40–51 cm; 19.9 in.) and the same coucal showing overall similarity in profile. From O. Heinroth, "Die Flügel von *Archaeopteryx*," *Journal für Ornithologie* 71 (1923): 277–83. Courtesy Franz Bairlein for *Journal für Ornithologie* and the Deutsche Ornithologen-Gesellschaft (DO-G).

is the ratio between wingspan and the cord or breadth of the wing, and birds such as galliforms with short, broad wings have a lower aspect ratio than birds that are more aerodynamically efficient. As one can see from the figure illustrating aspect ratio versus relative wing loading in various birds, *Archaeopteryx* fits almost exactly in the middle of the range.

Among the most insightful comparisons of the urvogel to its modern ecomorphological equiva-

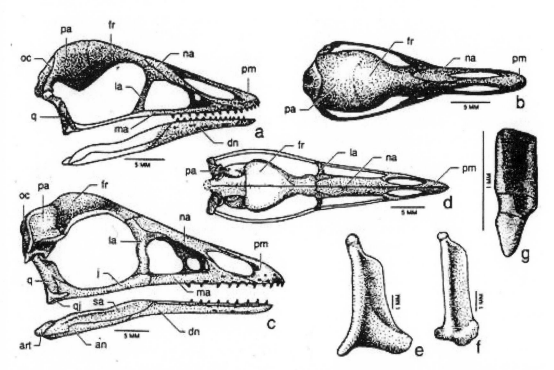

Fig. 3.10. Reconstruction of skull structures of *Cathayornis* and *Archaeopteryx*: (a) lateral view of *Cathayornis*; (b) dorsal view of *Cathayornis*; (c) lateral view of *Archaeopteryx*; (d) dorsal view of *Archaeopteryx*; (e) quadrate of *Archaeopteryx*; (f) quadrate of *Cathayornis*; and (g) labial view of tooth from the maxillary of *Cathayornis*. Abbreviations: an, angular; art, articular; dn, dentary; fr, frontal; j, jugal; la, lacrimal; ma, maxillary; na, nasal; oc, occipital; pa, parietal; pm, premaxillary; q, quadrate; qj, quadratojugal; sa, surangular. From L. D. Martin and Z. Zhou, "*Archaeopteryx*-Like Skull in Enantiornithine Bird," *Nature* 389 (1997): 556. Copyright 1997. Reprinted by permission from Macmillan Publishers Ltd.

lents was a study undertaken by the famous German biologist Oskar Heinroth.[31] Aside from his ornithological interests, Heinroth was among the first to apply his methods of comparative morphology to the study of animal behavior, and he is known as a founder of the field of ethology. Among his many accomplishments, he rediscovered and promoted the study of imprinting and was the mentor of Nobel Laureate Konrad Lorenz. In his study of *Archaeopteryx* Heinroth compared it to modern analogues, specifically the living magpie (*Pica*) and the coucal (*Centropus*). As one can see from the figure on page 66, the Berlin specimen of *Archaeopteryx* is a remarkably close match in size, proportions, and wing profile to living magpies, and especially to the pheasant coucal (*Centropus phasianinus*) of Australia and New Guinea, which inhabits thickets and is primarily terrestrial. Unlike *Archaeopteryx*, however, and like many other

predominantly ground-dwelling birds, the coucal has a loosely constructed tail, frayed at the tip, and loosely constructed wing feathers (as seen at the tips), with less wing ellipticity. The salient difference, however, is that, unlike *Archaeopteryx*, which shows no tail fraying due to terrestrial locomotion and has arboreally adapted, strongly curved pedal claws, the inner claw arc curvature is contained within the range of living ground-dwelling birds.[32]

Did *Archaeopteryx* Represent a Tetrapteryx Stage in Bird Evolution?

Paleontologist Nick Longrich restudied the Berlin specimen and his detailed examination revealed that, like some enantiornithine birds and microraptors from the Early Cretaceous of China, the urvogel possessed hind-limb plumage, characterized by flight feathers extending from the anterior or cranial surface of the tibiae and from the rear or caudal margins of the tibiae and femora.[33] These feathers had been noted by Per Christiansen and Niels Bonde, but they did not attempt an evaluation of the function of these hind-limb feathers.[34] The hind-limb feathers exhibit vane asymmetry, curved shafts, and a self-stabilizing overlap pattern. Such a hind-limb wing would facilitate lift generation in both the wings and in the tail, and this suggests that the hind-limb feathers acted as airfoils. Longrich concluded that the hind-limb wings formed about 12 percent of the total airfoil area. He stated that the "presence of a four-winged planform in both *Archaeopteryx* and basal Dromaeosauridae [Chinese microraptors] indicates that their common ancestor used fore- and hindlimbs to generate lift."[35] More recent studies using ultraviolet light, however, appear to show that *Archaeopteryx* was endowed with hind-limb "trousers," as in a number of modern birds of prey, and not hind-limb wings, as are present in microraptors (discussed in chapter 4).[36] The new findings from the Chinese Early Cretaceous microraptors suggest that arboreal parachuting and gliding preceded the evolution of flight in birds, and if so, *Archaeopteryx* would represent a more derived stage in avian flight or, being a mosaic of primitive and advanced features, may have lost the fully developed rear wings.

As we shall see, the feet and legs of *Microraptor*, a basal dromaeosaurid, originally described as a "four-winged dinosaur," also bear elongate, asymmetrical flight feathers, increasing in asymmetry toward the end of the respective limb, as in modern bird wings. As with *Archaeopteryx*, these hind-limb feathers of *Microraptor* apparently functioned in gliding, and perhaps in powered flight. Interestingly, William Beebe in 1915 described what he called a hypothetical tetrapteryx stage in the ancestry of avian flight, proposing that avian flight evolved via a four-winged gliding stage, the reconstruction of which is virtually identical to *Microraptor*.[37] Beebe's model was supported by what he perceived to be feather impressions on the legs of the Berlin specimen of *Archaeopteryx*, and he proposed that a "tetrapteryx" stage in avian evolution possessed "hindlimb wings" that served as passive parachutes. Commenting on the discovery of the four-winged *Microraptor*, Richard Prum remarked that the specimen "looks as if it could have glided straight out of the pages of Beebe's notebooks" and that it "provides striking support for the arboreal-gliding hypothesis of the origin of bird flight."[38]

A Miscellany of Avian Cranial Characters in *Archaeopteryx*

Many other characters of the *Archaeopteryx* skeleton have turned out to be more avian than previously imagined. Kenneth Whetstone, for example,

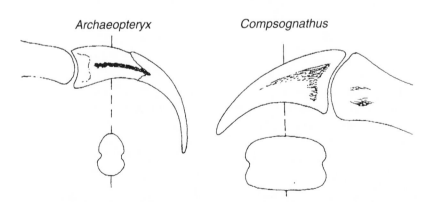

Fig. 3.11. The ungual phalanx and claw of the third (middle) toe of *Archaeopteryx* (left) compared to that of *Compsognathus* (right), in lateral view, with diagrammatic cross-section. Drawn by Derek Yalden from high-fidelity casts in the Natural History Museum, London. From D. W. Yalden, "Climbing *Archaeopteryx*," *Archaeopteryx* 15 (1997): 107–8. Courtesy Derek Yalden and the Jura-Museum, Eichstätt, Germany.

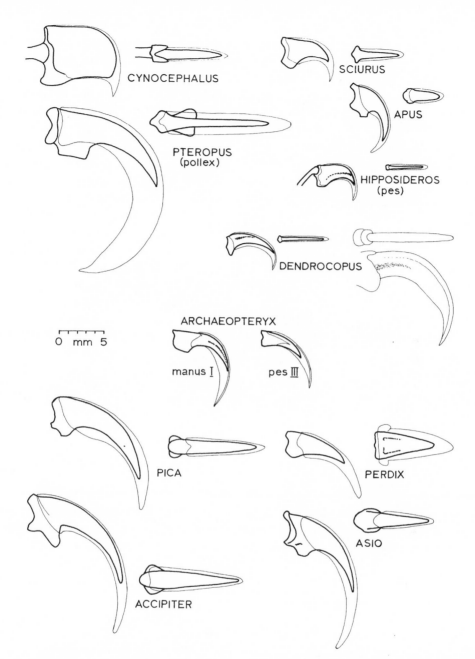

CYNOCEPHALUS

SCIURUS

PTEROPUS
(pollex)

APUS

HIPPOSIDEROS
(pes)

DENDROCOPUS

ARCHAEOPTERYX

0   mm   5

manus I

pes III

PICA

PERDIX

ASIO

ACCIPITER

Fig. 3.12. Derek Yalden's scale drawings of the ungual phalanges (heavy line) and horny claw (thin line) of a range of birds and mammals, in lateral and extensor views. *Archaeopteryx* claws are shown for comparison. The selection includes a perching bird that forages on the ground (magpie, *Pica pica*); two avian predators that use their feet to grasp prey (note the broader base) (long-eared owl, *Asio otus*; sparrowhawk, *Accipiter nisus*); a cliff-nesting bird (alpine swift, *Apus melba*); two trunk-climbing woodpeckers (lesser spotted woodpecker, *Dendrocopus minor*; white-backed woodpecker, *D. leucotus*); two trunk-climbing mammals (gray squirrel, *Sciurus carolinensis*; flying lemur, *Cynocephalus volans*); two bats (fruit bat, *Pteropus* sp.; leaf-nosed bat, *Hipposideros* sp.); and a ground-dwelling galliform (gray partridge, *Perdix perdix*). The extremely laterally compressed, strongly curved claws with needle-sharp points are characteristic of trunk climbers and do not fit the model of an earthbound predator. From D. W. Yalden, "Forelimb Function in *Archaeopteryx*," in *The Beginnings of Birds*, ed. M. K. Hecht et al. (Eichstätt: Freunde des Jura Museum, 1985), 91–97; courtesy Derek Yalden and the Jura-Museum, Eichstätt, Germany.

Fig. 3.13. Claw (including the bony ungual and the horny sheath) of the third finger (digit IV of ornithologists and most embryologists), right manus of Teyler specimen of *Archaeopteryx*: note the extreme curvature, lateral compression, and needlelike point. Actual straight length of claw, 10.5 mm (0.4 in.). Courtesy of the late John Ostrom.

reexamined the braincase of the London specimen after it was newly prepared and showed that it differed from that of theropods in the morphology of both the quadrate, prootic regions and the occiput: "The skull is much broader and more bird-like than earlier interpreted."[39] Alick Walker studied the otic region of the London specimen and found that it was "of a primitive, basically avian type."[40] The late German functional morphologist Paul Bühler studied the skulls of three specimens of *Archaeopteryx* and discovered that, like modern avians, the quadrate was movable against the skull and the upper jaw was movable in the avian prokinetic manner, in which the upper jaw moves independently with respect to the braincase. Bühler also discovered that the forebrain and cerebellum were enlarged in relation to the braincase and that the optic lobes were separated by the cerebellum, which indicated "that *Archaeopteryx* developed a much more birdlike level than has been assumed."[41] In 1988, Bernd

Haubitz and his colleagues used computed tomography on the skull of *Archaeopteryx* and concluded that "contrary to earlier interpretations of these specimens, the results . . . show the presence of an avian-like double-headed quadrate bone in this earliest true bird."[42] Finally, in describing the skull of the seventh specimen, Andrzej Elzanowski and Peter Wellnhofer found the palatine bone to be distinctively avian and "different from the bone of theropods and early dinosaurs. . . . The palatine structure supports strongly that *Archaeopteryx* is a primitive bird rather than a feathered preavian archosaur."[43] Elzanowski has also published an up-to-date summary of the anatomy of *Archaeopteryx* and his interpretation of its paleobiology.[44] However, in contrast to Elzanowski's view of the palatine bone, Gerald Mayr and colleagues found that, as in the so-called nonavian dinosaurs, *Archaeopteryx* had a plesiomorphic tetraradiate palatine bone, which, in combination with other characters, "blurs the distinction

Fig. 3.14. Above, dorsal view of and right wing and manual claws of the Berlin specimen, which exhibit claws closest in morphology to those of trunk-climbing birds, especially Neotropical woodcreepers (Dendrocolaptidae). Below, 1–5, claws of *Archaeopteryx bavarica*. (1) detail of shoulder girdle area showing the proximal parts of both humeri, scapulae, coracoids, the sternum (arrow), and right digits; × 2.3 natural size. (2) detail of (1), showing first right digit with bony claw, right coracoids, ribs, and gastralia; × 3.9. (3) horny claw of first right digit, lying dislocated at the lower margin of the slab; × 3.5. (4) claw with horny sheath of second right digit; × 3.7. (5) claw with horny sheath of third right digit; × 3.2. Above, Courtesy Daniela Schwarz-Wings, Museum für Naturkunde, Berlin; copyright Christoph A. Hellhake, Munich, Germany; below, photographs by Franz Höck, courtesy of Peter Wellnhofer.

of archaeopterygids from basal deinonychosaurs (troodontids and dromaeosaurs)."[45]

The Avian Claws and Hallux

Derek Yalden of the University of Manchester demonstrated in a 1985 paper that the claws of *Archaeopteryx*, specifically the manual claws, are like

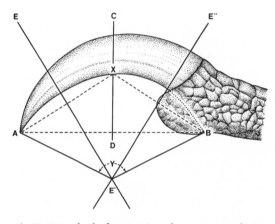

Fig. 3.15. Method of measuring claw geometry by inner arc; the claw is measured here in terms of relative degrees of a circle, with the angle (Y) the measure of the degrees of the arc. From A. Feduccia, "Evidence from Claw Geometry Indicating Arboreal Habits of *Archaeopteryx*," *Science* 259 (1993): 790–93. Reprinted with permission from AAAS.

those of trunk-climbing birds (and mammals) and that the pedal claws are similar to those of perching birds.[46] The manual claws of *Archaeopteryx* also resemble those of trunk-climbing or clinging mammals such as the flying lemur (*Cynocephalus*), various gliding marsupials of Australia, flying squirrels, and fruit bats. Indeed, the manual claws most closely resemble those of the woodhewers of the family Dendrocolaptidae, a group of trunk-climbing, or scansorial, Neotropical passerines that superficially resemble a large version of the better-known Northern Hemisphere treecreepers (Certhiidae) and provide an outstanding example of convergent evolution.

By using a geometry metric of inner claw curvature, one can show that claw arc in general distinguishes ground-dwelling birds from perching birds, and trunk-climbing birds have an even greater curvature, as well as laterally compressed claws. As a general rule ground-dwellers have relatively flat claws, and this is what one sees also in theropods such as *Compsognathus*, which exhibits flat, almost hooflike claws, totally unlike those of *Archaeopteryx*. Most interestingly, the curvature of pes claws of *Archaeopteryx* fossils falls completely outside the range of ground dwellers, and that of the manus claws falls within the range of trunk climbers.[47] Aside from arc curvature, the curved claws of trunk-climbing birds including

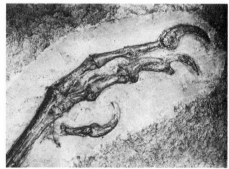

Fig. 3.16. Left, morphology of the entire foot of, left, the lyrebird (*Menura novaehollandiae*), a predominantly ground-dwelling bird, exhibiting the distinctive relatively straight claws of a ground-dweller, compared with that of, right, a bowerbird (*Chlamydera nuchalis*), a predominantly perching bird, which shows curved claws like those of *Archaeopteryx*. Right, in *Archaeopteryx* (left foot of London specimen) the

hallux is somewhat shorter and slightly elevated, but has a highly recurved hallucal and anterior claws, fitting into the perching bird category; it is a foot adapted for grasping branches. Left from Feduccia, "Evidence from Claw Geometry Indicating Arboreal Habits of *Archaeopteryx*," *Science* 259 (1993): 790–93; reprinted with permission from AAAS; right courtesy S. Tarsitano.

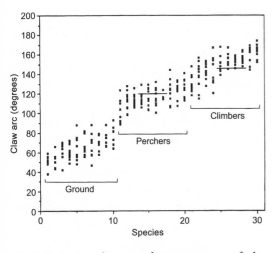

Fig. 3.17. Scatter diagram of measurements of claw arcs of thirty species of living birds showing degree of curvature. Note the almost complete segregation of ground-dwellers. The line in the center of the perchers indicates the mean value for *Archaeopteryx* pes claws, and the line in the middle of the climbers (to the right of the diagram) indicates the mean for manual claws. Recently discovered skeletal specimens fit within the parameters shown here. From Feduccia, "Evidence from Claw Geometry Indicating Arboreal Habits of *Archaeopteryx*," *Science* 259 (1993): 790–93. Reprinted with permission from AAAS.

Archaeopteryx differ from those of predatory dinosaurs in that they exhibit extreme lateral compression and have needlelike points.[48] Manual claws designed for predatory grasping may not be distinguishable from those designed for climbing solely by the angle of the claws, but they are easily distinguished by their thicker base, tapering toward the tip.[49] In addition to illustrating the similarity of the claws of trunk-climbing mammals and birds to those of *Archaeopteryx*, Yalden's figure illustrates the danger of using the bony unguals to estimate claw curvature, as was done by John Ostrom, accepted and perpetuated by Luis Chiappe, and attempted again in a paper by Christopher Glen and Michael Bennett.[50]

In an experimental analysis of dromaeosaurid claws, Phillip Manning and colleagues of the University of Manchester used a robotic limb to test the hypothesis that theropods such as *Deinonychus* used these claws to disembowel their prey. Their modeling supported the view that, contrary to the "disembowelling model," the distinctive claw may have aided in prey capture by being used to grip the hides of prey. They further suggested that, given their extreme size and curvature, dromaeosaurid unguals could have been used as climbing crampons instead of as slashing cutlasses. Comparing the claw curvature to those of arboreal birds, they show that the pedal digit II ungual of *Deinonychus* had an inner arc measurement of 160°, "supporting a climbing function for this structure."[51] In a more recent biomechanical study, using X-ray microtomography and other techniques, Manning and colleagues showed that the dromaeosaurid *Velociraptor* possessed highly recurved claws on the manus and on digit II of the pes that were well-adapted for multiple functions, including locomotion (walking, running, and climbing) as well as prey capture and grappling. They concluded that "enhanced climbing abilities of dromaeosaurid dinosaurs supports a scansorial phase in the evolution of flight," and more recently, David Burnham and colleagues studied the extremely curved claws of the small, basal dromaeosaurid *Microraptor*, concluding that tree climbing was a primary adaptation in birds, used for climbing tree trunks and performing little function in prey capture.[52] Adding to this scenario, Larry Martin and Jong-Deock Lim studied the wing feathers of the Berlin *Archaeopteryx* and discovered evidence for an avian prepatagium and postpatagium. Their study revealed an essentially avian hand with outer and middle fingers united nearly to the claw, with a long postpatagium extending along the back margin of the hand and arm, making it likely that the middle and outer fingers were also united with long primary feathers in other forms such as *Microraptor*, *Caudipteryx*, and *Confuciusornis*. The importance of this observation lies in the fact that *Archaeopteryx* and *Confuciusornis* especially are typically restored with separated fingers, in the former often reconstructed in a predatory function. As mentioned, they also found evidence for a propatagium, a triangular flap of skin that is bounded by a tendon that extends from the proximal humerus to the carpus and provides a considerable flight surface. The propatagium constrains the extension of the forearm, arguing for modern wing folding in the urvogel; and all specimens of *Archaeopteryx* preserve the avian folding pattern in at least one wing. Importantly, such a hand lacks significant grasping ability and would have been of little use in capturing prey; rather, the fingers and claws were used primarily for trunk climbing.[53] Given the relatively small size of these dinosaurs, there is certainly no reason they could not have scaled trees and used their claws for a variety of activities. With much less relative claw curvature, large varanids and iguanids climb trees with little effort, and numerous sizable mammals such as large cats, bears, and porcupines are frequent tree inhabitants.

Another uniquely avian anatomical feature of *Archaeopteryx* is a reversed hallux with a highly curved ungual, an adaptation for perching in which the hind toe is opposed to the front three toes for grasping branches. Precise determination of the degree of the reversal of the hallux in *Archaeopteryx* is difficult. In fact, Gerald Mayr and his colleagues argued that the tenth skeletal specimen exhibited a hallux that was not fully reversed, and they concluded that *Archaeopteryx* did not have a perching foot.[54] However, full reversal of the hallux is not required for a perching foot, and the orientation of the hallux in modern birds ranges from a high of 180° to a low of 65° in some totipalmate, pelecaniform birds (which nest in and are capable of perching in trees).[55] Many modern birds with a hallux less than fully reversed, such as galliform birds, or without a hallux, such as shorebirds, can easily perch in trees. Because the embryos of many extant birds have a mesially oriented hallux, it is difficult to imagine that any posterior reversal could be associated with anything other than the perching habit. Therefore, regardless of the degree of reversal of the urvogel hallux, it was clearly an adaptation for grasping branches.

In an apparent attempt to show that some theropods were birdlike, it has been argued from time to time that the small, Late Jurassic *Compsognathus* exhibits a reversed hallux as in modern birds, but photographs show that this is not the case.[56] There are well-preserved feet in a number of birdlike theropods, such as *Deinonychus*, *Velociraptor*, and *Coelophysis*, and they have no indication of a reversed hallux, although in the recently discovered Early Cretaceous Chinese microraptors (small

dromaeosaurs), there is a reversed hallux, and they show all the morphological signs of having lived in trees. As we shall see, however, the question is whether microraptors are true theropods or hidden birds, a remnant of the early avian radiation. In *Coelophysis* and *Compsognathus*, Late Triassic and Late Jurassic theropods, respectively, the first metatarsal and first toe were medial and not reversed. Moreover, there is no reason for an earthbound, cursorial theropod to have a reversed hallux, since it could only be a hindrance in running. As Walter Bock, proponent of the arboreal theory for the origin of flight, states: "Without a doubt the hallux in *Archaeopteryx* is reversed, and its foot functioned as a grasping one. Specialized ground-dwelling birds tend to reduce and lose the reversed hallux, and . . . it is difficult to conceive of aspects of terrestrial-dwelling that would have provided the selective agents needed for the evolution of a reversed hallux."[57]

Likewise, the right foot of the Munich specimen of *Archaeopteryx* is clearly preserved in a grasping position, with superimposed claws of the hallux and fourth toe.[58] This fact, combined with the lack of clearly definable cursorial adaptations in *Archaeopteryx*, would tend to argue in favor of a more arboreal existence, not to exclude activities on the ground. We will never know exactly the daily activities of the urvogel, but as Elzanowski correctly noted, "Paleobiological . . . reconstruction of *Archaeopteryx* clearly speaks against the cursorial origin of avian flight."[59]

Claws and Habitat

A good summary comes from a study of the claws and digits of the urvogel by P. J. Griffiths of the University of Wolverhampton, in which he noted that the most common plants in the Solnhofen habitat were the conifers *Brachyphyllum* and *Palaeocyparis*, which grew to about 3 meters (10 ft.), and brushland formed by Bennettitales, interspersed with open plains.[60] Of course, the lack of fossils is difficult to assess and certainly cannot be used as conclusive evidence for a lack of trees. Indeed, if no streams entered the ancient saline lagoon, as is often the case today for lagoons in arid regions, there would be no preservation or evidence of nearby trees. What we do know is that the flora did not change appreciably during the course of the Jurassic, much less the Late Jurassic, and there is ample evidence for forest inland from the arid lagoons. The slightly older Nusplingen deposits, about 250 kilometers (155 mi.) to the southwest of Solnhofen, when combined with the Solnhofen flora, provide a broader spectrum of plants, including numerous seed ferns (pteridosperms, including *Cycadopteris*), conifers (coniferopsids, including *Brachyphyllum* and *Palaeocyparis*), Bennettitales (*Zamites*), and ginkgoes (*Ginkgo* and *Baiera*), any of which could have provided perches and trunks to climb. Although most of the plants preserved from the Solnhofen lagoon are bushlike conifers, plants such as *Cycadopteris* grew perhaps 10 or so meters (33 ft.), and *Zamites* were substantial, arborescent plants. Regretfully, most of the gingko material was lost during World War II; yet ginkgoes (including the genus *Ginkgo*) are well known from Europe during the Jurassic, and these trees, such as today's *Ginkgo biloba*, grow to some 20 meters (66 ft.), so ancient ginkgoes could have provided suitable branches for perching, as well as substantial trunks for climbing. Peter Wellnhofer has favored a view of *Archaeopteryx* foraging in the interior of the German and Bohemian landmasses, primarily in the bushy and arborescent conifers, mainly araucarias, ginkgoes, seed ferns, and Bennettitales, and perhaps along the shores of pools and streams.[61]

It has often been noted that the fauna at Solnhofen was blown into the lagoons by strong offshore winds and storms. The fine sediments of the hypersaline lagoons would have sealed the fate of flying insects, pterosaurs, and the occasional *Archaeopteryx* in the forming Plattenkalk, or platy limestone: "Gales blowing from the land during an exceptionally low tide simultaneously explain the presence of both marine and flying animals in the sediment as well as the formation of the sediment itself."[62] The diverse insect fauna of Solnhofen included mayflies, thirteen species of dragonfly (the most of any fossil deposit), cockroaches, water skaters, locusts and crickets, bugs and water scorpions, cicadas, lacewings, beetles, bees and wasps,

and caddis flies and dipterans, in combination with abundant pterosaurs and *Archaeopteryx*, indicating a volant fauna that was preserved near the shore. The myriad insects could not have been too distant from land, and many could not have been far from fresh or brackish water to lay their eggs. The mere lack of fossil logs at Solnhofen cannot be used as evidence for their absence, for there were no rivers that could have transported them to the open sea. As the paleontological adage goes, lack of evidence is not evidence of absence.

Plant fossils from the Solnhofen limestone deposits have been interpreted as representing a semiarid, occasionally wet, forested environment with shrubs, small trees, and Bennettitales, but large conifer forests characterized by tall trees similar to modern araucarias that grow up to 60 meters (132 ft.) tall were no doubt present. As David Burnham has noted, "Araucariaceae had a cosmopolitan distribution throughout the Jurassic. . . . Undoubtedly, conifers formed part of the vast structured forests covering the landmasses to the north and east of the Solnhofen basins similar to the Yorkshire Jurassic Flora. . . . A forest with a tall, coniferous canopy was probably the habitat for *Archaeopteryx* as for other continental organisms found in the Solnhofen Formation."[63]

We may know now as much as we ever will about the habitat, but one thing is clear: the claws of *Archaeopteryx* are those of a predominantly arboreal bird, showing the telltale arc curvature with needlelike points and extreme lateral compression characteristic of climbing mammals and scansorial modern birds. The manual claws were used in trunk climbing and clinging to trunks, and perhaps in clinging to branches, because *Archaeopteryx* had not yet achieved the balance that is characteristic of modern birds and its center of gravity was more reptilian. As Griffiths has noted: "The structure of the foot of *Archaeopteryx* does not suggest a predominantly cursorial mode of locomotion. *Archaeopteryx* has an opposable reversed hallux which is an adaptation for an arboreal habit, and which does not resemble the hallux of *Compsognathus*. Although the structure of the foot shows *Archaeopteryx* would have been able to walk on the ground . . . the presence of the hallux and the type of claw suggest much of the time was spent in vegetation. . . . As the structure of the claws most closely resembles those of tree climbing modern animals and not those of predators, this suggests they were used primarily for climbing, but may have played a secondary role in predation as well as other activities."[64]

Storrs Olson, who has worked on the evolution of island birds, made the important observation that "perhaps not enough has been made of the fact that the known individuals [of *Archaeopteryx*] were inhabitants of islands."[65] Given that these urvogels inhabited the Solnhofen archipelago, a series of islands, the rather dramatic size and morphological differences that are apparent in the ten known skeletal specimens could be the result of divergence of island populations rather than reflective of a growth series. Populations isolated on islands often undergo adaptive radiation; the Hawaiian honeycreepers and Galápagos finches are vivid examples of this process. Microevolutionary events may be exacerbated in island situations, and climatic and vegetational changes that may appear relatively minor for broadly adapted continental species can have dramatic effects in an insular world such as Solnhofen. Although the urvogel shows little in the way of cursorial adaptations and has a number of strictly arboreal features, it may be that some populations, perhaps exemplified by the Solnhofen specimen, were becoming more terrestrial. Elzanowski pointed out that the pes of the larger Solnhofen specimen, which he considered a distinctive genus, is more symmetrical and suggestive of somewhat more cursorial habits. Could the size variation in the ten specimens regarded as stages in a growth series in reality represent a Solnhofen archipelago adaptive radiation?

Several recent studies have concentrated more on the morphology and proportions of the digits as opposed to the claws. Although his conclusions were not unequivocal, James Hopson of the University of Chicago provided an extensive statistical analysis of phalangeal proportions in birds, living and fossil, and theropods, finding that both *Archaeopteryx* and the Chinese Early Cretaceous *Confuciusornis* cluster with birds, such as pigeons and galliforms, that forage both in trees and on the

ground. Many of the more advanced Cretaceous species were clearly aligned within the arboreal or terrestrial column. In hind-limb proportions, *Archaeopteryx* clusters with pigeons as being "equally at home on the ground or in trees." Hopson, who has a history of favoring a cursorial, theropod origin of birds and their flight, suggested that since the proportions of the pedal phalanges and of the hind limb of *Archaeopteryx* overlap with those of the least cursorial terrestrial modern birds, "basal birds may be secondarily modified from more cursorial theropod ancestors."[66] In contrast, Zhonghe Zhou of the Institute of Vertebrate Paleontology and Paleoanthropology of the Chinese Academy of Sciences in Beijing and James Farlow of Indiana–Purdue University further analyzed the digits of *Confuciusornis* and concluded that it was arboreal and a powered flier, and that with respect to it and *Archaeopteryx*, "an opposable hallux of the foot is still one of the few characters unique to birds. There is no evidence indicating that this character was developed for any adaptation other than arboreal life. . . . Both *Archaeopteryx* and *Confuciusornis* were probably not much different in lifestyle from the protobirds that pioneered life in trees."[67]

It is difficult to accurately portray any fossil organism, but perhaps the best illustrations of *Archaeopteryx* have come from paleontologist and artist Manfred Reichel of the Natural History Museum of Basel, Switzerland. Among his many accomplishments, he rendered a series of beautiful and highly accurate pen-and-ink drawings of the urvogel and a number of pterosaurs in lifelike poses that are shown here modified in silhouette surrounding the Heilmann portrait of *Archaeopteryx* from his 1926 book *The Origin of Birds*.[68]

Peter Wellnhofer, who has studied all of the *Archaeopteryx* specimens, has written an extensive treatise on the earliest known, *Archaeopteryx: Der Urvogel von Solnhofen*.[69] Illustrated in detail and printed in an oversized format, the book discusses known specimens, quarries, and depositories, as well as people involved in the discoveries, and includes an extensive bibliography. Now being produced in an English edition, this book will no

Fig. 3.18. Adaptation of Gerard Heilmann's well-known reconstruction of a male *Archaeopteryx*, surrounded by silhouettes of Manfred Reichel's drawings of the urvogel in different life poses. Adapted from G. Heilmann, *The Origin of Birds* (London: Witherby, 1926); modified after M. Reichel, *Manfred Reichel, 1896–1984, Dessins* (Basel: Geological Institute of Basel University, 1984).

Fig. 3.19. Photograph of the Australian pheasant coucal (*Centropus phasianinus*), a bird remarkably similar in size and form to *Archaeopteryx* except in the possession of terrestrially adapted foot claws, shown next to a reconstruction of female *Archaeopteryx*. Photograph from *Wikimedia Commons*, by Aviceda, taken in southeastern Queensland, Australia, in 2003, licensed under the Creative Commons Attribution ShareAlike 3.0 License. Inset, female *Archaeopteryx*, adapted from drawing by G. Heilmann, *The Origin of Birds* (London: Witherby, 1926).

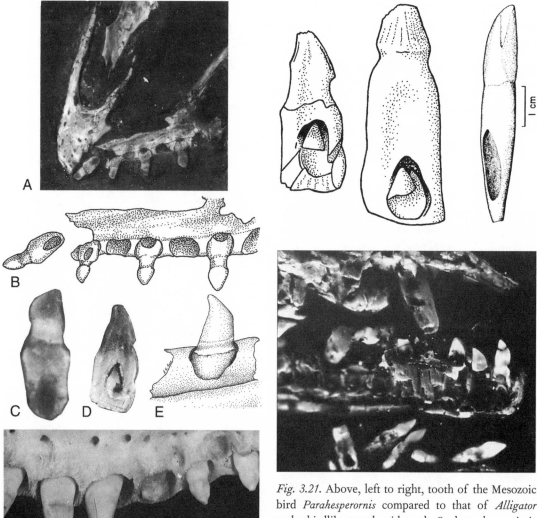

Fig. 3.20. Lingual views of the premaxillary and maxillary teeth of *Archaeopteryx lithographica* von Meyer, London specimen: (A) left premaxilla and right maxilla; (B) maxilla and isolated tooth; (C) isolated tooth (from right premaxilla?); (D) *Parahesperornis alexi,* left lower tooth; (E) drawing taken from photograph of a tooth of the seventh specimen of *Archaeopteryx,* showing similarity to sockets in the London maxillary; (F) right, lingual view of an alligator maxilla showing similarity of tooth and socket formation to (A) and (B). From L. D. Martin and J. D. Stewart, "Implantation and Replacement of Bird Teeth," *Smithsonian Contributions to Paleobiology* 89 (1999): 295–300. Courtesy Larry Martin.

Fig. 3.21. Above, left to right, tooth of the Mesozoic bird *Parahesperornis* compared to that of *Alligator* and a birdlike troodontid tooth. Such teeth are similar to those of *Archaeopteryx* and most other Mesozoic birds and probably represent the primitive condition for birds. Below, ultraviolet photograph detail of left dentary, rostral portion of premaxilla, and teeth of the Chinese Early Cretaceous bird *Yanornis,* showing *Archaeopteryx*-like teeth. Above, from L. D. Martin and J. D. Stewart, "Implantation and Replacement of Bird Teeth," *Smithsonian Contributions to Paleobiology* 89 (1999): 295–300; courtesy Larry Martin. Troodontid tooth redrawn from P. J. Currie and X.-J. Zhao, "A New Troodontid (Dinosauria, Theropoda) Braincase from the Dinosaur Park Formation (Campanian of Alberta)," *Canadian Journal of Earth Sciences* 30 (1993): 2231–47. Below, from S. Czerkas, ed., *Feathered Dinosaurs and the Origin of Flight* (Blanding, UT: Dinosaur Museum, 2002); copyright Stephen A. Czerkas.

doubt remain a leading treatise on this topic for years to come.

The Dentition of *Archaeopteryx*

Little is made in recent publications of the teeth of *Archaeopteryx* and other early birds, perhaps because they do not fit the pattern of a group that is directly derived from theropod dinosaurs. It is indeed no less than astounding that so little attention has been directed toward trying to understand their structure and nature of implantation, especially when one considers that one of the primary bases of mammalian paleontology involves tooth morphology.[70] The teeth of Mesozoic birds (such as are seen in *Archaeopteryx*, *Hesperornis*, *Parahesperornis*, *Ichthyornis*, *Cathayornis*, and others) are simple and remarkably similar to those of

Table 3.1. Comparison of morphology and other features of the dentition of Mesozoic birds and typical theropods. Maniraptoran tooth morphology is rather more diverse than this table allows, with troodotid teeth quite similar to those of *Archaeopteryx*. A transition from the dentition of Mesozoic birds to that of theropods is quite feasible, but the opposite transition, the current view (see Chiappe, *Glorified Dinosaurs*, 2007), would appear problematic.

Archaeopteryx, typical Mesozoic birds, and some troodontids	Typical theropods
Peglike, isodont, waisted teeth with expanded root crowns; devoid of surface ornamentation.	Bladelike, recurved, ziphodont teeth, with non-expanded roots; smooth, nonwaisted transition from tooth crown to root.
Posterior teeth with exaggerated, expanded roots.	Serrated ridges on front and back edges, consisting of closely spaced denticles, each squarish or oval in outline, separated by "slots" that cut meat fibers.
Subthecodont mode of insertion in deep, craterlike sockets; teeth begin in a groove, and sockets form around the root.	Teeth regimented along bony shelf, attached by cancellous bone.
Vertical tooth family; replacement teeth migrate labially just after initiation and develop under crown of predecessor.	Horizontal tooth family; replacement teeth form rows on lingual side of mature tooth, with up to three generations of teeth ranked side by side.
Oval resorption pit on lingual aspect of root surrounds developing crown (resorption pit closed at base).	Oval lingual scar on lower root (not resorption pit), left by lingually migrating replacement tooth.
Tooth covered with cementum and attached by periodontal ligament that may rot and allow tooth with root to fall out and be found isolated as fossils.	Teeth almost always found as tooth crowns only, having broken off at roots. (*Troodon* with rather coarse serrations, somewhat reminiscent of prosauropods; *Dromaeosaurus* with characteristic "twist" of anterior serrated carina, as in carnosaurs.)
Teeth designed for holding and eating whole prey.	Teeth designed for cutting and slicing flesh.

Sources: Adapted from A. Feduccia, "Birds Are Dinosaurs: Simple Answer to a Complex Problem," *Auk* 119 (2002): 1187–201; P. J. Currie, J. K. Rigby Jr., and R. E. Sloan, "Theropod Teeth from the Judith River Formation of Southern Alberta," *Canadian Journal of Earth Sciences* 30 (1990): 2231–47; and Martin and Stewart, "Implantation and Replacement of Bird Teeth," *Smithsonian Contributions to Paleobiology* 89 (1999): 295–300.

crocodilians. They are flattened, with an unserrated crown that becomes constricted, or waisted,
as it approaches the juncture of crown and root.
The root is expanded and covered with cementum,
and it is at least as broad as the crown, and usually
broader. Given the crocodylomorph morphology of Mesozoic bird teeth, it seems less remarkable that the Late Jurassic *Lisboasaurus*, hailed at
its discovery as a "missing link" between birds
and dinosaurs, was later shown to be a crocodylomorph.[71] Although it is not difficult to envision a
transition from the simple teeth of Mesozoic birds
to the laterally flattened, serrated teeth of theropod dinosaurs, the opposite transition, from typical theropod teeth to those of Mesozoic birds, is
both less likely and more problematic. Much has
been made of the presence in *Archaeopteryx* of
small triangular bones along the jaw known as interdental plates, which are also present in theropod
dinosaurs, but their presence in basal archosaurs
such as the Triassic *Euparkeria* renders the feature
of little phylogenetic significance. Some of the
recently discovered Cretaceous troodontids and
dromaeosaurs (microraptors) have some of the
posterior teeth devoid of serrations. Indeed, the
smooth-crowned teeth of the urvogel would appear to have a similar replacement pattern as those
of newly discovered troodontids.[72]

The Bird Brain of *Archaeopteryx*

A cast of the brain of the London specimen was
made by Cyril Walker in the mid-1980s and it
was considered at that time to be typically avian.
Then, in 2004 Patricio Domínguez Alonso and
colleagues studied the brain of the London specimen, using more sophisticated methods than were
available to Walker, and uncovered exciting new
information on both the brain and sense organs.[73]
Alonso and his colleagues were able to reconstruct
the brain and inner ear (the organ for hearing and
equilibrium) of the London specimen. Their reconstruction revealed that the brain and inner ear
of *Archaeopteryx* were much more birdlike than
had previously been imagined; in fact, they were
hardly distinguishable from that of modern birds,
although the brain was somewhat smaller. The

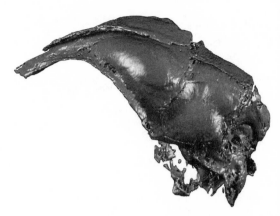

Fig. 3.22. Three-dimensional reconstruction of the
brain of *Archaeopteryx*, produced by P. D. Alonso
and colleagues using computed tomography. The
brain and inner ear, they concluded, showed that *Archaeopteryx* was probably equipped for flight. The reconstruction is about 20 mm (0.8 in.) in length. From
L. M. Witmer, "Inside the Oldest Bird Brain," *Nature*
430 (2004): 619–620. Copyright 2004. Reproduced
by permission from Macmillan Publishers Ltd. and
P. D. Alonso. Image by Patricio Domínguez Alonso.

areas responsible for movement were enhanced,
visual centers were enlarged, and in almost all
aspects the brain was that of a visually oriented
animal. The canals of the inner ear were more like
those of modern birds than those of reptiles, again
demonstrating that the urvogel had attained a high
level of agility and motor acuity.

To test whether these features might be associated with flight, comparison was made with the
results of a similar study performed by Lawrence
Witmer and his colleagues on similar structures in
a pterosaur. Their computed tomography analysis revealed expansion and reorganization of the
brain and the canals of the inner ear, similar to
that revealed by the Alonso group in *Archaeopteryx*. Moreover, the ratio of brain size to body
size was almost identical in the pterosaur and in
Archaeopteryx. Such independently evolved similarities suggested that there are fundamental neural requirements for flight, and Alonso and his
group argued that *Archaeopteryx* was well adapted
from a neurological point of view for flight. They
concluded that the brain of *Archaeopteryx* "closely
resembled modern birds in the dominance of the

sense of vision and in the possession of expanded auditory and spatial sensory perception in the ear" and that it represented "a stage further towards the modern bird pattern." They further emphasized that the urvogel "had acquired the derived neurological and structural adaptations necessary for flight, and . . . an enlarged forebrain suggests that it had also developed enhanced somatosensory integration with these special senses demanded by a lifestyle involving flying ability."[74] The flightless ratites, which Huxley thought were descended directly from dinosaurs through a non-flight stage, retain the neuroanatomy required by flight, and this turned out to be among the most important features revealing their ancestry from volant birds.

In a more recent paper, Stig Walsh and colleagues, including Witmer, studied the inner ear anatomy as a proxy for deducing auditory capability and behavior. Using microcomputed tomographic analysis, they measured the length of the duct of the bony part of the inner ear (the endosseous cochlear duct, or lagena) and showed that the length correlates with the mean hearing frequencies and even the complexity of calls. Their conclusion was that the urvogel *Archaeopteryx* was a good match for the emu, a modern ratite, which was evolved from a flighted ancestor.[75] Such approaches and other similar innovative ideas may take on even more importance in the future in assessing the "neo-flightless hypothesis" in avian evolution, that many of the so-called feathered dinosaurs are in fact secondarily flightless birds.

It has often been stated that were it not for the presence of feathers in *Archaeopteryx*, the fossil would have been identified as a small coelurosaurian dinosaur; this is an exaggeration. Many times the story has been related that the Eichstätt urvogel specimen was for years misidentified as a specimen of the small coelurosaur *Compsognathus*, but certainly this proposition has little to do with the issue at hand, and does not convince. Ignored are the misidentification of the Tyler specimen as a pterosaur; further, pterosaurs have been misidentified as birds.[76] An astute paleontologist would immediately notice the salient avian features of *Archaeopteryx*, features that show beyond doubt that it was a

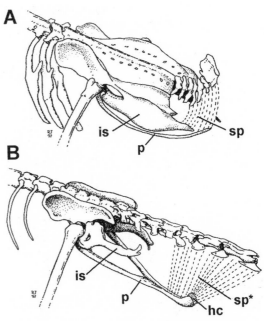

Fig. 3.23. Another birdlike structure of *Archaeopteryx* is seen in the pubis. Pelvic and tail skeleton, and suprapubic musculature of (A) modern perching birds (such as a pigeon) and (B) *Archaeopteryx* and the enantiornithine birds. In both, marked dorsal position and the projection of the distal pubis posterior to the ilium and ischium are associated with suprapubic musculature rotation of the pelvis and tail during arboreal roosting. Also, unlike the pubic boot characteristic of theropods, sauriurine birds have a hypopubic cup (or spoon), in the form of "cupped hands" to accommodate the pubic musculature, as shown. Abbreviations: is, ischium; p, pubis; hc, hypopubic cup (of pubis); sp, suprapubic muscles; sp*, probable suprapubic muscles. From J. Ruben et al., "Lung Structure and Ventilation in Theropod Dinosaurs and Early Birds," *Science* 278 (1997): 1267–70. Reprinted with permission from AAAS.

bird in the modern sense. As we have seen, among the more salient are the anisodactyl foot, with three anterior toes and the opposing hallux designed for perching; the avian skull and quadrate; avian teeth (like those of other Mesozoic birds, discovered a few years after *Archaeopteryx*); an avian hand designed for accommodating flight feathers; claws designed for climbing; an avian humerus; an avian scapula articulating with the coracoid at the avian 90° angle as in all modern volant birds, designed for reducing the angle in the recovery stroke (in flightless birds, the angle is obtuse); an opistho-pubic pubis ending in an avian hypopubic cup (as opposed to the theropod pubic boot); and the lack of a broad sacrum and any supra-acetabular shelf as in the theropods, although such structure is also absent in some maniraptorans.

It has also been extraordinarily difficult at times to assess the validity of many of the so-called dinosaurian characters because they are often de-scribed based on the "fact" that *Archaeopteryx* was a dinosaur. Among the prime examples is the pubis, which back in the 1970s was invariably de-scribed with a vertical orientation as in most thero-pod dinosaurs, with a pubic foot or boot.[77] As time progressed and the discoveries of more and more maniraptorans were unveiled with backward-pointing, or opisthopubic, pubes, the orientation of the *Archaeopteryx* pubis suddenly changed to the true avian opisthopubic orientation, and it was revealed that the pubes actually end in a hypopu-bic cup or spoon. Likewise in recent years it has become a common practice in paleontology to de-cipher not only character features but also mode of life, biomechanics, and even flight origin based on a cladogram rather than by independent assess-ment and testing.

Illustrating the complexity of the evolutionary process, one not readily apparent from modern cladograms, Gavin de Beer perceptively coined the term "mosaic evolution" in 1954 with refer-ence to *Archaeopteryx*, which exhibited a strange combination of avian and reptilian characters in different parts of its anatomy: "All these are char-acters which would not be in the least out of place if found in any reptile. On the other hand, there are a number of features in *Archaeopteryx* which are absolutely characteristic of birds." De Beer thought he had uncovered a new mode of organic evolution that might have widespread application, and today it is recognized in diverse groups, in-cluding the evolution of mammals and hominids. He added, "A necessary consequence of mosaic evolution and of the independence of characters evolving at different rates is the production of animals showing mixtures of primitive and spe-cialised characters."[78] Aside from the urvogel *Ar-chaeopteryx*, we shall see in the next chapter that this same theme carried over into the Cretaceous with the newly discovered Chinese aviary.

Ground Up or Trees Down?

The urvogel was far advanced aerodynamically beyond the genesis of flight in birds, but whether legitimate or not, *Archaeopteryx* has always played a crucial role in discussions of flight origins. As noted before, there have classically been two basic theories (and variations thereon) for the origin of flight: flight originated from the trees down, the arboreal theory, or from the ground up, the curso-rial theory. Much has been written on this topic, but to be brief, O. C. Marsh of Yale was the first, aside from Darwin on bats, to argue for an arbo-real origin of flight: "[The] power of flight prob-ably originated among small arboreal forms of reptilian birds. How this may have commenced, we have an indication in the flight of the *Galeo-pithecus* [flying-lemur], flying squirrels *Pteromys*, the flying lizard (*Draco*) and the flying tree frog (*Rhacophorus*). In the early arboreal birds which jumped from branch to branch, even rudimentary feathers on the forelimb would be an advantage as they would tend to lengthen a downward leap or break the force of a fall."[79] The enigmatic aspect of Marsh's proposal was that, following Huxley, he favored a dinosaurian origin of birds, which would seem almost incompatible with an arboreal origin of flight, as all known dinosaurs then were exclusively obligate, terrestrial bipeds.

Even John Ostrom, who never relented on his insistence that flight originated from the ground up, appreciated the biological and physical sim-plicity of an arboreal origin of flight: "The logic of the arboreal scenario is difficult to refute . . . an ancestral ground-dwelling quadrupedal reptile (sometimes identified as a 'thecodont' or primitive archosaur . . . leaping between branches . . . leap-ing between trees . . . parachuting between perches . . . gliding from high perches to lower trees or the ground . . . active, flapping flight. . . . Walter Bock . . . showed that a rational and logical explanation based on the accumulation of numerous microevo-lutionary changes over many generations could account for the seemingly macroevolutionary dif-ferences between a four-legged ancestor confined to a ground-existence and our high-flying feathered friends."[80]

Ostrom was an accomplished scholar, paleontologist, and biologist, and seeing inherent problems with a cursorial flight origin, he admitted doubts concerning it: "For obvious reasons, the arboreal theory has been favored ever since the cursorial theory was offered . . . as an alternative to . . . the more easily envisioned arboreal theory. Obviously, it is more difficult to accept a 'beginning flier' fighting against gravity as it struggles to become airborne. It is so much easier to accept that 'beginning flier' coasting down from an elevated perch under the power of gravity."[81]

It has been stated many times: flight in vertebrates has two prerequisites, small size and high places. Small size is essential because otherwise any incipient flight structures would have no effect in breaking a fall in jumping or parachuting; high places are essential because such sites provide the cheap energy offered by gravity. There are simply no other reasonable scenarios for the evolution of flight.

Colin Pennycuick of the University of Bristol notes: "A nonflying animal, destined to evolve the power of flight, will have some structures that are not yet wings, and do not, initially, perform well as such. To have a high chance of developing flight, [this creature] must be in a region of high potential diversity for flying animals. A nonflying animal over 1 kg [2.2 lbs.] is very unlikely to evolve flight, because only a limited range of specialized animals can fly successfully at such large sizes. To have the best chance of success, the ancestor should be in the mass range 10–100 g [0.4–3.5 oz.], where oscillating maladapted limbs at inappropriate frequencies can still produce acceptable results."[82]

Pennycuick's arguments are typical of those of the school favoring an arboreal origin of avian flight, which is biomechanically the only currently acceptable model. His calculations indicate that a large or even medium-sized bipedal theropod, for example, one the size of Ostrom's *Deinonychus*, would be an unlikely candidate for the evolution of flight, and interestingly, even *Archaeopteryx* weighing in at about 250 grams (less than 9 oz.), is larger than would be optimal for the evolution of flight.

Forgoing an extensive review, suffice it to say that the first proposal for a cursorial origin of flight was made in 1879 by Samuel Wendell Williston of the University of Kansas (named dean of medicine in 1899).[83] However, the first well-articulated proposal was made in 1907 and reiterated in 1923 by Baron Franz Nopcsa of Hungary, a flamboyant Hungarian nobleman and self-trained Transylvanian paleontologist, whose biography includes spying in World War I and volunteering his services as heir-designate to the vacant Albanian throne. He was the first to suggest that dinosaurs had complex social behavior and cared for their young, and he was the first to use the term "paleophysiology," which was later replaced by "paleobiology." Nopcsa was a leading European dinosaur paleontologist, and the government of Romania has reconstructed the Nopcsa castle to house a new dinosaur museum. Nopcsa published two papers on the subject of the origin of flight, detailing how a "protowing" developed as a propeller (not a wing), adding thrust to the powerful running legs of a bipedal "proavis": "We may quite well suppose that birds originated from bipedal long-tailed cursorial reptiles which during running oared along in the air by flapping their free anterior extremities."[84]

Among his more progressive ideas, Nopcsa thought that *Archaeopteryx*, though not an accomplished flier, was quite far removed from the early stages leading to flight, continuing: "The rounded contour of the *Archaeopteryx* wing, together with the feebly developed sternum, show us that *Archaeopteryx*, though perhaps not an altogether badly flying creature, can on no account have been a soaring bird, but a bird that was yet in the first stage of active flight."[85]

Despite the overwhelming biophysical considerations against a cursorial origin of flight, Ostrom envisioned the ancestral bird as an animal close in appearance to *Deinonychus*, although smaller, and was thus saddled with a cursorial origin of flight. So, in the early 1970s, John Ostrom proposed a very different version of Nopcsa's theory, which has been termed the "insect net" theory. Ostrom recognized that in order for flight to have originated from the ground up, there would have to be some selective pressure for the elongation of the

Fig. 3.24. Two illustrations showing the improbable ground-up model for the evolution of bird flight from earthbound dinosaurs. Above, the run-jump-fly scenario in birds. In these running scenarios, the origin of feathers would produce drag, which would tend to slow down a running animal. Then, once the animal is aloft, where would the energy come from to maintain flight? Below, right to left, *Compsognathus,* *Avimimus, Archaeopteryx,* pigeon, showing the near impossible lineage that would lead to avian flight from the biophysically unlikely ground-up genesis of flight. Above, from U. M. Norberg, *Vertebrate Flight* (Berlin: Springer, 1990); courtesy Ulla M. Lindhe Norberg. Below, reproduced with permission, © The Natural History Museum, London.

wing flight feathers, first in a nonflight context, and the scenario he used was that the wings were used to trap insects, elongating until they were of sufficient size to be aerodynamic.[86] His idea never really gained much momentum outside of the paleontological community, and at about the time of its origin, physical anthropologist Matt Cartmill, then at Duke University, wrote me a letter in which he outlined a challenge he had made to John Ostrom, a duel of sorts: a bottle of *Drosophila* would be opened in a room, and each combatant would receive ten cents for each fruit fly trapped. Ostrom would be given two bird wings and Cartmill would have a flyswatter!

Still another cursorial explanation was provided by Phillip Burgers and Luis Chiappe, who proposed that avian powered flight arose as a by-product of running and rowing, with wings generating additional thrust.[87] Their work suffers from serious flaws, among which are that no suitable phylogeny can be generated through comparison with analogous organisms and that most galliform birds, used in many of their comparisons, invariably hold the wings tightly against the body when running. More important, the generation of thrust in a running urvogel assumes major rotation of the humerus, and such movement would be impossible in the shoulder joint of *Archaeopteryx* because both the glenoid and the humeral head are elongated. Their thesis, however, was that, by flapping its wings to run faster, *Archaeopteryx* generated "residual" lift that enabled it to take off. Yet this increased lift would invariably cause a loss of traction—a major problem for all cursorial models for the origin of flight. A more pertinent consideration, however, is that *Archaeopteryx*, as we have seen, showed considerable aerodynamic advancements, so one could rightly question whether the urvogel can tell us anything about actual flight origins. Indeed, Sankar Chatterjee and Jack Templin, in a study of the flight of *Archaeopteryx*, noted that it "lacked both the powerful flight muscles and complex wing movements necessary for ground takeoff." They concluded, "Both anatomy and phylogeny strongly suggest that *Archaeopteryx* was an arboreal bird." Their flight simulation model suggests that it would have been more efficient and cost-effective to take off from a perch than from the ground. Further, they suggest that the urvogel could make short flights between trees using a method described as phugoid gliding, in which a volant animal can travel from treetop to treetop without expending much muscular energy. For example, when flying squirrels and crows take off from a tree, they at first lose height, expending little energy, and then swoop up to the next perch. There is an initial loss of height at an increasing speed; lift then increases as speed squares; and undulation, known as phugoid oscillation, follows. Phugoid gliding may be a realistic description of the flight path of *Archaeopteryx*, al-

though some degree of powered flight may have also been involved.[88]

As an aside, Kevin Padian, as a graduate student at Yale, ran into the same problem when tackling the question of flight origins in pterosaurs. His cladogram dictated that pterosaurs were the sister group of dinosaurs; therefore, pterosaurs for him were obligately bipedal and developed flight from the ground up. Luis Chiappe proclaimed the same "rules" for deciphering the origin of flight in birds: "The ancestral mode of life of birds was that of a cursorial biped. Inferences about the habits of *Archaeopteryx* should be made within this framework and not the reverse."[89] More recently, he noted: "The origin of flight from ground-dwelling maniraptorans that run while flapping their wings is plausible. . . . All we can say with confidence is that the functional and aerodynamic considerations required for terrestrial animals to become airborne were present in the theropod forerunners of birds, and that gravity-assisted explanations are not necessarily the only approach to explaining the beginnings of birds' most notable function."[90] Likewise, Padian has stated: "Because theropod dinosaurs were cursorial predators, the cursorial hypothesis is consistent with theropod phylogeny."[91] And Padian and Chiappe together have asserted that "some of the closest relatives of birds include the sickle-clawed maniraptoran *Deinonychus* that Ostrom had so vividly described. . . . The cursorial hypothesis is strengthened by the fact that the immediate theropod ancestors of birds were terrestrial. They had the traits needed for high liftoff speeds."[92] For these paleontologists all speculation concerning the origin of flight must follow the cladogram that dictates a theropod origin of birds and flight from the ground up.

When we follow Ostrom and his student Bakker through time, we see little change in the ground-up theory. Ostrom steadfastly maintained that protobirds and *Archaeopteryx* were terrestrial cursors and that flight originated from the ground up:

> *1970.* "Unique preservation of the horny sheaths of the manus claws provides new evidence that may be relevant to the question of the origins of avian flight. Tentative

interpretation suggests a cursorial rather than [an] arboreal origin."

1974. "I conclude that *Archaeopteryx* was not capable of powered flight other than that of fluttering leaps while assaulting its prey. *Archaeopteryx* was a ground-dwelling, bipedal, cursorial predator . . . probably not very different from its contemporary, *Compsognathus.*"

1975. "In spite of its very birdlike appearance, *Archaeopteryx* was closely related to certain small dinosaurs and could not fly."

1978. "*Archaeopteryx* supports two theories: warm-bloodedness in dinosaurs and dinosaurian ancestry of birds. The theropods . . . and especially the smaller kinds like *Compsognathus, Deinonychus,* and the struthiomimids, I suspect may have been true endotherms. And add to that the evidence that *Archaeopteryx* and other birds evolved from a small theropod dinosaur."

1985. "It is my contention that *Archaeopteryx* was still learning to fly—from the ground up—and that avian flight began in a running, leaping, ground-dwelling biped. . . . This animal [*Archaeopteryx*] was a highly adapted bipedal and cursorial ground-dwelling predator. That lends powerful credence to the cursorial theory of the origin of bird flight."

1986. "There is no evidence—and I underline the word *evidence*— that bird flight began from the trees down. What evidence there is all points to a highly adapted bipedal cursor and a 'ground-up' origin of flight. . . . The actual physical evidence . . . points very strongly toward a cursorial origin of avian flight."

1994. "It does appear that 'predatory hand movements' such as grasping action by swift-moving bipedal cursors was one of the key stages in the evolution of avian powered flight—somewhere in its pre-flight ancestry. . . . However this scenario plays, it does dictate the certainty of a bipedal pre-avian, probably predaceous organism. The best candidate (if not the only candidate) is a coelurosaurian theropod—all of which were cursors (necessarily so) after prey."[93]

In the final analysis, what we know is that many groups of terrestrial vertebrates, ranging from frogs, snakes, and lizards to pterosaurs, bats, and even rodents and gliding marsupials, have evolved some form of flight, parachuting, or gliding. They have done so by the simplest method possible, using the cheap energy provided by gravity. Such a context for the origin of flight is devoid of ad hoc hypotheses involving locomotor modules and other complex scenarios such as insect nets, and requires only small size and high places.[94] Small size is a necessity, so that flight membranes, scales, or feathers will have an effect on breaking the fall; high places are necessary to exploit gravity. As Ulla Norberg succinctly put it: "While all classes of land vertebrates have evolved some level of flight, it has demonstrably always been achieved in accordance with the extremely simple trees-down model, utilizing small size coupled with cheap energy provided by gravity, that is, high places."[95] In jumping or parachuting the animal must be lightly built and able to extend it appendages laterally so as to present as much surface area as possible. In addition, small size is a correlate of specialization for arboreal life, provides a favorable mass-surface area relation for vertical climbing, and lessens the effect of impact should the animal fall or jump to the ground. Smaller size also enables the animal to move through the medium of air at a lower speed, with smaller wings, in contrast to a cursorial life, which favors an augmentation in size to favor increased running speed. If the ancestors of birds were small arboreal archosaurs, there would also be little need to consider a bipedal ancestral form as dictated in the diagram by Walter Bock on page 87. Instead, the scheme might proceed directly from a terrestrial quadruped to an arboreal quadrupedal archosaur, and bipedalism in birds would have evolved in relation to releasing the forelimbs for flight, not as a result of having evolved from terrestrial cursors with already foreshortened forelimbs.

In *Vertebrate Flight,* Ulla Norberg suggested that maximization of net energy gain by foraging in trees would augment selection pressure for increased gliding, and that gliding from one tree to

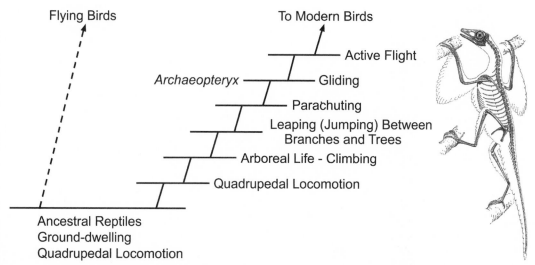

Flying Birds

To Modern Birds

— Active Flight

Archaeopteryx — Gliding

— Parachuting

Leaping (Jumping) Between
Branches and Trees

— Arboreal Life - Climbing

— Quadrupedal Locomotion

Ancestral Reptiles
Ground-dwelling
Quadrupedal Locomotion

Fig. 3.25. Walter Bock's well-known and simple diagram of the arboreal theory for the origin of avian flight followed closely the views of the arboreal model of O. C. Marsh. The model (right) illustrates stages that can be seen analogously in living animals and represents a reasonable pseudophylogeny. The ground-up scheme (left) caricatures the cursorial hypothesis, now generally discarded because of its magical nature. On the right, if birds arose from a common basal archosaur ancestor with dinosaurs, the bipedal stage would be eliminated, because arboreal life would involve a quadrupedal, small, arboreal archosaur. At far right is an adaptation of Gerhard Heilmann's illustration of a hypothetical "thecodont"—an arboreal, quadrupedal basal archosaur—as a protobird. Modified from W. J. Bock, "The Role of Adaptive Mechanisms in the Origin of the Higher Levels of Organization," *Systematic Zoology* 14 (1965): 272–87. Drawing from G. Heilmann, *The Origin of Birds* (London: Witherby, 1926).

another and then climbing back up the next tree during foraging activities maximizes net energy gain. Put simply, it takes less energy for an animal to climb a tree and then glide to the next tree than for the same animal to climb up and down a tree and then run to the next. Once the glide surface has evolved, the foraging efficiency would be dramatically increased and the time demands for locomotion during foraging would be drastically reduced. Parachuting and then gliding to escape predation may have also played an important role. Yet the beauty of the arboreal theory is that every microstage in the process is fully adaptive and fully processed by selection for a more efficient aerodynamic structure. Any extension of the feathers on the posterior forearm (brachium) and arm would convey an immediate advantage in parachuting and gliding. Indeed, as an analogous form, the Triassic archosaur *Longisquama* illustrates just that: a small, parachuting reptile, with elongate feather-like scales or parafeathers projecting along the posterior aspect of the forelimbs. Such a scenario for flight origins has no bearing on avian ancestry, as such a model would fit nicely for any other putative candidates, either basal archosaurs (or dinosauromorphs) or maniraptorans.

One should be cautious of classic scenarios that use *Archaeopteryx* and its Solnhofen habitat as the starting point in all discussions on the origin of birds and their flight. True, it is all we had, but even Huxley was unwilling to accept the urvogel as transitional, viewing it instead as an "intercalary" bird. Whatever *Archaeopteryx* was capable or incapable of biomechanically may have little to do with the major questions concerning the origin of birds and may even be misleading, since the urvogel no doubt existed well beyond the actual origin of flight. Despite this, virtually every study on the origin of birds and their flight has concentrated on the Solnhofen urvogel and the urvogel's

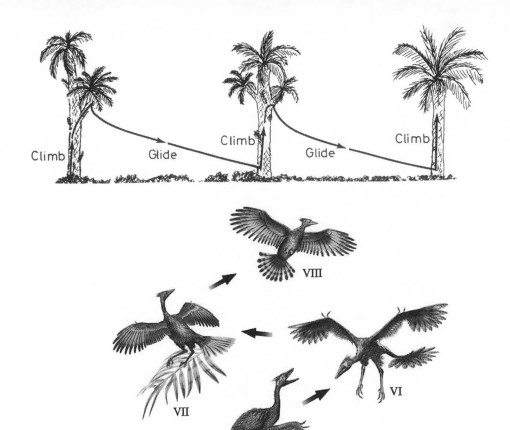

Fig. 3.26. Above, energy-saving mode of locomotion for a protobird: It costs less for an animal to climb a tree and then glide to the next tree than to climb up and down in a tree and then run to the next (as modeled by R. Å. Norberg). Below, drawings to illustrate the adaptively facile trees-down theory of flight origin as envisioned by Evgeny Kurochkin and depicted by Olga Orekhova-Sokolova. Given the recent discovery of the Early Cretaceous four-winged micro-raptors, one might modify stage VI with the addition of hind-limb wings, the tetrapteryx stage of William Beebe. Above, from U. M. Norberg, *Vertebrate Flight* (Berlin: Springer-Verlag, 1990); courtesy Ulla Lindhe Norberg. Below, from E. N. Kurochkin and I. A. Bogdanovich, "On the Origin of Avian Flight: Compromise and System Approaches," *Theoretical Biology* 35 (2008): 1–11; courtesy Evgeny Kurochkin.

ecomorphology. As we shall see, the new discoveries from the Early Cretaceous of China and earlier have opened up a new avenue for the study of flight origins, but the evidence still overwhelmingly favors a trees-down genesis of flight.

Many workers have tended to portray fossil animals according to specific analyses of single character complexes, not considering the behavioral plasticity of animals in general. As far as we can ascertain, in most evolutionary transitions, behavior typically precedes morphology to a large extent, and this is probably true in the early avian lineage. It is an important lesson to not place too much importance on biomechanical models that portray a limited set of behaviors of a fossil organism. Examples of behaviors that could have never been predicted if only fossils existed include but are certainly not limited to:

- the walking catfish (*Clarias*), which with no obvious adaptations for "walking," trek across land;
- female pythons endothermically brooding eggs;
- crocodiles galloping in a near erect posture;
- female turtles (such as *Pseudemys*) digging an egg chamber with a narrow opening and large chamber using their hind legs;
- a dog born with no front legs yet capable of efficient bipedal locomotion;
- elephants swimming; and
- a corn snake going up a tree (I first collected one heading up a pine trunk in Port St. Joe, Florida, in the late 1950s).

Thus, although *Archaeopteryx* is the classic urvogel, it is quite well developed as a bird, was capable of a great variety of behaviors (see the illustration on page 77 showing diverse behaviors), and was clearly too advanced and occurred too late in time to tell us a great deal about the ultimate origin of flight. At a minimum we can say that it was a flying bird with a bird wing and many of the flight attributes of modern birds and that it was certainly, considering the nature of the feet and claws, efficiently adapted for life in trees.

Yet periodically, individuals promoting a strict theropod origin of birds and a ground-up flight origin attempt to discover features of *Archaeop-*teryx that render it closer to the ground. A recent example is a study of the rachises of primary feathers of the basal bird *Confuciusornis* and the classic urvogel. Rubert Nudds and Gareth Dyke attempt to show that the rachises of these early birds were "much thinner and weaker than those of modern birds, and thus the birds were not capable of flight."[96] The study appeared unfounded on the observable facts and myriad flight adaptations for both taxa and was quickly rebutted both by a team of Chinese paleontologists led by Xiaoting Zheng and independently by Gregory Paul. The Chinese workers studied *Confuciusornis* fossils from the Shandong Tianyu Museum of Nature, which holds 536 specimens, pointing out that only four preserve clear impressions of the rachis of the primary feathers and that these measure about twice as large as those reported by Nudds and Dyke. They also point out that the primary feathers possess asymmetric vanes and curvature and that there is a right angle between the scapula and the coracoids, a feature of living flying birds. For his part, Paul argues that Nudds and Dyke overestimated the mass of the subjects and underestimated the strength of the primary shafts. His conclusion, like that of most workers, is that "the total biology of the birds indicates that they could achieve flapping flight."[97] Despite the massive evidence that urvogels flew, *Science News* featured an article entitled "Earliest Birds Didn't Make a Flap," while the author had written in 2004 a piece, "Bird Brain? Cranial Scan of Fossil Hints at Flight Capability."[98] When we cut through all the hyperbolic rhetoric, one thing is certain: *Archaeopteryx* flew!

Storytelling by Cladogram

Because the cladograms produced by the methodology of phylogenetic systematics dictate an origin of birds from advanced theropod dinosaurs, Lowell Dingus and Timothy Rowe linked the theropod ancestry of birds with the origin of flight from the ground up and the so-called thecodont (basal archosaur) hypothesis with the origin of flight from the trees down: "Our map [of avian relationships] suggests that flight evolved from the ground up, but exactly how this happened is another question altogether."[99] Their statement vividly illustrates the problem: birds are living theropods, theropods are earthbound, obligate bipeds, so flight must have originated from the ground up, but such a proposal strains credulity. So begins the "story-telling," but for many practitioners of cladistics, it does not qualify as storytelling because it represents what can be called cladistic inference, a scenario dictated by the cladogram. As Walter Bock commented, concerning Dingus and Rowe's assertion, "If the origin of birds and the origin of flight are tightly linked in this fashion, then the available discussion of all specialists in vertebrate flight is that the origin of avian flight from the ground up is exceedingly improbable, which would fatally weaken the dinosaur ancestry of birds."[100] Bock correctly notes that the origin of avian flight falls under the heading of historical-narrative explanation and that ever since Darwin, the classic and best approach in developing an evolutionary historical-narrative explanation is to establish a pseudophylogeny using known analogous organisms and hypothetical forms.[101] Each successive evolutionary step must be small, the organism at each stage must constitute a functional and adaptive whole, and an analogous known organism or viable hypothetical form must represent each stage in the pseudophylogeny. Such criteria do not apply to the various ground-up models for flight origins, which are evolutionarily complex, difficult to explain, and involve exaptations—features that evolved for a specific function and were later pre-adapted to another.

The WAIR Hypothesis: The Latest Ground-Up Model

Kenneth Dial, a functional morphologist at the University of Montana, is a long-term adherent of the dinosaurian origin of birds and the ground-up origin of flight. In a study of chukar chicks, he observed that they need to run quickly up very steep slopes, using flapping wings, presumably to escape from predators.[102] The ontogeny of flight in these specialized quail, he felt, could be a model for the origin of flight. Dial and colleagues called this theory the WAIR hypothesis (wing-assisted inclined running).

The chukar (*Alectoris chukar*), or Indian hill partridge, is a galliform bird. Denizens of semi-arid and arid, hilly, stony slopes with short grass, usually sparsely covered with low shrubs, chukars nest in a thinly lined ground scrape and lay usually ten to twenty eggs. Widely distributed across central Asia into eastern Europe, this game bird has been introduced into many parts of the world, including the United States and Great Britain. Gallinaceous birds, as noted earlier, are unique among living birds in several characteristics: although they are primarily ground-dwellers, many roost in trees; they have an extremely precocious development of the flight apparatus; and they are exceptionally strong fliers, capable of flight bursts from the ground and able to fly strongly for short distances. Galliforms never give rise to derived flightless species through paedomorphosis (arrested development) because of their precocious development of their flight musculature and associated bony architecture, which accounts for about 30 percent of their body weight. Indeed, their exceptionally large pectoral mass is the reason they are so highly prized as game birds.

Unlike *Archaeopteryx*, chukars have limbs that are highly adapted for a ground-foraging lifestyle and terrestrially adapted feet that feature a greatly reduced hallux and scant claw curvature on either the anterior or the posterior toes. Obviously, a well-developed hallux with a strongly curved claw, as in the Solnhofen urvogel, would hamper terrestrial locomotion. Despite its wide acceptance among paleontologists, the WAIR model fails to take into account the fact that early theropods had forelimbs already reduced to half the length of the hind limbs and that early theropods had a very small pectoral muscle mass. The flight muscle mass for *Archaeopteryx* is estimated to have been no more than about 9 percent of the body mass (as compared to about 30 percent in galliforms), based on a total body mass of 200 grams (7 oz.), and the urvogel lacked the wing stroke of a modern bird.[103] Finally, galliforms have a very well developed, broad sacrum with a formidable antitrochanter, absent in any known sauriurine birds, notably *Archaeopteryx*, Chinese Early Cretaceous urvogels, and Chinese four-winged microraptors.[104]

In a theoretical sense, the model harkens back to Haeckel's failed Biogenetic Law, in which, presumably in a highly derived modern bird, a feature, this one behavioral, is a recapitulation of behavior present in the adult protoavian. In an interview with *National Geographic News* in January 2008, Dial said, "The way in which vulnerable young birds use their wings while transitioning into adult bodies could be a model for how their ancestors developed the ability to fly."[105] Complicating matters, Dial alluded to a quasi-recapitulation in the supplementary information to another paper: "All birds . . . have the inclination to ascend to an elevated refuge. This includes *hatchlings that climb quadrupedally.*"[106] This statement would require that the ancestor, instead of being a "dinosaur," *would have to be a basal archosaur*, the current competing hypothesis for avian origins! Moreover, given that hind-limb wings, discovered in Jurassic birds and Early Cretaceous archaic birds, are thought to be primitive for Aves, it is unclear how early birds could have been significantly cursorial. Finally, the selection force for WAIR is the apparent need to run quickly up steep slopes and even tree trunks to escape predation. By this scenario the birds would need a downforce to give their feet an increased grip, but early birds, including *Archaeopteryx*, lacked the shoulder mechanism possessed by modern avians to produce the swift and powerful wing upstrokes. Thus, since the downforce on which the WAIR model is dependent must be generated by upstrokes, it is apparent that early birds would be incapable of WAIR.[107]

You Can't Teach an Old Dogma New Tricks (Dorothy Parker)

In a recent book on evolution, Donald Prothero of Occidental College, trying to make a rock-solid case for the current bird-dino dogma, states: "The fact that a small dinosaur [*Compsognathus*] and *Archaeopteryx* could be so easily mistaken for each other was a revelation for Ostrom. He revived Huxley's hypothesis." Yet as we have seen, Ostrom did not revive Huxley's hypothesis, that birds were derived through the flightless ratites from dinosaurs; Ostrom's idea was rather that birds were derived directly from coelurosaurian

theropods, from an ancestor closely resembling the dromaeosaurid *Deinonychus*. In the same paragraph, however, Prothero notes that the Teyler [Haarlem] specimen "had been misidentified as a pterosaur in 1855."[108] What are we to make of Prothero's analogy? If the misidentification of the Eichstätt specimen as a theropod somehow constituted evidence for a dinosaur origin of birds, why not a pterosaur origin, since a pterosaur was equivalently misidentified?

Prothero later disingenuously homogenizes dissenting scientists with Creationists: "They resemble the creationists, who attack one tiny detail of the subject without addressing all the rest of the evidence. . . . These arguments of the creationists, as well as the 'birds are not dinosaurs' minority, . . . are now rendered entirely obsolete. . . . Scientists still have to abide by the rules of science . . . not 'just-so' stories." Although this jab is presumably an allusion to Rudyard Kipling's classic *Just So Stories*, among them "How the Whale Got His Throat," "How the Camel Got His Hump," and "How the Leopard Got His Spots," Prothero then proceeds to tell readers his own "just-so" story, "How the Bird Got His Wings," in a nonflying, earthbound theropod dinosaur![109] His take on the WAIR model illustrates a lack of critical analysis: "Ken Dial (2003) showed that their emphasis on the 'trees down' origin of flight is misguided. Many birds, such as chukar partridges, use the lift of their wings to help them run up steep inclines but seldom use them for flight. It is easy to see how dinosaurs (which already had feathers for insulation) [!] . . . could have adapted these structures to make climbing steep inclines easier, and from there began to develop short glides and eventually true flight."[110]

And so it goes.

Walter Bock and others have pointed out the critical flaws in theories of the cursorial origin of flight, notably that most models postulate that flapping or rowing the wings would produce additional forward thrust.[111] There is just no recent analogue for this behavior in modern birds, and even birds that flap their wings when running do not use them to increase running speed. Most important, outstretched wings would increase drag on the running animal, necessitating additional force to overcome the increased resistance. In addition, the lift created by flapping the wings would counter the pull of gravity and would decrease the downwardly directed contact force between the animal and the ground. The result would be that the animal could not achieve as much forward thrust with its feet by running. As Bock notes, "Attempts to increase running speed by flapping the forelimb would be disadvantageous as they would increase air resistance and would decrease traction."[112]

In the final analysis, the WAIR model for avian flight origins qualifies admirably under Storrs Olson's "ground-down" category and is a classic example of cladogram-dictated storytelling. Instead of being a "just-so" story, it must inevitably fall into the "just-not-so" column.

Conclusion

For the past 150 years, *Archaeopteryx* has occupied center stage in all discussions of the origin of birds and, unjustifiably, in theories on the origin and evolution of avian flight. Yet it was already a bird, albeit with numerous reptilian features, and so whatever anatomical features it yields are unlikely to elucidate such earlier evolutionary events. That *Archaeopteryx* was a bird in the modern sense, however, has nothing to do with its ultimate ancestry, from either theropod dinosaurs or a common descent with theropods from an earlier ancestry. The transition to "birds" must have occurred much earlier, perhaps in the Late Triassic or Early Jurassic as many, including Huxley, have opined, and certainly from the trees down, whatever the actual ancestor. We will see how the recent discoveries from China, especially from the Early Cretaceous, have greatly enhanced our understanding of the early avian radiation, and how the discovery of a number of more primitive taxa, which can be called "basal birds," are in reality urvogels, leading to and from the more basal Solnhofen *Archaeopteryx* to the Cretaceous avifauna.

4

MESOZOIC CHINESE
AVIARY TAKES FORM

....

The Jehol Biota currently represents our best chance of viewing the composition and dynamics of an intact Early Cretaceous terrestrial ecosystem.

Z. Zhou, P. M. Barrett, and J. Hilton, "An Exceptionally Preserved Lower Cretaceous Ecosystem," 2003

The Jehol Biota: A Window on the Early Cretaceous

In the past three decades avian paleontology has been reinvigorated and literally revolutionized by the discovery of extraordinary new material from both Cretaceous and Tertiary rocks, nowhere more so than by the breathtaking and electrifying fossil discoveries from China. Yet, as is often the case in paleontology, these new data have raised as many questions as they have answered and many interpretations are speculative. While dazzling discoveries have poured out of the Chinese Early Cretaceous, other important discoveries have also greatly advanced the field.[1] In the early 1980s Paleocene and Eocene fossils of lithornithids, volant paleognaths related to and likely ancestral to the large flightless ratites, were discovered. These fossils provided critical evidence that the ratites were not hapless passengers on drifting Mesozoic continents but in fact part of a post-Cretaceous, explosive radiation, or "big bang," of birds and mammals beginning at about sixty-five million years ago, and that the ancestral ratites flew rather than drifted to their present homes.[2] Fossils of Paleogene (early Tertiary) birds, especially from the Early Eocene of the Green River Formation

of western North America, Early Eocene London Clay of England, and Middle Eocene Messel near Frankfurt, Germany, have exponentially increased our knowledge of that radiation.[3] Yet the Mesozoic history of birds until recently had been only sparsely documented, and even then mainly from the Late Cretaceous, particularly the well-preserved fossils of the now iconic *Hesperornis* and *Ichthyornis* from the Niobrara seaway of western North America.

A New Subclass Is Discovered

Then, in the early 1980s, British paleontologist Cyril Walker, who was curator of amphibians, reptiles, and birds for some thirty years at the Natural History Museum, London, discovered a completely unknown and distinctive subclass of Mesozoic birds known as the opposite birds, or enantiornithines (*einantios*, "opposite"): "Perhaps the most fundamental and characteristic difference between the Enantiornithes and all other birds is the nature of the articulation of the scapula and the coracoids, where the 'normal' configuration is completely reversed."[4] Although their pectoral girdle produces a structure morphologically different, it is equivalent in function to the triosseal canal of modern, ornithurine birds. Also reversed

Fig. 4.1. "Opposite birds," or enantiornithines, were the predominant land birds of the Mesozoic. Here *Iberomesornis* from the Early Cretaceous of Spain is shown near a shoreline. The nuthatch-sized opposite bird had strongly curved foot claws, equivalent to those of modern trunk-climbers. From A. Feduccia, *The Origin and Evolution of Birds*, 2nd ed. (New Haven: Yale University Press, 1999); drawing by John P. O'Neill.

is the fusion order of the three tarsal elements, fusing from proximal to distal (or inward to outward). In addition, these archaic birds exhibit a distinctive anatomy of the furcula and sternum. Walker's discovery of a completely unknown clade of Mesozoic birds was the most dramatic discovery of the century, an initially muted epiphany, being based on some isolated long bones, but his find would eventually send the entire field of avian evolution back to the drawing board. Then, in 1985, Russian paleontologist Evgeny Kurochkin described a Lower Cretaceous, pigeon-sized bird, *Ambiortus*, with a fully developed sternum and flight apparatus.[5] It was clearly a modern-type ornithurine bird, unexpectedly living much earlier in time than previously thought.

The discoveries of Walker and Kurochkin firmly established the existence of two major groupings of birds—archaic, sauriurine birds and archaic but modern-type ornithurine birds—together in the Cretaceous. In the wake of these discoveries, José Sanz of Madrid and his Argentine colleague José Bonaparte, described a well-preserved enantiornithine from the Lower Cretaceous of Spain, naming the beautiful fossil *Iberomesornis* in 1992.[6] That same year, Paul Sereno of the University of Chicago and his Chinese colleague Chenggang Rao described a sparrow-sized enantiornithine from the Lower Cretaceous of China, and named it *Sinornis*.[7] Similar to the Spanish bird in size and morphology, *Sinornis* was clearly a volant, perching bird, with a fully reflexed hallux, exceptionally curved pedal claws, and a well-developed wing. A second Lower Cretaceous Spanish enantiornithine was discovered shortly thereafter and named *Concornis*; it was twice the size of *Iberomesornis*, displayed all the known features of the opposite birds, including a strange, elongate pygostyle (fused tail vertebrae, also called a plowshare bone) and a sternum that was quite distinctive from that of a modern bird. These early discoveries were followed by others, including an Early Cretaceous Spanish enantiornithine exhibiting the earliest known alula, or bastard wing.[8] Many of the newly discovered enantiornithines were placed in separate genera and even orders, some of which, although geographically disjunct, appear morphologically quite similar, and thus a thorough review of the taxonomy of these birds will be in order.

In 1992, at the meeting of the Society of Avian Paleontology and Evolution in Frankfurt, Germany, a young Chinese graduate student of Academia Sinica's Institute of Vertebrate Paleontology and Paleoanthropology, Zhonghe Zhou, appeared for the first time and dazzled the conference participants with his latest discoveries of what could only be interpreted as a major adaptive radiation of opposite birds from the Chinese Early Cretaceous, represented by twenty individual specimens belonging to at least three distinctive morphological types; this was the first clear evidence of a major adaptive radiation of the enantiornithine birds.[9]

Fig. 4.2. The sparrow-sized *Sinornis*, the first well-pre-served enantiornithine bird described from the Chinese Early Cretaceous, presaging the massive discoveries of the last decade. *Sinornis* had strutlike coracoids, a carpometacarpus, and such primitive features as gastralia, ribs without uncinate processes, and unfused metatarsals. (A) Reconstruction of *Sinornis santensis*; cross-hatching indicates missing parts. (B) Skeleton as originally preserved (the figure combines information from the slab and counterslab). Abbreviations: I–IV, digits 1–4; dc, distal carpal; f, frontal; fe, femur; fi, fibula; fu, furcula; il, ilium; m, maxilla; pf, pubic foot (actually a hypopubic cup); py, pygostyle; sc, scapula; st, sternum; ti, tibia; ul, ulna; ule, ulnare. Scale bar = 1 cm. From P. C. Sereno and C. Rao, "Early Evolution of Avian Flight and Perching: New Evidence from the Lower Cretaceous of China," *Science* 255 (1992): 845–48, fig. 2. Reproduced with permission from AAAS.

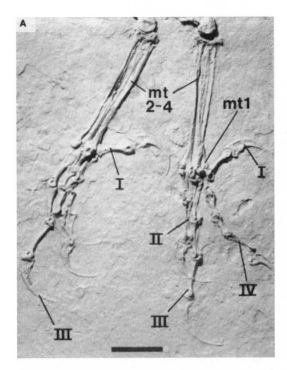

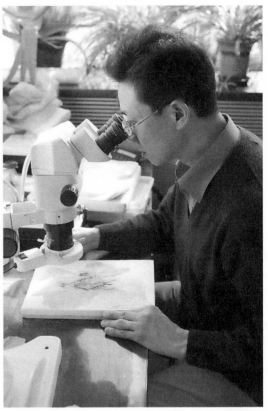

Fig. 4.4. Zhonghe Zhou in 2007 studying Chinese Early Cretaceous fossil bird in the laboratories of the Institute for Vertebrate Paleontology and Paleoanthropology, Beijing. Courtesy of Zhonghe Zhou.

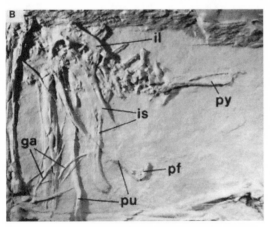

Fig. 4.3. Postcranial skeleton of *Sinornis* (epoxy cast from natural mold). (A) Feet showing unfused metatarsals (mt), retroverted digit 1 (I), and highly recurved unguals; scale bar = 5 mm. (B) pelvis and tail showing the erect ilium (il), blade-shaped ischium (is), pubis (pu) with pubic foot (pf), actually an avian hypopubic cup, large pygostyle (py), and gastralia (ga). From P. C. Sereno and C. Rao, "Early Evolution of Avian Flight and Perching: New Evidence from the Lower Cretaceous of China," *Science* 255 (1992): 845–48, fig. 4. Reproduced with permission from AAAS.

The discoveries occurred in September 1990, when Zhou found two small, articulated fossil birds while excavating paddlefish (*Polyodon*) fossils from lacustrine (lake) shales in Liaoning Province in northeastern China. Zhonghe Zhou came to the University of Kansas, received his PhD in 1999 under the direction of Larry Martin, and returned to China, where he has had a meteoric career rise, receiving two National Natural Science awards from the Chinese government. He is now director and senior research fellow of the Institute of Vertebrate Paleontology and Paleoanthropology of the Chinese Academy of Sciences, where he leads an outstanding research group of about ten scientists.

The Early Cretaceous Jehol Biota and Its Preservation

It was Zhonghe Zhou's initial discovery that presaged what would become one of the most significant periods for Lower Cretaceous paleontology anywhere. The extent of the Early Cretaceous radiation in China was fully appreciated only with the discovery of the *Jehol* Biota, and the ensuing discoveries would rank among the most dramatic breakthroughs in the study of avian evolution since the discovery of the urvogel, *Archaeopteryx*.[10] Indeed, the Jehol Biota, for the first time, provided a reasonably complete window on the avian world of the Early Cretaceous, and with diverse forest environments, the Jehol Biota contains a high percentage of arboreal forms, including pterosaurs, birds, and feathered "flying dromaeosaurs."[11]

A wide range of geologic age estimates have been assigned to the various Mesozoic deposits in China, but current best estimates for the Jehol Biota (from the bottom up: Dabeigou, Yixian, and Jiufotang Formations) indicate a late Early Cretaceous age, and consequently it was a late Hauterivian to early Aptian age fauna. Radiometric dating (^{40}Ar–^{39}Ar dates) of the entire Jehol Biota, including fossils from the younger Jiufotang Formation to the lower Yixian Formation, range from about 120 to 131 (Dabeigou Formation), and the mean average of the overlying lava layers and intrusive volcanics is about 121.[12] The fossil beds of Liaoning during this depositional period were characterized by massive volcanism that catastrophically and repeatedly covered the area with fine-grained siliceous ash that preserved the intricate details of an entire fauna from the forests, lakes, and marshlands of the region. Taphonomic studies based on the composition of volcanic tuffs of the Jehol Group have provided much additional information on the preservational environment. These tuffs are rich in volatiles such as sulfur, fluorine, and chlorine, supporting an interpretation that gaseous emissions associated with volcanic activity played an important contributory factor in the massive mortality of the Jehol organisms. Environmental catastrophes include gaseous emissions

of hydrofluoric acid, volatile chlorine compounds, sulfur dioxide, and hydrogen sulfide, which would also have contributed to acid rain.[13] In summary, the Jehol Biota lasted for at least 11 million years during the late Early Cretaceous, 131–120 million years ago, and the faunal development appears to have experienced three major phases, the most significant being the second, between the Barremian and Aptian, about 125 million years ago.

Liaoning Province, like most of rural China, is poor, and once the locals discovered that the abundant fossils were of monetary value, slabs with fossil specimens began to emerge by the hundreds, many coming from large quarry operations. Even today, only a small fraction of the fossils are discovered by professional paleontologists; most are uncovered by locals, many of them subsistence farmers. Although the illicit transport of Chinese contraband fossils has largely been halted, in the late 1990s not only contraband but fake fossils

Fig. 4.5. Map of China's Liaoning Province showing the primary collection sites for the Early Cretaceous Jehol Biota. Adapted and modified from various sources; see S. H. Hwang et al., "New Specimens of *Microraptor zhaoianus* (Theopoda: Dromaeosauridae) from Northeastern China," *American Museum Novitates* 3381 (2003): 1–44.

were appearing at an alarming rate all over the world, including the infamous *Archaeoraptor*, the fraudulent "missing link" discussed earlier, which combined the hindquarters of a microraptor with the forequarters of an ornithurine bird. Now,

. . . .

Table 4.1. Geological Time Scale, in millions of years, showing significant dates of fossils discussed in text; dates are for beginning of Period (rounded off) or Epoch. *Archaeopteryx*, the oldest known bird, is from the Late Jurassic Tithonian, which begins at 150.8 million years ago. Most of the Chinese birds are some 20–25 million years younger than *Archaeopteryx*, occurring during the middle Early Cretaceous (120–125 million years), but some are slightly older.

Cenozoic		
	Pleistocene (Ice Age)	1.8
	Paleocene	65.5
Late Cretaceous		
	Maastrichtian	71
	Campanian	84
	Santonian	86
	Coniacian	89
	Turonian	94
	Cenomanian	98
Early Cretaceous		
	Albian	112
	Aptian	125
	Barremian	130
	Hauterivian	136
	Valanginian	140
	Berriasian	146
Late Jurassic		
	Tithonian	151
	Kimmeridgian	156
	Oxfordian	161
Middle Jurassic		176
Early Jurassic		200
Triassic		251
Permian		299

Source: Dates from F. Gradstein, J. Ogg, and A. Smith, eds., *A Geological Time Scale* (Cambridge: Cambridge University Press, 2004).

. . . .

however, Chinese law forbids exports of such fossils, real or fake, although numerous fossils still appear for sale at various gem and mineral shows in the United States and elsewhere.

Early Ornithurines Were Also Present

Interestingly, the first Early Cretaceous bird discovered from China was not an opposite bird (suborder Enantiornithes) but an archaic ornithurine (subclass Ornithurae), the oldest known of which, *Gansus yumenensis*, was named for its discovery locality, Gansu Province in northwestern China. It was discovered in 1981 and described by Professor Lian-hai Hou, pioneering Chinese paleornithologist, and his colleague Z. Liu in 1984.[14] At the time it was probably the best material known from any Early Cretaceous bird, and to their credit, Hou and Liu were able to identify the fossil correctly as a water bird or shorebird-type ornithurine, even though the fossil consisted of only an incomplete hind limb.

In 2004, a collaborative expedition of researchers from the Carnegie Museum of Natural History and the Chinese Academy of Geological Sciences revisited the Xiagou Formation (about 10 million years younger than the Yixian and Jiufotang Formations), which had earlier yielded the single specimen of *Gansus*. They recovered a remarkable collection of partial to nearly complete avian skeletons of *Gansus* as well as three specimens (probably two types) representing enantiornithine birds; most important, the complete skeletons confirmed

Fig. 4.6. Pioneering Chinese avian paleontologist Lian-hai Hou, collecting fossils at Sihetun locality, October 1992. Courtesy L.-H. Hou.

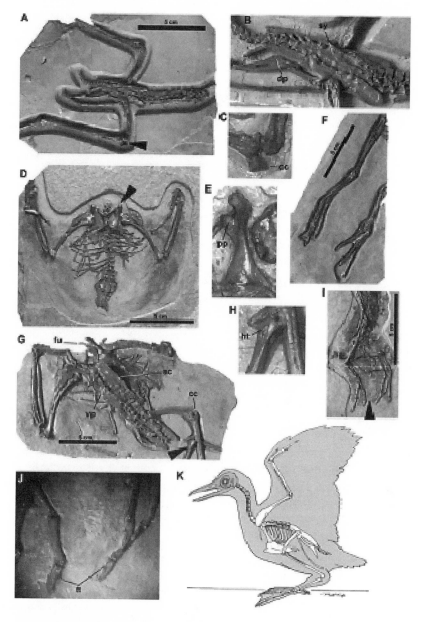

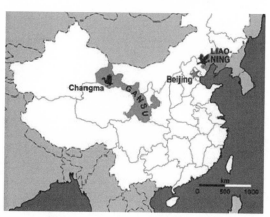

Fig. 4.7. Above, New fossil specimens of *Gansus yum-enensis*, a grebelike ornithurine from the Chinese Lower Cretaceous, along with skeleton and life reconstructions by Mark A. Klinger, Carnegie Museum of Natural History. Left, map of China with Changma locality in Gansu Province and fossil bird–producing deposits of the Jehol Group in Liaoning Province, some 2,000 km (1,242 mi.) to the east. From H.-L. You et al., "A Nearly Modern Amphibious Bird from the Early Cretaceous of Northwestern China," *Science* 312 (2006): 1640–43. Reprinted with permission from AAAS.

Hou and Liu's earlier identification and further revealed the bird to be a grebelike ecomorph.[15] The fifty-some-odd bird fossil specimens recovered from the lake deposits included nearly complete skeletons of *Gansus*, one consisting of two lower legs and feet superimposed, showing remarkable evidence of webbing between the digits and curious, and distinctive, two-pronged toe claws. The fossils show a bird comparable in size to the least tern (*Sternula antillarum*), with a wingspan about 30 centimeters (12 in.) and a short tail that terminated in an ornithurine-style pygostyle. Most important are the specializations of the hind limbs, which include adaptations for swimming and diving seen convergently in modern water birds such as loons, grebes, and diving ducks. Unfortunately, a skull has yet to be discovered. The recent *Gansus* material, in conjunction with other finds, adds support to the view that early ornithurines, unlike their land bird cousins, the enantiornithines, were denizens of near-shore habitats, and that Mesozoic ornithurine birds were predominantly present and may have originated in littoral niches.[16]

The Enantiornithines, or Opposite Birds

The first bird described by Zhou and his colleagues was *Cathayornis*, a tiny, nuthatch-size bird. According to Paul Sereno and colleagues, *Cathayornis yandica* may be the same as *Sinornis santensis*, illustrating that much work remains to be done on these recently described enantiornithines.[17] But remarkably, this small bird possessed a skull very similar to that of *Archaeopteryx*, though with only four teeth in the premaxilla.[18] This finding seemed to confirm Larry Martin's earlier view, originally published in the early 1980s, that *Archaeopteryx* and the enantiornithines form a monophyletic taxon, the sauriurine birds (Sauiurae), as distinct from the other modern-type ornithurine birds (Ornithurae); however, the skull type may simply be plesiomorphic (primitive) and therefore not indicative of an actual clade but rather a model of early avian skull architecture.[19] We can at least conclude that there was a fundamental dichotomy in birds by the Early Cretaceous: the ornithurines and the enantiornithines.

Regardless, the worldwide distribution of the enantiornithines and their adaptive radiation into

Fig. 4.8. *Cathayornis yandica*, discovered in 1990, had an *Archaeopteryx*-like skull (see chapter 3) and was approximately 29 mm (1.1 in.) in total length. The anterior part of the premaxillary is wide, with four teeth on the ventral margin, and the maxilla exhibits some fragmentary teeth. Two teeth are found on one separate thin-walled dentary. The conical teeth, typical of other Mesozoic birds, are slightly constricted at the bases of the crowns. The synsacrum is composed of eight free vertebrae, posterior to which are eight free caudals followed by the pygostyle, which is 15 mm (0.6 in.) long, tapering posteriorly. Courtesy of Zhonghe Zhou.

diverse ecological zones indicate that they were the dominant land birds of the Mesozoic.[20] New genera of Chinese enantiornithines have been steadily described, including among the twenty-odd forms such well-known names as *Sinornis*, *Cathayornis* (may be synonymous with *Sinornis*), *Boluochia*, *Eocathayornis*, and the longirostrine (long-jawed) enantiornithines *Longipteryx*, *Largirostrornis*, and, more recently, *Rapaxavis* ("grasping bird," in reference to its perching foot) from the Jiufotang Formation of Chaoyang in eastern Liaoning Province; *Eoenantiornis*, *Longirostravis*, *Shanweiniao*, and the small, juvenile bird *Liaoxiornis* from the Yixian Formation of Beipiao, Yixian, and Linyuan in Liaoning; and *Vescornis*, *Protopteryx*, and *Jibeinia*, from the Yixian Formation of Fengning in northern Hebei Province.[21] Many new forms will be described in the future, but in general they display a rather uniform morphology. Advancing the systematics of the group, the distinctive taxon Euenantiornithes was erected in 2002 by Luis Chiappe and Cyril Walker to comprise the vast majority of known enantiornithines, excluding *Iberomesornis* and *Noguerornis*.[22] In 1981 Walker perceptively hypothesized that the enantiornithines would prove to be both a widespread, globally distributed group and the dominant Mesozoic birds, but restricted to the Cretaceous: he was correct on both points. By the Late Cretaceous the enantiornithines had been collected on all continents except Antarctica and had undergone a remarkable adaptive radiation, producing large and flightless forms.[23] However, their flight anatomy was highly constrained, as in modern birds, presumably because of physically restrictive aerodynamic parameters, and measured forelimb proportions do not fall outside the range of extant birds, showing that enantiornithines had not evolved any unique set of forelimb proportions.[24] The taxonomy will be in flux for some time, but when one considers that the number of new species of Mesozoic birds discovered during the past decade is more than triple the number of those described during the past two hundred years, it is clear that we have recently added a remarkable new chapter

to the overall picture of avian evolution, spanning some sixty-five million years, from the Early Cretaceous (about 130 million years ago) to the end of the Cretaceous (sixty-five million years ago).

The Earliest Beaked Bird

From 1994 to 1999, there was an explosion in the number of discoveries of Early Cretaceous birds from the Yixian Formation in Liaoning and Hebei Provinces in northeastern China. These fossils included the first beaked bird, which was neither an enantiornithine nor an ornithurine. I had the privilege of joining Lian-hai Hou, Zhonghe Zhou, and Larry Martin in first describing *Confuciusornis sanctus* in a letter to *Nature* in 1995.[25] The previous year we had convened at the University of Kansas and studied the enigmatic specimen; only a laterally crushed skull and some leg and wing elements were preserved (see the resulting attempt at a life reconstruction below). Yet despite this limited material, it was clear that we had for the first time found a beaked bird from the Early Cretaceous. Over the next decade numerous additional

Fig. 4.9. Left, life reconstruction of *Confuciusornis sanctus* by John P. O'Neill, based only on the original 1994 material used to describe the species. Right, painting by Douglas Pratt of *Confuciusornis sanctus* in 1998, based on complete material. From A. Feduccia, *The Origin and Evolution of Birds* (New Haven: Yale University Press, 1996; Pratt painting from 2nd ed., 1999).

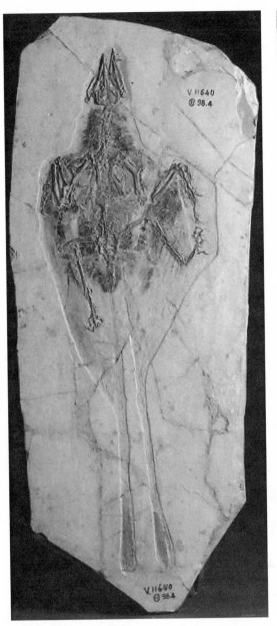

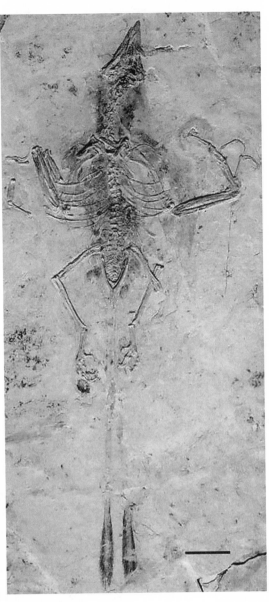

Fig. 4.10. Left, one of the thousands of specimens of the crow-sized *Confuciusornis sanctus*, a colonial inhabit-ant of forests surrounding the Early Cretaceous Liaoning lakes. The long, streamerlike tail feathers appear to have been sexually dimorphic. Right, *Confuciusornis feducciai*, main slab. Facing page, closeup of skeleton and distal tail feathers. The largest known species of the genus, differing also in major morphological fea-tures, it is the first definitive evidence for diversification in a Mesozoic bird. Courtesy of Zhonghe Zhou and Zihui Zhang.

specimens were discovered, exponentially rising by 2003 to estimates of over one thousand, and now literally numbering in the thousands, making *Confuciusornis sanctus* easily the most common fos-sil in the entire avian fossil record.[26]

As Zhonghe Zhou and his colleague Fucheng Zhang have pointed out, the exceptional abun-dance of diverse Early Cretaceous birds from these lacustrine deposits can no doubt be attributed to the frequent volcanic activity and the presence of

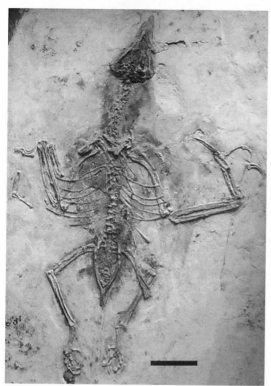

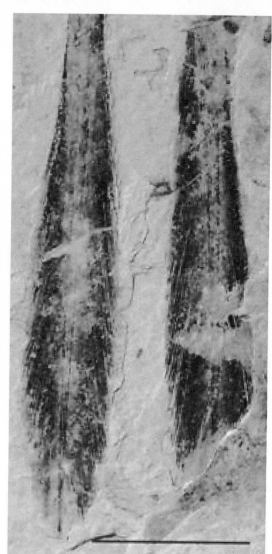

Fig. 4.10 (below). Interpretive life reconstructions of *Confuciusornis*. Left, reconstruction from Kevin Padian and Luis Chiappe, illustrating how ecomorphological reconstruction is dictated by the cladogram and not by anatomy. The hand is anatomically incorrect, with predatory fingers, and the feet are anatomically incorrect. In all, the diagram appears to be based on John Ostrom's insect-net model for bird origins from an earthbound dinosaur. Right, pen and ink image from Douglas Pratt's 1998 life reconstruction of *Confuciusornis*. Left, modified from painting by K. Sano, from K. Padian and L. M. Chiappe, "The Origin of Birds and Their Flight," *Scientific American* 278 (1998): 38–47.

partially isolated intermontane lakes in the region.[27] The exceptional preservation of a large number of specimens of *Confuciusornis sanctus* as well as juvenile individuals of other species of birds in taphonomically favorable lacustrine environments was, as noted earlier, the result of mass mortality caused by volcanic eruptions. Since *Confuciusornis sanctus* is by far the most numerous fossilized avian species from these deposits, it almost certainly gathered in selected areas in large social flocks.

Confuciusornis specimens vary in having either two elongate tail feathers, assumed to be males, or short tail feathers, presumed to be females. *Confuciusornis* may have been like many modern species, including paradise kingfishers, birds of paradise, manakins, and many tropical hummingbirds, that exhibit sexually dimorphic elongate tail feathers in males. This assumption should be treated with some caution, however, given the large size range within the species and the fact that in a number of modern birds, such as motmots and the pheasant-tailed jacana, both sexes sport elongate tail feathers. Still, most workers interpret the elongate

rectrices as functioning for display and sexual selection, arguing that the tails provide the earliest evidence of a sexually dimorphic avian species. One well-known and widely illustrated slab in fact preserves two individuals, one with and one without elongate tail feathers, although a morphometric analysis of more than a hundred skeletons failed to show any correlation between the presence or absence of the bladelike tail feathers and any size distribution classes.[28] Feathers were often preserved attached in lifelike position to the skeletons, or separately, and in unusual cases nearly complete plumage was preserved without skeletal elements. With their reversed halluces, strongly recurved claws, and shortened tarsus (much like that of coraciiform birds), the crow-sized *Confuciusornis* and the closely allied *Changchengornis* and *Eoconfuciusornis* were clearly arboreal. Their well-developed flight apparatus and strutlike coracoids indicate further that they were strong powered fliers.[29] As Zihui Zhang and colleagues show, the diversification of these birds at both the genus and species levels is probably the first well-

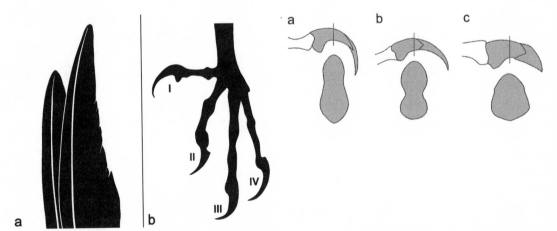

Fig. 4.11. Left, flight feathers and feet of *Confuciusornis* provide evidence for an arboreal, powered flier. The highly asymmetrical flight feathers (a) are those of a strong, capable flier. Note the extreme curvature (b) of all the pedal claws and the fully reversed hallux with equally curved claw. Right, the diagram compares cross-sections of the first pedal digit of *Confuciusornis sanctus* (a), with that of an arboreal bird, the house sparrow, *Passer domesticus* (b), and a predominantly terrestrial bird, a turkey, *Agriocharis ocellata* (c); not to scale. Lateral compression characterizes the pedal claws of arboreal birds. Left, drawn from photographs from L. M. Chiappe, S. Ji, Q. Ji, and M. A. Norell, "Anatomy and Systematics of the Confuciusornithidae (Theropoda: Aves) from the Late Mesozoic of China," *Bulletin of the American Museum of Natural History* 242 (1999): 1–89. Right, adapted with permission, from Z. Zhou and J. O. Farlow, "Flight Capability and Habits of *Confuciusornis*," in *New Perspectives on the Origin and Early Evolution of Birds*, ed. J. Gauthier and L. F. Gall (New Haven: Peabody Museum of Natural History, Yale University, 2001), 237–54.

documented, demonstrable indication of an avian adaptive radiation in the Cretaceous.[30]

One could say that *Confuciusornis* is now the best known of all Cretaceous birds, but many authors, notably Kevin Padian and Luis Chiappe, ignoring critical morphological evidence, especially from the feet and pedal claws, depict the species as a ground-dwelling predator, apparently in an attempt to bolster the theropod, ground-up theory for flight origins.[31] In addition to being anatomically incorrect in many aspects, their life restoration portrays *Confuciusornis* in a pose reminiscent of John Ostrom's insect-net model for the evolution of wings and flight feathers in a nonflight context, in accordance with their well-known bias. This reconstruction elicited the following reaction from Storrs Olson: "*Confuciusornis* is depicted like some medieval rendition of a dyspeptic phoenix that had just dismounted from a horse."[32]

Chiappe and Norell, with Chinese colleagues Ji and Ji, followed the same line of reasoning in their monograph on the genus.[33] Again, Storrs Olson set the record straight in a review of their paper: "The authors, steeped in cladistic fundamentalism, have been among the more insistent proponents of the origin of birds from theropod dinosaurs, with its attendant corollaries, such as the origin of flight from the ground up. . . . This paper will stand as an exemplar of manipulation of information to conform to preconceived ideas, but it is otherwise insufficiently credible or comprehensible to constitute a lasting addition to knowledge."[34]

One could add that although Chiappe and Norell reconstructed *Confuciusornis* as an earthbound predator, in their own paper they publish photographs of the feet and claws, illustrating what could only be described as a highly advanced perching foot, with highly recurved pedal claws, including the hallucal claw, with a fully reversed hallux. These recurved pedal claws, characterized by marked lateral compression as found in modern arboreal birds and *Archaeopteryx*, in combination with the relatively short tarsus and modern avian strutlike coracoids, could only mean that *Confuciusornis* was endowed with powerful flight capability and was a predominantly tree-dwelling bird.[35]

Another interesting avenue of inquiry relevant to *Confuciusornis* as well as to enantiornithine birds has been to study the histology of bone tissue in an attempt to reveal the rate of growth of these early birds and thereby derive conclusions about their metabolism and growth patterns. By examining cross-sections of long bones of enantiornithines the South African paleontologist Anusuya Chisamy and colleagues discovered that the bones are characterized by up to five growth rings that appear as concentric circles. As in tree growth rings, lines of arrested growth indicate a reduction in the rate of bone formation, a pause in growth.[36] Historically such studies have been attended by difficulties of interpretation, but growth rings are generally indicative of periodic bone deposition and, though not conclusive evidence of metabolic parameters, are nevertheless characteristic of such ectotherms as crocodilians, lizards, and turtles, in which bone is usually deposited cyclically, during the animal's growth phase each year. In the case of the enantiornithines, it is clear that these birds experienced not only reductions in their rate of bone growth but probably periodic cessation of growth. In contrast, both modern and Cretaceous ornithurine birds (including *Hesperornis* and *Ichthyornis*) typically lack growth rings in their long bones, indicating a rapid, unaltered growth pattern characteristic of determinate growth and endothermy, which we know is true for modern birds. The archaic enantiornithines probably had lower metabolic rates, may have been quasi-ectothermic, and likely had indeterminate growth, just as the ten skeletal specimens of *Archaeopteryx* show in their range of size. In an examination of cross-sections of the long bones of *Confuciusornis*, modern birds, and *Alligator*, Fucheng Zhang and colleagues concluded that *Confuciusornis* was more like extant ornithurine birds than enantiornithines and suggested that the genus could be endothermic.[37] However, they also suggested that because *Confuciusornis* is phylogenetically more basal than the enantiornithine clade, its endothermy may have evolved independently, as is certainly the case with its avian beak and loss of teeth. In a more comprehensive study, the osteohistology of some eighty thin sections of bones indicated that *Confuciusornis* had a growth rate "longer than in most small living birds but commensurate with larger birds or relatively slowly growing birds such as timami."[38] Yet tinamous are clearly endothermic. The reader is referred to appendix 1 for a more detailed discussion of problems related to the interpretation of bone histology in fossil birds.

Unlike enantiornithines and *Confuciusornis*, which were perching, arboreal forms, the Chinese Early Cretaceous ornithurines were denizens of shoreline and water habitats. Some perhaps were waders, as indicated by their long toes and by the relative proportions in the pedal skeleton; even webbing was preserved in *Gansus*. In contrast, all but one of the enantiornithines are known from lacustrine or terrestrial habitats. Ornithurines may have been adapted to the harsh, physiologically demanding shoreline environments by the acquisition of endothermy.[39] The physiologically advanced ornithurines, unlike enantiornithines, probably had strong flight and migratory ability, and specimens of the Late Cretaceous *Ichthyornis* were discovered well out to sea. As an unexpected corollary, ornithurines as a group survived the Cretaceous-Paleogene extinction event, while primitive enantiornithines, like the dinosaurs, met their demise. Whatever the case, the evidence appears overwhelming that "Early Cretaceous bird evolution highlights a distinctive dichotomy between enantiornithines and ornithurines, the two major avian groups of the Mesozoic."[40]

Chinese Basal Birds

Confuciusornithids, along with the more recently discovered "basal birds" *Jeholornis*, *Sapeornis*, *Zhongornis*, and *Zhongjianornis*, are among the nicely preserved, most basal of Chinese Mesozoic birds, but they possess a mosaic distribution of characters and do not clearly belong to either enantiornithine or ornithurine clades. They may be considered the sister groups of all major groups of extinct and extant birds, perhaps branching before the Mesozoic avian dichotomy. *Jeholornis* is the most similar to *Archaeopteryx*, having toothed jaws (although with a reduced number of teeth) and a long, comparatively unreduced tail.[41] The flight apparatus of *Jeholornis* was more advanced than that of *Archaeopteryx*. Its caudal vertebrae had long chevrons and prezygapophyses similar to those of *Archaeopteryx* but not as elongated as those in dromaeosaurs.[42] The stomach contents of *Jeholornis* included dozens of beautifully preserved ovules of unknown plants, showing it to be a seedeater, adding to the evidence for an extensive adaptive radiation of Mesozoic birds

Fig. 4.12. Jeholornis, a basal bird and seedeater, was less advanced than any other known bird except for *Archaeopteryx*. Additional specimens reported in 2003 show two skeletons and the fan-shaped distal tail feathers. Courtesy of Zhonghe Zhou.

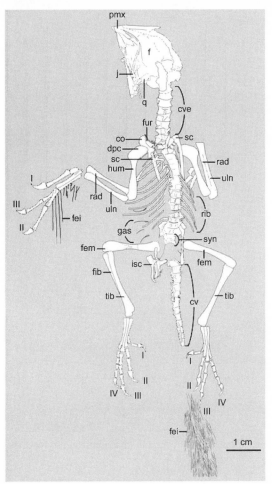

Fig. 4.13. Holotype of *Zhongornis haoae*. Left, photograph of main slab under normal light. Right, interpretive drawing of *Zhongornis skeleton*. Abbreviations: co, coracoid; cv, caudal vertebrae (c1–c13); cve, cervical vertebrae; dc, distal carpal; dpc, deltopectoral crest; dv, dorsal vertebrae; gas, gastralia; f, frontal; fei, feather impressions; fem, femur; fib, fibula; fur, furcula; hum, humerus; isc, ischium; j, jugal; mcI–III, metacarpals I–III; mtI–IV, metatarsals I–IV; pmx, premaxilla; q, quadrate; rad, radius; rib, thoracic ribs; sc, scapula; syn, synsacrum; tib, tibia; uln, ulna; I–IV, digits (manual or pedal) I–IV. Adapted from G. Gao et al., "A New Basal Lineage of Early Cretaceous Birds from China and Its Implications on the Evolution of the Avian Tail," *Palaeontology* 51 (2008): 775–91. Courtesy Ghunling Gao.

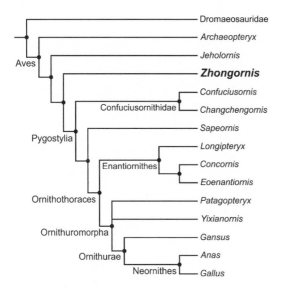

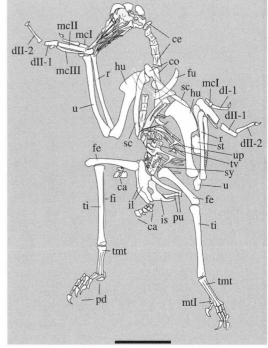

Fig. 4.14. Cladogram of Chinese basal birds. The most parsimonious cladogram resulting from a cladistic analysis using 242 variables and fifteen taxa (Dromaeosauridae plus fourteen avians, the former used as outgroup) is shown. The results of the cladistic analysis place *Zhongornis* as the sister taxon of Pygostylia, thus representing a critical stage in the evolution from basal, long-tailed birds to their short-tailed descendants. Adapted from G. Gao et al., "A New Basal Lineage of Early Cretaceous Birds from China and Its Implications on the Evolution of the Avian Tail," *Palaeontology* 51 (2008): 775–91. Courtesy Ghunling Gao.

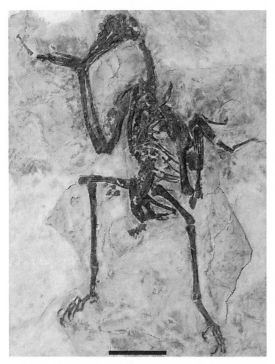

Fig. 4.15. Left, photograph of skeleton of *Zhongjianornis vangi* (scale bar = 5 cm). Right, line drawing of the holotype of *Zhongjianornis*. Abbreviations: ca, caudal vertebra; ce, cervical vertebra; co, coracoid; dI-1, first phalanx of alular digit; dII-1, first phalanx of major digit; dII-2, second phalanx of major digit; fe, femur; fi, fibula; fu, furcula; hu, humerus; il, ilium; is, ischium; mcI, alular metacarpal, mcII, major metacarpal; mcIII, minor metacarpal; mtI, metatarsal I; pd, pedal digits; pu, pubis; r, radius; sc, scapula; st, sternum; sy, synsacrum; ti, tibia; tmt, tarsometatarsus; tv, thoracic vertebra; u, ulna; up, uncinate process of rib. Scale bar = 5 cm. From Z. Zhou, F. Zhang, and Z. Li, "A New Lower Cretaceous Bird from China and Tooth Reduction in Early Avian Evolution," *Proceedings of the Royal Society of London Part B* 277 (2010): 219–27, figs. 1, 2. Courtesy of Zhonghe Zhou and the Royal Society of London.

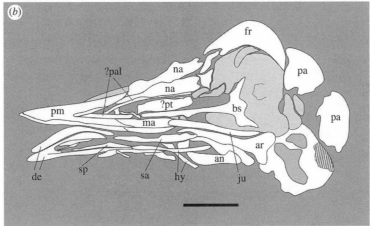

Fig. 4.16. (a) Skull of *Zhongjianornis yangi*, and (b) line drawing of *Zhongjianornis yangi*. Abbreviations: an, angular; ar, articular; bs, basisphenoid; de, dentary; fr, frontal; ju, jugal; ma, maxilla; na, nasal; pa, parietal; ?pal, ?palatine; pm, premaxila; ?pt, ?pterygoid; sa, surangular; sp, splenial bone. Scale bar = 1 cm. From Z. Zhou, F. Zhang, and Z. Li, "A New Lower Cretaceous Bird from China and Tooth Reduction in Early Avian Evolution," *Proceedings of the Royal Society of London Part B* 277 (2010): 219–27, fig. 3. Courtesy of Zhonghe Zhou and the Royal Society of London.

across diverse ecological zones.[43] *Sapeornis* had no lower jaw teeth, showing perhaps a trend toward the elimination of teeth, and the tail was reduced to a pygostyle. However, its short, robust coracoid closely resembles that of the more primitive *Archaeopteryx*, and it is the largest known bird from the Chinese Early Cretaceous, exceeding *Archaeopteryx* in size. Zhou and Zhang suggested that its extremely elongated forelimbs might indicate that *Sapeornis* was capable of soaring flight.[44]

A new basal lineage of Early Cretaceous birds emerged in 2008 with the discovery of *Zhongornis*, which provided critical new evidence relating to the transition between primitive long-tailed birds and their short-tailed counterparts, with a short tail ending in a fused pygostyle.[45] This new genus shows that at least in one lineage a decrease in relative tail length and number of caudal vertebrae preceded the distal fusion of caudals into a pygostyle, and in *Zhongornis* the four distal-most caudals appear to form a continuous lateral flange, which gives the appearance of an incipient pygostyle. Among the so-called pre-pygostylians, *Archaeopteryx* has twenty-one to twenty-three free

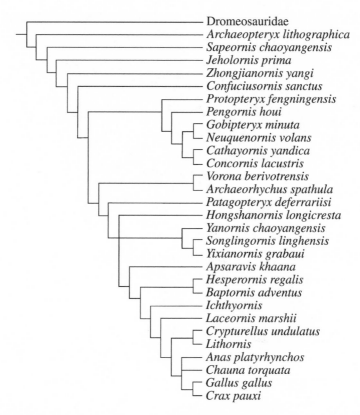

Fig. 4.17. Phylogenetic analysis of *Zhongjianornis yangi,* based on the strict consensus of the four most parsimonious trees. From Z. Zhou, F. Zhang, and Z. Li, "A New Lower Cretaceous Bird from China and Tooth Reduction in Early Avian Evolution," *Proceedings of the Royal Society of London Part B* 277 (2010): 219–27, fig. 6. Courtesy of Zhonghe Zhou and the Royal Society of London.

caudal vertebrae and the more advanced *Jeholornis* has an even greater number, thus showing no trend in tail abbreviation. The skull of *Zhongornis* is crushed but reveals edentulous jaws, and the orbit is large and the rostrum short, similar to that seen in a number of enantiornithines, especially reminiscent of *Eoenantiornis buhleri.* The metatarsals are unfused, but there is a reversed hallux that is shorter than the other digits; and there are no uncinate processes. Only faint traces of feathers, preserving long, straight parallel shafts, are associated to the right hand. Also of interest is that *Zhongornis* has a strutlike coracoid characteristic of volant birds and a manual phalangeal formula of x-2-3-3-x (assuming a hand with digits II, III, IV, as in modern birds), showing the avian trend toward phalangeal reduction from that of *Archaeopteryx,*

and yet it has a hand with an unreduced (clawed) digit IV, although reduced to three phalanges.

In 2009 Zhonghe Zhou and colleagues Fucheng Zhang and Zhiheng Li provided the description of still another remarkable basal bird, *Zhongjianornis,* from Lower Cretaceous lacustrine deposits of the Jiufotang Formation in Liaoning.[46] This remarkable new pigeon-size taxon is exceptionally well preserved, predates the basic ornithurine-enantiornithine dichotomy, and is the most basal bird to exhibit upper and lower edentulous jaws with a pointed snout. Phylogenetically close to the well-known beaked, toothless bird *Confuciusornis,* and slotting cladistically between *Jeholornis* and *Confuciusornis, Zhongjianornis* exhibits a robust, boomerang-shaped furcula resembling that of *Archaeopteryx* and other basal birds and a large deltopectoral crest of the humerus (and sternal plates), suggesting that this early flight architecture, like that found in *Jeholornis*, *Sapeornis*, and *Confuciusornis*, was later replaced by an enlarged sternal keel in more advanced avians. The new genus is relatively large, like other basal birds, confirming that during the evolution of more advanced avian flight there was a substantial size decrease, and the early absence of teeth may be a response to selection for weight loss. In addition, the loss of teeth in *Zhongjianornis* provides compelling evidence for multiple loss of teeth in early avian lineages.

Diverse Opposite Birds

The majority of Chinese Mesozoic birds are Enantiornithes, the archaic opposite birds. As noted earlier, they represent the dominant land birds of the Mesozoic, the Age of Reptiles, which became extinct at the close of the Cretaceous Period, along with archaic ornithurines, including the well-known toothed, foot-propelled divers, the iconic hesperornithiforms, the ternlike *Ichthyornis* and allies, and the dinosaurs. The enantiornithines were almost all toothed and were usually of small to moderate size. Some had extremely recurved claws and a well-developed hallux with a similarly recurved claw. They exhibit a well-fused carpometacarpus, reduced manual digits, and a morphology indicative of powered, flapping flight. Yet they had unique flight architecture, analogous in function but different structurally from the triosseal canal of modern birds, the short sternum was not deeply keeled, and the furcula had a long hypocleidium. They were almost entirely arboreal, and some may have been trunk-climbers, or scansorial.

The most primitive known enantiornithine was *Protopteryx*, a small bird with two elongate tail feathers that are generally similar to those of *Confuciusornis* but appear to have been sheathed (as is true of some *Confuciusornis* specimens), without the typical vanes composed of separate barbs and barbules as in modern feathers.[47] Nevertheless, *Protopteryx* exhibits an advanced flight architecture, including a well-developed alula, seen here in the most basal of the Chinese enantiornithines, and absent in *Archaeopteryx* and *Confuciusornis*.

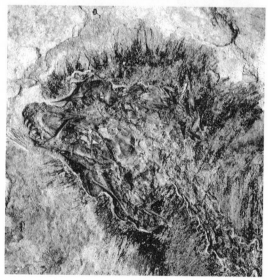

Fig. 4.18. Eoenantiornis buhleri, a moderate-sized enantiornithine with a short and deep rostrum and a particularly broad and deep skull. Courtesy of Zhonghe Zhou.

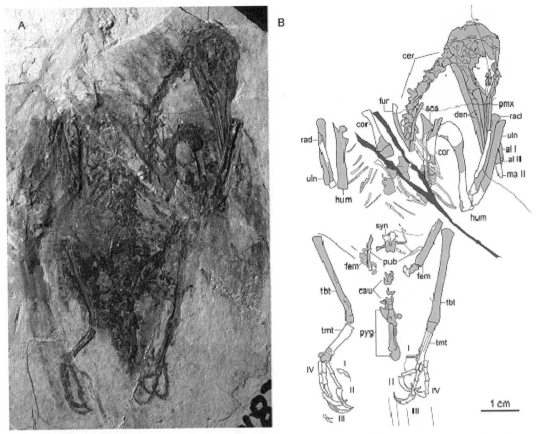

Fig. 4.19. Shanweiniao cooperorum, an enantiornithine of the family Longipterygidae. From J. K. O'Connor et al., "Phylogenetic Support for a Specialized Clade of Cretaceous Enantiornithine Birds with Information from a New Species," *Journal of Vertebrate Paleontology* 29, no. 1 (2009): 188–204. Copyright 2009 Society of Vertebrate Paleontology. Reprinted and distributed with permission of the Society of Vertebrate Paleontology [http://www.vertpaleo.org].

Another enantiornithine, *Longipteryx*, sported short hind limbs and long wings, suggestive to Zhou and Zhang of a lifestyle akin to that of a modern kingfisher, perhaps perching in trees near water and sallying out to make soaring sweeps over the lake surface.[48] When they first described *Longipteryx*, Zhang and Zhou noted that it differed from other known enantiornithines in the possession of uncinate processes, elongate jaws, relatively long wings, and short hind limbs. At the time of the description *Longipteryx* represented a new adaptive type of enantiornithine. Interestingly, in terms of taphonomy, although it was clearly a fully volant bird with a well-developed flight architecture, the flight feathers or wing remiges were

not preserved, thus pointing to the fact that body feathers can be preserved at Jiufotang when wing feathers are not. And most intriguing was the fact that the next long-billed enantiornithine to be discovered, *Longirostravis*, similar in many details to *Longipteryx*, was preserved with downy and body contour feathers but, unlike *Longipteryx*, with primary and secondary remiges as well as rectrices.[49] The long-billed *Longirostravis* further indicated that enantiornithines had undergone a more extensive adaptive radiation than previously thought, the authors suggesting that the bill was probably best adapted for probing feeding, perhaps in mud, like today's charadriiforms. Yet the discovery of another longirostrine enantiornithine, *Rapaxa-*

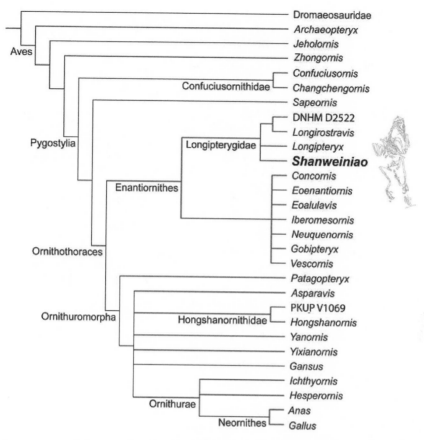

Fig. 4.20. Strict consensus cladogram showing the phylogenetic position of *Shanweiniao* within the enantiornithines (figure corrected). From J. K. O'Connor et al., "Phylogenetic Support for a Specialized Clade of Cretaceous Enantiornithine Birds with Information from a New Species," *Journal of Vertebrate Paleontology* 29, no. 1 (2009): 188–204. Copyright 2009 Society of Vertebrate Paleontology. Reprinted and distributed with permission of the Society of Vertebrate Paleontology [http://www.vertpaleo.org].

vis, with a well-developed grasping, arboreally adapted foot, has shown that these birds are likely arboreal and were possibly more like woodcreepers or hoopoes in their feeding behavior, probing in bark or other arboreal crevices.

The discovery of still another long-billed form, *Shanweiniao* ("fan-tail bird"), from the Yixian Formation, highlighted the existence of a diverse clade of trophically specialized opposite birds that Jingmai O'Connor and colleagues named Longipterygidae in 2009. The family, which includes *Longipteryx*, *Longirostravis*, *Rapaxavis*, and the new genus, comprises small to medium-sized enantiornithines in which the rostral portion of the skull equals or exceeds 60 percent of skull length and in which the dentition is restricted to the premaxilla and the rostral portion of the lower jaw (dentary).[50] In addition, and unique to *Shanweiniao*, the specimen preserved a fan-shaped, feathered tail, with four preserved rachises indicated, and long and slender rectrices, suggesting the evolution of advanced flight capabilities in the tail of enantiornithines.

Most intriguing, and indicative of a much more extensive adaptative radiation than previously envisioned, is the report of a putatively heterodactyl bird from the Chinese Early Cretaceous, as well as substantial documentation of a zygodactyl, roadrunner-like bird (*Geococcyx*) and an associated ichnofossil (imprint of the foot track).

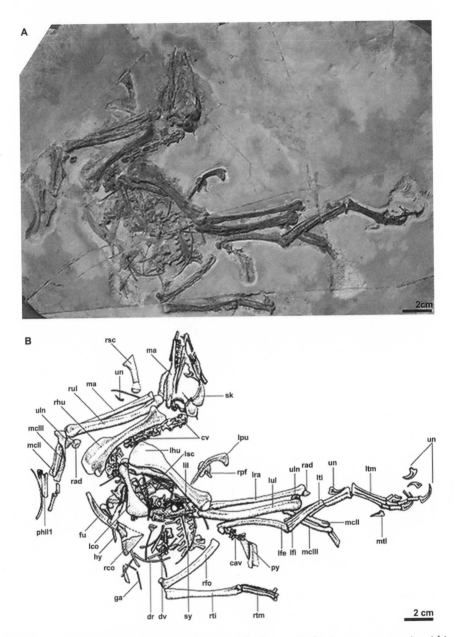

Fig. 4.21. Pengornis houi, a primitive enantiornithine, and the largest Early Cretaceous enantiornithine. Photograph by Zhonghe Zhou; from Z. Zhou, J. Clarke, and F. Zhang, "Insight into Diversity, Body Size and Morphological Evolution from the Largest Early Cretaceous Enantiornithine Bird," *Journal of Anatomy* 212 (2008): 565–77. Reprinted with permission from Blackwell Publishing, Ltd.

This predates the skeletal evidence of that type of avian foot by some 50 or more million years and indicates "a previously unknown degree of Cretaceous avian morphological and behavioral diversity that presaged later Cenozoic patterns."[51]

In 2008 Zhonghe Zhou, Julia Clarke, and Fu-cheng Zhang described a new enantiornithine, *Pengornis houi* ("Peng" is a Chinese mythical bird), named for Lianhai Hou, "a pioneering palaeo-ornithologist." The remarkably well pre-

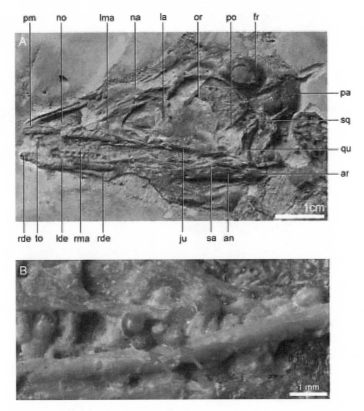

Fig. 4.22. (A) Skull of *Pengornis*. (B) Detail of the posterior dentary teeth. Abbreviations: an, angular; ar, articular; fr, frontal; ju, jugal; la, lachrymal; lde, left dentary; lma, left maxilla; na, nasal; no, nasal opening; or, orbit; pa, parietal; pm, premaxilla; po, postorbital; qu, quadrate; rde, right dentary; rma, right maxilla; sa, surangular; sq, squamosal; to, tooth. Arrows in-dicate possible wear facets. Photograph by Zhonghe Zhou; from Z. Zhou, J. Clarke, and F. Zhang, "In-sight into Diversity, Body Size and Morphological Evolution from the Largest Early Cretaceous En-antiornithine Bird," *Journal of Anatomy* 212 (2008): 565–77. Reprinted with permission from Blackwell Publishing, Ltd.

served specimen provides evidence of a relatively large enantiornithine, significantly larger than all previously described Early Cretaceous taxa, in-cluding *Longipteryx*, and approximately the size of the basal avian *Confuciusornis*. "With the discovery of *Pengornis*, there are similarly sized basal avians (e.g. *Confuciusornis*), Enantiornithes and stem Ornithurae . . . in the Early Cretaceous."[52] *Pen-gornis* exhibits a beautifully preserved skull sport-ing typical avian teeth on both upper and lower jaws, characterized by being conical, with nearly unrecurved crowns, with a constricted waist and elongate root with resorption pits, similar to those reported in *Archaeopteryx*. Cladistic analysis sug-gests *Pengornis* as a basal divergence within the

Enantiornithes and significantly changes our view of the characters thought to be uniquely diag-nostic of the Ornithurae, which is not surprising, given its basal position.

Much more descriptive work remains to be done, but all in all the Chinese Early Cretaceous enantiornithines provide evidence of a spectacular adaptive radiation, and when combined with the radiation of Cretaceous ornithurines, the Creta-ceous adaptive radiation of birds is comparable in many respects to the rich diversity exhibited by the modern bird radiation, which began in full swing in the Paleogene, some sixty-five million years ago, following the Cretaceous-Paleogene extinc-tions. Clearly, the enantiornithines represented

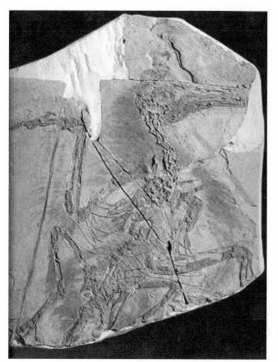
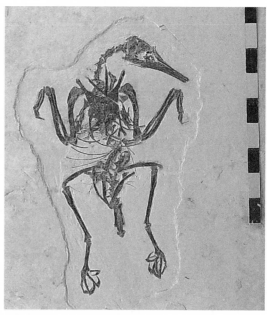

Fig. 4.23. Left, skeleton of *Longipteryx*, an enantiornithine with a bill designed for probing feeding behavior that was previously unknown for these birds, somewhat the ecological equivalent of modern coraciiform birds such as kingfishers. Importantly, with its long, strong wings and "flight architecture," indicating that it was not flightless, the specimen shows that feathers can be preserved without simultaneous preservation of the remiges. Right, *Rapaxavis*, a somewhat similar enantiornithine, but with claws adapted for grasping. *Longipteryx* and *Rapaxavis* vividly illustrate the scope of the dramatic adaptive radiation of archaic enantiornithine birds. Courtesy Zhonghe Zhou, Chunling Gao, and Zihui Zhang.

Fig. 4.24. Map showing the approximate geographical distribution of known fossil enantiornithine birds. Modified after various sources, especially G. J. Dyke and R. L. Nudds, "The Fossil Record and Limb Disparity of Enantiornithines, the Dominant Flying Birds of Cretaceous," *Lethaia* 42, no. 2 (2009): 248–54.

the most species-rich group of Cretaceous birds known, and the newly discovered trophically diverse forms indicate that they had undergone an adaptive radiation in many respects equivalent to that of modern land birds.

Newly Discovered Ornithurines

A number of Chinese ornithurines had been known from somewhat fragmentary material, including *Liaoningornis* and *Chaoyangia* from the Jehol Biota and the highly adapted water bird *Gansus* from Gansu Province. More recently, Zhonghe Zhou and Fucheng Zhang described two nearly complete skeletons of ornithurines, *Yanornis* and *Yixianornis* (and a third, *Songlingornis*). *Yixianornis* was preserved with fish in the stomach, consistent with a near-shore lifestyle. In addition, they and Zhi-Heng Li described *Jianchangornis* as a basal, toothed ornithurine bird with a predentary bone, now a feature known to be characteristic of the Ornithurae.[53] *Apsaravis*, from the nearby Late Cretaceous of Mongolia, is of uncertain taxonomic status, but it may be an ornithurine.[54] Another recently described ornithurine is the well-preserved *Hongshanornis*, from lacustrine deposits of the Lower Cretaceous Jehol Group in Inner Mongolia. It was a small bird with strong flight capability and a low aspect ratio wing, illustrating again the antiquity of sophisticated avian flight architecture.[55] *Hongshanornis* was probably a wader that fed in shallow water or marshes. Like modern shorebirds, these early ornithurine birds had rhynchokinetic skulls, which are adapted for probing in soft substrates in pursuit of shoreline worms and arthropods. Most interestingly, *Hongshanornis* lacked teeth and showed conclusively that the early ornithurines possessed a predentary bone, a synapomorphy also used to define the herbivorous dinosaurian clade Ornithischia. The authors concluded, "This find confirms that the aquatic environment had played a key role in the origin and early radiation of ornithurines, one branch of which eventually gave rise to extant birds near the Cretaceous/Tertiary boundary."[56]

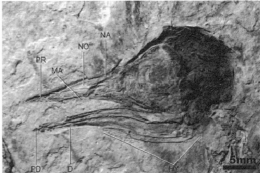

Fig. 4.25. Hongshanornis longicresta (left, main slab; right, skull), an ornithurine from the Lower Cretaceous of Inner Mongolia, China. From Z. Zhou and F. Zhang, "Discovery of an Ornithurine Bird and Its Implications for Early Cretaceous Avian Radiation," *Proceedings of the National Academy of Sciences* 102 (2005): 18998–9002. Courtesy of Zhonghe Zhou and National Academy of Sciences.

Note that although teeth were present in most Early Cretaceous birds, there are clearly exceptions. *Confuciusornis* was a beaked bird, without teeth, *Sapeornis* had lost teeth in the lower jaw, and the seed-eating *Jeholornis* had just three teeth in the lower jaw. All of the Early Cretaceous enantiornithines possessed teeth, and the only known genus devoid of teeth was the Late Cretaceous Mongolian *Gobipteryx*, although the teeth of the Early Cretaceous *Longirostravis* were restricted to the rostral end of the upper and lower jaws.[57] Clearly, the loss of teeth must have occurred independently in several lineages in the early evolutionary history of birds, in early ornithurines and in the extant neornithines, as well as independently in such birds as the Late Cretaceous *Gobipteryx* and the basal *Confuciusornis* and *Zhongjianornis*, the most primitive birds with a horny beak.

Another important discovery was reported by Zhou and Zhang in 2006, with the discovery of the beaked basal ornithurine *Archaeorhynchus* from the Lower Cretaceous Yixian Formation in Liaoning Province.[58] *Archaeorhynchus*, with a rhynchokinetic skull and toothless jaws, preserved numerous gizzard stones, perhaps suggesting an herbivorous diet, but its most interesting feature is the unfused distal tibiotarsus and well-preserved astragalus with a long ascending process. As the name implies, *Archaeorhynchus spathula* had a spatulate bill like that of some modern shorebirds. The discovery of *Archaeorhynchus* conforms to the existing evidence that early ornithurines were shore-dwelling birds, as were later Mesozoic ornithurines, such as the Late Cretaceous hesperornithiforms and ichthyornithiforms. The discovery of these nearly complete skeletons of ornithurines is one of the more unexpected and dramatic avian discoveries from the Chinese Early Cretaceous and demonstrates conclusively that modern-type birds occurred not only coeval with but from the same deposits as opposite, or enantiornithine, birds. The Chinese ornithurines displayed the reduction in size characteristic of the modern lineages of birds. Interestingly, most of the Chinese Cretaceous ornithurines were toothed, like most

other Mesozoic birds, and this feature continued to a large extent into the Late Cretaceous, where *Hesperornis* and *Ichthyornis* occurred.

The Chinese discoveries have gone a long way toward clarifying patterns in the evolution of birds from the gap between the urvogel *Archaeopteryx* and the birds of the Late Cretaceous. Among the surprises are that the horny avian beak appeared in the Cretaceous basal birds *Confuciusornis* and *Zhongjianornis* and that a pygostyle first appeared

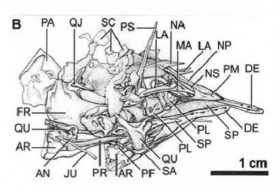

Fig. 4.26. Skull of *Archaeorhynchus spathula*. (A) specimen; (B) line drawing. Abbreviations: AN, angular; AR, articular; DE, dentary; FR, frontal; JU, jugal; LA, lachrymal; MA, maxilla; NA, nasal; NP, nasal process of premaxilla; NS, nostril; PA, parietal; PL, palatine; PF, pneumatic foramen; PR, prearticular; PS, parasphenoid; QJ, quadratojugal; QU, quadrate; SA, surangular; SC, sclerotic bones; SP, splenial. From Z. Zhou and F. Zhang, "A Beaked Basal Ornithurine Bird (Aves, Ornithurae) from the Lower Cretaceous of China," *Zoologica Scripta* 35 (2006): 363–73. Courtesy Blackwell Publishing, Ltd., and Zhonghe Zhou.

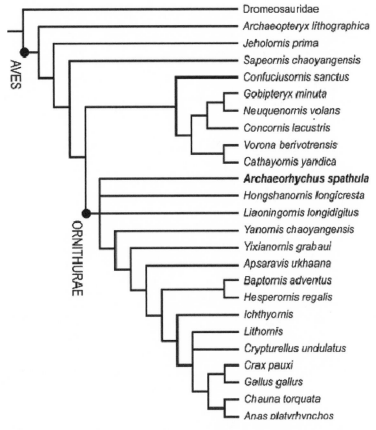

Fig. 4.27. Phylogenetic relationship of *Archaeorhynchus spathula* and other birds (strict consensus cladogram using 202 morphological characters obtained from four equally parsimonious trees). Phylogenetic analysis was conducted using PAUP * Version 4.0b10. From Z. Zhou and F. Zhang, "A Beaked Basal Ornithurine Bird (Aves, Ornithurae) from the Lower Cretaceous of China," *Zoologica Scripta* 35 (2006): 363–73. Courtesy Blackwell Publishing, Ltd., and Zhonghe Zhou.

in the basal bird *Sapeornis*. The alula first appeared in *Protopteryx* (and was also preserved in the later *Eoenantiornis*), as did a well-developed triosseal canal (of enantiornithine architecture). A modern type of sternal keel appeared in the early ornithurines, as did an emarginated sternum with four notches, presaging the condition in modern birds. Finally, in contrast to the enantiornithines, which were predominantly of small size, the Early Cretaceous ornithurines attained considerable size, and they were no doubt already endothermic. This being the case, the age-old correlation of feathers and endothermy, almost axiomatic in paleontology, is probably a misconception. Yet the most basal ornithurines, *Archaeorhynchus*, *Hongshanornis*, and

Liaoningornis, are smaller than the more derived ornithurines *Yanornis* and *Yixianornis*, illustrating a trend toward increasing body size among Cretaceous ornithurines. The fossils of the Early Cretaceous of China reveal that the Cretaceous avifauna was composed of a number of lineages basal to the two major clades, the Enantiornithes, the archaic land birds that underwent an extensive adaptive radiation, and the archaic, modern-type birds, the primitive ornithurines, limited primarily to near-shore or aquatic environments. The ancient avian radiation produced astounding and previously inconceivable morphological diversity, analogous in many respects to the diversity of the modern avifauna.

Feathered Dinosaurs from China

> With these discoveries [Chinese dinosaurs], the notion that feathers had cloaked the bodies of the theropod relatives of birds would become as solid as a rock.
>
> L. M. Chiappe, Glorified Dinosaurs: The Origin and Early Evolution of Birds, 2007

In the spring of 2008 I visited the North Carolina Museum of Natural History in Raleigh to view the American Museum of Natural History's traveling exhibit *Dinosaurs: Ancient Fossils, New Discoveries*. I was quickly amused by some of the new designations, now firmly engrained in the field. There were the usual "nonavian dinosaurs" and birds, the latter presumably "avian dinosaurs." What an unusual manner of naming beasts, to tell an observer what an animal is *not*, rather than what it is. And there were the now accepted "feathered, nonavian dinosaurs," but if birds are presumed to be living dinosaurs, one can only imagine that they are "feathered dinosaurs"! How did all this come to pass?

The feathered dinosaur odyssey began in 1996, when a black-and-white photograph of the first so-called feathered dinosaur was widely shown at the fifty-sixth annual meeting of the Society of Vertebrate Paleontology at the American Museum of Natural History. The photo showed a beautifully preserved specimen of a small coelurosaurian theropod, named *Sinosauropteryx* ("Chinese dragon feather"), recovered from the Early Cretaceous Yixian Formation. *Sinosauropteryx* was a typical theropod, an obligate biped with large hind limbs and greatly shortened forelimbs, similar in many details to the small Solnhofen *Compsognathus*. The fossil preserved details of the internal organs, including outlines of lungs, liver, and intestine. In addition, the specimen preserved tiny, clustered, but parallel fibers; these were immediately proclaimed to be protofeathers. With none of the usual scientific substantiation, not even microscopy, news of the first "feathered dinosaur" made the front page of the *New York Times*, writer Malcolm Browne claiming that the fossil provided the long-sought evidence that birds were descended from dinosaurs.[59] The article featured a pen-and-ink sketch by the well-known paleontological artist Michael Skrepnick of a small dinosaur clothed with downy feathers and carried the proclamation: "A fossil dinosaur has been found in China that shows traces of feathery down along its spine and sides—a sign that it may have been an early ancestor of birds." The article noted that "scientists crowded the hallways and meeting rooms to look at photographs of the Chinese specimen brought to the meeting by Dr. Philip J. Currie, who was chief of dinosaur research at the Tyrell Museum of Paleontology." The piece also observed that "rarely are scientific findings of this possible importance presented so casually." The specimen had been brought to the attention of Currie and Skrepnick just two weeks earlier in Beijing while they were leading a commercial "Dinotour." Currie, who is reported to be featured on Chinese trading cards, told the *Times*, "When I saw this slab of silt stone mixed with volcanic ash in which the creature was embedded, I was bowled over." Within a month, *Science* sported the same illustration of the "downy dino," noting that it "appears to have a mane of downy feathers," and the same story quickly appeared all over the popular press, even in *Audubon* magazine.[60]

Currie reported that when he first saw the meter-long (3 ft.) specimen, it appeared to be closely allied with the Solnhofen *Compsognathus*, but intriguingly, a dark line ran down its neck and backbone: "You could see at the nape of the neck something that looks like downy feathers with a central stem, or rachis," and "What I saw was breathtaking. It was a feathered dinosaur."[61] When Yale's John Ostrom first saw the photographs, he "literally got weak in the knees," because to him not only would the find help confirm his work on the dinosaurian ancestry of birds, but "an insulating covering might indicate the Chinese fossil is that of a warm-blooded animal."[62] While evidence for hot-blooded dinosaurs was declining, were *Sinosauropteryx*, a compsognathid, to be shown to exhibit primitive feathers, the conclusion among paleontologists would be that these unique structures evolved as an insulatory pelt rather than for flight.[63] Such a finding would bolster the idea advanced by Ostrom and Robert Bakker since the

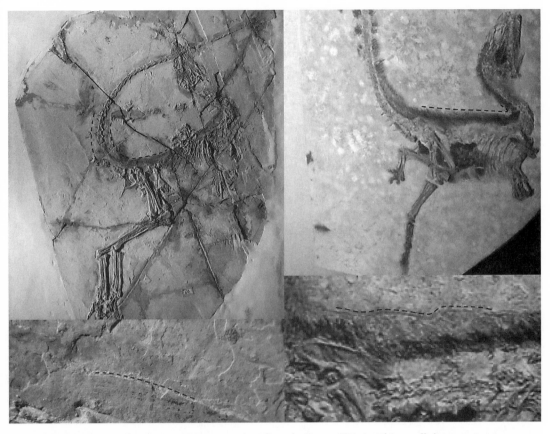

Fig. 4.28. The "downy dino" *Sinosauropteryx prima* (length 68 cm, 27 in.). This small dinosaur is a compsognathid, with forelimbs 40 percent the length of the hind limbs, as well as other typical theropod characters. Below, closeup of neck region showing the fibrous structures along the mid-dorsal line, now shown to be preserved collagen fibers. Note that the filamentous zone is clearly internal, as indicated by a demarcation of the body contour seen in all specimens. Photographs by John Ruben and Zhonghe Zhou (and the author).

early 1970s that feathers had evolved as insulation for incipiently endothermic dinosaurs rather than in an aerodynamic context. John Ostrom had earlier in his career searched extensively for feathers on the small Solnhofen theropod *Compsognathus* to no avail; now, however, the long-awaited discovery was at hand.

Unfortunately, these euphoric claims are not well supported by the data. There was not then, nor is there today, any evidence that the "integumentary" structures of *Sinosauropteryx* are feathers or are even featherlike. There was no light microscopy, transmission, or scanning electron microscopy, only the authoritative proclamation that a feathered dinosaur had been discovered. Additional discoveries of early flightless birds with

feathers, misinterpreted as theropods (to be discussed later), seemed to lend support to the idea. Complicated models for feather evolution from filaments emerged, with the concept that a downy covering insulated endothermic dinosaurs, despite both the fact that such a covering would be clearly maladaptive and the absence of any evidence for endothermy in dinosaurs. It stretches credulity to imagine a small dinosaur coated permanently with a downy pelt, as downy feathers in birds represent a temporary integumentary coating, abandoned as quickly as possible. It is during that juvenile stage that birds are most vulnerable, both to predation and to vagaries of weather. Baby ostriches, for example, if soaked by rain, will become hypothermic and die if they cannot seek the shelter of their mother's wings. With no compelling evidence for dinosaurian endothermy, there is no readily conceivable function for an otherwise maladaptive pelt that would become wet and mucky during rainy conditions.

Dream Team to China

Paleontologist Don Wolberg of the Academy of Natural Sciences in Philadelphia assembled a team of researchers, including John Ostrom, ornithologist Alan Brush, Larry Martin of the University of Kansas, and Peter Wellnhofer of the Bavarian State Museum in Munich, and photographer David Bubier of the Academy of Natural Sciences, to travel to China to study the specimen and report their findings. Reporting on the expedition on 31 March 1997, the *Philadelphia Inquirer* stated that no one on the team felt that the structures were bird feathers, and Peter Wellnhofer remarked, "We cannot call them bird feathers. . . . It's definitely something quite new and unusual. Whether it has anything to do with bird feathers, I don't know. Everything else is speculation." Describing their findings at the Academy of Natural Sciences in April 1997, the team concluded that the structures were not feathers, but John Ostrom also noted that they didn't understand what the fibers represented. Alan Brush speculated that they might be some type of protofeather, but with great perception, Larry Martin pointed out, "The big discussion is whether these fibers are under the skin or above the skin."[64] Martin had noted elsewhere that not only did the specimen exhibit scales with tiny rosettes but the body outline clearly showed that the fibers were contained within the outline. In addition, the obvious body outlines are not raised or sunken from the surrounding matrix and specimen, an indication that they are natural features and not an artifact of preparation. Paleontologists were quick, however, to jump on the "feathered dinosaur" bandwagon, and viewed the new find as adding key evidence to the already popular view of the origin of birds, confirming the preconceived notion that feathers evolved as insulation for endothermic dinosaurs.

Qiang Ji, invertebrate paleontologist and curator at Beijing's National Geological Museum, and S.-A. Ji initially described the newly discovered *Sinosauropteryx prima* in 1996 in a Chinese journal as the earliest bird fossil in China, the "Chinese dragon feather," even though nothing in the fossil remotely resembles either a bird or a feather.[65] Probably a relative of the Solnhofen *Compsognathus*, it has typical theropod features, including forelimbs some 40 percent the length of the hind limbs; a typical diapsid theropod skull with two large temporal openings; a long dinosaur tail and associated theropod architecture, including fan-shaped neural spines on dorsal vertebrae; a dinosaur pubis with pubic boot; and a broad sacrum with supra-acetabular shelf. *Sinosauropteryx* was a standard, cookie-cutter theropod.

After two years of hyperbole in the popular press following the 1996 announcement and after various writers had already made their superficially convincing case to the public for a feathered dinosaur, the fossil was described in *Nature* in 1998 by Pei-ji Chen, also an invertebrate paleontologist, from Nanjing, and his colleagues, and one would have to wait until 2001 for any reasonable description of the specimen or the structures.[66] Chen and colleagues were equally convinced that the structures had some relation to feathers and could be "previously unidentified protofeathers." "If small theropods were endothermic," they wrote, "they would have needed insulation to maintain high body temperatures." Chen and colleagues suggested that the "orientation and frequently sinuous lines of the integumentary structures suggest that they were soft and pliable, and semi-independent of each other."[67] (The fibrous structures range from about 4 to 5.5 millimeters [.16–.22 in.] on the skull to at least 21 millimeters [.83 in.] above the ends of the scapula, their length decreasing to some 16 millimeters [.63 in.] dorsal to the ilium.) Jennifer Ackerman unhelpfully asked in *National Geographic*, "Could those strange fibers be the earliest examples of bird feathers?"[68] Then, in 2003, paleontologist Xing Xu stated of *Sinosauropteryx*, "Although it was a primitive coelurosaurian, osteologically quite different from birds compared to non-avian coelurosaurians [which cannot be defined morphologically], *Sinosauropteryx* occupies an important position in understanding the origin of birds because of the presence of a hairlike covering on its body."[69] It should be added here that filamentous skin structures are also known in the therizinosauroid *Beipiaosaurus*, a basal tyrannosaurid, and the giant compsognathid *Sinocalliopteryx*, both from the Early Cretaceous of China, in a basal ceratopsian, *Psittacosaurus*, and in numerous dromaeosaurs, especially *Sinornithosaurus* (to be discussed later) and pterosaurs.[70]

Of the two compsognathids other than *Sinosauropteryx* known from the Lower Cretaceous Yixian Formation of China, *Huaxiagnathus* and *Sinocalliopteryx*, the latter genus (whose name means "Chinese beautiful feather") sports an array of very long integumentary filamentous structures as in *Sinosauropteryx*, and these, too, are clearly within the outline of the body and tail. Yet though no evidence links these structures to protofeathers or feathers, Ji and colleagues nevertheless state: "As in *Sinosauropteryx* . . . *Sinocalliopteryx* also bears the distinct filamentous integuments. This new fossil strengthens the presence of protofeathers . . . early at the base of Coelurosauria. . . . It is noteworthy that such integuments were also present in the areas of metatarsus. . . . As we know, the small dromaeosaurid *Microraptor gui* possesses the long and asymmetric vaned feathers on its metatarsus . . . suggesting the four-wing stage of the course from maniraptorans to birds. The protofeathers on metatarsus of *Sinocalliopteryx* could show this pattern appeared also as early as the basal coelurosaurs."[71] Apparently, since birds are dinosaurs, it simply follows that any integumentary filaments found with their skeletons are without question protofeathers.

Although Pei-Ji Chen and colleagues called the filaform, fibrous structures of *Sinosauropteryx* "integumentary," meaning epidermal derivatives of the integument, in fact there has never been any evidence for the existence of any kind of protofeather, and the real question is whether they were epidermal derivatives of the integument or part of the internal, integral integumental skin

architecture. Larry Martin and Stephen Czerkas note that a second specimen of *Sinosauropteryx* included a slab of scale impressions that had been removed from the body area, but this finding has not been published.[72] Another fellow member of the initial team, Peter Wellnhofer, remarked that the "protofeathers" of several Chinese taxa were also present in some pterosaurs and commented that the filaments might have nothing to do with protofeathers at all. As Dominique Homberger and K. N. de Silva of Louisiana State University have noted, an imbricating pattern of tuberculated scales does not provide a likely starting point for the complex musculature associated with feathers.[73] Most of the theropod dinosaurs discovered with well-preserved skin have scales, often the rounded pebble scales more familiar in the famed Komodo dragon, a giant varanid lizard. Certainly, skin is known from a host of ornithischian dinosaurs, particularly duckbills, as well as sauropods, and they all show a tuberculated scale pattern.

As noted, John Ostrom's careful examination of the Solnhofen theropod *Compsognathus* revealed no evidence of feathers:

> There are no feather impressions—nor any evidence whatever that is suggestive of feathers—anywhere on the *Compsognathus* slab. The reader can be sure that I made an exhaustive examination, under various lighting conditions, in search of evidence for feathers, but to no avail. If feathers had been present in *Compsognathus*, it is inconceivable to me that no evidence of them would be preserved, considering the complete and almost undisturbed manner in which the skeleton is preserved, the fine details of the skeleton, and the presence of portions of the one horny claw. But the fine-grained matrix shows nothing. Thus, I conclude that *Compsognathus* almost certainly was not feathered.[74]

It is therefore not surprising that the recently discovered small compsognathid *Juravenator*, described in a 2006 article in *Nature* by Ursula Göhlich and Luis Chiappe, has well-preserved integument

devoid of feathers.[75] The small carnivorous compsognathid, certainly very closely allied with *Comp-*
sognathus and *Sinosauropteryx*, was recovered from
the Late Jurassic deposits of Schamhaupten in
southern Germany. These deposits are part of the
same overall Solnhofen system, but slightly older,
at about 151 to 152 million years old, than those
containing the urvogel *Archaeopteryx*, which vary in
age but date to about 150 million years old. Cladistically slotting out as a basal coelurosaur, *Juravena-*
tor is of significance in the feather and protofeather
debate. Most important, the specimen preserved
large portions of integument as scaled skin around
the tail and hind-limb region. The integument of
Juravenator is formed of uniformly sized, smooth
tubercles that occur at a density of fifteen tubercles
per 25 millimeters (1 in.) of preserved tissue, quite
similar to the small, conical, nonimbricating tubercles of the skin of other known dinosaurs. Instead
of interpreting the fossil for what it actually exhibits, the authors contrarily concluded: "The absence
of feathers or feather-like structures in a fossil phylogenetically nested within feathered theropods indicates that the evolution of these integumentary
structures might be more complex than previously
thought."[76]

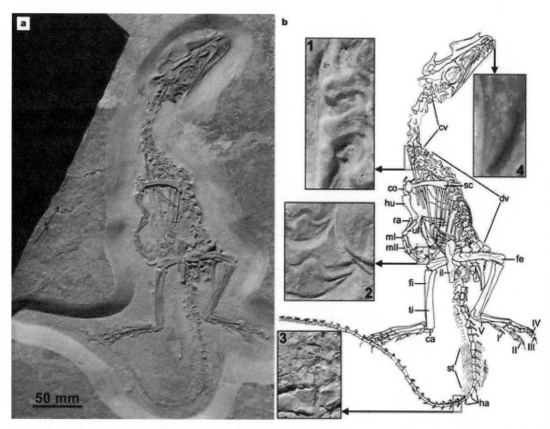

Fig. 4.29. The small Late Jurassic theropod *Juravena-*
tor starki. Serrated premaxillary teeth distinguish this
taxon from most other basal coelurosaurs. Note the
unbirdlike proximally high manual claws that taper
abruptly at their midpoint (2). Abbreviations: ca, cal-
caneus; co, coracoid; cv, cervical vertebrae; dv, dorsal
vertebrae; fe, femur; fi, fibula; ha, haemal arches; hu,
humerus; il, ilium; mI, metacarpal I; mII, metacarpal II; ra, radius; sc, scapula; st, soft tissue; ti, tibia;
ul, ulna; I–IV, pedal digits I–IV; V, metatarsal V.
From U. B. Göhlich and L. M. Chiappe, "A New Carnivorous Dinosaur from the Late Jurassic Solnhofen
Archipelago," *Nature* 440 (2006): 329–32. Copyright
2006. Reprinted by permission from Macmillan Publishers Ltd.

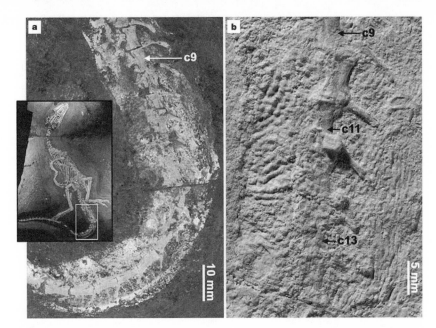

Fig. 4.30. Integument of *Juravenator starki*, showing typical tuberculated dinosaur skin: (a) specimen photographed under ultraviolet light; (b) specimen photographed under normal light; c9, c11, and c13 refer to numbered caudal vertebrae. From U. B. Göhlich and L. M. Chiappe, "A New Carnivorous Dinosaur from the Late Jurassic Solnhofen Archipelago," *Nature* 440 (2006): 329–32. Copyright 2006. Reprinted by permission from Macmillan Publishers Ltd.

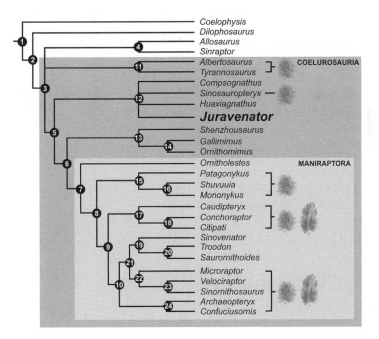

Fig. 4.31. Strict consensus cladogram, from Ursula Göhlich and Luis Chiappe, of the eight most parsimonious trees for twenty-eight theropod taxa, including *Juravenator*. According to the authors, in spite of the absence of feathers in the preserved integumentary portions, *Juravenator* is grouped with coelurosaur clades known for having feathery coverings. According to the authors, plumulaceous and/or pennaceous feathers have been discovered in taxa assigned to Tyrannosauroidea. U. B. Göhlich and L. M. Chiappe, "A New Carnivorous Dinosaur from the Late Jurassic Solnhofen Archipelago," *Nature* 440 (2006): 329–32. Copyright 2006. Reprinted by permission from Macmillan Publishers Ltd.

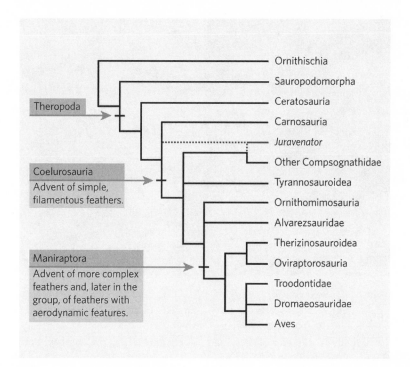

Fig. 4.32. Dinosaurs and feather evolution, as portrayed by Xing Xu. This evolutionary tree of dinosaur groups is compiled from several phylogenetic analyses and shows the possible occurrence of the main events in feather evolution. The tree shows alternative positions for *Juravenator* among the theropods. A compsognathid affinity for *Juravenator* is said to suggest that feathers evolved independently or were lost in some species early in coelurosaurian evolution, or that some coelurosaurs had limited feathery covering. If *Juravenator* turns out to be more basal than the known feathered dinosaurs, it would suggest that feather evolution started during the early history of the Coelurosauria. From X. Xu, "Scales, Feathers and Dinosaurs," *Nature* 440 (2006): 287–88. Copyright 2006. Reprinted by permission from Macmillan Publishers Ltd.

In an accompanying piece in the same issue, Xing Xu, in a tour de force, explains that "the presence of feathers in some extinct coelurosaurs and all living birds . . . suggests that all coelurosaurs, with the possible exception of *Tyrannosaurus rex*, are feathered."[77] He then employs "a robust theropod phylogeny" to reconstruct the sequence in which feathers evolved. We learn that filamentous feathers first evolved in basal coelurosaurs and that more complex feathers, with a thick central shaft and rigid vanes, appeared early in Maniraptora; finally, feathers with aerodynamic features appeared within the maniraptorans before the origin of birds. Xu's map for the evolution of feathers includes data from the discovery and description,

two years earlier, of a primitive tyrannosaurid from the early Cretaceous of China, *Dilong paradoxus* ("surprising emperor dragon"). *Dilong* was described by Xing Xu and colleagues, along with Mark Norell, in their paper "Basal Tyrannosauroids from China and Evidence for Protofeathers in Tyrannosaurids."[78] *Science* writer Erik Stokstad in an accompanying piece readily accepted the discovery without question, noting that the fossils were so well preserved that "one even shows a 'protofeather' fuzz covering the body" and remarking that "Norell proposes that smaller tyrannosaurs needed fuzz to stay warm but that their larger descendants . . . shed their insulation to keep from overheating."[79]

In 2006, following the description of *Juravenator*, French paleontologist Karin Peyer reexamined the larger specimen of *Compsognathus*, discovered around 1971 from the Upper Tithonian lithographic limestones of southeastern France.[80] The specimen measures about 1.4 meters (4.5 ft.) in length, comparable in size to the largest specimen of *Sinosauropteryx* (both specimens have a femur length of 108 millimeters, or 4.3 in.) and much larger than the original Bavarian *Compsognathus* (with a femur of 67 millimeters, or 2.6 in.). As in *Juravenator*, no feathers or featherlike structures were preserved; most important, however, the caudal vertebrae, up to the last preserved caudal, display patches of uniformly sized "bumpy structures" on their lateral sides, resembling the tubercles that form the *Juravenator* integument in shape, arrangement, and relative size.

As has been pointed out, and ignored, many times, skin impressions are not uncommon among dinosaurs and have been preserved in sauropods, ornithopods, stegosaurs, ankylosaurs, ceratopsians, and theropods; and some are preserved with internal organs. Yet in no case is there evidence of structures that could be remotely construed as feathers. Among the more notable theropods that preserve integument is the Lower Cretaceous Spanish *Pelecanimimus*, which exhibits an integument with a distinctive cross-hatching on the surface, representing wrinkling of the hide, but the surface of the skin is smooth, devoid of scales, much less hair or feathers.[81] In 1996, Alexander Kellner reported preserved skin in a Brazilian theropod, showing under scanning electron microscopy a very thin epidermis, formed mostly by irregular quadrangles bordered by deep grooves in a crisscross pattern. "No evidence," he writes, "was found of any structure covering the skin, such as dermal ossicles, scales or feathers, which should be preserved if they were originally present."[82] Then, in 1998, the cover of *Nature* featured the beautifully preserved specimen of the small Italian Lower Cretaceous maniraptoran theropod *Scipionyx*, perhaps the best preserved of all dinosaurs, showing details of soft anatomy, including internal organs, and even patches of preserved muscle

in the shoulder region and tail.[83] Yet, no evidence of feathers.

Novel Approaches to Feather Evolution

One of the most formidable problems of avian biology has been the evolutionary origin of feathers. Feathers represent the extreme in integumental complexity, both structurally and embryologically, and it is often noted that they are the most complex appendages produced by the vertebrate skin. In addition to the well-known thermal properties provided by the fluffy feather base, and the aftershaft, or hyporachis, a casual examination of the pennaceous feathers of a modern bird reveals

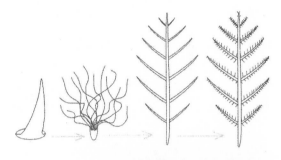

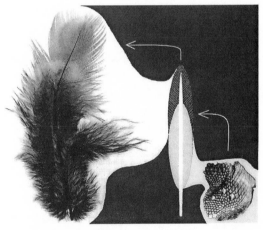

Fig. 4.33. Comparison of evolutionary models of feathers based on fossil evidence from *Protopteryx* (below) and supposed fossil evidence from *Sinornithosaurus* (above). Contrary to popular paleontological belief, there is no evidence of protofeathers from any Chinese fossils. From Z. Zhou and F. Zhang, "Origin of Feathers Perspectives from Fossil Evidence," *Science Progress* 84, no. 2 (2001): 1–18. Courtesy Zhonghe Zhou.

advanced aerodynamic design in almost every structural detail. Pennaceous feathers are characterized by their lightweight, unusually high strength-to-weight ratio, graded flexibility, possession of a reinforcing furrow on the ventral surface of the shaft, great resilience (when broken, they recompose in Velcro-like fashion, with barbs, and microscopic barbules and hooklets), and modular construction (which limits damage). Their unique design allows annual or biannual renewal of these critical flight structures through molting, and body contour feathers provide smooth, aerodynamic, laminar flow in flight, reducing drag as well as providing a waterproof surface, enhanced by the product of the oil, or uropygial, glands on either side of the pygostyle. The remiges (flight feathers), with their asymmetric vanes, create individual airfoil cross-sections, slotted wings, and wings with an overall airfoil cross-section; and their ordered arrangement as primary and secondary flight feathers allows a precisely folded wing, snugly fit to the sides of the body. As the distinguished Canadian biologist D. B. O. Savile put it, feathers "allow a mechanical and aerodynamic refinement never achieved by other means."[84] This of course prompts the question, why should such a complex integumentary structure, capable of producing such highly refined aerodynamic architecture, be evolved for insulation, when hair or a hairlike structure, simpler both embryologically and energetically, would be just as suitable? Nevertheless, the discovery of so-called protofeathers (now known as dino-fuzz) in Chinese dinosaurs was immediately embraced as definitive evidence that feathers evolved as an insulatory pelt for endothermic dinosaurs and were therefore exapted for their later aerodynamic function. "Exaptation" is an evolutionary term coined by Harvard's Stephen J. Gould and Yale's Elisabeth Vrba to denote structures that initially evolved for a specific function other than their later use; in other words, the initial structure has been coopted for a new, distinctive function.[85] New theories on the evolution of feathers, congruent with these views, quickly emerged.

In a 2002 article, Richard Prum and Alan Brush stated: "Feathers likely originated by selection for the growth of an integumentary appendage that emerged from the skin, without continuous investment in the dermis, resulting in the evolution of the novel tubular feather follicle. The tubular feather follicle and feather germ led to subsequent evolution."[86] In other words, the *tubular* nature of anything identified as a protofeather would be a sine qua non for its identification. Yet there is no evidence that the filaments so identified were hollow,[87] and worse, until recently there has never been any attempt to use the standard, normally required techniques of light and scanning electron microscopy to try to discover their structure. With no evidence that the filamentous structures were in fact protofeathers, Prum and Brush have built an entire model for the origin of feathers on the *fact* that they existed!

The two major models for feather evolution are illustrated here by figures (see pages 127 and 129) and will not be discussed in detail in this text. The two models can be characterized simplistically as the classic "scale-to-feather" model and variations thereon and the new "filament-to-feather" model introduced by Richard Prum.[88] Unfortunately, Prum dismisses alternative models derived from extensive evidence and ignores the well-known intriguing fact that feathers may develop from avian tarsal scutates under natural conditions and are easily induced to form on foot scutes by application of retinoic acid or bone morphogenetic protein.[89] In addition, in half-developed ostrich embryos a zone of more or less delimited scales can be observed near the dorsal border of the lateral apterium. Each scale's border displays a feather rudiment already sunk into a short feather sac near the posterior end. At the margin of the scale-covered area of the pectoral callosity, particularly in *Rhea*, the scales appear to be combined with feathers, at the top end of each scale.[90]

There are some scientific, ideologically unbiased reviews of the literature on this confusing topic, notably Lorenzo Alibardi and colleagues' "Evolution of Hard Proteins in the Sauropsid Integument in Relation to the Cornification of Skin Derivatives in Amniotes." The classic view is that feathers could be derived from archosau-

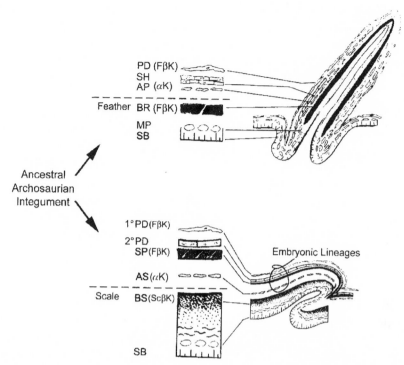

Fig. 4.34. Schematic representations of an embryonic scutate scale and feather, demonstrating the epidermal lineages of the scale (primary 1° periderm, 1° PD; secondary 2° periderm, 2° PD; subperiderm, SP; and alpha stratum, AS) and the feather (periderm, PD; sheath, SH; axial plate, AP; barb ridges, BR; and marginal plate, MP). In the scutate scale, the cells of the 1° periderm and subperiderm express the feather-type beta-keratins (FβK), whereas the cells of the secondary (2°) periderm and alpha stratum (AS) express alpha-type keratins (αK). After hatching, the embryonic epidermis of the scale is sloughed off (dashed line), and the beta stratum (BS), expressing scale-type beta-keratins (ScβK), remains above the stratum ba-sale (SB) and stratum intermedium along the outer surface of the scale. In the feather, the cells of the periderm and barb ridge express the feather-type beta-keratins (FβK), whereas the cells of the sheath, axial plate, and marginal plate express alpha-type keratins (αK). After hatching, only the structural elements resulting from the cells of the barb ridge remain protruding above the surface of the skin. Adapted and modified from R. H. Sawyer et al., "Origin of Feathers: Feather Beta (β) Keratins Are Expressed in Discrete Epidermal Cell Populations of Embryonic Scutate Scales," *Journal of Experimental Zoology (MDE)* 295B (2003): 12–24. Courtesy Loren Knapp.

rian scales as an alteration of the morphogenetic pattern of dermo-epidermal interactions. In this model, the rigid, corneous layer of archosaurian scales gradually became restricted to smaller and smaller areas, while the skin became nonspecialized and more pliable, giving rise to areas with no scales, known in birds as the apteria. The ancestral flat archosaurian scale became narrower and tuberculate, and the epidermis was nourished from a vascularized mesenchyme (mesoderm tissue), creating a condition resembling feather filaments.

Inductive signals with the epidermis then determined the formation of scales or feathers.[91] This model explains why feathers can be induced to develop under some natural and a number of experimental conditions from avian scutes. According to the filament model, the filamentous structures were hollow, and when epithelial folds appeared in the epidermis of these structures, the formation of feathers was initiated. Stages involved the formation of barb ridges and other structures. To add to the complexity of the problem, University of

South Carolina developmental biologist Roger Sawyer and colleagues have shown recently that feather beta-keratins are also expressed in embryonic scutate scales, thus illustrating the homology of scales and feathers at the level of their developmental origin.[92] Here we are concerned with the Chinese fossils, and we shall see that there are many problems with the assumption that the filamentous fibers on the Chinese specimens have any relation to feathers.

Dinosaur Protofeathers: Collagen Fibers?

> There is not a single close-up representation of the integumental structure alleged to be a protofeather.
>
> T. Lingham-Soliar, A. Feduccia, and X. Wang, "A New Chinese Specimen Indicates That 'Protofeathers' in the Early Cretaceous Theropod Dinosaur Sinosauropteryx Are Degraded Collagen Fibers," 2007

Theagarten Lingham-Soliar of the University of KwaZulu-Natal in Durban, South Africa, is a functional morphologist and paleontologist best known in the dinosaur world for his studies on extinct marine reptiles and predation in tyrannosaurs and for his adventurous discoveries of giant sauropod and theropod trackways in Zimbabwe. His interests are varied, however; he has used a variety of tools and methods to scrutinize structures and solve fundamental problems in vertebrate biology, ranging from the biology of mosasaurs and the dynamics of the fins in thunniform fishes to the fossil integument of Jurassic ichthyosaurs. This last area of research led him to the study of modern sharks, whereupon he noted that his high-resolution microscopic images of shark and ichthyosaur integument, with their complex meshwork of supportive collagen fibers, were similar to the integumental fibers discovered in the Chinese Sinosauropteryx and identified as so-called protofeathers.[93] In a series of papers he showed that the Chinese fibers can be matched not only with the collagen fibers of fossil ichthyosaurs but also with those of a number of diverse extant vertebrates, ranging from sharks to a variety of reptiles. Delving deeper into

fossil integumental structures, Lingham-Soliar explored what happens to animal tissue as an animal decomposes, stages that immediately precede fossilization. To see what happens to collagen during decay, he buried a dolphin carcass for a year, removing fragments of its skin tissue for investigation. High-power microscopic examination and chromatographic analysis showed that the degradation process led to compaction of the collagen fibers, the collapse of three-dimensional latticework of the blubber owing to lipid decay, and the separation of bundled fibers into smaller bundles and individual fibers. These degraded fibers presented almost limitless pattern permutations, in many instances showing featherlike patterns that are strikingly similar to many of those identified as protofeathers in the Chinese dromaeosaurs and including a wavy appearance of the allegedly external "appendages" reported in the fibers preserved with the Early Cretaceous dromaeosaur Sinornithosaurus millenii, which Richard Prum cited as illustrating a critical stage in his filament-to-feather evolution model.[94] Lingham-Soliar also proposed that the random orientation of some fibers in these Chinese fossils is likely attributable to breakages in the regimented, regular pattern of collagen fibers during decomposition.[95]

In examining a newer specimen of Sinosauropteryx, now in the Institute of Vertebrate Paleontology and Paleoanthropology, Beijing, Lingham-Soliar and colleagues showed that the integumental structures occur in more sheltered parts of the substrate as an orthogonal mesh. By contrast, in more exposed parts of the substrate the orthogonal mesh had degraded into isolated, short, occasionally sinuous integumental structures. Microscopy showed the regular close arrangement of the parallel fibers, and isolated integumental structures conformed precisely to collagen fibers. This specimen of Sinosauropteryx provides definitive evidence that the integumental structures previously described as protofeathers and proposed as support for a new hypothesis on feather morphogenesis are in fact degraded remains of a system of collagen fibers that reinforced the integument.[96] Also, the regular pattern of fibers, and

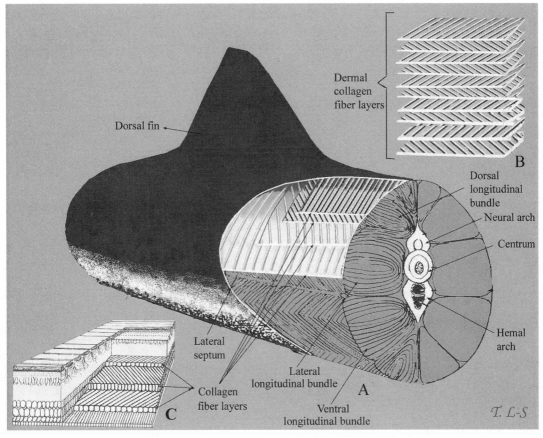

Fig. 4.35. Integumental collagen architecture from white sharks to worms. (A) Schematic view of the collagen fiber architecture of the dermis in the white shark (*Carcharodon carcharias*). (B) Exploded schematic view of ten layers of helically arranged collagen fibers in the stratum compactum of the dermis (varies from twelve to twenty layers) of *C. carcharias*, which form a complex architecture (as opposed to random networks of collagen that characterize the dermis of some animals). Note: the fiber bundles get progressively thicker deeper into the dermis, as previously hypothesized by Lingham-Soliar for those in the skin of the ichthyosaur, *Stenopterygius*. (C) Cuticle of the worm *Ascaris lumbricoides*, showing collagen fiber layers. Drawing by T. Lingham-Soliar (following R. B. Clark, *Dynamics of Metazoan Evolution* [Oxford: Clarendon Press, 1964]), from A. Feduccia, T. Lingham-Soliar, and J. R. Hinchliffe, "Do Feathered Dinosaurs Exist? Testing the Hypothesis on Neontological and Paleontological Evidence," *Journal of Morphology* 266 (2005): 125–66. Reproduced with permission from the Royal Society of London.

the line demarcating a body outline surrounding the tail and dorsum, suggested that *Sinosauropteryx* had a frill, similar to that of many modern lizards, running along its neck, back, and tail. This new evidence calls into question the validity of current theories about the evolution of feathers and their function that are premised on the theory that dinosaurs were feathered. To add to the confusion, the same fiberlike structures of *Sinosauropteryx* are found in a primitive ceratopsian (ornithischian) dinosaur, *Psittacosaurus*, and another unrelated ornithischian, a group thought to have no relation to the origin of birds, as well as a variety of pterosaurs from various ages and localities, including the Early Cretaceous Chinese rhamphorhynchoid pterosaur *Jeholopterus*.[97]

Collagen is a scleroprotein, the chief structural protein of the connective tissue layer of the skin, and one of the most ubiquitous proteins in all vertebrates, and other animals as well. Given

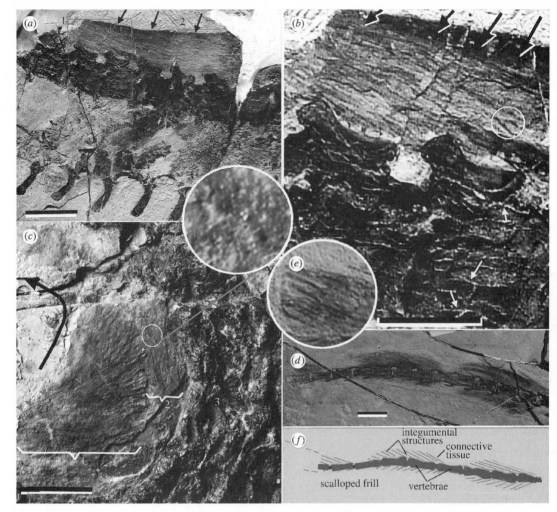

Fig. 4.36. Integumental structures of *Sinosauropteryx*. (a) Overview of the area with significant soft-tissue preservation within the body rather than coronal. Integumental structures occur in the tail recess and overlying the vertebrae; the chevrons have been displaced. (b) Detail of (a). Dark arrows show some isolated integumental structures; white arrows show integumental structures closely associated to give the impression of branching; white circle shows two closely associated integumental structures; arrows show the vertical part of the excavation in which the best preservation occurs. (c) Integumental structures at the juncture between the neck and body; detail in circle shows the angles of the beaded fibers; numerals 1 and 2 indicate integumental structures of the frill and skin/muscle, respectively; numeral 3 indicates the cervical vertebrae; the curved arrow shows the sharp backward curvature of the neck; arrowheads show straight fibers. (d) Integumental structures in the distal part of the tail showing gaps between preserved tissues. (e) Detail, showing beaded integumental structures. (f) Schematic of (d). Scale bars = (a, d) 2 cm and (b, c) 1 cm. T. Lingham-Soliar, A. Feduccia, and X. Wang, "A New Chinese Specimen Indicates That 'Protofeathers' in the Early Cretaceous Theropod Dinosaur *Sinosauropteryx* Are Degraded Collagen Fibers," *Proceedings of the Royal Society of London B* 274 (2007): 1823–29. Reproduced with permission from the Royal Society of London.

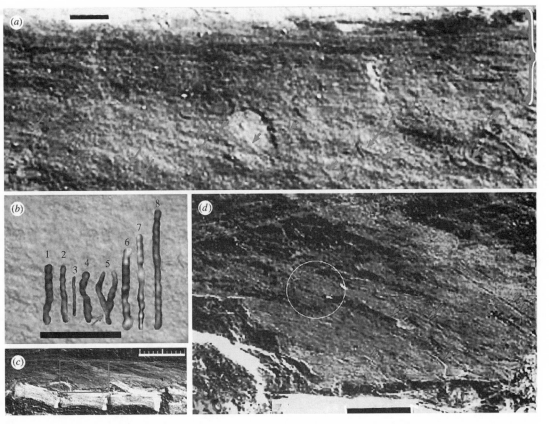

Fig. 4.37. Integumental structures in *Sinosauropteryx*. (a) Integumental structures in the proximal tail area; at the top, they are in regular parallel association; below the integumental structures are more random. (b) Isolated integumental structures from various parts of the preservation, reoriented for ease of viewing; b1–6, 8, from the integument within body; b7 represents integumental structure overlying well-preserved vertebrae. (c) Integumental structures in the last but three terminal caudal vertebrae preserved. (d) Detail of (c) showing integumental structures as part of a matrix of connective tissue at their lower half, while in the distal half, the individual structures are more evident (circled), becoming progressively more degraded toward the tips; circle shows midstage of regular, tight, parallel fiber associations. Scale bars = (a, b) 1 mm, (c) 1 cm, and (d) 2 mm. From T. Lingham-Soliar, A. Feduccia, and X. Wang, "A New Chinese Specimen Indicates That 'Protofeathers' in the Early Cretaceous Theropod Dinosaur *Sinosauropteryx* Are Degraded Collagen Fibers," *Proceedings of the Royal Society of London B* 274 (2007): 1823–29. Reproduced with permission from the Royal Society of London.

collagen's low solubility in water and its organization into networks or meshworks of tough, inelastic fiber, it is plausible that it would be preserved occasionally from flayed skin during fossilization. Yet the idea that the skin fibers preserved in the Chinese dinosaurs were collagen, by far the most likely explanation, was never considered by most paleontologists, since birds are dinosaurs and therefore any fibers in dinosaurs must represent protofeathers.

The corollary inferences derived from the alleged presence of filamentous protofeathers in theropods were not inconsequential: an entirely new model for feather evolution was conceived by authors who examined the fossils and readily embraced the view that the fibers represented primordial feathers. As noted, Richard Prum and Alan Brush promoted this view, writing in 2004: "Dinosaur fossils clearly show fully modern feathers and a variety of primitive feather structures. The

Fig. 4.38. Integumental structures preserved in an Early Cretaceous Chinese pterosaur *Jeholopterus* appear similar to the so-called protofeathers of birds and dinosaurs from the same deposits. Courtesy of Zhonghe Zhou.

conclusions are inescapable: feathers originated and evolved their essentially modern structure in a lineage of terrestrial, bipedal, carnivorous dinosaurs before the appearance of birds or flight."[98] In addition, they claimed, the first dinosaur discovered in the Chinese deposits, *Sinosauropteryx*, "had small tubular and perhaps branched structures emerging from its skin." Of course, virtually everything they wrote about these fossils has been refuted, and the fibers have been shown to be collagen fiber meshworks. Yet a model of feather evolution was constructed on this premise. To add to the confusion, Prum and Brush accept the view that the secondarily flightless bird *Caudipteryx*, with modern avian pennaceous feathers, was a dinosaur, and hence, not only did flight feathers originate in a nonflight context, but "the heterogeneity

of the feathers found on these dinosaurs is striking and provides strong direct support for the developmental theory."[99] Whatever miscellany of feathers they observed certainly cannot be confirmed, and there is no reason why feathers and collagen fibers would not be preserved on the same specimens. Both hair and collagenous fibers have been discovered in a beautifully preserved mammoth, and there is no evolutionary connection.[100] Moreover, as noted, similar skin fibers have been discovered in a number of fossils from lineages not hypothesized to have anything to do with the origin of birds, including an ornithischian dinosaur and a number of pterosaurs.

In 2004, the Society of Avian Paleontology and Evolution held its meeting in Quillan, France. I participated in a symposium on the homology of the avian digits and the evolution of feathers. A panel discussion rounded up the symposium, and questions were posed in writing. I asked the group, "What is the structural and/or biological evidence that the fibers discovered in the Chinese fossils are in any way related to feathers?" There was a pregnant silence, and it was apparent that no answer would be forthcoming. At last developmental biologist Paul Maderson stood up and emphatically proclaimed: "I can answer the question: *There is no evidence!*" It was noted that the only true answer was that "birds are living dinosaurs."

A sine qua non for the filament-to-feather model is the tubular nature of the filaments: "Feathers emerged only after a tubular feather germ and follicle formed in the skin of some species. Hence, the first feather evolved because the first tubular appendage that grew out of the skin provided some kind of survival advantage."[101] Yet there is no evidence that the structures are hollow. No research into the fine structure of the fibers precedes the work of Lingham-Soliar, which paleontologists have dismissed and seldom cite. The only speculation as to why the filaments might be hollow appears in a 2001 paper by Philip Currie and Pei-Ji Chen, in which they state that under magnification, "the margins of the larger structures are darker along the edges but light medially, which suggests that they may have been hollow."[102] But Lingham-Soliar has pointed out that collagen fibers in Jurassic ichthyosaurs also exhibit dark edges, a phenomenon related to mineralization from the inside out. Scanning electron microscopy by Terry Jones also produced no evidence of hollow structure, and his coworker and colleague John Ruben stated: "Detailed SEM scans of the superficial, fiber-like structures revealed no indication that they were either branched or hollow. In fact, they resemble nothing so much as frayed collagen fibers, similar to those seen associated with the dermis in many recently preserved herpetological specimens."[103]

In yet another portrayal of "feathered dinosaurs," Xing Xu and colleagues in 2009 described a new specimen of the basal therizinosaur dinosaur *Beipiaosaurus*, describing rodlike structures in the neck and elsewhere as "a new feather type in a nonavian theropod" that they called EBFFs, or "elongated broad filamentous feathers," with display rather than thermoregulation as their function.[104] Without citing Lingham-Soliar's work showing that the various filaments could be collagen fibers, they accept the Prum model as reflected in Chinese fossils: "All described feathers in nonavian theropods are composite structures formed by multiple filaments. They closely resemble relatively advanced stages predicted by developmental models of the origin of feathers, but not the earliest stage." They go on to report what they call a feather type in two specimens of *Beipiaosaurus*, in which they claim that "each individual feather is represented by a single broad filament." Somehow they view this morphotype as congruent with a "stage I morphology predicted by developmental models," and they claim that "all major predicted morphotypes have now been documented in the fossil record." Further, "This congruence between the full range of paleontological and developmental data strongly supports the hypothesis that feathers evolved and initially diversified in nonavian theropods before the origin of birds and the evolution of flight."[105] Of course, the only problem is that there is no evidence that the rodlike structures have anything to do with feathers; nor is there any evidence that the structures are hollow. Indeed, there is a widespread occurrence of ossified tendons in birds (particularly owls), dino-

saurs, and other vertebrates, and they cannot rule out any of a number of other possibilities, including fibers that support crests or frills.[106]

> Assuming an answer before the investigation has begun is an all-too-common human failing . . . an abandonment of the intellectual process.
> *Ralph D. Ellis*, Solomon, Falcon of Sheba, 2002

The epigraph above prefaces a reply to the proposed protofeathers in *Beipiaosaurus* by Theagarten Lingham-Soliar, in which he points out that in the entire debate on the existence of protofeathers, despite the importance of the hypothesis of protofeathers in dinosaurs that in all this time (from Chen and colleagues' paper in 1998 to Zheng and colleagues' 2009 publication), not a single close-up photo of a protofeather has been shown.[107] The proposed shafted feathers in an Early Cretaceous therizinosaur that "share some striking similarities with the filamentous integumentary structures seen in the ornithischian dinosaur, *Psittacosaurus*," mean that, if true, feathers would be widespread across dinosaurs. Given that Xu and colleagues suggested that there is "a potential primary homology among the integumentary feathers" of theropods such as *Beipiaosaurus*, the ornithischians *Psittacosaurus* and *Tianyulong*, and some pterosaurs, such a proposal would "push the origin of monofilamentous integumentary structures into the Middle Triassic at least." As has been noted, some pterosaurs share filamentous structures indistinguishable from the dino-fuzz present on so many of the Early Cretaceous fossils, and most recently Zheng and colleagues have included still another ornithischian, the heterodontosaurid *Tianyulong*, to the list of dinosaurs and archosaurs associated with feather origins, ranging from herbivores and small meat-eating coelurosaurs to the large, flesh-eating tyrannosauroids.[108] Without belaboring the point, none of these papers has provided any substantial evidence for the existence of any feather homologue in any dinosaur, and in contrast, in 2004 Wu and colleagues, using solid-state lasers as excitation sources, showed that the only way to distinguish collagen from keratin (the fundamental feather protein) is by using histology.[109]

Lingham-Soliar has reexamined the filamentous bristles of the ornithischian *Psittacosaurus* to verify evidence of their hollow nature. As described by Gerald Mayr and colleagues, "a dark stripe of varying width along at least a part of their midline . . . possibly indicates the presence of a hollow lumen inside these structures."[110] Lingham-Soliar shows that this hypothesis is negated by the presence of a degraded bristle in which "there is part of a black stripe with no surrounding material"; in other words, the black stripe represents nothingness and exists only with the solid outer tube. He continues, "The most reasonable explanation for the 'black stripe' is some form of diagenesis, i.e., chemical change to the 'bristle.'" Lingham-Soliar's paper neither supports nor refutes the hypothesis of protofeathers; rather, it rejects the alleged evidence to date for their existence. The overarching problem from a scientific point of view, however, is that there exists a field in which any wild proposal can be served up with no supporting evidence, and it is up to those who do not accept the notion to provide a refutation. When one considers that most of the papers supporting protofeathers have been published in such high-profile journals as *Nature*, *Science*, and the *Proceedings of the National Academy of Sciences*, the situation is no less than alarming. As Lingham-Soliar has aptly noted, "With the Chinese dinosaurs and the handling of the hypothesis of protofeathers, we [paleontologists] are in danger of once more being referred to as stamp-collectors— of opening Pandora's box."[111]

Despite the problems highlighted by Lingham-Soliar's exhaustive work, nothing has changed in the field, and feathered dinosaurs are still thought to have existed. In an interview following publication of the 2007 paper on the *Sinosauropteryx* fibers in the *Proceedings of the Royal Society* by Lingham-Soliar, myself, and Wang, paleontologist David Unwin of the University of Leicester was quoted as saying, "There's no need to panic. . . . Things may be more complex than we thought."[112]

More recently, a number of papers have attempted to develop an assay for the identification of fossil feathers and their color based on the presence of melanosomes, the pigment granules that provide tissues with color and are synthesized in skin melanocytes.[113] Melanosomes are large organelles (about 500 nanometers in diameter) and are normally visible under bright-field microscopy. Following their biogenesis and transport, melanosomes pass from the dendrites of melanocytes to neighboring keratinocytes, where they reside and provide color and pattern. In 2008, Jakob Vinther and colleagues published a paper in which they investigated the presence of melanosomes in fossil feathers. Applying the findings from the earlier work of Vinther, a team led by Michael Benton of the University of Bristol analyzed the structures associated with preserved feathers from the Chinese Early Cretaceous, including primarily *Confuciusornis*, and some other taxa, in an attempt to determine if the putative melanosomes were preserved in the feathers and apply their findings to theropod integumental filaments from the same deposits.[114]

They claimed to find both eumelanosomes and phaeomelanosomes preserved in life position within the structure of partially degraded feathers of *Confuciusornis*. Finding similar structures in the dromaeosaur *Sinornithosaurus* as well as in the region of filaments of the compsognathid theropod *Sinosauropteryx*, they claim to be able to reconstruct the actual color of these animals, which they infer for *Sinosauropteryx* as follows: "The dark-colored stripes on the tail . . . can reasonably be inferred to have exhibited chestnut to reddish-brown tones."[115] There are, however, major problems with the study, even if the structures can be confirmed to be melanosomes. First, the authors create massive confusion by including a true bird, *Confuciusornis*, in the same kettle with a microsaurid dromaeosaur, *Sinornithosaurus*, believed by many to represent a primitive offshoot of the early avian radiation, as discussed in chapter 1. *Sinornithosaurus* is very close to the four-winged *Microraptor*, which of course has true avian pennaceous feathers, and like *Microraptor* also had true feathers, so that claiming that the "filaments" in this fossil were loaded with microsomes ignores the fact that they are most likely dealing with true feathers or their ghosts. Conflation next becomes a major issue because they apply the same logic to the true theropod *Sinosauropteryx* and claim to have found only phaeomelanosomes in this fossil; but these melanocytes are not known in living reptiles. Further, they assert that they have located melanosomes within a filament, yet their micrographs do not substantiate this claim. In fact, there is no structure visible "within" single filaments, as we have seen from scanning electron microscopy of these filaments. More important, the consequences of compression and compaction during fossilization would make it virtually impossible to localize melanosomes within the tissues of the fossil. In living animals, melanosomes occur in the integument from the hyperdermis to the surface, as well as within avian feathers and mammalian hair bulbs. The authors do not explain why the description of the collagen profile of the filaments is incorrect, stating only that they have refuted "recent claims," but their arguments are vapid.

The most important flaw in Zhang and colleagues' argument is that they ignore the visible evidence, noticed by almost all who have examined the specimens, that the filamentous zone of *Sinosauropteryx* is internal. The body outline, especially along the tail, is clearly visible, with the filaments contained within, and the structure is best explained as a scalloped frill, as seen in many modern lizards. The beading profile of the fibers matches perfectly that seen in modern collagen fibers and is by far the best explanation for their identification. In the meantime, the authors proclaim: "Our work confirms that these filaments are probably the evolutionary precursors of true feathers." Yet more than a decade ago, I asserted: "The most obvious conclusion . . . is that these structures are collagen fibers beneath the skin that support some sort of midline skin flap or lizardlike frill that extended down the back to the tip of the tail and along the midline beneath the tail. The best preserved specimen appears to show a 'membrane line' immediately above the fibrous structures, indicating that the structures are internal, beneath the skin."[116] And in a review of Zhang and colleagues' paper on the color of *Sinosauropteryx*, Lingham-Soliar correctly noted: "A blanket view

that all sinuous structures in the Chinese dinosaurs were 'protofeathers,' is now being compounded by a blanket view that all micro-particles found in the same integumental structures are melanosomes." He produces a complex argument that the allegations for phaemelanosomes and therefore divining color in *Sinosauropteryx* are totally without merit.[117] For Vinther and colleagues' assertions on *Sinosauropteryx* to be correct, since the filamentous zone contacts the tail vertebrae, one would be forced to accept a reconstruction of the small compsognathid as having a ratlike tail, with bristly protofeathers emanating all over it. In reality, as noted, the filamentous zone is clearly internal and is most likely a meshwork of supportive skin collagen fibers providing the framework for a lizardlike frill.

It is worth noting again in this context that such a downy coating would be maladaptive in a small dinosaur. Young birds such as ratites are extremely vulnerable as downy chicks and must seek cover under the adult's protective wings in rainy weather to avoid hypothermia and death. Down feathers in a dinosaur would become mucky and soggy and would be maladaptive. The developmental goal of young precocial birds is to abandon the downy stage as quickly as possible. Given the uncertainty of many of the new proposals, the weak nature of the evidence, and the poor design of the research, discussion of melanosomes in fossil features will endure for years to come until more definitive technology can be applied to confirm or refute the assertions.

Identification of complex fossilized structures can be extraordinarily complex, as best illustrated by the "halo" of markings called "dendrites" that surround most of the Late Jurassic Solnhofen fossils. As unique as snowflakes, these lovely fernlike engravings surround the fossils and leave an impression of being of plant origin. In fact, however, they are inorganic in origin, the result of solution of iron or manganese from inside the strata, and reprecipitation as oxides along joint planes and discontinuities such as fossils.[118] Who would have guessed? What is needed is a thorough investigation of the taphonomy and nature of the "halos" surrounding the Chinese fossils. Like similar phenomena discovered in Paleogene birds, the halos are extraordinarily complex, comprising everything from bacterial biofilm to skin and muscle fibers, pennaceous flight feathers, and other types of feathers in all stages of decomposition and disarray. *Sinornithosaurus*, a dromaeosaurid with pennaceous avian feathers, might display bacterial biofilm, skin muscle, and collagen fibers, as well as degraded feathers of various types, but none of these might be protofeathers.

In beautifully executed experiments, Robert Sansom and colleagues observed the decay of two lowly branches of the chordate-vertebrate assemblage, the lancelet *Branchiostoma*, a chordate, and a larval lamprey, *Lampretta*, a member of the jawless fishes, the Agnatha, the first vertebrates to appear. They discovered that decomposition in both occurs in about the same sequence, features of the head tending to be lost before those of the trunk, including the notochord and muscle blocks. Most important, the overall appearance of the carcass transforms by decay until it resembles the much simpler morphology of the ancestral or stem chordate, with only the ancestral morphology remaining. In other words, the fossils fall into a low position on the evolutionary tree and would therefore mislead investigators. The authors use the term "stem-ward slippage" to describe this phenomenon, which of course severely compromises the usefulness of such fossils and provides a cautionary note to paleontologists. Since decay is a normal part of the fossilization process, the authors caution that stemward slippage "may be a widespread but currently unrecognized bias in our understanding of the evolution of a number of phyla." Could it be that these Chinese Cretaceous structures called protofeathers in primitive birds and feathered dromaeosaurs are little more than simple structures that have developed through degradation? It may be in fact that the halos of non-avian and non-dromaeosaurid fossils represent a great variety of filaform structures that are unrecognizable because of taphonomic degradation.[119] These myriad degraded structures can lend themselves to any variety of fanciful interpretations, and one can only imagine the interpretations from the Early Cretaceous of China if the head of

an ostrich were preserved. Surely every conceivable type of "protofeather" would be present!

Longisquama: A Curious Triassic Archosaur with Featherlike Scales

Abutting the western border of China is Kyrgyzstan, with its extensive Ferghana Valley containing an array of exposed Triassic sediments, and passing through Ferghana (Fergana) is the ancient Northern Silk Road, the northernmost of the silk roads leading to the ancient Chinese capital of Xian more than 2,600 kilometers (1,560 mi.) to the east. The Triassic sediments of this region were explored in the 1960s by the late Russian paleontologist Aleksandr Grigorevich Sharov, who was interested primarily in insects and pterosaurs.[120] Fortunately, Sharov discovered a small, mouse-sized (10–12.5 cm long; 5 in.) archosaur, *Longisquama insignis* ("long scale") in lacustrine sediments of the late Triassic (Norian, about 220 million years old), describing it in 1970 as a pseudosuchian thecodont. (Basal archosaurs, or thecodonts, are the stem group of reptiles that gave rise to dinosaurs, pterosaurs, crocodilians, and birds, but the term "thecodont" is now considered obsolete, referring to a paraphyletic collection of Triassic archosaurs.)

Following Sharov's careful description, Hartmut Haubold and Eric Buffetaut restudied the specimen in 1987 and concluded that it possessed a unique gliding adaptation, a series of double elongate integumental appendages that unfolded in butterfly fashion to form a parachuting or somewhat gliding wing.[121] I studied the specimen in Moscow in 1982 and again in 1999 at the University of Kansas and, like most who studied it, found little fault with Sharov's prior analysis. Based on 1982 Moscow observations with ornithologist and feather expert Peter Stettenheim, as well as later observations, under low, raking illumination, the featherlike scales of *Longisquama* exhibit transverse thickenings in series; these may be the result of repeated hypertrophy of germinal tissue. Terry Jones and colleagues, including myself and Russian paleontologist Evgeny Kurochkin, who had been working on additional

Fig. 4.39. Longisquama, the small arboreal, putative "thecodontian" reptile from the Late Triassic of Kyrgyzstan, is the only known Mesozoic reptile with featherlike scales or parafeathers. It is best interpreted as having had a unique gliding adaptation, with a double series of long, featherlike appendages (parafeathers) that were developed along its back. These could be unfolded in butterfly fashion (A) to form a continuous wing, which could be folded at rest (B). From H. Haubold and E. Buffetaut, "Une nouvelle interprétation de *Longisquama insignis*, reptile énigmatique du trias supérior d'Asie centrale," *Academie des Sciences, Comptes Rendus*, ser. 2A, 305 (1987): 65–70. Courtesy L'Academie des Sciences, Paris.

preparation of the tiny specimen, published on the single skeletal specimen of *Longisquama* in 2000.[122] The paper described the integumental appendages in more detail and noted that they attached to the body by a tubular or cylindrical structure resembling the calamus of a feather. The elongate scales, which I prefer to call parafeathers, are remarkably featherlike in a number of features, and unlike the imbricating scales of most reptiles, they are integumental appendages, each attaching segmentally by papillae to the upper back on either side of the vertebral column. That the parafeathers are integumental appendages is of exceptional interest in that they constitute the only such structures aside from the feathers of birds in all the archosaurs; and it means that they could be individually molted, as evidenced by the occurrence of individual parafeathers and an isolated "wing" in the fossil deposit. Additionally, the parafeathers appear to have been tubular and, like pennaceous feathers, were transversely subdivided, with a hollow core. They exhibit central thickening along

Fig. 4.40. A medley of so-called avimorph thecodonts—basal archosaurs or even nonarchosaurs—have attained a variety of avian-like characters. Depicted here is the tiny mesotarsal archosaur *Scleromochlus taylori*, from the Late Triassic (Carnian) of Scotland, considered a basal avesuchian by Michael Benton. It was initially placed in the Dinosauria and later moved to the Pseudosuchia. Kevin Padian postulated a clade Pterosauromorpha for Pterosauria and *Scleromochlus*. Benton erected a new group name "Avesuchia," and his analysis showed that it was "the most basal member of the bird line [of archosaurs], neither a sister group of Pterosauria nor of Dinosauromorpha." From M. J. Benton, "*Scleromochlus taylori* and the Origin of Dinosaurs and Pterosaurs," *Philosophical Transactions of the Royal Society of London B* 354 (1999): 1423–46; drawing by John Sibbick. Courtesy the Royal Society and Michael Benton.

Fig. 4.41. Another enigmatic so-called avimorph thecodont. The small (14 cm; 5.5 in.) Spanish Middle Triassic *Cosesaurus aviceps* is represented by a negative impression of the skeleton with little anatomical detail, but its skull, braincase, and beaklike jaws give a birdlike impression. Others have suggested that this juvenile archosauromorph is an eosuchian or even best placed in the Prolacertiformes. Andrew Milner pointed out that the specimen shows "how progenetic dwarfing may have reshaped the typical archosaur skull to result in the avian skull." From P. Ellenberger and J. F. de Villalta, "Sur la présence d'un ancètre probable des oiseaux dans le Muschelkalk supérieur de Catalogne (Espagne): note preliminaire," *Acta Geológica Hispánica* 9 (1974): 162–68; see also A. Milner, "*Cosesaurus*: The Last Proavian? *Nature* 315 (1985): 544. Image reprinted with permission from *Geológica Acta* (formerly *Acta Geológica Hispánica*).

Fig. 4.42. Opposite top, Sharov's reconstruction of the Triassic archosaur *Longisquama*, with circle showing the position of the flat, elongated scales along the posterior aspect of the forearm (brachium) and humerus, which must have been aerodynamic and functioned in parachuting and steering. This mouse-sized, arboreal archosaur may provide an analogy to the early models of an arboreal origin of flight, in which it was proposed that elongation of scales would render advantage to jumping and parachuting animals. In *Longisquama*, the highly elongate feather-like scales (parafeathers) were most likely paired in butterfly fashion and served in parachuting. Bottom left, wing of *Longisquama*, found as an isolated fossil, showing the overlapping parafeathers, partially folded. Center left, olotype of *Longisquama* (inset, above left, skull and neck; arrow points to elongate imbricating scales on skull and neck; inset below, the left humerus, digitally reversed from the counterslab), arrow showing the elongate, flattened postaxial scales. Scale 1 cm. Right: isolated folded wing of *Longisquama*, showing overlapping distal ends of integumental appendages. Vanes were sufficiently delicate that the structure of underlying vanes remains visible (scale bar = 5 mm); asterisks indicate tubular bases of three successive shafts that inserted

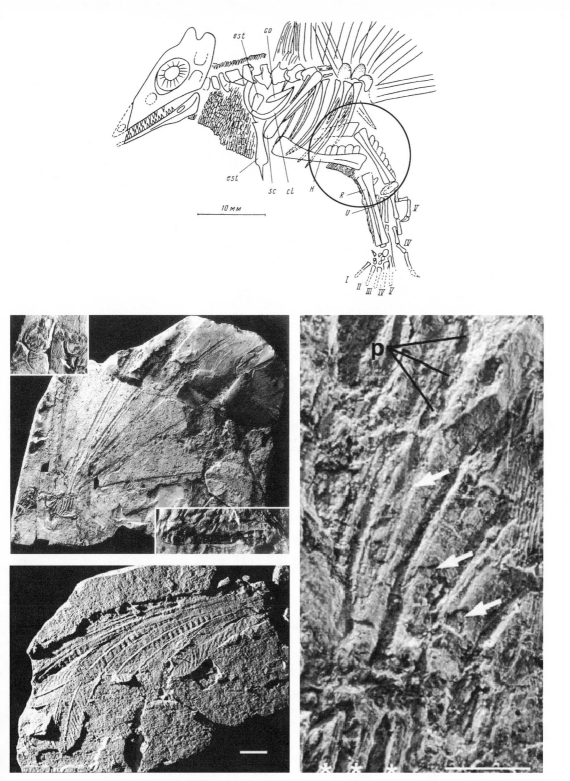

on the left dorsum above the ribs. Note the proximal tapering as well as the distinct, transversely subdivided, hollow core (indicated by arrows) of each tubular base. The morphology of these bases strongly implies that development of *Longisquama*'s elongate parafeathers took place within a follicle. Apparent pulp cavities (p) are also preserved (scale bar = 5 mm). Above left, modified from A. G. Sharov, "An Unusual Reptile from the Lower Triassic of Fergana [Russian]" *Paleontologiceskij Zurnal* 1 (1970): 127–30. Courtesy *Paleontologicheskii Zhurnal*. Photographs from T. D. Jones et al., "Nonavian Feathers in a Late Triassic Archosaur," *Science* 288 (2000): 2202–5. Reprinted with permission from AAAS.

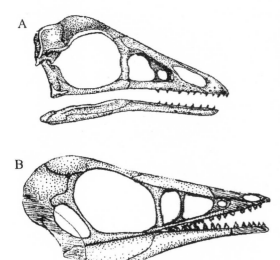

Fig. 4.43. Reconstruction of the skull of *Archaeopteryx* (A) compared to that of *Longisquama* (B). Right, photographs of the furcula of the Triassic archosaur *Longisquama* (upper) in ventral view, and that of the London specimen of *Archaeopteryx* (lower) in dorsal view. Note the ventral groove in the upper specimen, a feature also present in *Archaeopteryx* and Cretaceous enantiornithine birds. The possession of fused clavicles (furcula), a plesiomorphic (primitive) feature, in a basal archosaur shows that it is of little value in associating birds and theropods. Left, from L. D. Martin, "A Basal Archosaurian Origin for Birds," *Acta Zoologica Sinica* 50 (2004): 978–90. Courtesy Larry D. Martin. Right, courtesy Larry D. Martin.

the main axis that closely resembles the rachis of a feather and functionally supports these featherlike scales. Sharov distinguished this small thecodont from other Pseudosuchia "by the presence of long, featherlike appendages along the dorsum and concrescent clavicles [furcula]" and noted that "the structure of the dorsal appendages shows that they functioned as a kind of parachute, breaking the animal's fall as it jumped from branch to branch, or from the trees to the ground."[123]

Contradicting Sharov's original paper, Sebastian Voigt of the Geologisches Institut in Freiberg, Germany, and colleagues studied the specimen again in 2008, paying particular attention to the structure of the parafeathers in new finds discovered in 2007 in comparison with the original material: "We show that *Longisquama*'s appendages consist of a single-branched internal frame enclosed by a flexible outer membrane. Not supporting a categorization either as feathers or scales, our analysis demonstrates that the *Longisquama* appendages formed in a two-stage, feather-like developmental process, representing an unusual early example for the evolutionary plasticity of sauropsid integument."[124] Their interesting findings should lead to comparisons with the elongate, ribbonlike tail feathers discovered in the Chinese basal bird *Protopteryx* and others with such appendages. As Zhou and Zhang have suggested, such feathers "might actually represent a stage

Fig. 4.44. Evgeny Kurochkin at the *Longisquama* site in the Madygen Formation of Kyrgyzstan, in October 2006. Additional parafeathers but no skeletal remains were discovered. Courtesy of Evgeny Kurochkin.

from which elongated scales evolved into an early form of feather."[125] The question is whether there is a demonstrable similarity or even homology between these sheathed feathers and those of *Longisquama*, which have a "continuous outer surface . . . the outside of an enveloping membrane" that "surrounds a spacious interior."[126]

Regarding the function of the appendages, the authors oppose an aerodynamic habit for *Longisquama*, interpreting the dorsal appendages along the spine as a single row, instead of double (as interpreted by other workers). Such a view appears inaccurate both because it is impossible to envision how these curved, hockey-stick-shaped structures could possibly lie down along the back if in a single row and because the preserved isolated wing clearly shows the parafeathers partially folded in avian fashion, as they would be folded on each side. Because of their conclusion, the authors do not believe that *Longisquama* was a glider, but for this to be the case, one would have to explain the elongation of scales along the posterior arm surface as well as the aerodynamic nature of the parafeathers.

The elongate dorsal appendages are not the only modifications of the integument in *Longisquama*; a series of smaller, elongate flattened scales extend down the posterior edge of the brachium and upper arm. They are positioned much as the remiges of birds are and were no doubt flight surfaces. Presumably, the front limbs were used more in a steering capacity during the parachute glide, but the elongation of forelimb scales fits a general paradigm of a pseudophylogeny advocated by those promoting an arboreal flight origin, by which jumping and then parachuting animals gradually elongated scales on the posterior aspect of the forelimb to help break the fall, with continued elongation selectively adaptive at each minor stage.

The scales of *Longisquama* were not likely progenitors of avian feathers, but the specimen does illustrate that during the Late Triassic there was tremendous experimentation with featherlike scales in basal archosaurs presumably before the advent of feathers. It is striking that among Mesozoic archosaurs, feathers or featherlike scales are known only in birds and *Longisquama*, and it is also intriguing that *Longisquama* possessed a furcula similar to that of *Archaeopteryx* and a number of Cretaceous birds.

Larry Martin has argued that *Longisquama* is a potential avian ancestor: "There is nothing in *Longisquama*'s morphology that is inconsistent with a protobird."[127] Of course, others disagree, among them Richard Prum, who is said to have briefly studied the specimen and concluded that the structures were not featherlike, nor was the animal an archosaur.[128] Opinions on *Longisquama*'s systematic position vary widely, ranging from its original description as a pseudosuchian thecodont to an ancestor of pterosaurs and dinosaurs to an actual dinosaur or even a lepidosaur (a lizardlike glider). The counterslab, however, does show an antorbital fenestra and two temporal openings, consistent with it being within the archosaurian assemblage. Aside from that, we must await more material from the critical period of time represented by the Triassic.

Regardless of whether *Longisquama* has anything to do with bird origins, it was nonetheless a small, arboreal basal archosaur "thecodont" (perhaps best termed a Triassic archosaur, *incertae sedis*) with an *Archaeopteryx*-like skull and teeth that closely resemble those of Mesozoic birds (more thecodont than acrodont) and featherlike integumentary structures.[129]

Mesozoic Experiments in Flight

One of the interesting aspects of the *Longisquama* fossil is its age. During the Permian and Triassic Periods numerous, often bizarre arboreal forms experimented with the new aerial ecological zone for the first time. Notable among these experiments was another fossil from the *Longisquama* deposits, also discovered by Sharov, called *Sharovipteryx mirabilis* ("Sharov's miraculous wing," initially described as *Podopteryx*).[130] *Sharovipteryx* was a small gliding archosauromorph reptile, but unlike pterosaurs, it had short forelimbs, very long hind limbs, and a long tail. This strange reptile has been reconstructed with various conformations of a leathery patagial gliding wing; the most recent aerodynamic analysis, by Gareth Dyke of University College Dublin and colleagues, argued that it glided in a manner similar to a delta-wing jet.[131] The flight of *Sharovipteryx* was thus unlike the butterfly

Fig. 4.45. Life reconstruction of the 25-cm (9.8 in.) *Mecistotrachelos*, which glided through the forests of North America some 220 million years ago. This artist's reconstruction shows the small reptile soaring on skin-covered wings supported by delicate, elongated ribs. Ribs were hinged at the base and could be fanned out for gliding or folded against the body. *Mecistotrachelos* could steer by changing the angle and shape of its wings. From Fraser et al., "A New Gliding Tetrapod (Diapsida: Archosauromorpha) from the Upper Triassic (Carnian) of Virginia," *Journal of Vertebrate Paleontology* 27 (2007): 261–65. Image courtesy of Karen Carr. Copyright 2007 Karen Carr and Society of Vertebrate Paleontology; reprinted and distributed with permission of the Society of Vertebrate Paleontology [http://www.vertpaleo.org].

parachuting and gliding of *Longisquama*, the active flapping flight of pterosaurs, or the rib-supported (or rod-supported) gliding of various lineages of Late Permian and Triassic diapsid lizardlike lepidosaurs, including the recently discovered archosauromorph *Mecistotrachelos* from the Late Triassic of Virginia, which, like modern Southeast Asian lizards of the genus *Draco*, glided on membranous wings supported by elongated ribs that could spread laterally to provide a gliding surface. This new discovery also shows that the gliding habit arose in least three distinct tetrapod clades.[132]

The ease with which animals can take to the aerial zone is vividly demonstrated in a recent study by Ardian Jusufi and colleagues at the University of California at Berkeley, in which they examined the role of the house gecko's (*Cosymbotus platyurus*) tail in aerial descent and gliding.[133] Using high-speed video, they showed that

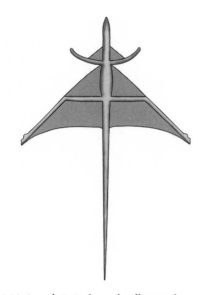

Fig. 4.46. Gareth J. Dyke and colleagues' reconstruction of the Triassic gliding reptile *Sharovipteryx*, which closely resembles a delta-winged jet fighter. In flight, *Sharovipteryx* would have controlled its pitch with a small front wing formed by skin flaps that stretched from body to forelimbs. Hind-leg position would have determined the aerial properties of the large, delta-shaped rear wing. This gliding reptile is from the same locality as *Longisquama*. Modified after G. J. Dyke, R. L. Nudds, and J. M. V. Rayner, "Flight of *Sharovipteryx*: The World's First Delta-Winged Glider," *Journal of Evolutionary Biology* 19 (2006): 1040–43.

at polar latitudes, with wood similar to that of extant conifers. Angiosperms had yet to evolve, so there were no vines or other similar plants. The importance of these trends in botanical evolution is clarified by consideration of the gliding creatures of rain forests in Borneo and of the structure of those rain forests. There are more than thirty gliding species in Borneo's rain forests.[134] These include four species of flying frogs that use loose flaps of skin on their limbs as well as long webbed fingers to parachute, flying geckos that use webbed toes and flanges of skin on the legs and tail for gliding, lizards that can expanded their rib cage for gliding, and ribbon-flat paradise tree snakes that flatten their bodies to propel themselves distances of up to 30 meters (98 ft.). Most of these creatures are more aerodynamically agile than is generally realized: the harlequin tree frog (*Rhacophorus pardalis*) is capable of true gliding, and with webbed hands and feet for airfoils, it can not only glide but also make 180-degree turns in midair. *Draco* lizards can make 30-meter (98 ft.) glides between trees, landing with precision in upright position on tree trunks. Then, among the mammals, are flying lemurs and fourteen species of flying squirrels ranging in size from the 15-centimeter-long (6 in.) pygmies to the red giant flying squirrel (*Petaurista petaurista*), which can exceed 1 meter (3 ft.) in length, including the tail.

a gecko falling with its back to the ground rapidly swings its tail to right its posture. Once righted to a sprawled gliding posture, the gecko uses circular tail movements to control yaw and pitch during its descent. Their results show conclusively how large, active tails can function as effective control appendages and illustrate how simple is the transition to gliding flight.

Patterns of botanical evolution during the Late Permian and the Triassic are perhaps more relevant to the evolution of widespread experimentation with arboreality and with forms of flight among reptiles than is generally realized; the first tall forests emerged during this geological time span. A rise in global temperatures accompanied the ascent of tall gymnospermous trees even

Why is the diversity of gliding organisms so much higher in the rain forests of Borneo than in equally rich Amazonian or central African rain forests? In Amazonian rain forests the only aerialists are the small squirrel monkeys that parachute from tree to tree while assuming the same spread-eagle posture typical of practically all gliders; there are also Central American parachuting frogs. One possible explanation is that the rain forests of Borneo, unlike those of the Amazon basin, Africa, and other similar regions, are dominated by giant dipterocarp trees, which fruit infrequently and unpredictably, making food sparser in the Borneo forests and forcing animals to expand the range of territory they must canvass for food.[135] Gliding offers an energetically cheap and efficient method of canopy travel for a variety of animals, sparing them a long trip down to the ground and back up again.[136] Another explanation is that the diptero-carp forests of Borneo are taller and have fewer lianas; they have a more discontinuous canopy than Amazonian rain forests. In dense Amazonian rain forest canopies, gliding is both more difficult and unnecessary; an organism can simply clamber from tree to tree.

By the end of the Permian the characteristic *Glossopteris* flora would be supplemented by larger conifers such as the tall, columnar *Araucarioxylon* (similar to the living *Araucaria*, the Southern Hemisphere monkey-puzzle), a tree growing to 80 meters (262 ft.). *Araucarioxylon arizonicum*, for example, grew in Arizona's Petrified Forest National Park, some 380 square kilometers (147 sq. mi.), a flat tropical forest in the northwest of the Pangaea supercontinent. These conifers, nearing 60 meters (200 ft.) in height and 60 cm (2 ft.) in diameter, would have provided, in terms of canopy habitat, an analogous ecosystem to that of the Bornean rain forest, with a high, open canopy devoid of lianas interconnecting trees. The modern rain forest of Borneo could suggest an explanation, albeit a general one, for evolutionary experimentation in the aerial ecological zone during the Triassic, a process that produced not only the pterosaurs but flying lizards and strange forms such as the delta-winged *Sharovipteryx* and the butterfly-glider *Longisquama*. Could this be the geologic period during which the precursors of birds first took to the air?

The Feathers of Protopteryx and Allies

As noted earlier, *Protopteryx* was a primitive enantiornithine bird with advanced flight architecture and a well-developed alula, but it also possessed a number of primitive characters, including a long manus and alular digit and an unfused carpometacarpus, all similar to *Archaeopteryx*.[137] The most interesting character of *Protopteryx* is not skeletal but integumental; the type specimen preserves two elongate feathers of unusual morphology. These feathers, never before seen, lack barbs and therefore show no branching at the proximal end (the distal ends are missing), which is attached to the pygostyle as in all enantiornithines. Distally, the vanes are undifferentiated between

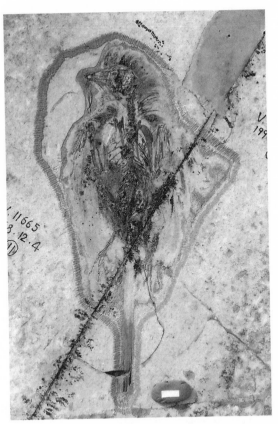

Fig. 4.47. *Protopteryx fengningensis*. The primitive enantiornithine *Protopteryx*, from the Lower Cretaceous Yixian Formation of China, provides evidence of a triosseal canal in early birds. Most interestingly, the two central, elongate tail feathers are scalelike, without branching within the vanes, suggesting to the authors that feathers evolved initially from scales rather than filaments. From F. Zhang and Z. Zhou, "A Primitive Enantiornitine Bird and the Origin of Feathers," *Science* 290 (2000): 1955–59.

the central rachis and the outer branched barbs. It is noteworthy that this type of undifferentiated, scalelike feather is also present in some specimens of *Confuciusornis* and at least four other enantiornithines and differs from the feathers of all other known fossil or extant birds. Fucheng Zhang and Zhonghe Zhou argue that these primitive feathers provide a reasonable model for the evolution of modern feathers according to the scale-to-feather model, through the following stages: (1) elongation of scales; (2) appearance of a central shaft; (3) differentiation of vanes into barbs, first dis-

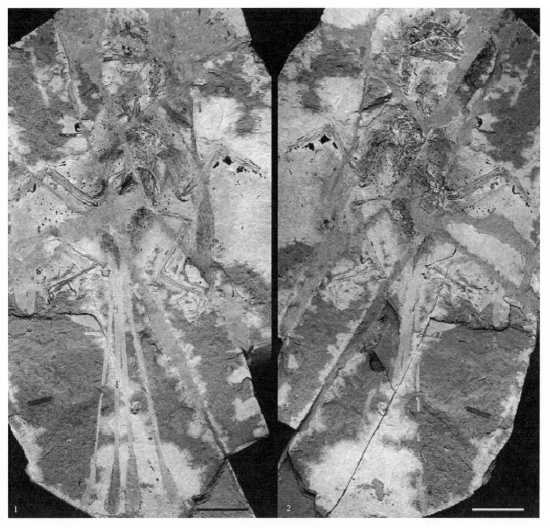

Fig. 4.48. Main slab (1) and counterslab (2) of the enantiornithine *Paraprotopteryx gracilis*. Scale bar = 2 cm. From S.-T. Zheng, Z.-H. Zhang, and L.-H. Hou, "A New Enantiornithine Bird with Four Long Rectrices from the Early Cretaceous of Northern Hebei," *China Acta Geologica Sinica* 81 (2007): 703–8. Courtesy S.-T. Zheng and L.-H. Hou.

tally; and (4) appearance of barbules and barbicles.[138] Thus, if one considers the undifferentiated *Protopteryx* feathers to be "protofeathers," such a view would suggest that modern feathers evolved through a process whereby scales elongated, developed shafts with undifferentiated vanes, and eventually developed barbs and barbules distally. Zhang and Zhou further suggest that flight and down feathers may have differentiated separately from elongated, nonshafted scales in an early stage in the evolution of feathers (see the illustration on page 127). In terms of early function, they write,

"The study of *Protopteryx* suggested that birds probably first developed the feather for [an] aerodynamic purpose."[139]

Another bizarre enantiornithine recently described had four elongate rectrices, unlike the more common doubles seen in *Protopteryx* and *Confuciusornis*. *Paraprotopteryx gracilis* was recovered from the same locality as *Protopteryx* and is morphologically similar to it in having elongate rectrices, an unfused carpometacarpus and distal tibiotarsus, and complete manual claws. It differs, however, in having four long tail feathers

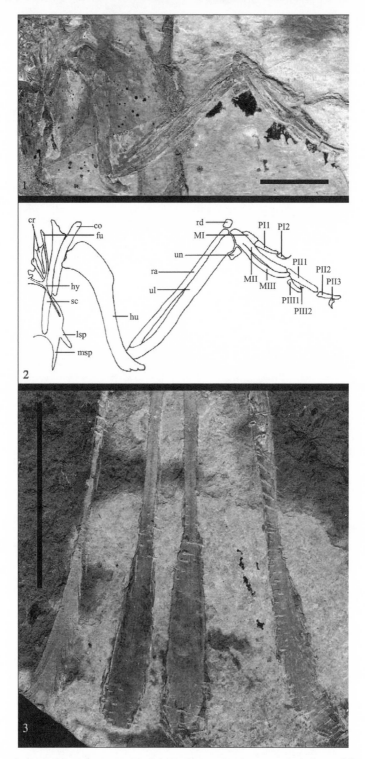

Fig. 4.49. Photograph (1) and line drawing (2) of the upper sternum, pectoral girdle, and forelimb of *Paraprot-opteryx*. Scale bar = 1 cm. Abbreviations: co, coracoid; cr, cervical ribs; fu, furcula; hu, humerus; hy, hypoclei-deum; lsp, lateral sternal process; msp, medial sternal process; MI, alular metacarpal; MII, major metacarpal; MIII, minor metacarpal; and phalanges of digits, as indicated; ra, radius; rd, radiale; sc, scapula; ul, ulna; un, ulnare. Distal ends of rectrices (3). Scale bar = 2 cm. From S.-T. Zheng, Z.-H. Zhang, and L.-H. Hou, "A New Enantiornithine Bird with Four Long Rectrices from the Early Cretaceous of Northern Hebei," *China Acta Geologica Sinica* 81 (2007): 703–8. Courtesy of L.-H. Hou.

attaching to a pygostyle, the central pair being the same length but slightly broader than the lateral pair, though otherwise similar in morphology.[140] The proximal four-fifths is ribbonlike, with no branching in the vanes; but the oval-shaped, enlarged distal end shows the occurrence of a shaft, barbs, and a pennaceous vane as in modern feathers. In *Protopteryx* the distal ends are missing. The authors conclude that the tail feathers of *Paraprotopteryx* may suggest an intermediate stage from elongate scale to feather, or perhaps an indicator of sexual dimorphism, as illustrated by the shaft-tailed whydah (*Vidua regia*), which has four elongated black tail feathers with expanded tips during the breeding season. The feathers may also

have provided a "functional advantage in supplementing the lifting surface to compensate the unskilled flight as in some extant birds."[141]

Scansoriopterids: Birdlike Archosaurs from the Jurassic

In 2002, still another interesting discovery was reported in several journals independently. A diminutive and peculiar archosaur, *Epidendrosaurus* ("upon-tree lizard"), was described by Fucheng Zhang and colleagues as a maniraptoran dinosaur, the group containing dromaeosaurs, oviraptorosaurs, and troodontids (and perhaps therizinosauroids and alvarezsaurids), although it apparently

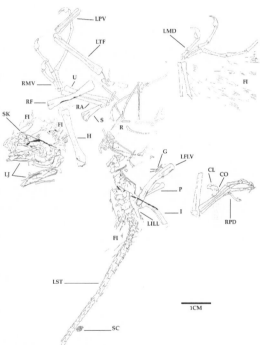

Fig. 4.50. Scansoriopteryx heilmanni (=Epidendrosaurus). Left, photograph of main slab. Stephen Czerkas and Chongxi Yuan note: "It would not be so surprising or unexpected to find such a primitive looking animal as *Scansoriopteryx* from much earlier periods of time dating from the Middle Triassic or even further back into the Permian." Right, interpretive drawing of *Scansoriopteryx*. Abbreviations: CL, clavicle; CO, coracoid; FI, filamentous impressions of wing feathers; G, gastralia; H, humerus; I, ischium; LFLV, left femur lateral view; LILL, left ilium lateral impression; LJ, lateral jaws; LMD, left manus dorsal impression; LPV, left metatarsals ventral impression; LST, impression of left side of tail; LTF, left tibia/fibula; P, pubis; R, area with ribs; RA, radius; RF, right femur; RMV, right manus ventral impression; RPD, right pes dorsal impression; S, scapula; SC, scales; SK, skull; U, ulna. From S. A. Czerkas and C. Yuan, "An Arboreal Maniraptoran from Northeast China," *Dinosaur Museum Journal* 1 (2002): 63–95. Copyright Stephen A. Czerkas. Reprinted with permission.

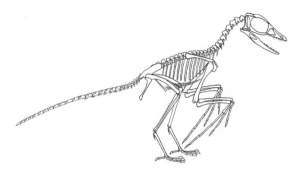

Fig. 4.51. Skeletal reconstruction of *Scansoriopteryx* by Stephen Czerkas showing, among other features, that the pelvis is still like that of a reptile (as opposed to a theropod). There is no pubic boot, there is an avian hallux, there is no furcula, but there are separate clavicles, the forelimbs are elongate, the scapula broadens distally, and the outer finger (probably IV) is totally unlike the equivalent digit III of theropods. From S. A. Czerkas and C. Yuan, "An Arboreal Maniraptoran from Northeast China," *Dinosaur Museum Journal* 1 (2002): 63–95. Copyright Stephen A. Czerkas. Reprinted with permission.

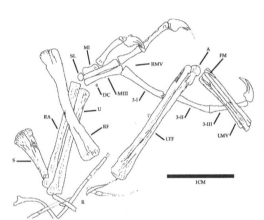

Fig. 4.52. Drawing of right manus and associated wing elements (left) of the small scansoriopterid next to photograph (right), showing the nonavian morphology of the scapula, lower left (S). To the right are leg elements, including the tarsometatarsus. Note also the nontheropod nature of the astragalus (A). From S. A. Czerkas and C. Yuan, "An Arboreal Maniraptoran from Northeast China," *Dinosaur Museum Journal* 1 (2002): 63–95. Copyright Stephen A. Czerkas. Reprinted with permission.

lacks any salient theropod synapomorphies.[142] Though not well preserved, *Epidendrosaurus* possessed feathers that resemble those of the dromaeosaurid *Microraptor*, lacked the fully perforated acetabulum characteristic of theropods, and had a strange elongate outer finger, which is the longest of the manus (in theropods the middle is the longest) and analogously resembles the digging finger of the tiny Malagasy primate aye-aye (*Daubentonia madagascariensis*), which uses its elongate middle finger as an insect-finding tool for probing in tree crevices. To date, a number of specimens have been described, and important questions are raised by this enigmatic group, not the least of which are just what they are and which groups they are most closely allied with. Overlooking the obvious

specialization of the elongate outer finger, it is difficult to justify classifying this small archosaur as a theropod. Are they simply basal birds or even basal archosaurs? Why are they called theropods?

Stephen Czerkas of the Dinosaur Museum in Blanding, Utah, and his Chinese colleague Chongxi Yuan independently described another specimen of the small *Epidendrosaurus* in 2002 and named it *Scansoriopteryx* ("climbing wing").[143] There was some confusion concerning priority because of the near-simultaneous descriptions of different specimens, but *Scansoriopteryx* is generally accepted as a junior synonym of *Epidendrosaurus*, although the family name is derived from the designation by Czerkas and Yuan (hence Scansoriopterygidae, "climbing wings").[144] These

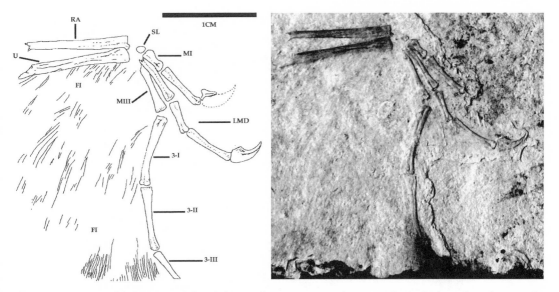

Fig. 4.53. Drawing of left manus (left) of the small scansoriopterid next to photograph (right), showing the elongate preaxial finger and the feathers emanating from the hand, in avian fashion. Abbreviations: SL, semilunate carpal element; RA, radius; U, ulna; FI, filamentous feather impressions. From S. A. Czerkas and C. Yuan, "An Arboreal Maniraptoran from Northeast China," *Dinosaur Museum Journal* 1 (2002): 63–95. Copyright Stephen A. Czerkas. Reprinted with permission.

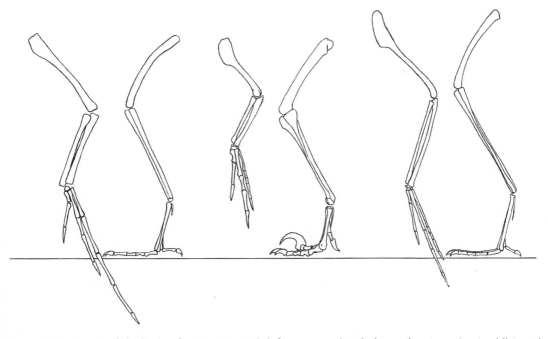

Fig. 4.54. Drawings of the limbs of scansoriopterid (left), compared with those of *Deinonychus* (middle), and *Archaeopteryx* (right), drawn to same femur length. Note that the length of the radius, ulna, and humerus in the scansoriopterid most closely approaches that of *Archaeopteryx*, and also that the third and fourth metacarpals are not elongate when compared to *Archaeopteryx* but are more similar to those of *Deinonychus*. The characteristic fourth digit (excluding the metacarpal) is nearly three times the length of that of *Archaeopteryx*. From S. J. Czerkas and C. Yuan, "An Arboreal Maniraptoran from Northeast China," *Dinosaur Museum Journal* 1 (2002): 63–95. Copyright Stephen A. Czerkas.

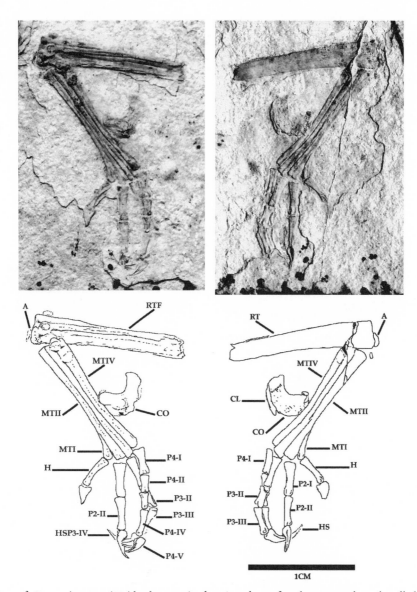

Fig. 4.55. Feet of *Scansoriopteryx* (*Epidendrosaurus*), showing the unfused metatarsals and well-developed reversed first toe or hallux, an adaptation for perching in trees for scansoriopterids. Photographs of right pes main slab (left) and right pes counterslab (right), with corresponding drawings below. Abbreviations: CL, clavicle; CO, coracoid; RT, right tibia; MTIV, fourth metatarsal; MTII, second metatarsal; MYI, first metatarsal; H, hallux; P refers to phalanges with appropriate numbers (note the highly curved nature of the claw, indicating arboreal habits). From S. A. Czerkas and C. Yuan, "An Arboreal Maniraptoran from Northeast China," *Dinosaur Museum Journal* 1 (2002): 63–95. Copyright Stephen A. Czerkas. Reprinted with permission.

authors were reluctant to assign this enigmatic, small arboreal archosaur to any specific taxon, but they noted that it might be of basal saurischian or pre-theropod status, and they argued that it showed that avian status was derived before the advent of theropods.

Scansoriopterids, unlike theropods, which are characterized by a completely open acetabulum, or hip joint, have a partially closed acetabulum, and there is no supra-acetabular shelf as in theropods. Scansoriopterids have a reversed hallux and highly recurved pedal claw that could only be interpreted

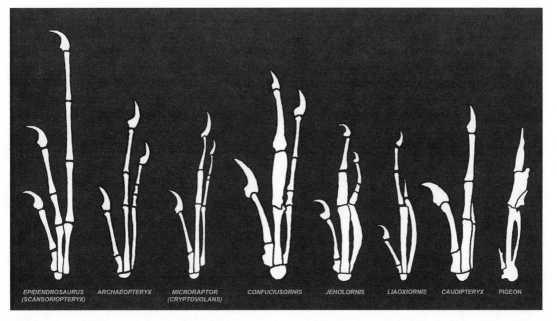

EPIDENDROSAURUS (SCANSORIOPTERYX) — ARCHAEOPTERYX — MICRORAPTOR (CRYPTOVOLANS) — CONFUCIUSORNIS — JEHOLORNIS — LIAOXIORNIS — CAUDIPTERYX — PIGEON

Fig. 4.56. A hypothetical evolution sequence of the evolution of bird hands, going from the scansoriopterid and *Archaeopteryx* hands, specialized for climbing, to hands with reduced finger size, followed by loss of the outer finger (IV). Copyright Stephen A. Czerkas.

as an arboreal adaptation. They have a greater relative arm length than any known theropod, nearly equal to that of *Archaeopteryx*, but with very different proportions. Excluding the elongate outer digit, the total arm length is between that of *Archaeopteryx* and *Deinonychus*. The most diagnostic features are the extraordinary length of digit IV and the relative lengths of the manual proximal phalanges to the penultimate phalanges; the phalanges are progressively shorter distally rather than longer. The progressive reduction of phalangeal length in the manus is primitive (plesiomorphic) within the Archosauria; it is not a character of theropods. As Jacques Gauthier has noted, theropods are united by having the second manual digit, not the outer one, as the longest.[145] In addition, there is little development of any ascending process of the astragalus, and the calcaneum appears to have a peglike articulation with the astragalus. The rodlike clavicles are separate, and the scapula is not like that of theropods, being expanded distally. The nearly closed acetabulum and associated head of the femur, which is offset from the shaft in nontheropod fashion, illustrates that

scansoriopterids had not achieved the fully upright, bipedal stance characteristic of the Dinosauria. In addition, and unlike *Archaeopteryx*, not only is the pubis directed forward but its short length and proportions are suggestive of that of basal dinosauromorphs such as *Marasuchus* (=*Lagosuchus*). The pubis is short, and the ischium is large and lacks a fully perforated hip socket (acetabulum); the ilia are widely set apart, and the pubic peduncle is very small and unexpanded. These are all plesiomorphic characters within the Archosauria, consistent with their classification as basal archosaurians or dinosauromorphs as opposed to true theropods.

More recently, Stephen Czerkas has discovered impressions of elongate feather shafts, set off at angles running across the metacarpals and metatarsals, indicating that scansoriopterids were also "four-winged birds." This finding confirms not only their flight ability but that the tetrapteryx stage was primitive for avians. Czerkas has also discovered impressions of elongate rods running along the rail vertebrae, like those of dromaeosaurs, suggesting that scansoriopterids are basal

Table 4.2. Pretheropod and avian characteristics in scansoriopterids.

Pretheropod characteristics:

- Outer digit of manus longer than other digits (in all theropods the middle manual digit is the longest)
- Outer metacarpal straight, more robust, and longer than the mid-metacarpal
- Phalanges of outer manus digit become progressively shorter distally (as in basal archosaurs)
- Acetabulum shallow and largely closed (fully opened in theropods and dinosaurs in general)
- Supra-acetabular crest absent or with only incipient development as a low rim
- Anteriorly directed pubic bones of short length reminiscent of the lagosuchid archosaurs *(Marasuchus)*
- Pubic peduncle very small and unexpanded
- Distal ends of pubes and ischia not fused distally
- Pubes lack pubic boot
- Ischia longer and more robust than pubis
- Femur without offset head (offset at right angle even in baby theropods); this indicates that scansoriopterids did not have a fully upright limb posture
- Femur lacks a distinct neck or rounded head
- Scapula expanded distally

Avian characteristics:

- Forelimbs long, greater than known theropods, nearly the length of that in *Archaeopteryx*
- Presence of semilunate carpal
- Primary feathers on manus present (as indicated by impressions of several rachides preserved along the metacarpals; note that the total length is unknown, and whether these feathers were fully asymmetrical and pennaceous is unknown)
- Anisodactyl perching foot with reversed hallux
- Tail with short anterior caudal vertebrae followed by elongated posterior vertebrae in *Scansoriopteryx/Epidendrosaurus* (note that the tail is short and resembles a pygostyle in *Epidexipteryx*)
- Elongate tendons overlapping two or more vertebrae in *Scansoriopteryx/Epidendrosaurus* (similar to dromaeosaurs)

to the dromaeosaurs. However, in addition, there is an avian-like semilunate carpal. Certainly the fact that scansoriopterids could spread the hind limbs outward in a splayed posture, more than in typical birds, indicates that a true upright stance was achieved only later and independently from true dinosaurs.[146] In summary, there are no apomorphic features that would unite scansoriopterids with theropods, except for the avian-like precursor to a semilunate carpal, also known in maniraptorans.

Czerkas and Yuan conclude by noting that while the scansoriopterid "represents an arboreal precursor of *Archaeopteryx*, in essence it also represents a 'protomaniraptoran' . . . and it represents 'an arboreal lineage of theropods,' or a 'pre-theropod' lineage of saurischian archosaurs [which the authors favor] which could climb."[147] Even Zhang and colleagues, who described *Epidendrosaurus* provisionally as an arboreal coelurosaur, note, "Phylogenetic analysis has shown that *Epidendrosaurus* is very close to the transition to birds." They do not go so far as to state that this animal is not a theropod, but they do state that the "climbing function in *Epidendrosaurus* was acquired before birds."[148]

Whatever their status, the scansoriopterids closely resemble the proavis from Gerhard Heilmann's 1926 book, and their anisodactyl (avian) perching foot indicates arboreal habits.[149] Given the characters of the manus, combined with the perching foot, these animals were well adapted for climbing. Although *Epidendrosaurus* was discovered from rocks of Late Jurassic age, Czerkas and Yuan also suggest that, based upon skeletal morphology, it would not be surprising to find such creatures from much earlier periods, dating back to the Middle Triassic or even earlier.

Enter Epidexipteryx

Another even more surprising discovery was the announcement in *Nature* of a small but "bizarre" Jurassic maniraptoran from China with elongate ribbonlike feathers named *Epidexipteryx hui*.[150] The type specimen is a feathered, pigeon-sized organism preserved on both slab and counterslab, from Daohugou, Inner Mongolia. Unfortunately, the age of the Daohugou deposits ranges widely

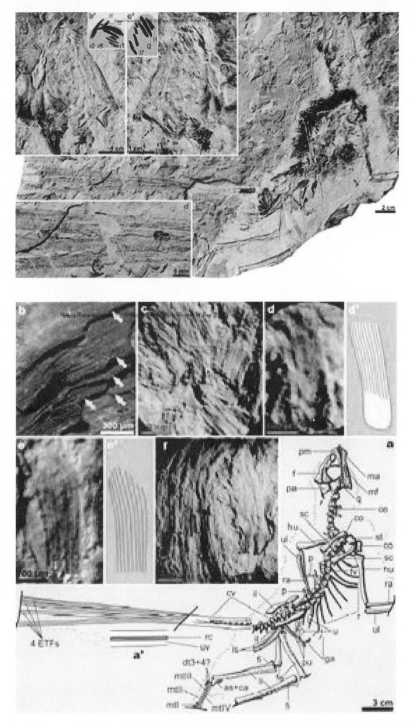

Fig. 4.57. *Epidexipteryx hui*. Upper, photographs of main slab and close-up photos; lower, line drawings and close-up photographs of feathers. (a) main slab and counterslab, (b, c) skull in main slab (b) and counterslab (c); (d) four elongate ribbon-like tail feathers; (b′, c′) line drawings of b and c, respectively, showing teeth. Lower, line drawings and close-up photos of *Epidexipteryx hui*. (a) skeleton and feather outline, showing that each shafted feather is formed by central rachis and two unbranched vanes (a′). (b, d, d′) proximal regions of nonshafted feathers; barbs are parallel and closely united as an unbranched membranous structure (d, d′), vanes are either layered, indicated by arrows (b), or arranged irregularly (c). (f) distal regions of nonshafted feathers, in which barbs appear loosely parallel. From F. Zhang, Z. Zhou, X. Xu, and C. Sullivan, "A Bizarre Maniraptoran from China with Elongate Ribbon-like Feathers," *Nature* 455 (2008): 1105–8. Copyright 2008. Reprinted by permission from Macmillan Publishers Ltd.

from the Middle Jurassic to the Early Cretaceous, although published radiometric dates range from 152 to 168 million years ago, from Middle to Late Jurassic. More work is needed to clarify the exact age of these deposits, with recent studies giving conflicting views ranging from assertions that the Daohugou Formation represents the earliest evolutionary stages of the Jehol Biota and "belongs to the same cycle of volcanism and sedimentation as the Yixian Formation" to calculations that it is of Middle Jurassic age.[151]

The authors describe the small "maniraptoran" (estimated to be 164 grams [5.8 oz.] in body mass) as a basal avialan, characterized by a combination of characters seen in several different theropod groups, particularly the Oviraptorosauria, which is thought by many to be a group of secondarily flightless birds. Their phylogenetic analysis slots *Epidexipteryx* with *Epidendrosaurus*, which together form a monophyletic Scansoriopterygidae, "a bizarre lineage at the base of the Avialae."[152] Interestingly, *Epidendrosaurus* also shows striking similarities with the oviraptorosaurs and, to a lesser degree, therizinosauroids, both very close to, if not of, avian status.

For example, in *Epidexipteryx* the last ten unfused caudal vertebrae form a distally tapering pygostyle-like structure similar to the elongate, incipient pygostyle in some basal birds. By comparison, the oviraptorosaur *Nomingia* has an actual pygostyle that may have supported a feather fan, as in *Caudipteryx*.[153] In 2008, Tao He and colleagues described a new oviraptorosaur, *Similicaudipteryx*, thought to be a caudipterid, with a daggerlike pygostyle formed by the last five caudals, making it the first known caudipterid with an actual pygostyle and indicating that such structures may have been more common within the group.[154] Typically, the pygostyle has been considered an indication of a flighted ancestry, which is the most reasonable explanation for its presence.

Epidexipteryx has strange, enlarged, and procumbent anterior teeth, much larger than the posterior ones. In combination with a short, high skull there is a resemblance to some basal enantiornithines as well as to oviraptorosaurs and basal therizinosauroids, groups that may have avian affinity.

Although the authors describe two types of feathers in the specimen, elongate tail feathers with undifferentiated vanes and nonshafted body feathers, the body feathers appear as filamentous parallel fibers arising from a putative membranous structure, and it is difficult to be certain that these features represent a type of feather. The authors argue that the presence of the ribbonlike tail feathers, combined with the absence of pennaceous

Fig. 4.58. Above, *Jinfengopteryx*, one of the most beautiful fossils recovered from the Chinese fossil gold rush, is of uncertain age but of great interest in that it sports feathers. The small, birdlike *Jinfengopteryx* is now thought to be a troodontid instead of an archaeopterygid, as originally described. It has an *Archaeopteryx*-like tail and unserrated avian-like teeth. Below, life reconstruction of *Jinfengopteryx* by Matt Martyniuk, from Wikimedia Commons, reproduced under the terms of the *GNU Free Documentation License*, Version 1.2. Photograph of specimen on display in Hong Kong Science Museum, 2007, from Wikimedia Commons, by Laikayiu, reproduced under the GNU Free Documentation License, Version 1.2. (Images not intended to support any concept presented in this text.)

wing feathers, indicate that *Epidexipteryx* was incapable of flight, and they may well be correct, but one should be cautious in making such assumptions. The well-preserved enantiornithine *Longipteryx*, in which feathers but not flight remiges were preserved, nevertheless could clearly fly, as shown by other similar fossils in which down, contour feathers, and wing flight and tail feathers were preserved. The authors argue that the presence in *Epidexipteryx* of long, ribbonlike rectrices without any apparently preserved flight feathers supports an early display function in the origin of feathers, but such a conclusion is premature, given current knowledge. They conclude that "*Epidexipteryx* is the oldest and most phylogenetically basal theropod known to possess display feathers, indicating that basal avialans experimented with integumentary ornament as early as the Middle to Late Jurassic."[155] However, given its sister group relationship to scansoriopterids that had flight feathers, and the fact that lack of preserved flight feathers does not always indicate their absence (for example, *Longipteryx chaoyangensis*), it remains speculative whether this basal avian was flightless.

The Birdlike Troodontids

Among the most birdlike maniraptorans is the clade of small, lightly built troodontids ("wounding tooth"), first named for the species *Troodon formosus* (3 m, 6 ft.; 50 kg, 110 lbs.), described by Joseph Leidy in 1856, from the Late Cretaceous of North America, although the family name was first applied in 1924 to a group of ornithischian dinosaurs. Known primarily from the Cretaceous of Asia and North America, these gracile animals are noted for their similarities to dromaeosaurs and birds in numerous features, including an enlarged, sickle-shaped claw on the second pedal digit and, especially, their exceptionally high encephalization quotients, which are comparable to those of living ratites, their keen sense of vision, and their well-developed middle ear. Their large eyes are oriented in a forward position to provide stereoscopic, binocular vision. Troodontids are distinctive in normally having a high number of maxillary and dentary teeth characterized by a distinct constriction between the tooth and crown,

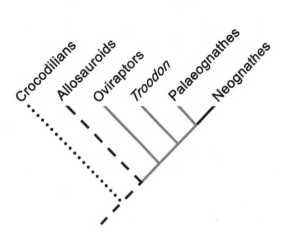

Fig. 4.59. Hypothetical cladogram showing the distribution of different methods of parental care. From left to right: crocodiles have maternal care; allosaurids are unknown or equivocal; oviraptorosaurids, troodontids, and palaeognaths have paternal care; and modern, neognathous birds typically have biparental care. Although the authors interpreted the evidence from troodontids (preserved males at nests) as indicating that paleognathous birds inherited their "ancestral" paternal care, it is equally likely that it is independently derived among paleognaths (secondarily flightless birds) and troodontids and oviraptorosaurs, which may also be secondarily flightless. Adapted and modified from D. J. Varricchio et al., "Avian Parental Care Had Dinosaur Origin," *Science* 322 (2008): 1826–28.

with coarse serrations, often only on the posterior carina (the teeth of *Byronosaurus* are devoid of serrations). Some teeth show the resorption pits characteristic of Mesozoic bird teeth, and like ratites, troodontids exhibit a tall ascending process on the astragalus. The characteristic troodontid skull ends in a U-shaped snout, narrow at the end.

Although poorly known for much of the twentieth century, troodontid fossils have been recovered recently with great regularity, including specimens in association with eggs and embryos, feathers, and juveniles. Today, some eight taxa are known from the Cretaceous of the Gobi Desert, including the well-known *Saurornithoides* and *Sinornithoides*, as well as the more recently discovered *Byronosaurus* and *Sinovenator*.[156] Much excitement has been generated by the discovery of two fully feathered troodontids in China. The basal troodontid *Sinovenator* ("Chinese hunter") is quite birdlike and has many similarities to the dromaeosaurs. Like most of the other Chinese Early Cretaceous theropods, it was a chicken-sized animal measuring less than 1 meter (3 ft.) in length, with a very birdlike braincase. It also shares numerous features with primitive dromaeosaurs and avialans, illustrating again a nexus of these maniraptorans and birds.[157] Among the most remarkable troodontid finds is *Mei* (Chinese "soundly sleeping"), a duck-sized animal from the Early Cretaceous of Liaoning discovered in 2004. With a snout nestled beneath one of the forelimbs, it was interpreted to have been preserved by instant burial in volcanic ash in an avian roosting posture.[158]

The most interesting work on this group has come from David Varricchio of Montana State University, who has extensively studied troodontid nests, eggs, and embryos. He and his coworkers based their analyses on the repeated discovery of "dinosaur" adults in close association with nests and egg clutches, particularly the classic troodontid *Troodon* (discovered on a clutch from the Upper Cretaceous of Montana), as well as the closely allied oviraptorosaurids *Oviraptor* and *Citipati*, the latter two from the Late Cretaceous Gobi Desert discovered sitting atop their egg clutches. They attempted to assess parental care by examining clutch volume and the bone histology of the brooding adults, discovering that the large clutch volumes of *Troodon*, *Oviraptor*, and *Citipati* scale most closely with a paternal bird care model. In addition, they discovered using bone histology that the adults fossilized at the nests lacked medullary bone, a feature of female birds, which lay down a layer of this spongy bone within the long bones during the reproductive period. Large egg clutch volumes can evolve in avian species without maternal care because females that provide no maternal care can devote more resources to eggs, and a "clutch" in such species is typically composed of eggs from multiple females (in the case of the ostrich, eggs are contributed by an average of three birds). Most living birds lay small clutches with small egg volume, and biparental care, with males and females sharing incubation duties and care of young, is the general rule. In contrast, the

paleognaths, along with the flightless ratites and the closely allied but volant tinamous, thought to be primitive within living birds, exhibit strictly paternal care, in which males incubate the eggs and care for the young alone.

The team of paleontologists concluded that the large clutch volumes in these Cretaceous maniraptorans, combined with a suite of avian reproductive features, favored an interpretation of paternal care, possibly within a polygamous mating system. "Paternal care in both troodontids and oviraptorids indicates that this care system evolved before the emergence of birds and represents birds' ancestral condition."[159]

Taking a different perspective, biologist Lee Kavanau of the University of California at Los Angeles notes that because ratites are without exception secondarily flightless birds and tinamous are reluctant, clumsy fliers, the evidence from the new dinosaur analyses strengthens the view that troodontids and oviraptorosaurids are secondarily flightless birds, as argued by any number of other scientists. He points out that although secondary flightlessness may favor paternal care with clutches of large egg volume, such care is unlikely to have been primitive and that there is other evidence pointing to the evolution, first of maternal, followed by biparental care in the earliest avian ancestors.[160] It should also be noted that although the Varricchio team did an outstanding job of uncovering these remarkable finds and analyzing their data, the fossils represent evidence only from the Late Cretaceous, toward the end of the reign of dinosaurs.

Jinfengopteryx

With the recent description of the chicken-sized, feathered troodontid *Jinfengopteryx elegans* ("elegant golden phoenix feather"), the question of the avian versus theropod status of various Chinese fossils has plunged the birdlike maniraptorans, the troodontids, into the heart of the discussion. *Jinfengopteryx* was first described in 2005 as a new genus of archaeopterygid and the most basal avialan. Luis Chiappe, however, pointed out that the fossil appeared to have more in common with troodontids, including an enlarged claw on the second toe, and *Jinfengopteryx* was later cladistically slotted with the birdlike troodontids, showing affinity with genera such as *Sinovenator*, *Byronosaurus*, and *Mei*.[161] Yet the tenuous cladistic allocation may well belie the uncertainty of its phylogenetic status, and it is most likely avian. Turner and colleagues pointed out that *Jinfengopteryx*, if correctly allocated, would be the first troodontid to preserve evidence of feathers.[162] A small maniraptoran, 55 centimeters (2 ft.) long, it was discovered in the Qiaotou Formation of Heibei Province, China, and is of uncertain age, but possibly from the Early Cretaceous. Known from a single nearly complete and articulated specimen, it sports impressions of pennaceous

Fig. 4.60. The red-legged seriema (*Cariama cristata*). Seriemas are long-legged, largely cursorial gruiform birds confined to southern South America. The two monotypic genera of the family Cariamidae represent relics of the ancient gruiform radiation that produced the flightless carnivorous birds the phorusrhacids, as well as the bathornithids and idiornithids. Seriemas prefer to remain on the ground and are formidable runners. When alarmed, however, they can fly, and their overall proportions are not much different from a number of the Lower Cretaceous Chinese fossils, such as *Protarchaeopteryx*, usually interpreted as flightless. Drawing by George Miksch Sutton, by permission of Dr. Dorothy S. Fuller.

feathers, but unlike related forms such as *Pedopenna* ("foot feather"), it lacks flight feathers on the hind legs.[163] If a true troodontid, it would be the first preserved with avian feathers, and it therefore takes on added importance, especially because in general troodontids exhibit cursorial adaptations and their wings are shorter than those of *Archaeopteryx*. Although the original describers suggest that *Jinfengopteryx* supports the cursorial origin of avian flight, it could just as well illustrate a form that was becoming secondarily flightless, since it exhibits a semilunate carpal element and an *Archaeopteryx*-like tail, and as noted earlier, there are many volant birds with reduced wings. Troodontids have long been recognized to be extremely birdlike in many morphological features and have even been considered secondarily flightless birds.

Most interestingly, however, troodontids have birdlike teeth, and they have recently been shown to exhibit the same type of tooth replacement as seen in early avians.[164] In fact, the only dichotomy between Mesozoic bird teeth and their vertical tooth replacement disappears when combined with the dromaeosaurids, troodontids, therizinosauroids, and alvarezsaurids. Many have unserrated or partially serrated teeth, constrictions between the crowns (waisted), expanded roots,

Fig. 4.61. Foot of the red-legged seriema, showing the enlarged claw of digit II reminiscent of the equivalent toe in microraptors. Although little is known about the life-history of these intriguing birds, there is some field evidence that the sickle claw performs a dual function in climbing and predation. Photograph by Matt Edmonds, from captive bird. Licensed under the GNU Free Documentation License, Version 1.2.

and tooth replacement pits closed at their bases, as well as vertical tooth replacement. In fact, the dromaeosaurids *Sinornithosaurus* and *Microraptor* have teeth devoid of serrations; but vertically erupting teeth have been described only in troodontids. When one combines tooth characters with features of the hand, feathers, and other characters, it becomes apparent that dromaeosaurids, troodontids, and Mesozoic birds especially are so closely allied that whatever microraptors are, so too are birds.

Anchiornis

The most recent fossil described from the treasure trove of Chinese Mesozoic fossils is *Anchiornis huxleyi* ("nearby bird," alluding to its being very closely related to birds), from the Middle-Late Jurassic Tiaojishan Formation of Liaoning, China, some 160 to 155 million years ago. The authors describe *Anchiornis* as the "smallest known non-avian theropod dinosaur," but their analysis recovered "a monophyletic Avialae with *Anchiornis* as the most basal avialan."[165] Feathers, though present on the original specimen, were too poorly preserved to provide critical information. However, other specimens show well-developed feathering, with avian wing remiges, asymmetric vanes, and hind-limb wings, rendering life on the ground, as originally suggested, most unlikely. Regrettably, *Anchiornis* is referred to today as a feathered dinosaur, though it is certainly a basal archaeopterygid and, with its avian wings, may have been an apt glider, somewhat like the New Zealand kakapo (*Strigops*), in which the flight feathers have much less vane asymmetry and fewer interlocking barbules distally than those of volant birds. Kakapos are weak fliers that climb up trees and glide to the ground. They are capable of significant descending glides during which they occasionally flap their wings, but they cannot attain significant-level flight.[166] On the ground, however, kakapos are not hindered by extensive hind-limb wings as seen in basal birds. *Anchiornis* probably spent its days clambering about branches in search of food, gliding occasionally to other trees and rarely to the ground, where it would rapidly ascend a tree trunk to avoid predation.

Estimated at a body length of 34 centimeters (13.3 in.) and about 110 grams (3.9 oz.) in body mass, this animal had a forelimb about 80 percent as long as the hind limb and was somewhat smaller than some basal birds, including *Archaeopteryx*. Certainly the length of the forelimbs is concordant with that of some basal birds, as is the presence of a semilunate carpal element, so *Anchiornis* could have well been volant. Among the more interesting features are the femur and sacral region. Most interestingly, the acetabulum, like that of scansoriopterids, is not fully open as in theropods, and it exhibits a "peripheral medial wall." Although the authors describe a "distinct supra-acetabular crest," it appears more as a slight rim and is certainly not like any typical theropods. In all, *Anchiornis* is more derived than scansoriopterids but less derived than *Archaeopteryx*, making it in all probability an intermediate between the true avian ancestors and the classic urvogel. Unfortunately, the fossil is from western Liaoning, China, from lacustrine deposits of "uncertain Jurassic-Cretaceous age" but new information indicates that the age is near the Middle-Late Jurassic boundary (Tiaojishan and Daohugou Formations, which probably have the same fauna), making the find the only fauna preserving feathers that predates *Archaeopteryx*.[167] Also, new finds show that the feathering of *Anchiornis* includes profusely feathered feet, illustrating also that it could not have been a terrestrial animal, as originally proposed. Xing Xu considers *Anchiornis* to be a basal troodontid, but in terms of its morphology there is little reason to regard *Anchiornis* as a nonavian theropod ("small maniraptoran dinosaur") rather than a basal bird as originally described.

More recently, in a paper entitled "Plumage Color Patterns of an Extinct Dinosaur," Quanguo Li and colleagues analyzed *Anchiornis* melanosomes. They concluded that the body was gray and dark, the face had rufous speckles, and the long limb feathers were white with distal black spangles. Despite the ensuing publicity this study received in the popular press, the results are premature, as there has yet to be a survey of melanosomes across a broad spectrum of living birds. Whatever the case, the term "dinosaur" as applied

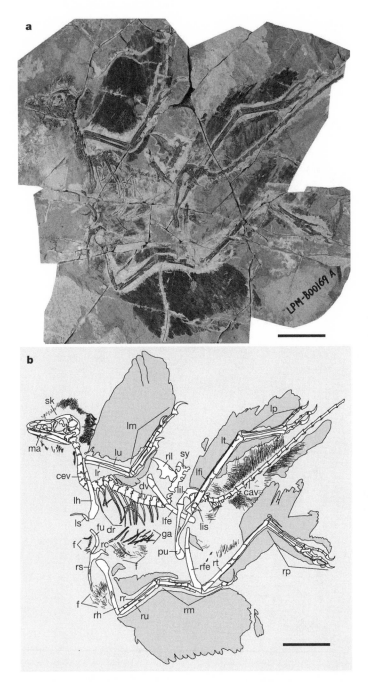

Fig. 4.62. Photograph (a) and line drawing (b) of *Anchiornis huxleyi* LPM-B00169 A (slab). Abbreviations: cav, caudal vertebra; cev, cervical vertebra; dr, dorsal rib; dv, dorsal vertebra; f, feather; fu, furcula; ga, gastralia; lfe, left femur; lfi, left fibula; lh, left humerus; lil, left ilium; lis, left ischium; lm, left manus; lp, left pes; lr, left radius; ls, left scapula; lt, left tibiotarsus; lu, left ulna; ma, mandible; pu, pubis; rc, right coracoid; rfe, right femur; rh, right humerus; ril, right ilium; rm, right manus; rp, right pes; rr, right radius; rs, right scapula; rt, right tibiotarsus; ru, right ulna; sk, skull; sy, synsacrum. Scale bar = 5 cm. Courtesy L.-H. Hou. From X. Xu et al., "A Pre-*Archaeopteryx* Troodontid Theropod from China with Long Feathers on the Metatarsus," *Nature* 461 (2009): 640–43. Copyright 2009. Reprinted by permission from Macmillan Publishers Ltd.

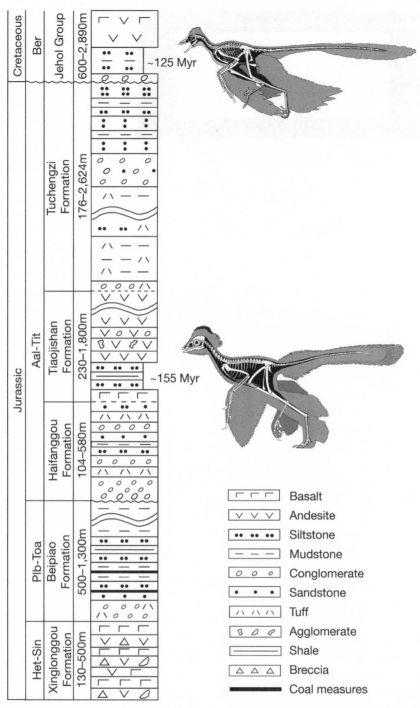

Fig. 4.63. Stratigraphic column of Jurassic and Lowest Cretaceous strata in western Liaoning, showing horizons from which feathered dinosaurs have been described. Two major horizons have produced such specimens. The Tiaojishan Formation has yielded *Anchiornis huxleyi* and dates to about 155 million years ago, whereas the Jehol Group has yielded *Microrap-* *tor* and other feathered dinosaurs and dates to about 125 million years ago. Courtesy L.-H. Hou. From X. Xu et al., "A Pre-*Archaeopteryx* Troodontid Theropod from China with Long Feathers on the Metatarsus," *Nature* 461 (2009): 640–43. Copyright 2009. Reprinted by permission from Macmillan Publishers Ltd.

to *Anchiornis* was hasty and misleading; most likely, the team described putative but still questionable color in a basal bird.[168]

Given that the origin of and interrelationships within the troodontids are poorly understood and that a large suite of avian characters are present, especially in *Anchiornis*, we must remain open to the view that they represent a primitive group of early avians.

The Nontheropod Sacral Region

A major question concerning the theropod status of many of these fossils, including the microraptors, scansoriopterids, *Epidexipteryx*, and *Anchiornis*, attends their nontheropod morphology in many aspects, but particularly of the sacral region and femur. Functional morphologist Fritz Hertel of California State University at Northridge and paleornithologist Kenneth Campbell of the Natural History Museum of Los Angeles County performed an extensive morphological and functional analysis of the sacral region and antitrochanter in birds, concluding: "The antitrochanter is a uniquely avian osteological feature of the pelvis that is located lateral to the postero-dorsal rim of the acetabulum. This feature makes the avian hip joint unique among all vertebrates, living and fossil, in that a significant portion of the femoral-pelvic articulation is located outside of the acetabulum."[169] Stated simply, the antitrochanter serves as a brace to prevent abduction or movement of the hind limb away from the main plane of the body. The antitrochanter-femur articulation is a "drum-in-trough-like form, and is a derived feature of birds that probably evolved as an aid in maintaining balance during locomotion." They conclude that the avian antitrochanter is unrelated to the so-called antitrochanter of dinosaurs, which probably serves as an area of massive muscle attachment, as opposed to an articular surface. Their observations confirm Alick Walker's conclusion that dinosaurs "have no true antitrochanter as is present in the bird."[170] And as Larry Martin and Stephen Czerkas point out, "*Archaeopteryx* lacks an antitrochanter in the acetabulum of the pelvis. In fact, there is no articular surface at the back of the acetabulum, nor a supra-acetabular shelf char-

acteristic of dinosaurs.[171] Hertel and Campbell show that not only is the antitrochanter uniquely avian but it evolved after the appearance of the earliest birds and that it was not present either in *Archaeopteryx* (confirming Martin and Czerkas's observation) or in the Early Cretaceous Chinese bird *Caudipteryx*, even though *Caudipteryx* was cursorial. Such observations coincide with those of Matthew Carrano of the Smithsonian Institution, who concludes that there was little similarity in hind-limb locomotion between theropods and birds.[172] As Hertel and Campbell state: "Because an antitrochanter does not occur in the oldest known birds, it must be assumed that it evolved subsequent to the establishment of the avian lineage. An antitrochanter appears in the Mesozoic Enantiornithes, and the presence of an antitrochanter in the late Cretaceous birds *Hesperornis* and *Ichthyornis*, and particularly the highly derived state of the antitrochanter in *Hesperornis*, argues for a fairly early appearance of this osteological feature within the avian lineage."[173]

Any conclusions regarding flight origins, however, cannot be based on what one discovers in a particular fossil or whether the wings are not strictly as long as those of *Archaeopteryx*. We are still far from anything resembling the actual avian ancestor, in terms of both geologic time and morphology. Whatever their systematic status, in terms of flight capability, most of the Early Cretaceous Chinese basal birds are generally compared with modern fully feathered wings. Yet there are other components to the production of lift, notably the forgotten propatagium (also prepatagium), which may have played a significant role in flight origins in archosaurs. Richard Brown and Allen Cogley used flight experiments with live birds and computer modeling to determine the aerodynamic contributions of the propatagium in avian flight. In house sparrows (*Passer*), they found that removal of the secondary feathers while leaving the six distal primaries and an intact propatagium had no noticeable effect on flight performance, and they discovered, surprisingly, that "the cambered propatagium is the major lift generating component of the wing proximal to the wrist."[174] The propatagium may thus have played an early

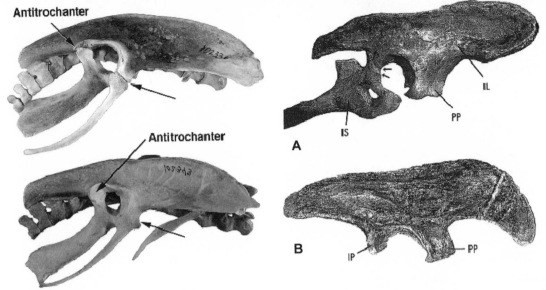

Fig. 4.64. Left, maturation of the pelvis in the brown kiwi (*Apteryx australis*); upper, fledgling pelvis showing the pelvic bones still unfused (arrows); lower, adult pelvis with complete fusion of all pelvic bones. Note the contribution of the ilium and ischium to the formation of the antitrochanter. Right, the ilium and ischium of (A) *Archaeopteryx* and the ilium of (B) *Caudipteryx* do not have any structures that could be interpreted as an antitrochanter (also absent in microraptors and scansoriopterids), providing further evidence that the avian antitrochanter evolved after the avian lineage was established. In the London specimen of *Archaeopteryx* (A), there was displacement of the ischium from the ilium, the amount of displacement being indicated here by the two arrows lying within the acetabulum. The top arrow indicates the ventral edge of the ischiadic peduncle of the ilium; the lower arrow indicates the line along which this edge would attach in life. Abbreviations: IL, ilium; IP, ischiadic peduncle; IS, ischium; PP, pubic peduncle. Ilia not to scale. Anterior to right. From F. Hertel and K. E. Campbell, Jr., "The Antitrochanter of Birds: Form and Function in Balance," *Auk* 124 (2007): 789–805. Courtesy of the editor of *Auk*.

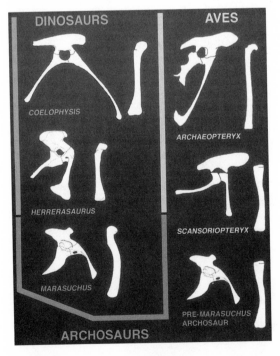

Fig. 4.65. The separate lineages of true theropods and birds are illustrated by the sacrum and femur, suggesting that each group evolved separately from basal archosaurs. In the scansoriopterids the structure of the femur is unlike that of dinosaurs in not having an offset head of the femur, a condition suggestive of a pre-*Marasuchus* (dinosauromorph) stage. In addition, the sacrum in scansoriopterids more closely resembles that of archosaurs than that of theropod dinosaurs. Reprinted with permission. Copyright Stephen A. Czerkas.

significant role in the evolution of flight and should be considered in addition to the preserved feathered avian wing.

In an important finding, Larry Martin and Jong-Deock Lim of South Korea, while studying the wing feathers of the famous Berlin *Archaeopteryx*, noticed that the hand and arms were surrounded by a natural depression. Their cast of the region revealed an essentially avian hand.[175] The outer and middle fingers were united nearly to the claw, and a posterior patagium extended along the margin of the hand and arm, a structure previously postulated to exist by Heilmann in 1926.[176] There was also an indication of a propatagium, which in modern birds constrains the extension of the forearm, so that the junction between the humerus and the ulna usually reflects the pattern of wing folding in fossil birds. In addition, as noted, the patagia provide a substantial lift component. They concluded that the urvogel's hand was essentially modern in conformation, lacking any significant grasping capabilities and therefore of little use for capturing prey but used primarily for climbing.

In summary, aside from the very birdlike troodontids, which may well be secondarily flightless birds (some may have been volant), given the phylogenetic slotting of small arboreal archosaurs between the clade of Troodontidae and Dromaeosauridae and archaeopterygids, the Scansoriopterygidae and its possible allies could just as easily be basal birds as anything else, not far removed from the basal dinosauromorph level of evolutionary development. Whatever they turn out to represent, they are devoid of any definitive theropod characters, and they vividly illustrate that the true avian ancestors should be sought in deposits of earlier geologic time.

A Jurassic Ceratosaur Clarifies Avian Digital Homologies?

Classifications vary, but ceratosaurian theropods are generally thought to be nested in the same group, the Tetanurae ("stiff tails") (including Maniraptora), within which birds are also nested, so the discovery of a Middle-Late Jurassic small, new ceratosaur from the Shishugou Formation of western China (approximately 156–161 million years ago) has been the cause of considerable excitement.[177] Named *Limusaurus inextricabilis*, the specimen represents the only known beaked, herbivorous Jurassic theropod, a small creature sharing cranial features with both coelophysids and other ceratosaurs, and also exhibiting such unusual features as a fully developed rhamphotheca, highly abbreviated forelimbs with very short hands, and a cursorially adapted, elongate hind limb.[178] What makes it of such great interest, however, is its basal phylogenetic position. Recall that the first dinosaur to be named was the English *Megalosaurus bucklandii*, considered now a basal tetanuran, and was the dinosaur that had impressed Thomas Huxley with its chickenlike hind limbs. Not only is the new find among the earliest known ceratosaurs, but its theoretical phylogenetic proximity to birds and apparent preservation of a full complement of digital elements makes it of considerable interest. Intriguingly, *Limusaurus*, ornithomimosaurs, and shuvosaurid suchians, such as the crocodylomorph *Effigia*, share remarkable similarity in having small heads with large orbits, edentulous jaws, long necks, and elongated hind limbs, thus providing a stunning example of convergence evolution among three higher archosaurian groups.

Given such a high degree of homoplasy, it would seem inappropriate to draw any overarching conclusions from the anatomy of such a specialized theropod, but the paper was published by *Nature*, and *Nature News* was quick to pounce on the discovery with the title "Dinosaur's Digits Show How Birds Got Wings."[179] While some paleontologists embraced the new find as having solved the age-old problem of the homologies of the avian and dinosaurian digits, many were far more skeptical, noting that the new aberrant ceratosaur may not be a major role player in a larger evolutionary picture and was simply a highly adapted, bizarre theropod enjoying an unusual lifestyle. Following this logic, Kevin Padian said, "It is equally reasonable that we are just dealing with another odd possibility of evolution." Yale's Günter Wagner, however, stated, in line with his "digital frame-shift" hypothesis (explained in appendix 2), that "the certosaur fossil may be

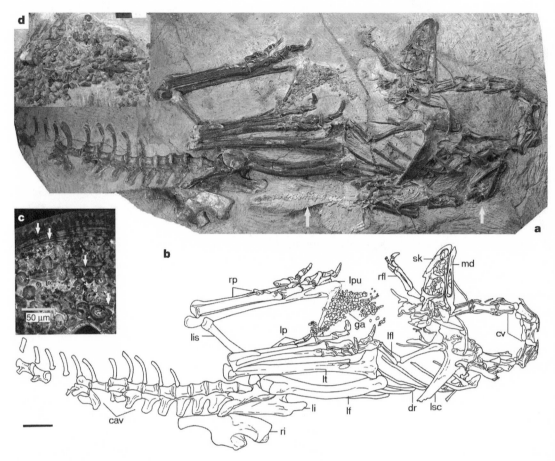

Fig. 4.66. The small herbivorous ceratosaur *Limusaurus*. (a) Photograph and (b) line drawing, scale bar = 5 cm; (c) histological section from fibular shaft (showing age to be approx. 5 years); (d) close-up of gastroliths, scale bar = 2 cm. Abbreviations: cav, caudal vertebrae; cv, cervical vertebrae; dr, dorsal ribs; ga, gastroliths; lf, left femur; lfl, left forelimb; li, left ilium; lis, left ischium; lp, left pes; lpu, left pubis; lsc, left scapulocoracoid; lt, left tibiotarsus. From X. Xu et al., "A Jurassic Certosaur from China Helps Clarify Avian Digital Homologies," *Nature* 459 (2009): 940–44. Copyright 2009. Reprinted by permission from Macmillan Publishers Ltd.

showing us a species in the midst of a digit identity shift, but whether the digits that we see in later theropods are the actual second, third and fourth digits or the first, second and third digits or the first, second, and third digits in the second, third and fourth positions, altered by gene bombardment to look like the second, third and fourth digits, is difficult to determine."[180] Obviously, there is no dearth of opinions on the evolutionary significance of *Limusaurus*.

One might wonder how a highly specialized, aberrant ceratosaur could possibly tell us much about bird digital homology. However, *Limusaurus* does slot out close to the putative position of birds in phylogenetic analyses, and the age is Jurassic. It is widely agreed that theropod dinosaurs possess a grasping, raking hand reduced to three digits identified as digits I, II, and III. This scheme has been accepted by paleontologists based on nicely preserved hands from Late Triassic and Early Jurassic forms ranging from *Herrerasaurus* to *Dilophosaurus*, which exhibit three elongated fingers with a semivestigial fourth finger and a fifth finger that is nearly absent. Given this observation, paleontologists have traditionally claimed that birds had to possess a hand with digits I, II, and III because they are thought to be living dinosaurs, and the phalangeal formula of 2-3-4 in the Late Jurassic *Archaeopteryx* has the same formula. Yet despite conflicting paleontological and new genetic evidence for both a I-II-III and a II-III-IV avian hand, nearly all developmental studies show birds to have a tridactyl hand comprising digits II, II, IV. Because of the genetic evidence on both sides, resolving this discrepancy continues to be an issue of interest in studies of bird origins. The importance of phalangeal formulae may have been promoted excessively, as phalangeal counts can vary even within species, and "digits I-III do not display a 2-3-4 phalangeal formula in any known ceratosaur, demonstrating that the conservatism of this formula is not absolute."[181]

To put it another way, theropods display a pattern of lateral digital reduction, in which digits have been reduced progressively from the outside inward (lateral digital reduction, LDR, or postaxial reduction), as an almost unique occurrence in tetrapods. In contrast, most tetrapods that exhibit digital reduction have bilateral digital reduction—that is, symmetrical reduction from both sides inward (bilateral digital reduction, BDR)—a reduction pattern that can be duplicated experimentally. Therefore, the discovery of bilateral digital reduction in *Limusaurus*, if correct, would indicate that the avian pattern might be more widely distributed among early theropods and that the ancestral avian theropod may well have had a hand with digits II, III, IV. The authors concluded that "if birds possess digits II-III-IV as most developmental studies indicate, the data strongly support the interpretation that all tetanurans have digits II-III-IV (-V)."[182] A similar conclusion was reached years before based on the developmental evidence of the architecture of the bird hand, combined with the morphological similarity of the avian hand to maniraptoran theropods.[183]

However, one major problem attends the argument on the hand of *Limusaurus* and basal theropods—namely, that the dinosaur hand (Saurischia and Ornithischia) all demonstrate manual lateral digital reduction, in the manus combined

with bilateral digital reduction in the pes, so obviously the situation is far more complicated than has been portrayed.[184]

Discussion of the problem of avian digital homology is a bit of a peripheral topic and an interruption thrown in the middle of a chapter on Chinese birds, but it is relevant to the topic of bird origins, and is considered of sufficient importance to include an expanded discussion of the topic as appendix 2.

Dinosaurs with Feathers?

> You can't depend on your eyes when your imagination is out of focus.
> *Mark Twain*, A Connecticut Yankee in King Arthur's Court, *1889*

Chinese paleontologist Xing Xu noted that, the year following the *Sinosauropteryx* discovery,

> two other feathered dinosaurs were found from the same area. One was named *Protarchaeopteryx robusta* and the other *Caudipteryx zoui*. . . . *Caudipteryx* has short arms like *Compsognathus* but other features suggest it is an oviraptorosaurian dinosaur. . . . This was the first time in history that feathers had been discovered on non-avian animals—living or extinct. . . . In 1999, two other feathered dinosaurs were again reported from western Liaoning. *Beipiaosaurus inexpectus* . . . is the biggest theropod yet discovered from Liaoning. It is more than 2 meters long. . . . *Sinornithosaurus* . . . the second species, is a close relative of *Velociraptor* but much smaller. . . . Actually *Sinornithosaurus* represents one of the most birdlike dinosaurs, and is more closely related to birds than the other species mentioned. . . . The third theropod reported in 2002 was a small arboreal dinosaur, *Epidendrosaurus*. . . . *Epidendrosaurus* has some features previously unknown in non-avian dinosaurs, including a fully reversed hallux. . . . A new feathered dinosaur *Microraptor gui* . . . was described early in 2003. . . . Surprisingly it has long pennaceous feathers not only on its forelimbs and tail, but also on its hind limbs (the so-called "four-winged dinosaur"). Furthermore, the

feathers were almost identical to those of living birds, with asymmetrical vanes, a feature associated with flight or gliding in extant birds. It is very likely that *Microraptor gui* is a gliding animal, representing an intermediate stage between flightless dinosaurs and the volant birds.[185]

The conflation was in full swing!

In 1998 I received my last letter from John Ostrom before his death in 2005; he was in poor health and under care in his last years. We had corresponded for many years, and John kindly invited me to meet him in Washington, DC, at the headquarters of the National Geographic Society, to view with him the newly displayed Chinese fossils, including the first truly "feathered dinosaur"

(*Caudipteryx*) and *Sinosauropteryx*, with its putative protofeathers. Regrettably, I had to decline his offer because of prior commitments. Soon after, a BBC crew contacted me about meeting in Chapel Hill, where I was chair of the Department of Biology at the University of North Carolina, for a filmed interview. They sought my opinion of a forthcoming cover article in *Nature* on these first "feathered non-avian dinosaurs." After carefully reading the article, I could only conclude that the taxa being described as feathered theropods were in fact secondarily flightless birds—Mesozoic kiwis, as it were—and this tentative conclusion framed the theme of my interview. At the time, I had not examined the specimens, and my view was thus somewhat speculative. Since then, having studied the specimens a number of times, I have seen nothing

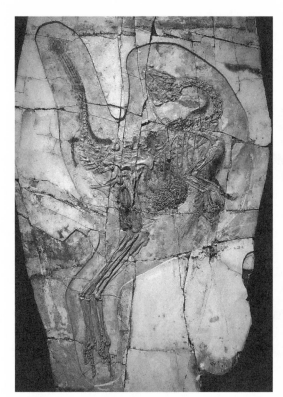
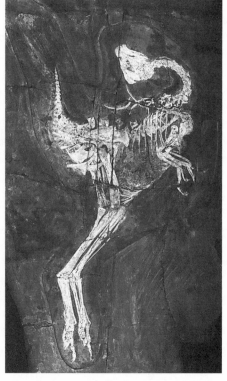

Fig. 4.67. The turkey-size flightless bird ("Mesozoic kiwi") *Caudipteryx*, an oviraptorosaur originally described as the first dinosaur with true avian feathers in a cover article in *Nature* in 1998. It has a rather typical, deep bird skull with birdlike teeth, a ventrally located foramen magnum, a short tail with a proto-pygostyle, a reduced femur, an avian foot, and many other birdlike characters. It had a terminal fanlike tail, typical also of other basal birds such as *Jeholornis*; note the mass of gizzard stones. Left photograph and right ultraviolet light image both courtesy of Stephen Czerkas. Copyright Stephen A. Czerkas.

to contravene that view. Interestingly, before I examined the specimens, I had a note from a colleague at the Smithsonian who had made the trek over to see the highly exalted specimens, and had the following report, which I paraphrase here: "Well, we went through the hall, and there was *Sinosauropteryx*, an obvious theropod dinosaur, and there was *Caudipteryx*, an obvious bird." Such was the sentiment of most ornithologists who viewed the specimens. At that time, it was claimed that *Sinosauropteryx* had protofeather filaments over the body, but the body outline clearly encircled the filaments over the back and provided a nice outline of the tail. The dino-fuzz filaments, which resulted in the small compsognathid being depicted as a downy covered creature, were incorrect, and the structures were internal within the integument and not epidermal integumental derivatives, and would later be shown to be supportive collagen fiber meshworks. It seems the only people who were having problems with the identifications were those who were trying to solidify a story of avian descent from ancestors that flourished in the Early Cretaceous.

Caudipteryx and *Protarchaeopteryx* were described in *Nature* in 1998 by Qiang Ji, Philip Currie, Mark Norell, and Shu-An Ji.[186] The findings were popularized by the press, and as noted in chapter 1, *Nature*'s editor Henry Gee proclaimed, "Birds are dinosaurs—the debate is over."[187] In the article the authors stated, "Although both theropods have feathers, it is likely that neither was able to fly. . . . These new fossils represent stages in the evolution of birds from feathered, ground-living, bipedal dinosaurs."[188] They then proclaimed that feathers were irrelevant in the diagnosis of birds and that the presence of feathers on flightless theropods indicated that the hypothesis that feathers and flight evolved together was incorrect. Gee would go on to write that "these creatures effectively close the debate on whether or not birds and dinosaurs share a close evolutionary heritage. The answer is a resounding 'yes.'" He further noted, "These fossils make it clear that feathers appeared in evolution long before flight."[189] Yet the debate is not, and never has been, about "whether or not birds and dinosaurs share a close evolutionary heritage," since both sides of the debate have almost always

agreed on that issue. The real question is whether Aves is nested within Theropoda and whether flight evolved from earthbound dinosaurs. Of course, the real problem was that the authors had actually described two secondarily flightless birds.

Following Gee, Lawrence Witmer proclaimed that "the presence of unambiguous feathers in an unambiguously nonavian theropod has the rhetorical impact of an atomic bomb, rendering any doubt about the theropod relationships of birds ludicrous."[190] This may well be another example of the phenomenon that I characterize by the phrase "the weaker the field of science, the more heated the rhetoric." First, as we shall see, these "feathered dinosaurs" are secondarily flightless birds. Second, *Protarchaeopteryx* likely could fly at least to some degree, as its wing proportions closely resemble those of a number of volant birds, especially the South American seriemas.

Ji and colleagues' "unambiguous evidence supporting the theory that birds are the direct descendants of theropod dinosaurs" was immediately challenged by a number of papers. These articles presented evidence that the oviraptorosaurs, including *Caudipteryx*, were secondarily flightless birds and debated the posture and morphometrics presented in Ji and colleagues' work.[191] Most prior phylogenetic analyses of this Cretaceous group placed oviraptorosaurs outside of Aves, recognizing as the sister taxa the Dromaeosauridae or Troodontidae, or a clade Deinonychosauria, which contains both. Interestingly, however, the two cranial characters used by Jacques Gauthier to define Coelurosauria, the pterygopalatine fenestra and ventral pocket in the ectopterygoid, are both absent in the oviraptorosaurs.[192] Also, *Caudipteryx* exhibits an array of manual digits with a formula of 2-3-2, similar to that of more advanced birds. Recall that the digital formula of 2-3-4, possessed by *Archaeopteryx* and theropods, is the only reason some have argued that *Archaeopteryx* has the hand of a theropod. Here the table is turned: the same people now argue that *Caudipteryx* is a theropod, despite its avian digital formula.

The most impressive analysis of the oviraptorosaurs was published in 2002 by a team of Polish paleontologists noted for their careful work,

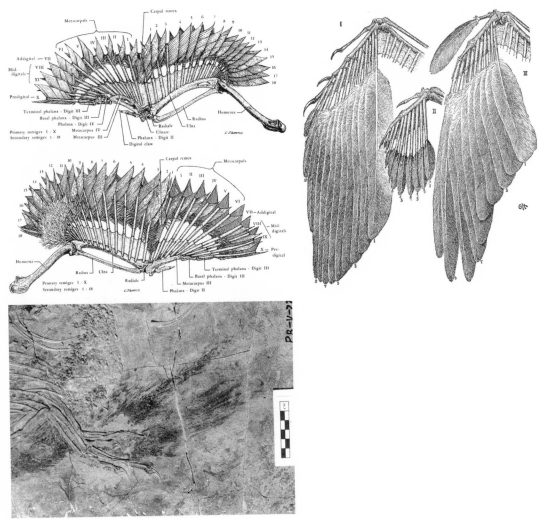

Fig. 4.68. Upper left, drawings (ventral view, upper; dorsal view, lower) showing the precise attachment of primary and secondary feathers in the avian wing (*Gallus*). Upper right (left to right), dorsal views of the wings of *Archaeopteryx*, baby hoatzin (*Opisthocomus hoazin*), and adult hoatzin. Note that the arrangement in *Gallus* is virtually the same as in *Archaeopteryx* and the hoatzin. Lower left, the wing of the Lower Cretaceous flightless bird *Caudipteryx*, showing feather attachment as in a typical flighted bird. Upper left from A. M. Lucas and P. R. Stettenheim, *Avian Anatomy: Integument*, 2 vols. (Washington, DC: US Government Printing Office, 1972); upper right from G. Heilmann, *The Origin of Birds* (London: Witherby, 1926); lower left courtesy of Zhonghe Zhou.

Teresa Maryańska, Halszka Osmólska , and Mieczyslaw Wolsan. Maryańska and colleagues performed an extensive phylogenetic analysis of the group using 195 skeletal characters, scored for four outgroups and thirteen ingroups.[193] Their analysis unequivocally places the Oviraptorosauria with the Avialae, in a sister-group relationship with *Confuciusornis*, *Archaeopteryx*, Therizinosauria, Dromaeosauridae, and Ornithomimosauria, which are successively more distant outgroups to the Oviraptorosauria. By their scheme, birds evolved from more primitive theropods, and oviraptorosaurs are birds that became flightless and, like the ratites, re-evolved some primitive features.

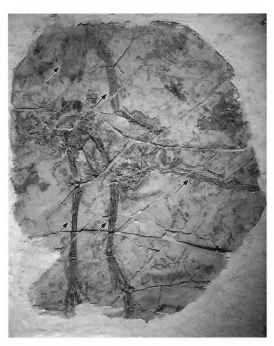

Fig. 4.69. Sculpture life restoration of *Caudipteryx* by Stephen Czerkas. Although most reconstructions show it as a predatory theropod, *Caudipteryx* was an herbivore, as attested to by the presence of numerous gizzard stones. The wings and hands were not used in predation but were the remnants of a "flight wing"— still locked up by vestiges of flight feathers—and the wing claws have greatly reduced curvature compared to those of trunk climbers such as *Archaeopteryx* and *Confuciusornis*. Additional specimens of *Caudipteryx zoui* reported in 2000 clarified many issues relating to the skeleton. The hallux was at least partially reversed, "suggesting that the ancestor of *Caudipteryx* had probably possessed the arboreal capability," and the manual digital formula of 2-3-2 was that of a bird, not a theropod; interpretation as a theropod would involve the loss of the distal two phalanges independently of similar reduction in the early evolution of birds. (Z. Zhou et al., "Important Features of Cau- dipteryx: Evidence from Two Nearly Complete New Specimens," *Vertebrata PalAsiatica* 10 [2000]: 241–54.) Copyright Stephen A. Czerkas.

This important analysis has been generally ig- nored by many paleontologists, who continue to assert that these taxa are the first clear evidence that avian plumulaceous feathers and the complex arrangement of primary and secondary wing flight feathers arose before the divergence of Aves (in the modern sense). Therefore they assert that the

Fig. 4.70. Protarchaeopteryx, one of the two "feathered dinosaurs" described in a cover article in *Nature*, is a Mesozoic bird. Not as well preserved as *Caudipteryx* and with longer wings, it was probably predominantly a ground-dwelling bird, but with some flight capabil- ity, like the living South American seriemas. Arrows, from left to right: upper, tail feathers, wing, and wing; lower, two legs. Courtesy of John Ruben.

avian "flight" hand and flight remiges arose in a nonflight context, in earthbound dinosaurs. Here we see how Chinese paleontologists have been concerned with the very avian-like morphology of caudipterids: "The two other feathered dinosaurs, *Caudipteryx* and *Protarchaeopteryx*, are different. They undoubtedly possess real feathers. . . . They are considered as feathered dinosaurs in this book. . . . However, it must also be pointed out that *Cau- dipteryx* has uncinate processes and reduced hands as in birds. Therefore, they might well be flightless birds that had reversed many of their bird charac- ters with the loss of flight." In addition, the more recently described *Similicaudipteryx* had avian- like feathers preserved in a molting stage: "The juvenile tail feathers of *Similicaudipteryx* are en- tirely consistent with the morphology of moulting feathers of living birds."[194]

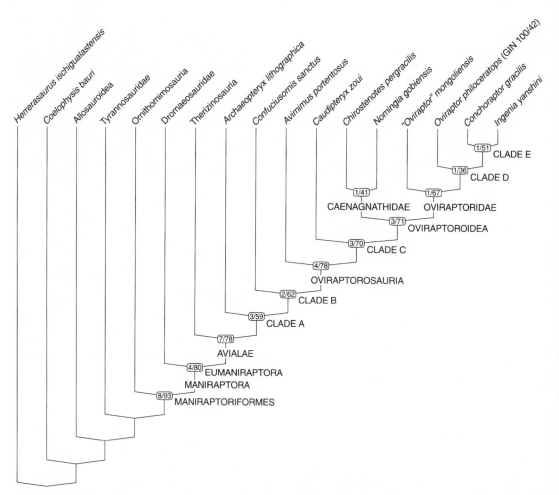

Fig. 4.71. The most parsimonious cladogram inferred from the cladistic analysis of Teresa Maryańska, Halszka Osmólska, and Mieczyslaw Wolsan showing that *Caudipteryx* is a bird, a secondarily flightless Mesozoic "kiwi." Reprinted with permission from T. Maryańska, H. Osmólska, and H. M. Wolsan, "Avialan Status for Oviraptorosauria," *Acta Palaeontoliga Polonica* 47 (2002): 97–116. Reproduced with permission from *Acta Palaetontologica Polonica*.

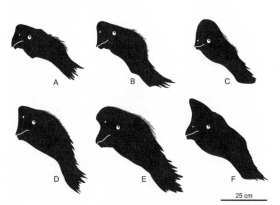

25 cm

Fig. 4.72. Adaptive radiation of Oviraptorinae with sufficient skull material to restore: (A) *Oviraptor philoceratops*, (B) *Nemegtomaia barsboldi*, (C) *Rinchenia mongoliensis*, (D) *Citipati osmolskae*, (E) *Citipati* sp., and (F) Unnamed sp. Silhouettes adapted from Martyniuk, Wikimedia Commons; licensed under the *Creative Commons Attribution 2.5 Generic* license. (Images not intended to support any concept presented in this text.)

....

Table 4.3. Avian features of *Caudipteryx* and of derived oviraptorosaurs, such as *Citipati* and *Ingenia* (assuming digits II, III, IV, as in birds)

Caudipteryx:

- Boxy skull with large expanded cranial vault and beaklike snout
- Nasal opening larger than antorbital fenestra
- Expanded frontal with supraorbital rim (as in primitive birds)
- Postero-ventrally located foramen magnum
- Teeth (restricted to front of upper jaw) constricted at the base "waisted" (as in Mesozoic birds)
- Uncinate processes (also in maniraptorans)
- Scapula articulates with coracoid at 90° angle (indicative of flight ancestry)
- Pedal digit I at least partially reversed (indicative of flight ancestry); and metatarsal I articulates with the postero-medial, rather than medial surface, of metatarsal II
- Tail greatly reduced; 22 unfused caudals (avian pygostyle present in the oviraptorosaur *Nomingia*); twelve avian rectrices attached to distal caudals
- Relatively short trunk and long neck; avian center of gravity (as in flightless ratites)
- Avian wing, with remnants of aerodynamic architecture
- Semilunate largest of three carpal elements
- Avian phalangeal formula 2-3-2
- Typical avian plumaceous feathers present, with rachis and vanes, with plumaceous barbs at the feather base
- Avian flight remiges with symmetrical vanes (as in flightless birds)
- Approximately fourteen primary remiges attach to middle metacarpal (III), phalanges 1 and 2
- Avian arrangement of tail feathers and avian molt pattern in a juvenile *Similicaudipteryx*
- Outer digit (IV) reduced, composed of two reduced phalanges (as in the Early Cretaceous enantiornithine *Eoenantiornis*; modern birds have only one)
- Outer digit abuts tightly on first phalanx of middle digit as in advanced birds

- Only two unguals retained (digits II and III); ungual of outer manual digit (IV) lost
- Antitrochanter absent (as in early birds and in *Archaeopteryx*)
- Tibia longer than femur (as in early birds)

Derived ovoraptorosaurs (e.g., *Citipati* and *Ingenia*):

- Fused prefrontals
- Reduced maxillae
- Extensively pneumatized narial region
- Shape of lacrimal "reverse C-shaped" as in *Confuciusornis*
- Contralateral communication between at least some tympanic diverticulae
- Fusion of the articular and surangular
- Articular surface for quadrate with development of lateral or medial process (or both)
- Pneumatic presacral vertebrae
- More than five sacrals
- Ossified uncinate processes
- Ossified sterna plates
- Costal facets on sternum
- Sternum with lateral process
- Anterior margin of sternum grooved antero-laterally for reception of coracoids

Source: F. D. James and J. A. Pourtless IV, "Cladistics and the Origin of Birds: A Review of Two New Analyses," *Ornithological Monographs* 66 (2009): 1–78.

....

Another interesting oviraptorosaurid is *Incisivosaurus gauthieri*, named for Jacques Gauthier of Yale, who pioneered the usage of phylogenetic systematics in archosaur paleontology. The genus name refers to the specimen's prominent, rodent-like front teeth, which appear to be those of a plant eater.[195] The numerous teeth, in contrast to those of advanced oviraptorosaurs, which are edentulous, indicate that *Incisivosaurus* was a basal member of the group, and a number of features show close affinity with the birdlike therizinosaurs, also herbivorous. Additional dinosaurs have been discovered from the Jehol Biota, including a number of ceratopsians (mostly *Psittacosaurus*), several ornithopods, and a small basal ankylosaur, but these are not discussed further here.[196]

New Evidence for Endothermy?

Regrettably, paleontologists have used the discoveries of the first "feathered dinosaurs," *Caudipteryx* and *Protarchaeopteryx*, as evidence for everything from flight from the ground up to feathers evolving for endothermy and to all avian flight architecture having evolved in an earthbound dinosaur. Here, for example, is Philip Currie on endothermy: "The warm-blooded proponents gained some support in 1996 by the discovery in China of a pair of smaller 'feathered' theropods. . . . The discovery of these specimens has given additional support to the hypothesis that theropod dinosaurs were the direct ancestors of birds, and that some theropods were endothermic."[197]

This statement is particularly disturbing because the evidence for endothermy in dinosaurs has been slowly dismantled over the years, especially by comparative physiologist and paleobiologist John Ruben of Oregon State University and Devon Quick.[198] It is known that dinosaurs grew rapidly as juveniles, with growth slowing at sexual maturity, and some paleontologists speculatively view such fast growth as a sign of some form of endothermy, at least in the young, with a switch to inertial homeothermy or gigantothermy when large size was attained.[199] Much of this reasoning, however, is based on the putative identification of protofeathers in numerous species of theropods from the Early Cretaceous of China, an identification, as we have seen, that has been seriously questioned or even proven incorrect. Also, as we have seen, archaic enantiornithines were fully feathered but exhibit ectothermic growth rings in their bones, indicating that they were likely quasi-ectotherms. It is as though every tiny bit of evidence for endothermy in dinosaurs, including bone histology, erect posture, brain size, feathers, and nesting behavior, independently has some limited credibility, and once added together, the totality of these tiny bits of inconclusive or incorrectly interpreted evidence produce a half-truth, Then somehow these half-truths become additive

and constitute a truth, which invariably translates into resurrecting endothermic dinosaurs. Yet the vast majority of papers on hot-blooded dinosaurs fail to clarify their biothermal semantics, and the assumption appears to be that any animal exhibiting homeothermy (constant body temperature) is necessarily endothermic (producing body heat endogenously), when all physiologists know that constant body temperature is not a reliable indicator of an animal's metabolic rate. As Zhonghe Zhou and Fucheng Zhang have correctly pointed out, "There is no compelling evidence that any dinosaurs were hot-blooded."[200]

For the reader who would like a detailed summary of this paleontological *Fantasia*, there is an entire symposium volume entitled *Feathered Dragons: Studies on the Transition from Dinosaurs to Birds*.[201] And Mark Norell and Xing Xu published a twenty-two-page review entitled "Feathered Dinosaurs" in 2005 in the *Annual Review of Earth and Planetary Sciences*.[202] With little supportive evidence, much of the speculation surrounding protofeathers and feather origins is making its way into respectable texts. Like Luis Chiappe in *Glorified Dinosaurs* (chapter 1), Donald Prothero in his 2007 book *Evolution* is guilty of selective citation, his references containing no reference to any view other than the current orthodoxy concerning bird origins, no citation of the papers by Lingham-Soliar, nor any mention of the evidence indicating that the first discovered "feathered dinosaurs" are actually flightless birds.[203] Not surprisingly, then, color plate 6 of *Evolution* shows "feathered nonflying dinosaurs from the Lower Cretaceous . . . of China."

In 2008–9, alas, the American Museum of Natural History's traveling exhibit, entitled *Dinosaurs: Ancient Fossils, New Discoveries*, had a life-reconstruction mount of *Caudipteryx*, with the notation: "First nonavian dinosaur found with modern feathers."

The Conflation of Feathered Dinosaurs and Birds

> The smallest dinosaur is the bee hummingbird . . . found only in Cuba.

M. A. Norell, E. S. Gaffney, and L. Dingus,
Discovering Dinosaurs in the American Museum of Natural History, *1995*

The epigraph above is a perfect summary statement of the current orthodoxy: birds have been defined out of existence; they do not merit their own taxonomic distinction; they are simply living dinosaurs. Stephen Jay Gould was perplexed by the popular concept that birds are living dinosaurs. In an interview he once stated that "a sparrow is not a *Tyrannosaurus,*" even if birds did branch off from a small group of theropods. Gould's concern was that such a statement could be interpreted in transformationist fashion—that is, that dinosaurs evolved into birds in an anagenic way, an extension of the linear, transformationist view that plagues the popular understanding of human evolution and even leads to racist notions of relative "advancement."[204] Yet from a purely scientific point of view, the most disturbing aspect of the paleontological approach is that, because it is assumed that "birds are dinosaurs," any filamentous integumental structures in theropod dinosaurs must therefore be protofeathers, and the burden of proof is placed on those skeptical of such identifications.

Feathered Dromaeosaurs: The Final Proof?

For most paleontologists the final proof that Aves is nested within Theropoda is the discovery of dromaeosaurs with pennaceous feathers. With each discovery of the small Early Cretaceous basal dromaeosaurid microraptors, the excitement has steadily reached a crescendo, reflecting the view that the ultimate avian ancestors have been uncovered. First, of course, was the discovery in 1996 of the first so-called feathered dinosaur and first named Jehol coelurosaurian, *Sinosauropteryx,* which turned out to be devoid of any structures related to feathers; it is a compsognathid closely allied with the Late Jurassic *Compsognathus* and *Juravenator,* both of which had typical tuberculated dinosaur skin. Then in 1998 the two "feathered dinosaurs" *Caudipteryx* and *Protarchaeopteryx* were featured on the cover of *Nature*; these, too, had their day of glory but were quickly viewed by many for what they were, secondarily flightless

birds, fitting nicely into the pattern of flightless paedomorphosis elucidated by Sir Gavin de Beer in ratites.[205]

In quick succession, in 1999 two other feathered dinosaurs were reported, again from western Liaoning. *Beipiaosaurus inexpectatus* was described as the largest theropod discovered from Liaoning, measuring over 2 meters (6 ft.) long and thought to have weighed some 85 kilograms (187 lbs.).[206] With tiny, lanceolate "prosauropod-like" teeth, a short tail, and unusually long, recurved claws, this primitive therizinosaur was most likely a plant eater. It was preserved with a halo of fibers, quickly proclaimed to be protofeathers, surrounding the entire body, but as usual, there was no evidence to support the view that the fibers were in any way allied with feathers. Xu and colleagues even suggested that the "feathers" of *Beipiaosaurus* represented an intermediate stage between those of *Sinosauropteryx* and more advanced birds. As noted earlier, with additional material of *Beipiaosaurus,* Xu and colleagues described what they believed were a new feather type, elongated broad filamentous feathers, or EBFFs, preserved in the head and neck region and along the posterior half of the trunk, and as noted earlier, with scant evidence they concluded that the structures were for integumentary display.[207] In addition, X.-T. Zheng and colleagues described an Early Cretaceous heterodontid dinosaur (basal ornithischian) also with elongate filamentous integumentary structures, noting their possible homology with filamentous structures in other dinosaurs.[208] They noted that "the discovery of filamentous integumentary structures in *Tianyulong* provides an unprecedented phylogenetic extension of archosaurian dermal [sic; they would of course be epidermal] structures, previously reported only in derived theropodan saurischian dinosaurs, to Ornithischia." Since then, however, Lingham-Soliar has provided a critical assessment of the possible link between these elongate fibers in these dinosaurs, as well as "tail bristles" described earlier from the basal ceratopsian *Psittacosaurus,* also from the Chinese Early Cretaceous.[209]

In 1999, a feathered dromaeosaur, *Sinornithosaurus millenii,* was described.[210] This dromaeo-

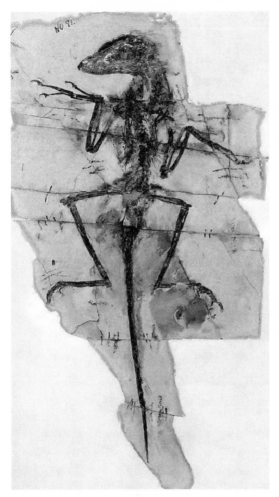

Fig. 4.73. The specimen dubbed "Dave," most likely the Chinese Early Cretaceous *Sinornithosaurus*. Although reconstructed as a flightless, feathered dinosaur, the birdlike hand and length of the forelimbs are concordant with it being a volant form that did not preserve its complete remiges, as was the case for *Longipteryx*. From Q. Ji et al., "The Distribution of Integumentary Structures in a Feathered Dinosaur," *Nature* 410 (2001): 1084–88. Copyright 2001. Reprinted by permission from Macmillan Publishers Ltd.

saur, the "Chinese bird lizard," smaller than but closely allied with the famous *Velociraptor*, created a sensation in the popular press and was considered more closely related to birds than any previously described theropod. The nearly complete type skeleton resides in the collections of Beijing's Institute of Vertebrate Paleontology and Paleo-anthropology, while a second specimen, collected

about 130 kilometers (78 mi.) from the original site, but also in Liaoning Province, resides in the National Geology Museum in Beijing. To date, several species have been identified, but the various specimens all appear to belong to the same genus. One incredibly well preserved specimen, dubbed "Dave," may represent a third species or simply a juvenile.[211] These microraptors exhibit the salient synapomorphies of the Dromaeosauridae: a stiffened tail with elongate prezygapophyses and chevrons that span several tail vertebrae, a retroverted pubis with a birdlike hypopubic cup (as opposed to a theropod pubic boot), and a modified second pedal digit with a sickle claw. Phylogenetic analyses show them to be basal within the Dromaeosauridae, and certainly very close allies of the four-winged *Microraptor*. Among the more interesting revelations was their occurrence in the Early Cretaceous, because before their discovery, dromaeosaurs were known only from later in the Cretaceous.

Various claims have been made concerning the integument of *Sinornithosaurus*, including that the preserved integumental structures represent a critical stage in the evolution of feathers from the more primitive type found in *Sinosauropteryx* to that found in modern birds. Yet the claims of an advanced stage of protofeathers in this genus have been firmly refuted by Lingham-Soliar, and the evidence is simply too weak to be considered seriously.[212] With little evidence, Richard Prum and Alan Brush stated in *Scientific American*:

The heterogeneity of the feathers found on these dinosaurs is striking and provides strong direct support for the developmental theory [their theory developed following the discovery of so-called protofeathers]. The most primitive feathers known—those of *Sinosauropteryx*—are the simplest tubular structures and are remarkably like the predicted stage 1 of the developmental model. *Sinosauropteryx*, *Sinornithosaurus*, and some other nonavian theropod specimens show open tufted structures that lack a rachis and are strikingly congruent with stage 2 of the model. There are also pennaceous feathers that obviously had differentiated barbules and coherent planar vanes, as in stage 4 of the model.[213]

The real confusion comes in when one considers that most of these basal dromaeosaurs, which as we shall see in chapter 7 are most likely early birds, actually have fully developed pennaceous avian feathers, and many of the specimens show the fibers (collagen fibers, so-called protofeathers) along with true bird feathers, creating conflated arguments related to feather origins. The Chinese rocks present a confusing mixture of true theropods like the compsognathid *Sinosauropteryx*, basal dromaeosaurs, which are likely birds, and secondarily flightless birds such as *Caudipteryx*, some forms with collagenous fibers, some with genuine feathers, and still others with both structures preserved in the same fossil. Small wonder there is total confusion when discussing issues of definition and the origin of feathers! As Prum and Brush note, "We now know that feathers first appeared in a group of theropod dinosaurs and diversified into essentially modern feathers."[214]

Still more problematic is the assumption that because protofeathers were discovered in earthbound theropods, feathers must have arisen along with a host of avian flight characteristics in cursorial animals. Yet careful examination of the morphology of *Sinornithosaurus* ("Chinese-bird-lizard") reveals nothing in its anatomy to preclude flight at some level. The reconstruction by the group at the American Museum of Natural History for their traveling exhibit on dinosaurs is that of a terrestrial cursor, but the hand appears very much like the flight-adapted hand of birds, showing definitively that it was derived from a flying ancestor, and the length of the arm is certainly within the range of many flying birds. In addition, the flight feathers coming off the hind limbs would have been a maladaptive hindrance for any terrestrial cursor and would militate against such a proposed lifestyle. Reconstructions like these are based on the cladogram that classifies them as terrestrial theropod dinosaurs and not on well-reasoned anatomical analysis. Indeed, the interpretation is reminiscent of John Ostrom's flyswatter model for the origin of bird wings. Despite

Fig. 4.74. Silhouettes of, from top to bottom, Late Jurassic *Anchiornis*, the microraptor *Sinornithosaurus*, and an oviraptorosaur, illustrating that the "insect net" model for the evolution of avian flight feathers and wings, introduced by John Ostrom, is alive and well today despite its improbability. By today's dinosaur-bird orthodoxy, all of the avian flight architecture, including the "flight hand" and flight feathers (remiges), are thought to have evolved in a non-flight context, an improbable scenario that is almost non-Darwinian, a stretch of credulity. In addition, the reconstructions of *Anchiornis* (originally described as an archaeopterygid and later transferred to Theropoda) and *Sinornithosaurus* are extremely unlikely, because the elongate "flight feathers" coming off the foot (and toes in *Anchiornis*) would be maladaptive, a hindrance in terrestrial locomotion. Such feathers are much more likely related to flight function, illustrating a volant ancestry. Yet all these creatures are reconstructed as earthbound predators to accommodate the cladogram.

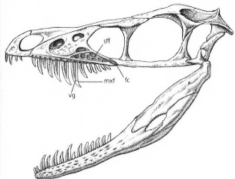

Fig. 4.75. Careful examination of this animal shows teeth and other anatomical features indicating that *Sinornithosaurus* was poisonous and probably used its elongate, fanglike teeth to penetrate its feathered prey, which were subdued with poison that flowed down grooves in the teeth. From E. Gong et al., "The Birdlike Raptor *Sinornithosaurus* Was Venomous," *Proceedings of the National Academy of Sciences* 107 (2010): 766–68.

the substantial evidence to the contrary, all the recent Chinese discoveries seem reconstructed to reflect Ostrom's model and flight origin from the ground up.

Phylogenetic analyses have shown the turkey-sized *Sinornithosaurus*, represented by at least two species, to be basal among the dromaeosaurs, and as might be expected, it shows characteristics, particularly of the skull and shoulder, that are very similar to *Archaeopteryx* and other avialians; in addition, it has fully developed avian pennaceous feathers. The skull is especially birdlike, and in 2009 a team of paleontologists headed by Enpu Gong of Northeastern University in Liaoning, and including Larry Martin, David Burnham, and Amanda Falk of the University of Kansas, discovered the presence of lateral grooves on the tooth crowns and other associated structures of the jaw region, including a maxillary fossa supradental channel, or duct with pits running along the tooth base, that led them to conclude that *Sinornithosaurus* was likely venomous, judging from similar morphologies in living and extinct reptiles. Poison delivery systems are also known in a fossil mammal, and some modern shrews have a venomous bite. The authors surmised that a gland produced venom that flowed through the duct and down the grooves of the teeth. They suggested that the poison was possibly similar to that of rear-fanged snakes and helodermid lizards in that it probably

did not kill the envenomated prey but rapidly placed it in a state of shock, with a "bite and hold" method of venom delivery.[215] *Sinornithosaurus* probably fed on the abundant birds of the Early Cretaceous Jehol forests, using its long, fanglike anterior maxillary teeth to penetrate the feather coating of its avian prey. With its birdlike features, flight wing, and skull, *Sinornithosaurus* was very closely allied with the four-winged glider *Microraptor*, discussed below, and was therefore within the early avian radiation. If terrestrial, it was most likely derived from volant ancestry, given its *Archaeopteryx*-like flight hand.

Four-Winged Dinosaurs

In 2000, Xing Xu, Zhonghe Zhou, and Xiaolin Wang of the Institute of Vertebrate Paleontology and Paleoanthropology described the first mature, nonavian dinosaur smaller than *Archaeopteryx*. Named *Microraptor zhaoianus*, this small dromaeosaur possessed many birdlike features, including more birdlike waisted teeth, with serrations only on the posterior carina (elevated edge) of the posterior teeth. In addition, it exhibited a birdlike ischium, uncinate processes, a semilunate carpal and bowed ulna, highly recurved manual claws, a birdlike foot with highly recurved pedal claws, and elongated penultimate phalanges, comparable only to those of arboreal, trunk-climbing birds. Most intriguing, and foreshadowing future discoveries, was the preservation of "integumentary structures" especially near the femur, where they extend almost perpendicular to the bone and have features of avian feathers with a rachis. With their new find, the authors proposed that "the theropod ancestors of birds may have passed through an arboreal phase."[216]

Shortly after the discovery of *Microraptor*, a team led by Xing Xu and Zhonghe Zhou collected an additional specimen in Liaoning, and the group purchased an additional five specimens recovered from the same deposits. They designated and described one as *Microraptor gui*, the now famous flying dinosaur described in *Nature* in 2003 in the article "Four-Winged Dinosaurs from China."[217] To the surprise of many, the new specimens clearly showed two sets of wings, on both fore- and hind limbs. The long, pennaceous remiges were true avian flight feathers, with primary feathers anchored to the hand and secondary feathers attached to the ulna, as in modern birds; there is also a birdlike alula. The body appears to be covered in typical contour feathers. The flight feathers on the arm, leg, and tail had asymmetric vanes, each an individual airfoil. The remiges include approximately twelve primaries and about eighteen secondaries, and the longest primaries are 2.7 times as long as the humerus. On the hind limbs the fourteen or so primary flight feathers were anchored to the upper foot bones and the lower and upper legs. In addition, the new specimens showed clearly that microraptors had a fairly typical, primitive avian flight hand, quite similar to that of *Archaeopteryx*, down to the semilunate carpal element, as well as the same regimented arrangement of primary and secondary flight feathers. The holotype, which appears typical of the species, is estimated to have weighed about 1 kilogram (2.2 lbs.) and measured about 77 centimeters (30 in.) in length, with an exceptionally long, bony tail bearing asymmetric feathers and ending in a fan of feathers. Otherwise, the specimen is indistinguishable from *Microraptor zhaoianus*, and some believe that all the microraptor specimens, including another described under the name *Cryptovolans pauli* by Stephen Czerkas and colleagues, belong to a single species, which would technically be *Microraptor zhaoianus*.[218]

Following the initial description of *Microraptor*, Sunny Hwang and colleagues described additional, nearly complete skeletons of this small arboreal dromaeosaur, and their analyses found that dromaeosaurs are monophyletic and that *Microraptor* is the sister taxon to other dromaeosaurs and, together with troodontids, forms a monophyletic Deinonychosauria, the sister group of the Avialae.[219] They also found that small size is primitive for the group. Interestingly, *Microraptor* is particularly similar to the troodontid *Sinovenator*, and each is a primitive member of its respective, closely related group and close to the deinonychosaurian split between the dromaeosaurs and troodontids. Since then, scores of these creatures have been discovered. More than thirty reside at the Institute of Vertebrate Paleontology and Paleoanthropology, but there are many more, perhaps hundreds of specimens.

The surprising occurrence of leg flight feathers, along with the presence in the Early Cretaceous of four-winged, arboreal microraptors with little cursorial ability led Zhonghe Zhou and Fucheng Zang to introduce a new variation on hypotheses for the origin of avian flight, the "Dinosaur-trees-down" hypothesis.[220] It also had become apparent that microraptors were an indication of a tetrapteryx stage in the early evolution of avian flight, an idea, as noted earlier, first proposed in 1915 by American naturalist William Beebe.[221] Beebe's hypothetical tetrapteryx derived from his studies of young chicks, in which he found what appeared to be vestiges of such a feather configuration on the legs. The microraptor finds lend considerable credence to Beebe's view, as do additional findings of hind-limb "trouser" feathers on *Archaeopteryx* (perhaps secondarily reduced), elongate pennaceous feathers on the hind limbs of an unnamed enantiornithine bird, and the discovery of what has been described as a feathered maniraptoran dinosaur.[222] However, *Archaeopteryx* and the enantiornithine differed from microraptors in apparently lacking the hind-limb metatarsal feathers. The maniraptoran, named *Pedopenna*, from the Middle to Late Jurassic Daohugou beds where *Epidexipteryx* was discovered, is represented by a leg with beautifully preserved elongate feathers,

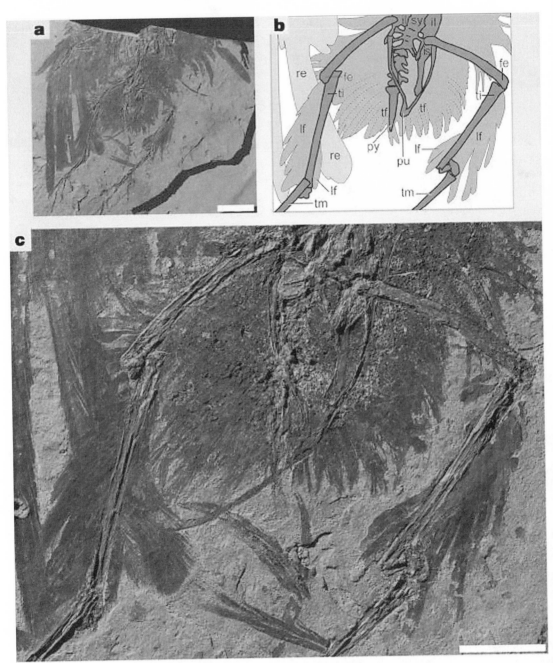

Fig. 4.76. Unnamed basal enantiornithine bird from the Early Cretaceous Yixian Formation showing the hind-limb skeleton preserved with aerodynamic feathers attached to its legs (lf) and weak, narrow and relatively short tail feathers (tf). (a) Photograph of the specimen in a single slab, showing the association of the preserved skeletal elements, almost all of which are articulated, with feathers. (b) Drawing of the specimen. Abbreviations: fe, femur; il, ilium; is, ischium; pu, pubis; py, pygostyle; re, remiges; sy, synsacrum; ti, tibiotarsus; tm, tarsometatarsus. (c) Close-up view of slab. Scale bars = (a) 2 cm, (b) 1 cm. From F. Zhang and Z. Zhou, "Leg Feathers in an Early Cretaceous Bird," *Nature* 143 (2004): 925. Copyright 2004. Reprinted by permission from Macmillan Publishers Ltd.

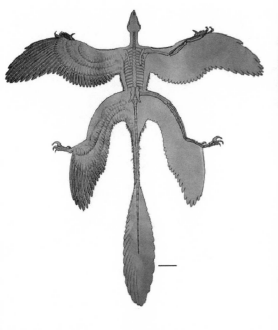

Fig. 4.77. Reconstructions of *Microraptor*. Left, reconstruction from the description by Xing Xu and colleagues. Upper right, foam model rendered by the Kansas group based on an actual skeleton, which performed nicely in wind tunnel tests. Lower right, the "tetrapteryx" stage in the evolution of flight predicted by naturalist William Beebe in 1914. Left, from X. Xu et al., "Four-Winged Dinosaurs from China," *Nature* 421 (2003): 335–40. Upper right, image of model made by David A. Burnham, courtesy David A. Burnham; model painted by Elizabeth Ebert. Inset photo from Shandong Tianyu Museum of Nature, Pingyi, Shandong, Director: Xiaoting Zheng; photo courtesy David A. Burnham. Lower right, from W. A. Beebe, "Tetrapteryx Stage in the Ancestry of Birds," *Zoologica* 2 (1915): 38–52.

and the skeletal elements show synapomorphies of both basal dromaeosaurs and basal birds, but with a less specialized second pedal digit than in dromaeosaurids. The authors have suggested that it is more closely related to Aves than any other known nonavian theropod but might have been volant, since "the presence of metatarsus feathers . . . [is] inconsistent with a cursorial habit."[223] In their 2003 paper on *Microraptor*, the authors made the same observation with respect to that taxon: "The metatarsus feathers are inconsistent with the suggestions that basal dromaeosaurs are cursorial animals because such long feathers on the feet would be a hindrance for a small cursorial animal. It is unlikely that a small dromaeosaur could run fast with such an unusual integument and this provides negative evidence for the ground-up hypothesis for the origin of avian flight."[224]

The discovery of a four-winged planform in basal birds and basal dromaeosaurs suggests that the fore- and hind-limb wings were a plesiomorphic, primitive condition present in a common ancestor and provides strong evidence that arboreal parachuting and gliding stages preceded powered flight in birds.

Yet considerable debate ensued concerning the positioning of the flight remiges. In a recent paper by David Hone, Helmut Tischlinger, Xing Xu, and Fucheng Zhang, the feathers of *Microraptor* were reexamined using ultraviolet light, revealing much more detail than was previously available.[225] They were particularly interested in the asymmetrical flight feathers on the manus and pes and were able to determine that these remiges attached to the bones as in modern birds. They discovered that the feathers have not disarticulated and

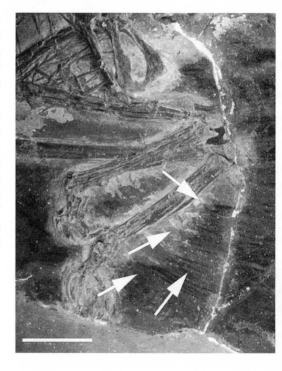

Fig. 4.78. The famous four-winged dinosaur *Microraptor gui*, as preserved on the main slab. Below, close-up of lower hind limb of same specimen under UV light, showing that the feathers do indeed penetrate the "halo" and extend to the actual bones; this shows that the hind-limb flight remiges are in their natural positions. Photograph of specimen courtesy of Zhonghe Zhou. X. Xu et al., "Four-Winged Dinosaurs from China," *Nature* 421 (2003): 335–40. Copyright 2003. Reprinted by permission from Macmillan Publishers Ltd. UV photograph from D. W. E. Hone et al., "The Extent of the Preserved Feathers on the Four-Winged Dinosaur *Microraptor gui* under Ultraviolet Light," *PLoS One* 5, no. 2 (2010): 1–7. Distribution under the Creative Commons License Deed, Attribution 2.5 Generic.

moved away from the body but extend into a halo surrounding the animal, reach the bones, and are preserved in their natural position. They show that the flight feathers are actually longer than previously thought, thus increasing the wing area over that calculated previously.

Their earlier work on *Archaeopteryx*, using ultraviolet light by Helmut Tischlinger of Stammham, Germany, and in association with David Unwin, revealed that the equivalent pedal feathers of *Archaeopteryx* were not equivalent to those of *Microraptor*, as had been previously suggested, but were best interpreted as "feather trousers" of many extant birds, including particularly birds of prey, and are exclusively attached to the tibia and femur. They concluded that since both troodontids and dromaeosaurs are preserved with feathers on the metatarsus, "the most parsimonious interpretation would suggest that the lack of feathers on the metatarsus of *Archaeopteryx* is derived."[226]

Of course, interpretation of the hind-limb wings in these taxa has been contentious. Xing Xu and colleagues in their initial cover article in 2003 in *Nature* reconstructed *Microraptor* in the most natural pose possible for an animal capable of gliding flight, with tandem wings spread horizontally as in Beebe's tetrapteryx, arguing, as noted, that because of its projecting long hind-limb flight feathers, it would have been quite clumsy on the ground. It was thought to have been a tree-dweller and a finely tuned glider. However, almost immediately, dinosaur paleontologists objected to Xu's interpretation; Sankar Chatterjee and colleague Jack Templin argued that the wing design envisioned by Xu and colleagues conflicted with the known articulation of theropod limb joints, namely, a parasagittal posture of the hind limb. In contrast, Chatterjee and Templin reconstructed an alternative planform of the hind limb, concordant with a normal theropod posture: "The wings of *Microraptor* could have resembled a staggered biplane configuration during flight, where the forewing formed the dorsal wing and the metatarsal wing formed the ventral one . . . its biplane wings were adapted for undulatory 'phugoid' gliding between trees, where the horizontal feathered tail offered additional lift and stability and controlled pitch. Like the Wright 1903 Flyer, *Microraptor*, a gliding relative of early birds, took to the air with two sets of wings."[227]

Shortly after the publication of the Chatterjee and Templin's *Microraptor gui* biplane model, *NOVA* produced an hour-long program entitled *The Four-Winged Dinosaurs*, in which the two quite different interpretations of the flight planform of *Microraptor* were on display.[228] One was developed by a team at the American Museum of Natural History; on their model, the flight planform of *Microraptor* was essentially that of a biplane, using

the standard dinosaurian anatomical interpretation, and it was based on an artist's reconstruction from photographs of numerous specimens. A different model was reconstructed by Larry Martin and David Burnham of the University of Kansas, in collaboration with David Alexander, expert in biomechanics and animal flight and author of a book on the topic.[229] The Kansas group reconstructed the pelvic and hind-limb bones based on casts of a specimen, and it supported the initial interpretation by Xing Xu and colleagues, showing a more primitive sprawled posture with the hind limbs splayed out laterally (as on the cover of *Nature*). Yet the program appeared to favor the dinosaurian model, and even held to the model for feather evolution based on their having evolved for insulation as the original function, despite the lack of evidence for hot-blooded dinosaurs. In contrast to almost all who studied the fossil, Mark Norell of the American Museum of Natural History said, "I wouldn't say that we know that *Microraptor* even was a glider, let alone a flyer."[230] He also noted, "I really get frustrated by this [argument on the origin of birds] because we shouldn't be having to deal with something which has been settled for 20 years. I don't know when Magellan got back from sailing around the world, if he was frustrated, too, by people who still said the Earth was flat."[231] Continuing, the *NOVA* program narrated, "*Sinosauropteryx* . . . 'first Chinese winged lizard.' . . . With fuzzy feathers . . . it is unlikely that it ever flew. The feathers could, however, help keep the dinosaur warm through cool nights in the temperate forest."[232]

But what flight model would best fit the anatomy of the microraptor? The best test model for the biplane model was parachuting, and despite the apparent certitude of the participants (Norell commenting that "the animal couldn't physically do it"), it was clear through wind tunnel tests that the sprawled-posture, splayed-out model was superior aerodynamically to its rival, and the MIT aerodynamic engineers who performed the wind tunnel tests were quite enthusiastic about the Kansas model, which exhibited classic airflow behavior; they noted that it "climbed steadily and more predictably than anything they had seen so far."[233]

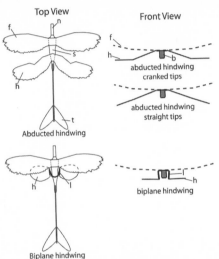

Fig. 4.79. Above, David Alexander, expert in biomechanics and the field of animal flight, testing the *Microraptor* flying model originally envisioned by Xing Xu, with hind-limb wings splayed laterally. Alexander's model flies well both in the field and in wind tunnel experiments. Below, different hind-wing configurations tested by Alexander. The bi-wing model actually flew but required a reorientation of the center of gravity that required a very heavy additional weight to the head region. The original interpretation by Xing Xu, and now based on direct casts of the fossil bones, shows the animal in a sprawled posture with the hind-limb wings splayed out laterally, an anatomical profile with superior performance in wind tunnel experiments. Above, photograph copyright David A. Burnham. Below, from D. E. Alexander et al., "Model Tests of Gliding with Different Hindwing Configurations Tested in Wind Tunnel Experiments for the Four-Winged Dromaeosaurid *Microraptor gui,*" *Proceedings of the National Academy of Sciences* 107 (2010): 2972–76. Copyright 2010 National Academy of Sciences.

The work by the Kansas group on the morphology of *Microraptor* should be given careful attention; aside from functional considerations, there are now specimens with the hind legs spread in a splayed position, which would not be possible if they were obligate bipeds such as typical theropods or chickens. It should be noted that anatomy is not always the most reliable indicator of the full range of motion or posture available to an organism; joints and articulations can seem more restrictive of motion and posture than they actually are. Many mammals and birds with upright posture can assume a splayed-out posture of the hind limbs, including carnivores like cats and dogs, bats, virtually all rodents (especially squirrels), primates (including the tropical American glider, the monk saki or volador *Pithecia monachus*, and the flying lemur *Cynocephalus*), and various gliding opossums. There is a vast anecdotal literature on cats falling from buildings, and in all cases the cats spread themselves out like flying squirrels. Notes from this field, now termed "feline pesematology," indicate that for cats falling from heights of up to thirty-two stories there is about a 90 percent rate of survival.

Following the *NOVA* program, the Kansas group, led by David Alexander, used their reconstruction of *Microraptor*, based on an actual skeleton cast in the round, to produce lightweight, three-dimensional physical models and performed flight tests with different anatomically reasonable hind-wing configurations. For example, hind wings were abducted and extended laterally (the hypothesized "tetrapteryx" of William Beebe), following the original reconstruction by Xing Xu. Other models were also tested, including the one favored in the *NOVA* program, with the biplane configuration, a model based on a bipedal dinosaurian reconstruction broadly at odds with the original reconstruction. The results showed that while the biplane model glided almost as well as the tetrapteryx, it required an extremely unlikely reorientation of the center of gravity, with a massively and ungainly, very heavy head for aerodynamic stability and stable gliding. The authors concluded, "Our model with laterally abducted hindwings represents a biologically and aerodynamically

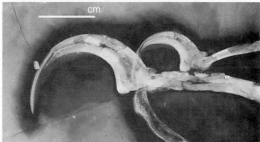

Fig. 4.80. Above, specimens of *Microraptor*, showing the legs in the splayed position, left, and complete skeleton, right. Given the position of the hind legs, the narrow sacrum, and the elongate "flight" feathers of the hind-limb wings, preserved on most specimens and illustrated here (below left), this animal was no terrestrial cursor, but is frequently so reconstructed to accommodate the cladogram. Below right, highly recurved, laterally compressed claws of *Microraptor*, approaching an inner claw curvature of 180°, indicate an arboreal, climbing function. Top left photographs taken in the public exhibits, Shandong Tianyu Museum of Nature, Pingyi, Shandong, Director: Xiaoting Zheng; courtesy of David A. Burnham. Top right photograph of complete specimen displayed in Hong Kong Science Museum, by Laikayiu, from Wikimedia Commons, licensed under the GNU Free Documentation License, Version 1.22. Other photographs by David A. Burnham, courtesy of David A. Burnham. Copyright David A. Burnham.

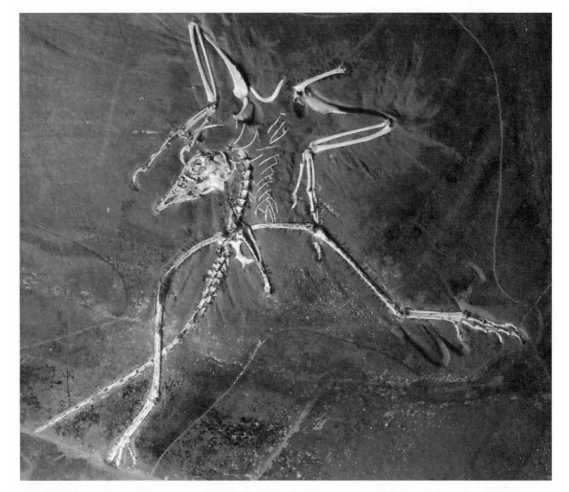

Fig. 4.81. The tenth skeletal specimen, "Thermopolis," of *Archaeopteryx* under ultraviolet-induced fluorescence photography to show the preserved bone substance. Preservation of this specimen proves that *Archaeopteryx* was not fully obligately bipedal with a parasagittal gait but rather was capable of a splayed position of the hind limbs. Note particularly the very narrow sacrum, typical of arboreal birds, and unlike the broad sacrum of a terrestrial bird such as a chicken or ratite. The Thermopolis specimen has been reconstructed as an earthbound predator, in keeping with the Ostrom "flyswatter" model, to accommodate the cladogram. From G. Mayr et al., "A Well-Preserved *Archaeopteryx* Specimen with Theropod Features," *Science* 310 (2005): 1483–86. Reprinted with permission from AAAS.

reasonable configuration for this four-winged gliding animal. *M. gui*'s feathered hindwings, although effective for gliding, would have seriously hampered terrestrial locomotion."[234]

In the final analysis, Xing Xu's original interpretation of the anatomy and flight profile of *Microraptor* would appear to be the best interpretation to date. Regardless, the current pace of discoveries of these amazing animals will offer further clarification of both their anatomy and their flight profile, but their lack of truly definitive theropod characters and their possession of numerous salient avian features, including a perfect arrangement of primary and secondary wing feathers, all point to their being of avian descent.

Feathers Are for the Birds

When the chicken-sized "feathered dinosaur" nicknamed Dave was discovered, it was hailed as among the finest, most beautifully preserved theropods yet discovered, displaying a full array of body feathering. Although probably a specimen of the previously described *Sinornithosaurus*, the specimen was nothing less than spectacular, and it created a sensation in the popular press.

In his book *Unearthing the Dragon*, Mark Norell describes how, when in Beijing, he and his American colleague Mick Ellison struggled to communicate to Chinese colleagues that the specimen was indeed a feathered dinosaur and not a bird.[235] Ellison is a photographer and illustrator at the American Museum of Natural History who has worked closely with Norell for more than a decade. Norell noted that although Qiang Ji certainly realized that the new animal was important, and not a bird but a feathered dinosaur, the rest of the Chinese crew was confused: "They had not moved on from the character-based definition that anything with feathers was a bird." Norell recalled that, being tired, the more he tried to explain, the more confused they got, and "Our cultural baggage and lack of a common language collide[d] and [blew] right past each other." After a nap and massage at the exclusive Xiyuan Hotel, Norell and Ellison dug into the minibar to reflect on their afternoon's conversation and agreed that the event could be summed up by a flashback to an old Cheech and Chong routine. For the uninformed, Richard "Cheech" Marin and Tommy Chong were a comedy duo of the 1970s and 1980s hippie, free-love, and drug culture movements. In one comic routine Cheech tries unsuccessfully to communicate with Chong about the whereabouts of a character named Dave. From their similar experience that day in China, Norell and Ellison decided to apply the colloquial name "Dave" to the new fossil, and their discovery followed its usual course, climaxing in another *Nature* paper, "The Distribution of Integumentary Structures in a Feathered Dinosaur."[236] Many who have been in the field of paleontology and have suffered through many of the highly speculative ideas these many years can readily sympathize with the apparently obtuse Chinese colleagues who were still back in the dark ages on "character-based" definitions; many of the rest of us are also still at that stage. Most would certainly have not been

spared the usual ridicule that is exacted on those not following the new phylogenetic systematics and its taxonomic lexicon, the phylocode. Yet like our Chinese colleagues who could not communicate in those terms, many Westerners still find the Linnaean system to be more than satisfactory, and until further credible evidence can be marshaled to destroy the time-honored view that birds are unique in having feathers, the following axiom still holds: if it has avian feathers, it is a bird.

Discoveries Continue

We have learned of a remarkable new world from the discoveries made in the rocks of the Chinese Early Cretaceous; they have opened a rare window on the past, equivalent perhaps to that offered by the Late Jurassic Solnhofen fauna of southern Bavaria, where *Archaeopteryx* was discovered. The Chinese discoveries without doubt provide the most remarkable picture of bird evolution for any specific period of the Mesozoic. Among the most remarkable discoveries was that the opposite birds, or enantiornithines, were the dominant land birds of the Mesozoic and had an extensive adaptive radiation, though most were arboreal. We see for the first time the appearance of the avian beak in the strange confuciusornithids, a group that is clearly advanced over *Archaeopteryx* and other sauriurine birds in any number of features but still not at a modern level of structure. Yet, as is apparent from *Confuciusornis* and early beaked ornithurines, the avian beak and loss of teeth must have evolved multiple times in birds. Luis Chiappe has historically tended to view the evolution of birds in a more linear, orthogenic fashion, proceeding from the urvogel *Archaeopteryx* through the enantiornithines to the ornithurines and modern birds, a view following a strict cladistic interpretation.[237] A less orthogenic view of the early evolution of birds, with a dichotomy between enantiornithines and other archaic birds on one hand and early ornithurines on the other, is validated by the discovery of complete skeletons of Early Cretaceous ornithurine birds at rocks of the same geologic age, including notably *Hongshanornis, Yanornis, Yixianornis,* and *Gansus*. That these were birds of the shoreline environment, as noted, suggests that early ornithurines were probably limited to near-shore environments, whereas enantiornithines and other archaic birds dominated terrestrial ecosystems. It is clear from the Chinese fossils that by the Early Cretaceous the two main lineages of Aves, the Sauriurae and Ornithurae, had split, the archaic sauriurines destined to become extinct at the close of the Cretaceous, while the ornithurines survived and gave rise explosively in the Early Tertiary to the modern avifauna. The other revelation of the Chinese avifauna is that the evolution of flightlessness, not surprisingly, is almost as old as the evolution of flight itself, and among the Cretaceous avifauna are forms that represent almost all stages of flight and flightlessness. The fantastic revelation of the Chinese Early Cretaceous must rate as perhaps the greatest boom in the modern era of vertebrate paleontology.

Despite these remarkable discoveries and the general agreement as to their relationships, Bradley Livezey of the Carnegie Museum of Natural History, and his associate Richard Zusi, using a cladistic analysis of some 2,954 morphological characters of 150 avian taxa, come up with a classification that groups *Archaeopteryx* and *Confuciusornis* in the Superorder Archaeornithes and all Mesozoic birds in the Parvclass Palaeoaves.[238] Such a conclusion can only generate additional distrust in a methodology that has already proved incapable of sorting out massively convergent groups, such as loons and grebes, and hawks and owls, which Livezey and Zusi, of course, include in a single clade. Aside from the annoyance of the misspellings of scientific names in the text and figures, numerous misquotations, and erroneous statements on fossil taxa, the very doubtful character scorings make the work more a catalog of detailed avian anatomy than any work from which a veracious phylogeny can be gleaned.[239] This work should at best be considered as a Sears and Roebuck catalog of detailed avian morphology, coded by human hands and therefore often through the eyes of the beholder, to produce a phenetic, ecomorphological association of the world's birds.

In contrast, careful cladistic approaches to discerning phylogenetic affinities of the Early Cretaceous birds have been quite successful, illustrating that at the level of species and genera the approach

works admirably and provides the most prudent approach available today for describing such fossils. However, when one ascends to the family, ordinal, and even higher levels, cladistics and other approaches often lose their ability to divine relationships because of multiple complexities, including primarily massive convergences and resulting homoplasy.

Conclusion

The earliest known adaptive radiation of birds is now well documented by the recent remarkable discoveries from the Early Cretaceous of Liaoning Province and neighboring areas in northeastern China. The fossils include primitive urvogels or basal birds, not far removed from the Solnhofen *Archaeopteryx*. Other specimens provide evidence of a dichotomous branching early in the evolution of birds, including the enantiornithines, or opposite birds, on one hand, and archaic modern-type birds, the ornithurines (which, surprisingly, occur coeval with the sauriurines), on the other hand. One must now ponder whether the ornithurines go back to the time of the urvogel *Archaeopteryx*, as they are certainly present shortly thereafter, fully developed, within a maximum time-frame of approximately twenty-five million years. Together both avian clades, Sauriurae and Ornithurae, show a great range in size, flight capability, diet, and ecological adaptive zones. A general decrease in body size appears to have characterized the initial trend toward more sophisticated flight architecture, and the Early Cretaceous ornithurines appear to have achieved near-modern flight capability. Early ornithurines apparently lived primarily in near-shore habitats and show all the signs of having been endothermic. The enantiornithines, which had a distinctive flight apparatus, less advanced than that of ornithurines, and a different, more primitive pygostyle, were probably physiologically not as advanced as their ornithurine counterparts. Enantiornithines were the dominant land birds of the Mesozoic and were predominantly arboreal or scansorial, well adapted to the flourishing forested regions of the warm Early Cretaceous. Many were herbivores, some were specialized seedeaters, and still others were insectivores or omnivorous, some perhaps feeding on fish. In addition, the volant fauna exhibited a significant radiation of pterosaurs.[240]

The greatest challenge for Mesozoic paleontologists now will be to tease out the avian lineage from that of true theropod dinosaurs. Certainly, the sophisticated avian flight hand, with its precise regimentation of primary feathers in juxtaposition to the secondary flight remiges of the forearm, or brachium, along with a reversed hallux and lack of dinosaurian sacral architecture, suggest that the microraptors are indeed derivatives and not close to the ancestry of the avian lineage. Yet they are almost certainly related to the other, earthbound members of the Dromaeosauridae, including the much later *Deinonychus*, *Velociraptor*, and lesser-known genera possessing the characteristic stiffened dromaeosaurid tail and hypertrophied second pedal sickle claw. Are the troodontids and oviraptorosaurids, as well as the alvarezsaurids and therizinosauroids, also part of this lineage? Is it not reasonable to hypothesize that these forms are also part of the avian lineage, albeit one that achieved flightlessness early in the Cretaceous Period? Workers have pondered why there were no ratites or equivalent radiation of flightless birds during the Mesozoic, but perhaps this radiation has been revealed in the fossil record in the form of hidden birds, secondarily flightless remnants of the early avian radiation, the true identities of which we are just now discovering.

Yet *Archaeopteryx* is still the classic urvogel—the oldest well-studied bird yet discovered, perhaps some 25 or more million years older than most of the Early Cretaceous Chinese fossils. As we saw in chapter 3, the Solnhofen urvogel is a mosaic of reptilian and avian features, a true bird, and the more it is studied, the more and more bird-like it is revealed to be. Ignoring the element of geologic time, however, many paleontologists have proposed that the Liaoning fossils provide evidence for all the stages of the evolution not only of birds and bird flight but also of feathers, from fiberlike protofeathers to pennaceous, asymmetrical flight remiges. Such a claim is remarkable and would be astounding in any fauna, but is especially so for a fauna so far temporally removed from the time of avian origins, presumably before

the Middle Jurassic and perhaps well back into the Triassic. University of Pennsylvania paleontologist Peter Dodson, remarking on the inadequacies of cladistic methodology, tells us: "To maintain that the problem of the origin of birds has been solved when the fossil record of Middle or Late Jurassic bird ancestors is nearly a complete blank is completely absurd. The contemporary obsession with readily available computer-assisted algorithms that yield seemingly precise results that obviate the need for clear-headed analysis diverts attention away from the effort that is needed to discover the very fossils that may be true ancestors of birds. When such fossils are found, will cladistics be able to recognize them? Probably not."[241]

As our very talented and distinguished Chinese colleagues Zhonghe Zhou, Fucheng Zhang, Xing Xu, and others progress in their discoveries of early birds, perhaps new quarries will be opened in earlier lacustrine environments of the Middle to Early Jurassic and even earlier in the Triassic. These sediments could reveal answers to many unresolved questions about the origin of birds, flight, and feathers, and one eagerly awaits the day of their discovery. Indeed, Xing Xu and colleagues recently reported on their discovery of the approximately 160-million-year-old, small-crow-sized, feathered *Xiaotingia*, close to *Anchiornis*, within Archaeopterygidae.[242] Not surprisingly, cladistic analysis, albeit tenuous, shows the urvogel *Archaeopteryx* closer to *Xiaotingia* and *Anchiornis* than to other birds, shifting these forms to the Deinonychosauria (based largely on skull characters), a view not fully embraced by some reviewers. Xu and colleagues number the three fingers of *Xiaotingia* and other maniraptorans as II-III-IV, a view endorsed here. Feather impressions and elongate femoral feathers indicate a four-winged, or tetrapterygian, condition, strikingly similar to that hypothetically and presciently envisioned by William Beebe in 1915, which he termed a "tetrapteryx" stage in avian flight. One can legitimately ask why *Xiaotingia*, *Anchiornis*, *Microraptor*, and other four-winged forms are not reconstructed as trunk-climbing, arboreal avian gliders instead of as cursorial theropod predators, the prevalent view among paleontologists? Are *Xiaotingia* and *Anchiornis* among the "hidden birds of China"?

5

BIG CHICKS

....

The Flightless Birds

The Child is father of the Man.

William Wordsworth, "The Rainbow"

The evolution of the flightless condition in birds has been among the most intriguing and most studied features of avian evolution, in part because of people's fascination with large animal size in general, harkening back to the discovery of dinosaurs and the giant New Zealand moa in the nineteenth century. In the early years after their discovery, flightless birds were thought by Thomas Henry Huxley and others to be direct derivatives from dinosaurs that led to modern volant birds. We now know that these views were erroneous, but the general, superficial appearance of the large flightless ratites (ostrich and allies) to dinosaurs has traditionally lured most ornithologists to view them as primitive within living birds and has created considerable confusion regarding the evolution of birds and their putative relationship to dinosaurs.

Small or Large Steps for Evolution

In *On the Origin of Species*, Charles Darwin made a point of arguing in favor of a view that he articulated with the phrase *"natura non facit saltum"* (nature does not make jumps, originally from Linnaeus).[1] This is essentially the concept of gradualism in biological evolution. In reviewing the book in 1860, aside from coining the term "Darwinism,"

Huxley expressed concern with Darwin's disregard for the possibility of saltation (jumping), or sudden macroevolutionary events in the scheme of biological evolution.[2] Darwin seemed to be stuck on the proposition of gradualism, perhaps in part because of his early association with Charles Lyell (1797–1875), England's most prominent geologist of the day. Lyell's first and most influential work, the three-volume *Principles of Geology* (1830–33), introduced the foundational doctrine of the geological sciences, uniformitarianism, summarized by the dictum, "The present is a key to the past."[3] Thus, the natural processes operating in past geologic time were the same as those observed today. Applying his uniformitarian principle, Lyell argued that extant geologic formations had formed through the steady accumulation of small changes over long periods of time, and this principle had an enormous influence on the young Darwin, who was given volume 1 of *Principles* by John Stevens Henslow just before HMS *Beagle* began its famous voyage in 1831. The influence is clearly seen when Darwin writes of seeing rock formations on Saint Jago, the first stop of the *Beagle*, "through Lyell's eyes." Ironically, Darwin received volume 2 of *Principles* while in South America, and in that volume, Lyell clearly rejected organic evolution and substituted instead a concept of "Centres of Creation," which was not too distant from the later view of Richard Owen.

The term "uniformitarianism" was actually coined by William Whewell in 1832, and he also was responsible for introducing the term "catastrophism," intending to convey the supernatural process of the earth's creation by a series of catastrophic events—in other words, by processes that could not be observed today.[4] Whewell was a philosopher, scientist, Anglican priest, and one of the Cambridge dons whom Darwin had encountered during his tenure there. Incidentally, Whewell invented the term "scientist." However, with respect to geology, whatever lingering influence there might have been on Darwin from Whewell was clearly overshadowed by that of Lyell and his famous *Principles*. Huxley and Darwin had many discussions on the topic of saltation, and in a letter to Lyell dated 25 June 1859, Huxley unveiled his position in stating that "the fixity and definite limitation of species, genera, and larger groups appear to me to be perfectly consistent with the theory of transmutation. In other words, I think transmutation may take place without transition."[5] Then, in a letter to Darwin dated 23 November 1859, Huxley expressed his joy at having read *Origin*: "a lucky examination having furnished me with a few hours of continuous leisure." He then brought up what he calls his "only objections," among which were: "That you have loaded yourself with an unnecessary difficulty in adopting *Natura non facit saltum* so unreservedly," and "It is not clear to me why, if continual physical conditions are of so little moment as you suppose, variation should occur at all."[6]

Yet how and why could such saltation occur? Perhaps the main mechanism known today is alteration of the timing of the developmental process. Such a change in the timing of the events of ontogeny, which may lead to changes in size, shape, and morphology of the adult organism, is termed "heterochrony," a word introduced by Huxley's close friend and scientific colleague, the famous German embryologist Ernst Haeckel, in 1875.[7] Modification of the developmental process can be accomplished through comparatively simple changes affecting control genes regulating the timing of the developmental process.

Heterochrony

The lexicon of heterochrony, simply stated as phyletic change in the timing of development, has been complicated and inconsistent, but many authors follow a modern nomenclature articulated by the late developmental biologist Pere Alberch and colleagues in 1979 and followed in *Evolutionary Change and Heterochrony*, edited by Kenneth McNamara.[8] Heterochrony is considered by many to be the most common mechanism of developmental change and thus of evolutionary change, and it takes on an ascendant role in the evolution of vertebrates.[9] The terminology is complex, but here we are concerned primarily with paedomorphosis, in which a descendant undergoes less growth during ontogeny than its ancestor. If the descendant undergoes more growth, the process is known as peramorphosis. These states can be achieved by varying the onset or offset timing or the rate of development. In birds, the most common process that produces paedomorphosis is perhaps neoteny, in which the actual rate of growth is slowed, and many ornithologists have thus referred to flightless birds as neotenic, more properly neotenous, following Gavin de Beer.[10] Whatever the case, most flightless birds are the result of paedomorphosis, with the underlying heterochronic mechanism being neoteny or some other undiagnosed candidate.[11] And as far as an understanding of the ornithological literature is concerned, most of the literature adheres to such definitions as we encounter in Stephen Jay Gould: "Paedomorphosis: The retention of ancestral juvenile characters by later ontogenetic stages of descendants," and "Neotony: Paedomorphosis . . . produced by retardation of somatic development."[12]

Man Is a Neotenic Ape

Heterochrony has been considered a key component in human development and hence in human evolution in reference to great apes and especially the chimpanzee. The notion of human paedomorphosis, that man was a neotenic ape, was recognized when juvenile pongids reached the zoos and museums of Europe in the nineteenth century and is patently obvious on examining a juvenile and adult chimpanzee; it is easy to see that the juvenile is much more similar to a human adult than to its own parent. The similarity is particularly apparent in the strong negative allometry of the brain and positive allometry of the jaws.[13] Louis Bolk (1866–1930), professor of anatomy at Amsterdam, promoted the concept that man was the result of a permanent juvenile state of apes—"Bolk's fetalization theory of human origins"—and listed similarities between adult humans and juvenile apes: "Our essential somatic properties, i.e. those which distinguish the human body form from that of the other Primates, have all one feature in common, viz they are fetal conditions that have become permanent."[14] In the case of human paedomorphosis, the adult morphology exhibits resemblances to the fetal state in the ancestral form, particularly evident in the enlarged brain and the flattened face. In other words, retardation of developmental maturation has led to the expression of features found in the embryos or juvenile stages of closely related primates. Importantly, many of the resultant features of the adult morphology are not adaptations produced by natural selection but rather represent nonadaptive consequences of a neotenic phenotype that is itself adaptive. Growth of the head and brain in the chimp starts at approximately the same

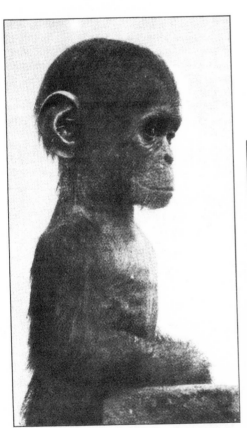

Fig. 5.1. Swiss zoologist and paleontologist Adolf Naef noted in 1926, "Of all animal pictures known to me, this is the most manlike." There is no better illustration than the much-reproduced photograph by Herbert Lang of dead specimens of juvenile and adult chimps, from a 1909 American Museum of Natural History expedition to the Congo. Even if we assume some postmortem posturing for photography, the head is clear evidence of verisimilitude of humans and chimpanzees, which, like the earlier embryo, is strikingly more similar to a human adult than to the adult chimp. A. Naef, "Über die Urformen der Anthropomorphen und die Stammesgeschichte des Menschenschädels," *Naturwissenschaften* 14 (1926): 445–52.

time as in humans but ends at birth; the growth of the head and brain in humans continues for several years following birth, and this accounts for the relatively larger human head and brain. Other notable paedomorphic features of man include the "embryonic" distribution of body hair (including only the head, armpits, and pubic regions), flattened face, short jaw, position of the toe, and degree of head flexion. As another expression of the juvenile nature of man, it is obvious to anyone that the late fetus of a chimp or gorilla more closely resembles adult man than adult ape. As the poet William Wordsworth expressed it, "The Child is father of the Man."[15]

Juvenilization

Given the role of heterochrony in human evolution, the topic is contentious. Although there are many subcategories of heterochrony, we are concerned here only with the phenomenon of paedomorphosis, or as some call it "juvenilization," in which development is arrested and the adults of a species retain characteristics of the juvenile or fetus. First described in 1922 by embryologist Walter Garstang, paedomorphosis today is considered a plausible means by which humans have domesticated everything from dogs (about twelve thousand or more years ago) to chickens and cattle, goats, and sheep, and even Asian elephants.[16] The salient characteristic of juvenility is a common feature, and interestingly, the phenomenon of docility in dogs through paedomorphosis has been suspected in a remarkable Russian forty-year red fox breeding program aimed at domestication that resulted in a similar phenomenon.[17] Through selective breeding, the Russian foxes behave in much the same manner as our dog companions, which have become man's best friend and protective ally. Would that the splendid red fox that I kept as a young boy (along with a red phase screech owl) had shown signs of domestication! Yet such domestication of animals carries heavy ethical burdens, and as the late animal activist Roger Caras proclaimed many times, "We are responsible for anything we domesticate . . . forever."

Fig. 5.2. During the late 1950s, Russian geneticist Dmitry Belyaev began a decades-long program in Siberia to breed domesticated foxes. He hoped to show that physical and morphological changes in domestic animals, particularly dogs, could have resulted from selection for a single behavorial trait, friendliness toward people. The continuing experiments appear to support his hypothesis. From L. N. Trut, "Early Canid Domestication: The Farm-Fox Experiment," *American Scientist* 87 (1999): 160–69. Courtesy Lyudmila N. Trut.

Amphibian Paedomorphosis

Paedomorphosis is best known in amphibians. For example, numerous ambystomid and protean salamanders undergo complex metamorphosis from an aquatic juvenile stage with external gills to air-breathing, land-dwelling adult forms. Perhaps the best known of the temperate zone neotenic urodele amphibians is the mudpuppy *Necturus*, a large salamander with external gills that lives in murky streams and oxbow lakes and is obligately neotenous and therefore permanently aquatic. There are also facultatively neotenic forms associated with aquatic habitats that may at times dry up periodically or become otherwise unfavorable, and these aquatic larvae are capable under such circumstances of undergoing metamorphosis to the adult state. However, the classic example is the Mexican axolotl (*Ambystoma*), a tiger salamander that usually reproduces aquatic larvae, and remain as gilled larvae, not undergoing metamorphosis. Yet, unlike some other neotenic salamanders such as the

North American mudpuppy (*Necturus*), axolotls can easily be induced to undergo metamorphosis by injection of iodine (which is used in the genesis of thyroid hormones) or by simply injecting thyroxine. Upon metamorphosis, the axolotl becomes a perfectly respectable salamander.

As noted, metamorphosis is dependent on thyroid hormones, first demonstrated by J. F. Gudersnatsch in 1912, when he fed horse thyroids to tadpoles of the frog *Rana temporaria*, resulting in their undergoing metamorphosis.[18] In addition to amphibians, paedomorphosis is known to occur in certain termites and cockroaches, as well as in protochordates; given recent genomic findings within the latter taxon, this evolutionary mechanism takes on considerable importance in vertebrate origins.

Possible Mechanisms of Heterochrony

Alistair Dawson and colleagues have been instrumental in elucidating the developmental mechanisms by which the evolutionary process of paedomorphosis has proceeded in birds.[19] Removal of a chick's thyroid gland soon after hatching will inhibit its growth and development, and thyroidectomy can also induce sexual maturation in postbreeding photorefractory European starlings. Dawson and coworkers performed an interesting procedure in which four-day-old starlings were thyroidectomized by injection of radioactive iodine, which resulted in slowed body growth (measured by sternal length), and the birds retained the short, wide bill, large protuberant eyes, and distended abdomen characteristic of young birds.[20] Skull sutures remained unfused in one-year-old birds, and juvenile plumage persisted. Behavior was also affected; the same birds did not learn to feed independently until fourteen weeks, compared to four weeks for controls. Experimental birds also did not develop fully effective thermogenesis and had to be maintained at 25°C (77°F) to prevent shivering. They concluded that "neonatal thyroidectomy . . . caused neoteny in a passerine bird," and they noted numerous similarities between neotenic characteristics of thyroidectomized starlings and those of ratites, suggesting

that hypothyroidism may have been a factor in ratite evolution, "possibly because their evolution occurred in an area deficient in iodine." Dawson and colleagues even speculated that living ratites might be hypothyroid. Such a conclusion would certainly not be extraordinary, as neoteny is well known in amphibians and is thyroid dependent. Undoubtedly the genetic change needed to produce hypothyroidism is very small, and it could have evolved independently in different avian groups.

As noted earlier, lancelets (amphioxus) are an ancient, relict lineage of chordates, and the recent work on their genome has emphasized once again that their close chordate cousins the sea squirts, or tunicates, are in terms of sequence homology most likely the group closest to the ancestry of chordates, even though their adult structure less closely resembles that of chordates than it does that of the lancelets. How can this counterintuitive conclusion be true? Marine biologist Walter Garstang (1868–1949) was the first to study marine invertebrate larvae in detail. He is particularly known for his vehement opposition to Ernst Haeckel's "biogenetic law," his theory of recapitulation, noted in an earlier chapter; and in an address to the Linnaean Society in 1922, he coined the now well known phrase, "Ontogeny does not recapitulate phylogeny; it creates it."[21] Garstang was particularly interested in the phylogenetic relationships of tunicates and chordates, and in the 1890s he had argued that vertebrates arose from some type of "protoechinoderm" through paedomorphosis with the echinoderm larva transforming into an adult tunicate or ascidian. Garstang's theory was that selective forces acting on the larvae promoted a motile larval form with a muscular tadpole-like tail for dispersal and the accelerated development of the gonads and sexual maturity, permitting the abandonment of the sessile, bottom-dwelling adult stage, a transformation analogous to the dramatic metamorphosis seen during the life history of most sea squirts (Urochordata). Through this process the active larval form became a quasi-fishlike, free-living adult. Garstang's paedomorphic larval theory is still alive today, and to bolster his view, there is a group of tunicates that closely conforms to his model: the Larvacea exist only as free-living "tadpole larvae." These so-called tadpole larvae have all the salient characteristics of chordates, including a notochord, a dorsal tubular nerve cord, pharyngeal pouches, and an endostyle (homologous with the thyroid gland). In addition, electron microscopy of the tail musculature has shown it to be identical to the striated muscle of vertebrates. Thus, although the ultimate echinoderm ancestor of vertebrates is a radially symmetrical creature with an exoskeleton of calcite (calcium carbonate) and a strange water vascular system with tube feet, along with other unusual features, evolution could proceed through the larval forms, completely bypassing the adult stage, and therefore major evolutionary change could be possible in a relatively short time.

Flightlessness in Birds

While the myriad examples of paedomorphosis are of great interest to biologists, of paramount interest to us here is the phenomenon of the evolution of flightlessness in birds, for flightlessness is an evolutionary event that may produce major morphological change with very little genetic modification, including avians that in profile bear a striking resemblance to dinosaurs.[22] Unfortunately, there is almost ubiquitous confusion, even among ornithologists, concerning the evolution of the flightless condition. For example, penguins are often referred to as flightless birds. While it is true that penguins do not fly through the air, and are in that sense flightless, they are adapted to literally fly through the dense medium of sea water, and as a consequence, they have an extremely well developed flight apparatus, with a large, keeled sternum and a massive pectoral muscle mass. They likely evolved from ancestral seabirds (analogous to, and perhaps not unlike, the modern diving petrels *Pelecanoides* of the Southern Hemisphere) that are volant but fly into the water to capture prey, swimming with their wings (wing-propelled divers) and then emerging into the air.[23] A number of aquatic avian types have acquired this type of pseudo-flightlessness, including the now-extinct plotopterids, penguin ecological counterparts in the Eocene to Oligocene of the Northern Hemi-

sphere (northern Pacific). This group of often huge, penguinlike birds were pelicaniforms, contained the largest diving birds known, and, like the giant penguins, became extinct by the Early Miocene, when seals and porpoises, which were undergoing an extensive adaptive radiation, occupied the niches for pelagic endotherms of that size worldwide. In addition to penguins and their plotopterid counterparts, the twenty-three species of Northern Hemisphere alcids are wing-propelled diving birds that propel themselves through the sea using strokes of half-folded wings; they include the extinct, flightless Lucas, or mancalline, auks and the famous great auk (*Pinguinus*) of the northern Atlantic. Aside from wing-propelled divers, numerous groups of birds have become foot-propelled divers, reducing the flight apparatus and increasing the mass of the hind limbs. Included in this group are the flightless Early to Late Cretaceous hesperornithiforms (with lobate webbing on the feet as in modern grebes), one living grebe (the Atitlán grebe), the Galápagos cormorant, a possibly flightless giant anhinga, ducks, and others.

The overarching point is that there are myriad flightless seabirds or water birds, and the reasons for the evolution of flight loss may be quite distinctive, but the seabirds and water birds have little to do with the current discussion on avian origins. We are here interested in the terrestrial setting, where land birds evolved, and here we discover a bewildering array of flightless forms ranging from hoopoes, parrots, and rails to giant elephant birds and carnivorous phorusrhacids. Many flightless birds still tell of their ancestry from flying types, but many are so highly modified that we cannot trace their line of descent. One thing is certain, however: all flightless birds have been derived from flying ancestors.

Tertiary Flightless Experiments

In 1876, Edward Drinker Cope discovered the giant terrestrial bird *Diatryma*, from the Paleocene and Eocene of the Northern Hemisphere, which stood about 2 meters (6.5 ft.) tall and had a head nearly 50 centimeters long (1.5 ft.); it is estimated to have weighed some 175 kilos (385 lbs.). These

Fig. 5.3. Skeletal and life reconstructions of the giant Paleogene *Diatryma*. From W. D. Matthew, W. Granger, and W. Stein, "The Skeleton of *Diatryma*, a Gigantic Bird from the Lower Eocene of Wyoming," *Bulletin of the American Museum of Natural History*, no. 37 (1917).

giant dinosaur-like, flightless birds (including the genus *Gastornis*) had a gigantic head and powerful neck, which identified it as a fierce, bipedal predator, superficially a diminutive version of *Tyrannosaurus rex*, both with small, more or less func-

tionless forelimbs but powerful hind limbs.[24] As Stephen Jay Gould noted, "*Diatryma* must have kicked, clawed and bitten its prey into submission."[25] Actually, the short, stout legs of this bird show that it was not a fast runner, but neither were the archaic mammals of the time. The late dean of modern vertebrate paleontology, Alfred Sherwood Romer, wrote: "The presence of this great bird at a time when mammals, were, for the most part, of very small size (the contemporary horse was the size of a fox terrier) suggests some interesting possibilities—which never materialized. The great reptiles had died off, and the surface of the earth was open for conquest. As possible successors were the mammals and the birds. The former succeeded in the conquest, but the appearance of such a form as *Diatryma* shows that the birds were, at the beginning, rivals of the mammals."[26]

Both *Diatryma* and *Gastornis* were gone from North America and Europe by the Late Eocene; they were transitory successes in an attempt by birds to take over the niche of a large, bipedal carnivore, left vacant by the demise of the theropod dinosaurs, but they were ultimately unable to compete in the rapidly evolving world of Cenozoic mammals. South America, in its splendid isolation through most of the Cenozoic, and therefore in geographic seclusion from advanced North American carnivorous, eutherian mammals, produced a radiation of large, swift, predaceous birds known as the terror birds, or phorusrhacids. Phorusrhacids ranged in size from about 1–3 meters (3.3–9.8 ft.) and were represented by as many as eighteen or so species.[27] Known primarily from Argentina and Brazil (one is known from Florida, and there is another group from the Cenozoic of Europe), many were flightless or nearly so, and they were cursorial denizens of the open grassland ecosystems. Their large heads armed with a powerful hooked beak made them highly adapted for killing prey and tearing flesh.[28] In terms of their ecological roles, the adaptive radiation produced three lineages normally placed in separate families: cursorial predators, graviportal scavengers, and cursorial predators capable of limited flight, such as their living close relatives the seriemas (Cariamidae). Most of the phorusrhacids became extinct

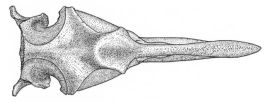

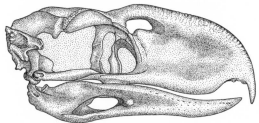

Fig. 5.5. Skull of *Patagornis marshi*, in dorsal and lateral views, showing the powerful, laterally compressed and strongly hooked bill. Length, 35 cm (14 in.). Redrawn from F. Ameghino, *Revista Argentina Historia Natural* 1, no. 4 (1891): 441–45.

Fig. 5.4. An early life restoration of a large Miocene phorusrhacid (*Phorusrhacos longissimus*) from South America, by Charles R. Knight. From Frederic A. Lucas, *Animals of the Past* (New York: McClure, Phillips, 1901).

Fig. 5.6. The large Lower Miocene phorusrhacid *Paraphysornis braziliensis*, shown next to Brazilian paleornithologist Herculano Alvarenga. Courtesy of H. Alvarenga.

when placental carnivores reached South America with the opening of the Panamanian land bridge, which resulted from a buildup in polar ice caps and a concomitant drop in sea level by about 50 meters (164 ft.), approximately 2.5 million years ago. Yet following the opening of the Middle American land bridge and the extinction of South American phorusrhacids, a behemoth species, *Titanis walleri*, persisted in Florida and southern Texas until about 1.8 million years ago.[29] *Titanis walleri* was originally described by University of Florida paleornithologist Pierce Brodkorb as being larger than an ostrich and more than twice the size of South American rheas. The phorusrhacid-like birds from the Eocene and Oligocene of Europe did not achieve exceptionally large size and were likely weak fliers; many paleontologists today believe that they represent a paraphyletic lineage, and they show similarities to a number of gruiform families.[30]

Aside from the unusual carnivores discussed above, and although slightly less than 2 percent of avian species are flightless, the total number and diversity of flightless species is nevertheless astounding, illustrating the phenomenon as pervasive within birds. Flight loss has occurred in eleven nonpasserine and at least two passerine families, depending on exactly how one defines flightlessness.[31] This diversity includes everything from the fleet-footed ostrich to a now-extinct flightless hoopoe. Yet of all the flightless species alive, none has attracted more attention than the large ratites, named from the Latin *ratis*, raft, referring to the flat, unkeeled sternum. This term sets them in contrast to all other living birds, the carinates, from the Latin *carina*, meaning the keel of the ship, in reference to the keeled sternum for the attachment of the major flight muscles.

Ratites: Large Flightless Birds

The enigmatic Ratitae, which comprises the living African ostrich, the South American rhea, the Australian emu and cassowary (also of New Guinea), and the New Zealand kiwi, also includes a long list of extinct forms, notably the dozen or so species of New Zealand moa and an undetermined number of species of elephant birds from Madagascar. Australia also featured a major extinct group called the mihirungs, or dromornithids, which resembled ratites in size and morphology and were long thought to be of that group but, following the discovery of their bizarre, massive skulls, were shown to be anseriform (ducklike) derivatives.[32] The morphological diversity of ratites is matched

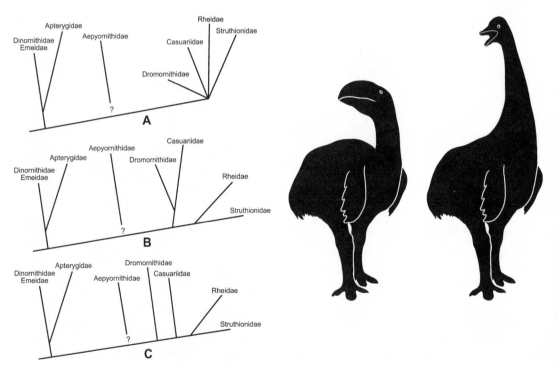

Fig. 5.7. Left, various attempts using phylogenetic systematics to assess the relationships of dromornithids. However, only the discovery of the skull demonstrated that the mihirungs were of anseriform rather than ratite affinity. Right, silhouettes of dromornithids: left, after discovery of their skull and anseriform affinity; and right, before the skull discovery, when they were considered ratites. Adapted and modified from P. Rich, "The Australian Dromornithidae: A Group of Large Extinct Ratites," *Los Angeles County Museum of Natural History, Contributions to Science* 330 (1980): 93–103.

only by the number of theories advanced to explain their origins.

Charles Darwin, in the first edition of *On the Origin of Species*, in near Larmarckian ("use and disuse"—a long-discarded theory of evolution) fashion, promoted the view that the ratites were derived from flying ancestors through "disuse" of the wings and the concomitant increased use of the hind limbs:

> There is no greater anomaly in nature than a bird that cannot fly; yet there are several in this state. . . . As the larger ground-feeding birds seldom take flight except to escape danger, I believe that the nearly wingless condition of several birds, which now inhabited or have lately inhabited several oceanic islands, tenanted by no beast of prey, has been caused by disuse. . . . We may imagine that the early progenitor of the ostrich had habits like those of a bustard, and that as natural selection increased in successive generations the size and weight of its body, its legs were used more, and its wings less, until they became incapable of flight.[33]

Yet it was Darwin's primary competitor in the race for a theory of evolution by natural selection, Alfred Russel Wallace, who appeared to have a fully comprehensive view of ratite evolution through arrested development: "Since, therefore, the struthious birds [ostrich and allies, or ratites] all have perfect feathers, and all have rudimentary wings, which are anatomically those of true birds, not the rudimentary forelegs of reptiles, and since we know that in many higher groups of birds—as the pigeons and the rails—the wings have become more or less aborted, and the keel of the sternum greatly reduced in size by disuse, it seems probable that the very remote ancestors of the rhea, the cassowary, and the apteryx [kiwi], were true flying birds."[34]

Darwin's longtime adversary Richard Owen, whose views on evolution were very uncertain, referred to the "Cursores" or ratites, as "those birds in which the power of flight is abrogated," coming surprisingly close to a view that ratites were descended from birds that were formerly volant.

Owen explained the loss of flight through "arrested development of the wings unfitting them for flight."[35] In other words, Owen was advocating a paedomorphic origin for the flightless condition in ratites, which was subsequently supported by numerous paleontologists and ornithologists including Max Fürbringer (1888), Robert Broom (1906), J. E. Duerden (1920), Erwin Stresemann (1927–34), and William King Gregory (1935).[36] Owen's clear view of ratite evolution predated by some ninety years the splendid work of Sir Gavin de Beer, in a monograph published in 1956, and it was the first comprehensive biological explanation for the general evolution of flightlessness in birds.[37]

Here we are using the term "paedomorphosis" to denote this important mechanism by which macroevolutionary change may occur with very minor genetic change, perhaps involving only the timing of developmental events. As noted before, considerable caution must be exercised in using these terms, because although the initial stages in the evolution of a novelty such as flightlessness may be initiated by heterochronous developmental events (attested to by lack of anatomy associated with late stages of development), other features, such as hypertrophy of hind limbs and associated architecture for cursorial locomotion, may be due to subsequent selection for a specific biological role and not heterochronous development.

The first recognition of heterochrony for any avian species is from H. E. Strickland and A. G. Melville's 1848 description of the Mauritius Island dodo, *Raphus cucullatus*:

> These birds were of large size and grotesque proportions, the wings too short and feeble for flight, the plumage loose and decomposed, and the general aspect suggestive of gigantic immaturity. . . . We cannot form a better idea of it than by imagining a young Duck or Gosling enlarged to the dimensions of a Swan. It affords one of those cases . . . where a species, or a part of the organs in a species, remains permanently in an underdeveloped or infantine state. . . . The dodo is (or rather was) a *permanent nestling*, clothed with down instead

of feathers, and with the wings and tail so short and feeble, as to be utterly unsubservient to flight.[38]

Rails: Models for the Evolution of Flightlessness

A dozen or so groups of living birds have independently evolved flightless or weak-flighted derivatives. Of all living birds, rails (Gruiformes: Rallidae) are by far the most likely to lose flight, and it would appear that if selection can favor that condition, the rails have a tremendous proclivity to abandon their flight architecture.[39] More than one-fourth of all rails, living or recently extinct, totaling more than sixty species, have lost the ability to fly. All flightless rails are island forms, the islands ranging in size from the tiny, austere Laysan to the large, topographically diverse New Guinea and New Zealand. In Hawaii some twelve species of flightless rails developed, and "it is likely that all islands in the Pacific were inhabited by one or more species of flightless rail before the arrival of humans."[40]

Rails therefore provide a living laboratory for the study of the developmental, structural, and physiological changes that occur in the evolution of flightlessness. By examination of the chicks of rails and galliforms (chickenlike birds), one can easily see the two extremes in the relative development of the pectoral flight apparatus and its associated muscle mass at hatching, as well as the relative ease with which the developmental mechanism of paedomorphosis can result in the evolution of flightlessness. Rails, hatched with almost no development of the sternal keel and associated flight apparatus, can easily arrest development of the sternal keel and give rise to flightless species, whereas galliform birds exhibit a near fully developed flight anatomy, and they cannot easily develop flightless species. For galliform species, or birds like them that hatch fully formed, embryological development would have to be arrested far in advance of hatching, when other organs have not yet developed, and the results might well be lethal.

Storrs Olson vividly demonstrated in a 1973 paper the precise changes in the posthatching de-

velopment of the sternum in a purple gallinule (*Porphyrio martinica*), a flying rail, illustrating that in early stages of its development the keel, or carina, of the sternum corresponds to the shape of the keel in two flightless species.[41] The downy chick at about a week from hatching still has a completely cartilaginous sternum, closely resembling that of the flightless New Zealand rail, the weka (*Gallirallus australis*). At a later stage, when the chick is fully feathered but not quite volant, the sternal development resembles that of the flightless Guam rail (*Gallirallus owstoni*). Using cleared and stained specimens of the king rail (*Rallus elegans*), Olson further showed that going from seventeen to forty-seven days after hatching, the articulation of the scapula and coracoid progresses from an obtuse angle in the younger rail to an acute angle in the older bird; the acute angle shortens the distance through which the dorsal elevators of the wing must act in affecting the recovery stroke of the wing, thus producing greater power.

When almost any bird becomes flightless, it resembles an earlier stage of development of a volant member of the group, and other changes universally include an immediate reduction of the muscle mass and bones of the wing and pectoral girdle. As noted, the keel of the sternum is reduced or lost along with the pectoralis and supracoracoideus flight muscle mass. Because most flightless birds are no longer under selective pressure for a lightened skeleton for flight, they tend to increase in size, often dramatically, as in the ratites. In other words, as birds are similar in possessing the ability to fly, they likewise become structurally similar in the loss of flight. The flight feathers, designed as aerodynamic structures, with asymmetric vanes to form individual airfoils, revert to symmetry. Contour feathers, which produce smooth body contours for laminar flow in flight, tend to become loosely constructed. If there is not continued selection to warrant the energy expenditure for the embryogenesis of these complex structures, whether muscles, bones, or feathers, they will tend to be lost in flightless birds. Birds are essentially flying machines, and the disappearance of the mitochondrial rich and energetically expensive flight muscles is a watershed in terms

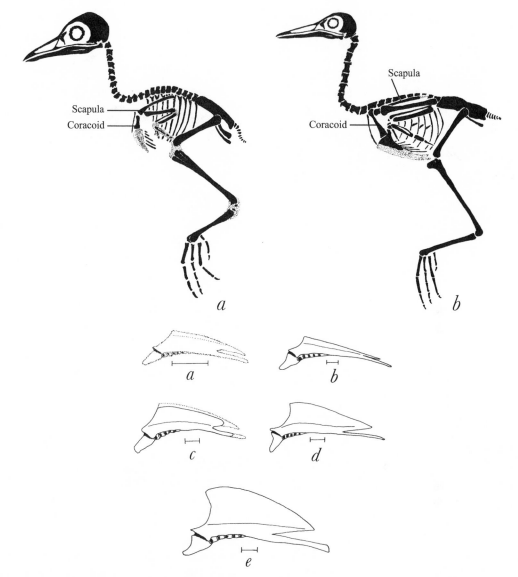

Fig. 5.8. Above, cleared and stained skeleton of the king rail (*Rallus elegans*), a flying rail: (a) at seventeen days after hatching; (b) at forty-seven days, reduced so that the femur lengths in the two drawings are equal. Stippled areas represent cartilage. Note that the articulation of the scapula and coracoid forms an obtuse angle in the younger form but an acute angle in the older form. Below, development of the sternum of the purple gallinule (*Porphyrio martinicus*), a flying rail (a, c, e), showing how the carina in its early stages corresponds to the shape of the carina in two species of the flightless rails (b, d): (a) *P. martinina*, downy chick about a week old; the sternum is entirely cartilaginous but has nearly the same conformation as: (b) the weka (*Gallirallus australis*), a flightless rail, adult; (c) *P. martinica*, an immature that is fully feathered but not quite volant; the sternum is still partly cartilaginous and now resembles: (d) the Guam rail (*Gallirallus owstoni*), adult; (e) *P. martinicus*, adult. Scale bars = 5 mm. Dotted lines indicate cartilage. From S. L. Olson, "Evolution of the Rails of the South Atlantic Islands (Aves: Rallidae)," *Smithsonian Contributions to Zoology* 152 (1973): 1–53. Courtesy Storrs Olson.

Fig. 5.9. The weka (*Gallirallus australis*) of New Zealand. Wekas are aggressive, chicken-sized, flightless rails. Formidable predators, they have added introduced mice and rats to their culinary repertoire. Photograph by the author, Stewart Island, New Zealand.

Fig. 5.11. The avian order Galliformes, represented here by the Malleefowl (*Leipoa ocellata*), includes in addition the curassows, guans, and chachalacas (Cracidae), guineafowl (Numididae), and pheasants, quail, grouse, and turkeys (Phasianidae). Galliform birds are mischaracterized as being exclusively ground-dwellers. In fact, they are among the strongest avian fliers, capable of explosive-burst takeoffs from the ground, and the vast majority roost in trees. Drawings by George Miksch Sutton, reproduced with permission by Dr. Dorothy S. Fuller.

Fig. 5.10. Chicks of the clapper rail (*Rallus longirostris*), left, and the great currasow (*Crax rubra*), a galliform, right, represent the two extremes in relative development of the pectoral flight apparatus of the downy young. Rails, born with essentially no development of the sternal keel and associated flight apparatus, easily arrest development and give rise to flightless species. Galliform birds, by contrast, are endowed with nearly fully developed flight architecture at birth. For them to give rise to flightless species, development would have to be arrested long before hatching, when other organs have not yet developed, and the results might be lethal. Left, modified after D. Ripley, *Rails of the World* (Boston: Godine, 1977); right, adapted and modified after J. Delacour and D. Amadon, *Curassows and Related Birds* (New York: American Museum of Natural History Press, 1973).

of energy savings. As an illustration of the magnitude of such change, the flight muscles can account for as little as 7.8 percent of body weight in the white-throated rail (*Laterallus albigularis*), about the same as that estimated for the urvogel *Archaeopteryx*, to a high of some 36.7 percent in Cassin's dove (*Leptotila cassini*). Another measurement is that flight muscles, combined with wing muscles, bones, and sternum, combine to produce an average of 20–25 percent of total body weight in modern birds. Naturally, any reduction in these energy-consuming flight muscles confers tremendous energy savings.[42] Jared Diamond put it beautifully: "The energy saved by flightlessness can be put to other purposes. A winged rail on a predator-free island is like a 60-kg backpacker forced eternally to carry 15 kg of bricks and to regurgitate half of each meal."[43]

The simplest way to jettison the flight apparatus is through arrested development, or paedomorphosis. All known birds possess the characteristic features of flightless birds at some stage of their development in the egg, including a disproportion-

Fig. 5.12. Living ratites ("Cursores" of Richard Owen) are a diversified assemblage of flightless birds whose relations and biogeography have been controversial. Bottom right, red-winged tinamou (*Rhynchotus rufescens*), a putative close ally of the early Eocene lithornithids, thought to be volant ancestral forms. It is postulated that various members of Lithornithidae flew to various continents and islands and gave rise to the living flightless ratites through neoteny. Top right, going counterclockwise, emus (*Dromaius novaehollandiae*, Australia); cassowaries (*Casuarius unappendiculatus*), Australia and New Guinea; kiwis (*Apteryx australis*), New Zealand; rheas (*Rhea americana*), South America; and ostriches (*Struthio camelus*), Africa. Although from Huxley's time to the present it has been argued that ratites are ancient, there is no evidence for such a proposal; they are exclusively Cenozoic and evolved independently from volant forms that flew to Australia, New Zealand, Africa, and South America. From A. Feduccia, *The Origin and Evolution of Birds* (New Haven: Yale University Press, 1996). Drawings by George Miksch Sutton, with permission from Dr. Dorothy S. Fuller.

ately large pelvic region and hind limbs, greatly re-duced wings, and a greatly reduced sternum with little or no keel or flight muscle. By altering one or several developmental regulatory genes—those controlling timing of the developmental period and therefore arresting perinatal development—flightlessness can easily be achieved in the adult, and the change can be quite remarkable, resulting in an individual often quite distinctive from its im-mediate ancestor. These ontogenic events involv-ing heterochronous developmental alterations are the probable factors involved, as noted earlier, in major events in vertebrate history, including chor-date origins, the amphibian radiation, and even the origin of man from apes.

Obviously, when development is arrested and affects a particular organ, other organs will also be arrested at that particular stage of the developmen-tal process, and therefore such an arrest of devel-opment can take place only after essential organs have developed sufficiently to sustain the normal processes of life. The salient point is that because many types of birds have a precocial development of the pectoral muscle mass and flight apparatus, they are unable to develop flightless species. By examination of chicks one can determine if the group is a major candidate for the origin of flight-less species. For example, in contrasting chicks of the clapper rail (*Rallus longirostris*) and a gal-liform, the great curassow (*Crax rubra*), we see two of the extremes in the relative development of the flight apparatus. Because the flight apparatus de-velops late in rails, there is a strong propensity for flightless species to develop, whereas in galliforms, developmental arrests would have to occur far in advance of hatching, when other organs would not be developed enough to be functional, and the results might therefore be lethal.

Gruiforms (rails and allies), pigeons (and doves and dodos), and grebes are groups that have late development of the sternum and flight apparatus and that thus show a penchant for the evolution of flightless species. Postponed devel-opment of the flight apparatus and sternum in these groups allows paedomorphosis, resulting in the loss of the flight without also resulting in the loss of the bird.

Flightless gruiforms also have such neo-tenic features as a large pelvis and hind limbs and greatly reduced wings, exactly the features we also see in the ratites and most flightless birds. But in addition, associated with flightlessness and gigan-tism there tends to be an increase in longevity, a decrease in metabolic rate, and a consequently im-proved capacity for fasting, which might be a con-siderable advantage for island frugivores, where there is seasonal productivity of fruit.

It is also interesting to note that a distinctive set of characters recurrently appears in secondarily flightless birds from whatever group they evolve. These characters, including but not limited to a keel-less sternum, an obtuse angle of scapula and coracoid, reduced wings, and degenerate feathers, appear more or less the same in everything from flightless rails, to geese, ibises, grebes, and passer-ines. Yet many of these features have been used in cladistic analyses of the ratites in attempts to dem-onstrate relationships. It is the lack of recognition by those who practice phylogenetic systematics of paedomorphosis as a macroevolutionary mecha-nism, by which an iterative evolution from differ-ent ancestry can give rise to quite similar forms, that has led to errors in evaluating the phylogeny and biogeography of the living ratites.

Biogeography by the Cladogram: Ratites Are Tertiary "Big Chicks"

The modern practice of phylogenetic reconstruc-tion calls for a cladistic analysis that provides an adequate beginning point for further inquiry but should not be considered a final authority. Un-fortunately, such cladograms tend to become in-violate dogma, and the practice dictates that unless someone uses the same methodology to refute the original tree, it remains the preferred hypothesis, with its attendant phylogenetic inferences con-cerning everything from physiology and behavior to biogeography. Orthodoxy aside, errors *do* oc-cur, and the so-called cladistic inferences derived from such cladograms can therefore produce mis-leading conclusions. Use of cladistics for evalu-ating the phylogeny of large flightless birds has proved extraordinarily difficult because of the suite of juvenile characters that distinguish such taxa and that are quite similar in all forms, from rails to ratites, regardless of their individual ancestry. Consequently, corollary inferences based on cladistic analyses of flightless birds have tended to produce grossly misleading conclusions.

One such error is the inference that the large flightless ratites are ancient and shared a common flightless ancestor before the breakup of the major continents.[44] According to this hypothesis, the putatively flightless ancestors of extant ratites were passengers on drifting continents, leaving the rhea on South America, the ostrich on Africa, and the emu and cassowary on Australia. The term "vicariance" is used to describe the seemingly mysterious disjunctive distribution patterns of related organisms caused by geographic barriers, often oceans, that split the range of ancestral species, or geologic events that separated once continuous ranges. Perhaps the most widely proclaimed example involved the splitting of the southern continent Gondwanaland, giving rise to South America, Antarctica, Africa, and Australia. According to this so-called vicariance view, the pattern of the avian fossil record is extremely biased. The rich Tertiary record of birds exists because sediments of that age are extremely abundant compared with those of the Cretaceous, and what Cretaceous fossils are found are discovered mostly in the Northern Hemisphere because that is where most paleontologists search for fossils. Over the past three decades much of the biogeographic distribution of major avian groups has been attributed to continental drift or plate tectonics, and therefore to what has been called vicariance biogeography. In particular, Joel Cracraft of the American Museum of Natural History has used vicariance biogeography to attempt to attribute most disjunct avian distribution patterns to drifting continents. The application of vicariance biogeography to the analysis of the origin of ratites is merely the best known example of Cracraft's general strategy for explaining much of avian biogeography.

Although the ratites can only be traced to the Early Paleogene, and mainly the Eocene, there is simply no evidence whatsoever that any ratite existed in the Mesozoic, not even in the Late Cretaceous, although there are flightless birds, some of substantial size, from that period. Thus, there is no reason to believe that ratites were distributed by any tectonic events. Even so, Yale University's Charles Sibley and colleague Jon Alquist in 1990 and other publications became enamored with Cracraft's hypothesis for a Gondwanaland origin of ratites and used the split of the continents to calibrate their molecular clock.[45]

Lithornithids: Volant Paleognaths

Our understanding of the evolution of ratites changed dramatically when Peter Houde of New Mexico State University began to discover fairly complete fossils in nodules from the Eocene of North America. Houde's breakthrough in the seemingly intractable problem of the origin of ratites came in 1981 with the description by Houde

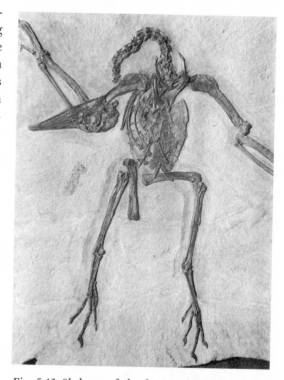

Fig. 5.13. Skeleton of the flying paleognath *Pseudocrypturus cercanaxius*. From P. Houde, *Paleognathous Birds from the Early Tertiary of the Northern Hemisphere* (Cambridge, MA: Nuttall Ornithological Club, 1988). Courtesy of Peter Houde.

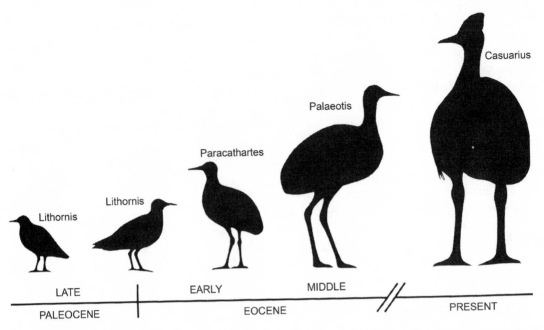

Fig. 5.14. Grades of ratite evolution as represented by the series of fossils *Lithornis celetius, L. promiscuus, Para-cathartes howardae, Palaeotis weigelti,* and a modern cassowary (*Casuarius*), to scale. From P. Houde, *Paleognathous Birds from the Early Tertiary of the Northern Hemisphere* (Cambridge, MA: Nuttall Ornithological Club, 1988). Courtesy of Peter Houde.

and Storrs Olson, published as a cover article in *Science*, of fossil material of lithornids, fully volant, midsized paleognathous carinate birds from the Paleogene of North America.[46] These remarkable birds, now known from the Paleogene of western North America and Europe, had well-developed wing and pectoral architecture, and many fossils exhibited a well-preserved paleognathous palate. Actually, lithornithid fossils had been sequestered in specimen drawers in various museums for more than a century, masquerading under the names of disparate families of modern neognathous birds, but the telltale skull with preserved palate had not then been discovered. Houde and Olson concluded, "The new fossil birds reported here are probably remnants of what may have been a diverse radiation of paleognathous carinates. . . . Tinamous and ratites may have descended independently from various families or orders within this radiation of paleognaths, or some of the ratites may have evolved secondarily from neognathous birds through neoteny." The order Lithornithiformes differs from the only living volant paleognaths, the Central and South

American tinamous (Tinamiformes), in some specific characters, especially the sternum, and unlike the terrestrial tinamous with their flat claws, the lithornithids were arboreal, with long and strongly curved claws, adapted for perching in trees. Despite the differences, Houde further wrote, "I cannot overemphasize the similarity of the Lithornithiformes and the Tinamiformes," and "Analysis of the Lithornithiformes is most appropriately made . . . with the Tinamiformes, the group to which they are phenetically most similar and from which they presumably diverged."[47]

The important point for the present discussion is that although the tinamous are ground-dwellers capable of short bursts of powerful flight from the ground, in the manner of galliforms such as quail, the lithornithids were not particularly cursorial, were adapted for life in trees, and were capable of sustained flight. The controversy surrounding the monophyly or paraphyly of the ratites need not be of concern here; what is striking is that the lithornithids occur in the right time period for dispersal of paleognaths all over the world and then became extinct in the Eocene. This contradicts

Cracraft's model of a single flightless ancestral ratite being widely distributed in Gondwanaland during the Cretaceous (for which no evidence exists) that gave rise to various lineages of ratites, as described here by Houde: "Populations of this flightless pluripotent ratite ancestor already present in Gondwanaland . . . were isolated by ocean barriers as they arose. Ratites are viewed as passive passengers on rifting continents; populations isolated and diverged as daughter continents were born. The present-day distribution of ratites is due only to the distribution of this flightless ancestor in Gondwanaland and the subsequent breakup of this ancient supercontinent to form the existing Southern Hemisphere landmasses."[48]

The euphoria for vicariance biogeography has been dampened not only by Houde's discoveries of the volant paleognath lithornithids but also by the absence of any fossil evidence of ratites before the Cretaceous-Paleogene boundary and by the discovery of a crane-sized, flightless paleognathous ostrich from the Middle Eocene of Germany. This last revelation showed that ratites were not only present in the Northern Hemisphere but evolved there long after the breakup of Gondwanaland and only subsequently emigrated from

Europe to Africa. Although one fragmentary fossil is assigned to the Rheidae from the Paleocene of South America, the oldest true fossil ostrich is Miocene; fossils of New Zealand moa are from the Miocene at the oldest; and there are no confirmable Tertiary fossils of Malagasy elephant birds, even though good Cretaceous deposits there have produced avian remains. To view the elephant birds, moas, and kiwis as passengers on ancient drifting landmasses of Madagascar and New Zealand, respectively, strains credulity. One must also consider that even with respect to the southern continents, their separation at the Late Cretaceous was no more than the approximate distance separating mainland Ecuador from the Galápagos Islands, and midway there was a series of volcanic islands. The ancestors of elephant birds and the moas and kiwis (which, based on DNA comparisons, are of separate origins), like those of the other ratites, must have flown to their current homes and subsequently become flightless.[49]

The Ease of Avian Dispersals

A view of the world map of avian migration patterns, as well as the flight capabilities of even the

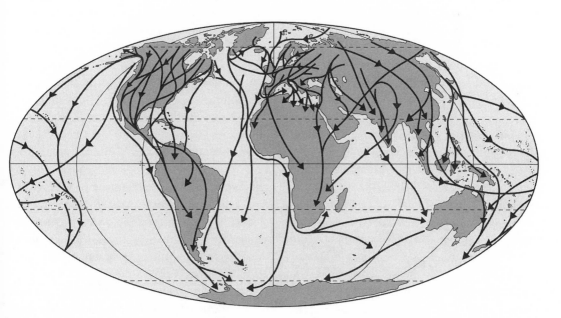

Fig. 5.15. World map showing migration patterns of a number of birds, illustrating the incredible dispersal abilities of birds, often underestimated in drawing biogeographic conclusions. Redrawn from various sources.

"weak" fliers such as rails, in which only a relatively small percentage of total body weight is devoted to the flight apparatus, reveals that birds do get around. The rails are among the most widespread of all avian families and are capable of performing exceptionally long migrations over land and extensive stretches of ocean. Rails have colonized innumerable South Pacific and many other oceanic islands. The large purple swamphen (*Porphyrio porphyrio*), a clumsy-looking, awkward flier, ranges from the Mediterranean basin and Africa across to Southeast Asia, on to Australia, New Zealand, and Micronesia, and in New Zealand is known to cross mountain ranges! Given the flying abilities of birds, combined with new discoveries and lack of any ancient fossils, the most likely explanation is that no birds owe their current distribution to ancient, continental vicariance biogeography. Many of the most puzzling modern distribution patterns, including the pantropically distributed trogons (Trogonidae), and even West Indian todies (Todidae), once thought to be endemic to the islands, have now been found as Paleogene fossils in Europe and North America, illustrating that they once were widespread and are now restricted to a relict distribution pattern.

The Appeal of the Vicariance Model

Despite the mountain of contrary evidence, the vicariance view of ancient flightless ratites and drifting continents has proven appealing to the popular press and to writers of textbooks. Illustrating this point, Richard Dawkins accepted the vicariance model for ratites in his book *The Ancestor's Tale*, in which he tells us that "the ratites reached their present separated homelands without benefit of flight. How did they get there? . . . They walked. All the way. . . . What we now know as separate continents were joined together, and the great flightless birds walked dry shod."[50] Dawkins devotes fifteen pages to the whole laborious story. Regrettably, this hypothesis, though appealing, is wrong.

One of the most overlooked aspects of this discussion is that although the continents split during the Cretaceous, the real biogeographic issue concerns not the split but the distance that separated the continents during the Early Tertiary. For example, the separation of the southern continent, Gondwanaland, during the Paleocene was likely no more than 1,600 kilometers (1,000 mi.), and with island arcs dotting the mid-Atlantic ridge, midway between South America and Africa, the Atlantic Ocean does not appear as formidable a barrier as it is today. The same is true of the separation of New Zealand and Australia and of Madagascar and Africa. Birds migrate! The eastern kingbird (*Tyrannus tyrannus*) nests in North America as far north as central Canada and migrates thousands of miles to winter in South America from Peru to Bolivia. When one considers that the tiny vermillion flycatcher (*Pyrocephalus rubinus*) nests along the Gulf Coast of North America, winters in Argentina, and has managed to find its way to the Galápagos Islands, volcanic islands some 960 kilometers (600 mi.) off the Ecuadorian coast, the barrier for lithornithids dispersing in the Early Tertiary seems quite surmountable.

Fortunately, science marches on, albeit in a punctuated fashion, and a recently published study by John Harshman of the Field Museum of Natural History and colleagues, from polygenomic evidence from ratites, provides compelling evidence for multiple loss of flight in ratites. The authors present evidence from twenty unlinked nuclear genes that unequivocally places the flighted tinamous within the ratite assemblage, making the ratites polyphyletic and strongly suggesting multiple losses of flight. According to this analysis, the most plausible hypothesis would require at least three losses of flight and would thus demand a fundamental reconsideration of the proposals that relate ratite evolution to continental drift. By this model, and considering that the ratites are "big chicks," the many morphological and behavioral similarities within the group originated by parallel or convergent evolution. As the authors appropriately conclude, "Perhaps the impact of our phylogeny should be viewed as yet another example of the phenomenon that Huxley called 'the great tragedy of science—the slaying of a beautiful theory by an ugly fact.'"[51]

The Remarkable Flightless Birds of Hawaii Provide Important Lessons

Among the most recent remarkable studies of the evolution of flightlessness is the work of paleornithologists Storrs Olson and Helen James of the Smithsonian's Museum of Natural History. Their 1991 monograph on the Late Holocene nonpasserine birds from the Hawaiian Islands describes a remarkable avifauna that was extirpated by the Polynesians shortly after their arrival, around AD 300. Interestingly, the original avifauna probably numbered more than fifty species and therefore outnumbered the extant native Hawaiian avifauna.[52] Recall that the Hawaiian Islands and Galápagos Islands are similar in age, at about three million years old, and as such not only provide a showplace of evolution and adaptive radiation but also illustrate the rapidity of the evolutionary process. Aside from the nine or ten species of flightless rails, other exciting discoveries were at least two flightless ibises (genus *Apteribis*), and most pertinent here is that the first leg bones of these ibises that were found were so much shorter and stouter than those of typical ibises their identity was uncertain until additional material was uncovered. "The hindlimb . . . is so modified from that of typical ibises as almost to defy identification. At first we had only the femur, tibiotarsus, [and] tarsometatarsus to work with, and of all modern birds the proportions of these elements most closely approached those of the kiwis . . . and it was not until we received the associated material from Maui . . . that our suspicions were confirmed."[53] The Hawaiian flightless ibises were restricted to Maui Nui and "could not have colonized these islands more than 1.8 million years ago, the age of the oldest rocks found on Molokai, the oldest of the Maui Nui group."[54] These ibises probably foraged by probing the forest floor litter in the manner of the kiwi.

Bizarre gooselike ducks, the moa-nalos (Hawaiian *moa*, "fowl," and *nalo*, "lost/vanished"), were also discovered. These extinct, flightless "geese" (*Thambetochen*, "amazing goose") were in fact ducks that became gooselike in both morphology and size: "The big flightless 'goose-like' birds of the Hawaiian Islands may actually have been derived from something more like a mallard (*Anas platyrhynchus*) and certainly were not derived from true geese, although in their terrestrial,

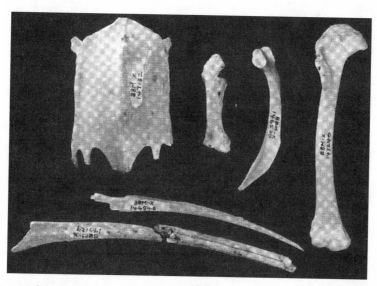

Fig. 5.16. Fossil bones of the flightless Hawaiian ibis *Apteribis glenos*. From left to right and top to bottom: sternum (note lack of keel), coracoid, scapula, humerus (note reduced size), rostrum fragment, and two mandibular fragments. From S. L. Olson and A. Wetmore, "Preliminary Diagnoses of Extraordinary New Genera of Birds from Pleistocene Deposits in the Hawaiian Islands," *Proceedings of the Biological Society of Washington* 89 (1976): 247–58. Courtesy Storrs Olson; photograph by Victor E. Krantz.

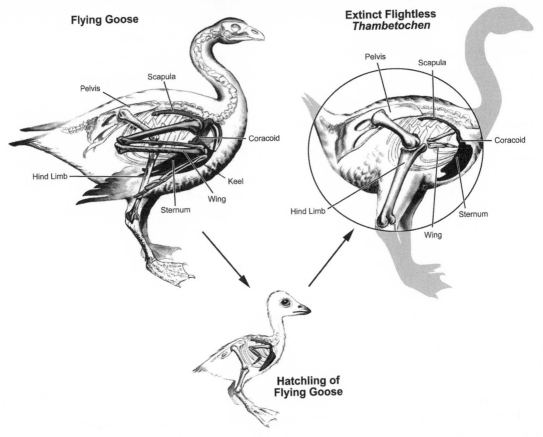

Flying Goose

Pelvis
Scapula
Coracoid
Hind Limb
Keel
Wing
Sternum

Extinct Flightless
Thambetochen

Pelvis
Scapula
Coracoid
Hind Limb
Sternum
Wing

**Hatchling of
Flying Goose**

Fig. 5.17. Compared with a flying goose, the extinct flightless Hawaiian "goose" *Thambetochen* had stout hind limbs, a massive pelvis, small and unsculptured wing bones, and a small sternum with no keel, much like that of ratites. The angle between the coracoid and scapula is obtuse, a trait common among flightless birds but never found in adult flying birds. Some of the skeletal features of adult flightless birds result from the retention of juvenile characters, a process called paedomorphosis, or arrested development. The hatchling of a flying goose, below, is too young to fly. Notice the open, obtuse angle between coracoid and scapula, reduced wings, and absence of a keel. Flying birds, like the modern goose above, have well-developed scapulae and coracoids that meet at a sharp, approximately 90° angle and help brace the wing bones. They also have a sizable keel for the attachment of the major flight muscles. Modified from drawing by Douglas Cramer, from H. F. James and S. L. Olson, "Flightless Birds," *Natural History* 92 (1983): 30–40. Courtesy of Helen James.

Fig. 5.18. Lateral view of the skeleton of the small, flightless (or nearly so) suboscine passerine the ash-colored tapaculo (*Myornis senilis*) of Colombia, a furcula reduced to two clavicular splints, and absence of a sternal keel. US National Museum specimen 610474. Photograph by the author.

herbivorous habits they were undeniably 'goose-like.'[55] The four species of moa-nalos were the ecological equivalents of the giant tortoises of the Galápagos and Indian Ocean Islands. In both cases they evolved in the absence of any large herbivorous birds and perhaps in response to the absence of herbivorous mammals. The strangest of all, *Thambetochen chauliodous*, from Molokai, had a beak with toothlike projections, and one genus, *Chelychelynechen* ("turtle-jawed goose"), was named for the turtlelike aspect of its beak. Most interesting were their skeletal features, which were quite distinctive from any living anseriform (duck- and gooselike birds). Like many large flightless birds, moa-nalos possessed robust hind-limb elements, reduced wings and pectoral girdles, and a sternum that lacked a keel and was unfused along the midline of the posterior half. The scapula articulated with the coracoid at an obtuse angle, and there was a weak furcula (clavicles) with widely divergent rami, or splints. They were quite similar in size and proportions to the smaller moa of New Zealand, and were it not for their skulls, these moa-nalos could have been identified as ratites!

Relationships of the Hawaiian Gooselike Waterfowl

Hawaii's large waterfowl encompass two distinctive groups, as noted above, the flightless moa-nalos, consisting of four species of large flightless birds that are thought to be gooselike duck derivatives; and the true geese (tribe Anserini), which comprise at least three species exhibiting a wide range of morphology, the smallest of which is the living nene. Although the nene is capable of strong flight, its extinct relatives, including the "nene-nui" from Maui, were at best weak fliers, and there is a yet-undescribed giant flightless goose from Hawaii, as well as additional undescribed species from Kauai and Oahu.[56]

Two recent studies of extinct Hawaiian geese illustrate the inadequacy of character-based cladistic analyses, which form the entire basis of phylogenetic reconstruction in the fossil record of vertebrates. Michael Sorenson of the Smithsonian's Molecular Genetics Laboratory and col-

leagues studied the relationships of the extinct moa-nalos, based on sequenced mitochondrial DNA extracted from fossil bones. They determined that these massive, gooselike birds are most closely allied not with geese but with the dabbling ducks (tribe Anatini). Their conclusions should be a warning to all who use standard morphology-based cladistics: "The evidence of phylogeny in morphological characters was obscured by the evolutionary transformation of a small, volant duck into a giant, terrestrial herbivore."[57] In another study, Ellen Paxinos of the Smithsonian's Genetics Program and colleagues performed a phylogenetic analysis of mitochondrial DNA sequences from fossils of the Hawaiian true geese, including the living nene (*Branta sandvicensis*) and its extinct, less volant relatives, and discovered that the group is nested phylogenetically within the living species the Canada goose (*Branta canadensis*) and produced a radiation of at least three species that differed in body size and flight ability. The Hawaii bird was closely related to the nene but some four times larger and flightless; it is estimated to have evolved on the island of Hawaii in fewer than five hundred thousand years. Interestingly, Paxinos notes, "The giant Hawaii goose resembles the moa-nalos, a group of massive, extinct, flightless ducks that lived on older Hawaiian islands and thus is an example of convergent evolution of similar morphologies in island ecosystems."[58]

It is highly unlikely that definitive relationships of these two Hawaiian anseriform radiations could have been resolved by morphology alone. Only through analysis of DNA, both recent and fossil, were such conclusions possible.

Evolution Can Be Very Rapid

Most tend to view the evolutionary process as requiring vast periods of geologic time, as we see in the case of lineages, like horses, that evolved from the Eocene (about fifty million years ago) to the present. However, evolution often proceeds by explosive bursts, and there are often long periods in certain lineages with little or no evolutionary change, times known as evolutionary stasis. One

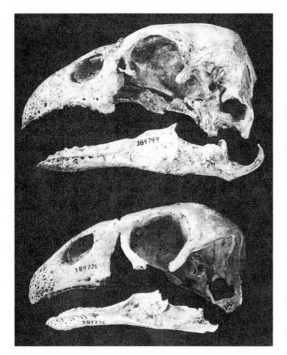

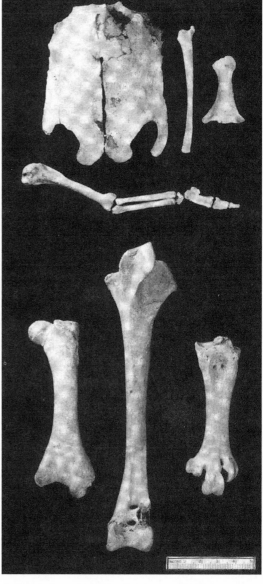

Fig. 5.19. Bones of extinct Hawaiian, flightless goose-like ducks. Left, skulls of two genera of "toothed" moa-nalos. Above, *Thambetochen chauliodous*; below, *Ptaiochen pau*. Right, Ratitelike fossil bones of *T. chauliodous*. Left to right and top to bottom, sternum (note lack of keel), scapula, coracoid, right wing (note drastic reduction), femur, tibiotarsus, and tarsometatarsus. The pelvis and hind limbs were extremely robust, and the wing and pectoral girdle was abbreviated. In addition, the sternum lacked a keel and the posterior half was unfused along the midline; the furcula was weak, with widely divergent rami. In all but the head, the moa-nalos closely resembled ratites convergently. Note that the bones have come to resemble convergently those of the large ratites in loss of the sternal keel, reduction of the wing to a vestige, and the greatly increased size of the leg bones. Scale bar = 5 cm. Skull photographs courtesy of Storrs Olson. Left, from S. L. Olson and A. Wetmore, "Preliminary Diagnoses of Extraordinary New Genera of Birds from Pleistocene Deposits in the Hawaiian Islands," *Proceedings of the Biological Society of Washington* 89 (1976): 247–58. Courtesy of Storrs Olson; photo by Victor E. Krantz.

of the most remarkable periods of explosive evolution in birds and mammals involved the period of time following the Cretaceous-Paleogene extinction event, sixty-five million years ago, when the dinosaurs, plesiosaurs, mosasaurs, ammonites, and many other groups became extinct.[59] A limited number of avian and mammalian groups emerged from the cataclysmic "slate-cleaning" that ended the Cretaceous Period, and they began to evolve into the innumerable newly available niches. Among the best-documented cases of rapid evolution to a truly derived morphology involved the whales, or cetaceans, which evolved from land ungulates in some eight to ten million years, during the Early Paleogene.

As for the time required for flightlessness to evolve, opinions have varied throughout the debate on their evolution, but the time required is quite small. Huxley, contrarily, thought the ratites were a remnant of an ancient lineage, "waifs and strays" that evolved directly from dinosaurs and had changed little since then; they were the "scanty modern heirs of the great of ornithoid creatures which once connected Birds with Reptiles."[60]

As for rails, Storrs Olson has demonstrated that "the span of time needed to evolve flightlessness in rails can probably be measured in generations rather than in millennia."[61] Take, for example, moorhens that occur some 400 kilometers (250 mi.) apart in the south Atlantic. On Tristan da Cunha and Gough Island there are flightless populations of two nearly identical species (or subspecies) of the modern common moorhen (*Gallinula chloropus*) that must have colonized the islands independently from small founder populations of moorhens that initially flew to the islands, and they must be of very recent origin. Coots of the genus *Fulica* on both Mauritius and Reunion are very similar flightless subspecies that evolved independently and recently from a common ancestor, and there are flying and flightless races or subspecies in a single rail species, *Dryolimnus cuvieri*. One final and impressive example is the flightless rail *Atlantisia elpenor*, present only on Ascension, a South Atlantic island whose oldest rocks date at just 1.5 million years. It is difficult to know just how long it would take for these flightless rails to

become unrecognizable as to their ancestry, once they developed gigantism or bizarre adaptations, as is the case in the Hawaiian moa-nalos, which resembled geese rather than their duck ancestors, or the flightless ibis, whose hind limb elements most closely resembled those of kiwis. The Hawaiian Islands' oldest rocks date to less than 6 million years, and the ibis from Molokai could be no older than 1.8 million years, the oldest dated rocks on Molokai.[62] The Galápagos Islands, dating to a maximum age of some 3 million years, was home to the fourteen or so species of giant Galápagos tortoises and the flightless Galápagos cormorant.

In an important study bearing on time involved in loss of flight, Beth Silkas, Storrs Olson, and Robert Fleischer, using mitochondrial sequence data to estimate phylogenetic relationships and ages of four species of flightless insular rails in the genus *Porzana*, concluded that "loss of flight appears to have evolved rapidly in these insular rails, based on both sequence divergence values and data on the ages of the islands. In the case of the Laysan Rail . . . divergences including loss of flight probably evolved in less than 125,000 years."[63] Finally, as noted concerning the radiation of geese in Hawaii, there was a giant Hawaii goose closely allied with the living nene. The study by Paxinos and colleagues, based on

Fig. 5.20. Silhouettes (not to scale) of a small New Zealand moa, left, and that of a Hawaiian moa-nalo, right, showing the overall similar morphology, acquired through convergent evolution. Bones of moa-nalos very closely resemble those of large extinct flightless birds, and some attained their moalike morphology in approximately a million years or so.

local rate calibration from their DNA data, suggested that it split from an ancestor with the nene less than 500,000 years ago; and interestingly the giant Hawaii goose is restricted to an island that is only ~430,000 to 500,000 years old. Their results: "supports the hypothesis that its gigantism, flightlessness, and robust cranial morphology evolved during that time frame on the island of Hawaii."[64]

Given the data, there can be no doubt that the time needed for the evolution of flightlessness and bizarre morphologies can be quite brief.

Sir Gavin Rylands de Beer

Each ontogeny is a fresh creation to which the ancestors contribute only the internal factors by means of heredity.
Gavin de Beer, Embryos and Ancestors, *1940*

The brilliant British scientist and embryologist Gavin de Beer (1899–1972) was among the first to bring the old German school of embryology led by founders Karl Ernst von Baer and Ernst Haeckel into a more modern perspective. As an Oxford undergraduate, de Beer researched under the direction of T. H. Huxley's grandson Julian Huxley (1887–1975), one of the most influential biologists of his day. The Oxford zoology department was then headed by the great descriptive embryologist Edwin Goodrich (1868–1946). Together with Huxley and de Beer, Goodrich defended Darwinian evolution, which before the First World War was in danger of "eclipse."[65] In summarizing de Beer's contributions, Oxford's Tim Horder argued that his theories of embryology played a crucial role in evolution's modern synthesis and that his enduring work still influences how we think about the genome, evolution, and developmental biology.[66]

In addition to being director of the British Museum, president of the Linnaean Society, and a fellow of the Royal Society, de Beer was knighted in 1954 and received the Royal Society's Darwin Medal in 1957. In what is perhaps his best-known book, *Embryos and Evolution* (1930; later editions were entitled *Embryos and Ancestors*), he attacked Haeckel's theory of recapitulation (biogenetic

law) as being inconsistent with the modern evolutionary synthesis, and he stressed the importance of heterochrony in evolution, especially paedomorphosis.[67] De Beer also formulated the concept of clandestine evolution, which was at odds with Darwin's gradualist doctrine, but that de Beer thought would explain abrupt changes in the fossil record. Stephen Jay Gould summarized his work nicely: "In a series of remarkable books that established the synthetic theory of evolution, Gavin de Beer's *Embryology and evolution* was the first and the shortest (1930; expanded and retitled *Embryos and ancestors*, 1940; 3rd ed. 1958). In 116 pages de Beer brought embryology into the developing orthodoxy. . . . For more than forty years, this book has dominated English thought on the relationship between ontogeny and phylogeny."[68]

Aside from his biological contributions, de Beer served with the Grenadier Guards in World Wars I and II, wrote several publications on Darwin, and published extensively on Switzerland

Fig. 5.21. Ernst Haeckel, eminent German biologist and embryologist, and German champion of Darwin's theory, known today for his formulation of the Biogenetic Law. Photo from National Institutes of Health, US Department of Health and Human Services; published in *Photographische Gesellschaft* (1906).

and the Alps, with a particular interest in Hannibal. But, regarding the present discussion, de Beer made important contributions to paleornithology, and he was the first to elucidate the concept of mosaic evolution as exemplified by the urvogel *Archaeopteryx*.[69] He was also the first to fully apply the concept of heterochrony to avian evolution and its role in the evolution of flightlessness in the ratites.[70]

As a historical digression, it was Ernst Heinrich Philipp August Haeckel (1834–1919), the renowned German physician, biologist, embryologist, philosopher, and naturalist, who initially developed the field of embryology in relation to phylogeny.[71] Following studies at the University of Jena under Carl Gegenbaur (both Haeckel and Gegenbaur supported Huxley's dinosaurian origin of birds), he became professor of comparative anatomy at Jena, where he worked for forty-seven years. In addition to discovering and describing literally thousands of new species and mapping one of the first genealogical trees, Haeckel was responsible for many common words in the biological lexicon, including "ecology," "phylum," and "phylogeny"; and he was the first to use the term "First World War." In addition he introduced the concept of heterochrony. Haeckel promoted Darwin's work in Germany, although he had a modified version, combining some elements of Lamarckism, and in 1866 he met with Charles Darwin, Thomas Huxley, and Charles Lyell at Darwin's home, Down House. A flamboyant figure, Haeckel is best known today for advancing his "recapitulation theory," which became the so-called biogenetic law. As generally interpreted, the "law" stated that not only was the ontogeny of an individual a recapitulation of its phylogeny but more strictly it passed through, in succession, the adult stages of its primitive ancestors. The concept has been widely refuted, although a more moderate interpretation, that the ancestral embryonic developmental process is repeated and built upon, is clearly not only valid but extremely useful in evolutionary studies. Haeckel has been criticized for taking artistic license in his drawings of embryos, but again, the general concept behind the drawings remains valid, and in summary, his work, though an oversimplification of a complicated process, nevertheless represented a brilliant starting point for future research.[72] Actually, the first recapitulationist was Johann Friedrich Meckel (1781–1833), who described the parallelism between the embryonic stages of higher animals and permanent states of lower animals: "The development of the individual organism obeys the same laws as the development of the whole animal series; that is to say, the higher animal in its gradual development essentially passes through the permanent organic stages that lie below it."[73]

However, before Haeckel, the first solid biological principles of the relationship of ontogeny and phylogeny were formulated by Karl Ernst von Baer (1792–1876), a founding father of embryology who discovered the blastula stage and the notochord. Von Baer's formulation of the relation between ontology and phylogeny was later condensed into what is known as "von Baer's laws." Stated simply: the more general characters of a group appear earlier in development than the more specialized features; that is, more general (conservative) structures are formed earlier than more specialized structures. Thus, instead of passing through adult stages of ancestral forms, contrarily, the embryo separates itself from them. Therefore, the embryo of a "higher form" resembles the embryo of ancestral forms, not the adult. In this context the study of embryos of birds and other organisms has been an invaluable tool, as we shall later see, in studying the process of evolution. As Stephen Jay Gould appropriately pointed out, recapitulation, properly understood, is simply a fact that cannot be eliminated by attacks on it.[74]

Percy Roycroft Lowe: Font of Contemporary Dinosaurian-Bird Controversy

Another prominent figure in the controversy concerning ratites was Percy Lowe (1870–1948), who studied medicine at Cambridge University, served as a surgeon in the Second Boer War, and while in South Africa became interested in ornithology. Lowe became curator of ornithology in 1919 at the British Museum (Natural History) and was president of the British Ornithologists' Union from

1938 to 1943. He produced a series of papers on ratites that were in stark contrast to those of de Beer but coincided largely with the views of Huxley. Although largely dismissed by today's ornithological community, Lowe's views, reflecting his position as a strict recapitulationist (following Haeckel), resurfaced within the works of advocates of the theropod hypothesis for the origin of birds in the 1970s, especially flight genesis from the ground up, though his contribution was not acknowledged.

In 1867, Thomas Henry Huxley was the first to make a truly comprehensive study of the avian palate with the goal of identifying a well-defined character complex to aid in achieving a rational classification of living birds, and he grouped the living ratites based on their common possession of a dromaeognathous palate.[75] In 1900, William Plane Pycraft renamed the ratite bony configuration the paleognathous palate to denote their presumed reptilian and hence primitive nature.[76] (As mentioned earlier, there is a living family of volant, paleognathous birds, the Central and South American tinamous, as well as the extinct lithornithids.)

The name Palaeognathae has had ramifications for the future status of the ratites, even down to the present, for almost everyone has regarded them as primitive within living birds, perhaps largely because of their appearance, although there is really no solid evidence to support this conclusion, and certainly none from the fossil record. Like most modern birds, the oldest ratite fossils date from the Paleogene, and there is nothing remotely resembling a ratite from the Late Cretaceous or earlier. So, although the paleognathous birds may be primitive, the ratites evolved in the Paleogene and successively later on. If ratites were ancient, Huxley's "waifs and strays" of an ancient radiation of birds, there should be abundant fossils from the Cretaceous. But they are absent: Where are the Mesozoic fossils of ratites, none even from the Latest Cretaceous, where numerous avian fossils are found? Surely with their large size and heavy bones, they would be preserved as part of the Mesozoic aviary, and interestingly, a recently discovered giant flightless bird from the Late Cretaceous

of France is not even a neornithine bird but a large flightless enantiornithine.[77]

Recall, however, that Huxley viewed the ratites as having been descended directly from dinosaurs without passing through a flying stage. The antiquity of the ratites has been argued on the basis of the similarity of the ratite arm and hand to that of small dinosaurs like *Ornitholestes* and *Struthiomimus* (the ostrich mimic) and retention of rudimentary claws on the fingers. Yet rudimentary claws are known in an endless array of modern birds ranging from hawks to ducks and rails, and one can even see rudimentary claws on the chicks of rails such as the purple gallinule.

Let us return to Percy Lowe, who wrote a series of papers between 1928 and 1944 in which he proposed that the ratites and small coelurosaurian dinosaurs actually shared a common ancestor, a minor departure from the straight-line evolution, or orthogenesis, proposed by Huxley.[78] Like Huxley, however, Lowe argued that the ratites had evolved from creatures that had never acquired flight. Lowe even viewed *Hesperornis* as a "swimming ostrich" and included it among the "aquatic or swimming palaeognaths," writing: "as reptilian as it was possible to be without losing its claim to be avian . . . an aquatic palaeognath, just as the Ostriches (Struthiones) were cursorial palaeognathes . . . there is little reason to think that either had volant ancestors."[79]

Lowe, in typical recapitulationist fashion, interpreted the retained cranial sutures in ratites as "a belated manifestation of a reptilian character in the 'Ratite' skull." Further, he wrote, "the adults of the existing Struthiones are clothed in prepennal down and have not reached a more advanced stage of development than the downy chick of a fowl."[80]

Reflecting his view that birds descended through a nonflight stage, the predominant view today among paleontologists, Lowe concluded that flightlessness in ratites was "primary," noting that "the Struthiones represent a perfectly natural group descended from some common ancestor which left the main avian stem before flight had been attained."[81] Thus, the polarity of avian evolution was for Lowe the same as for the current dinosaurian-bird orthodoxy: "In the Struthiones the entire make-up of the adult is chick-like and reflects the beginning of the story of avian evolution instead of the end."[82] Then, in 1942, Lowe also made it clear that he did not adhere to the concept of ratite neoteny, which would be in stark contrast to his views: "The stage of development in the wing . . . *considerably antedates in point of evolution any mere juvenile phase in the Neognathae.*"[83]

It is interesting to note that Lowe considered the flightless condition in all birds not to have arisen from volant ancestry but to be primary: "The rails, as a group, may in the past have been slow to acquire the power of full flight, or even of flight in moderate perfection, while some, like the fossil forms . . . [*Aphanapteryx* and *Diaphorapteryx*], may not have acquired it at all. Rails seem, in fact, to have been primarily and constitutionally either bad fliers or bad 'triers.'"[84]

Lowe's views never gained much momentum, and eventually a mountain of anatomical evidence

made it clear to most workers that the ratites, like *all* other flightless birds, are derived from flying predecessors and that heterochrony had played an important role.[85] Lowe's arguments had gone off the deep end (literally!) when he proclaimed that penguins derived from primarily terrestrial, nonflying ancestors. As we have seen, penguins are flying birds, and because they fly through the unusually dense medium of sea water, their flight architecture and associated muscle mass are among the most well built, most exceptional of all living birds. Lowe's views on both ratites and penguins were attacked by numerous individuals, particularly George Gaylord Simpson, and today his model for the origin of flightlessness is considered untenable.[86] In the case of ratites, one need only look at the "flight" features of these birds to show that they were derived from flying ancestors, an observation first elucidated by de Beer (and later authors), who cited adaptations to flight in ratites that could only be indicative of flight ancestry:

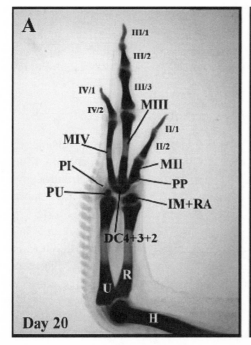
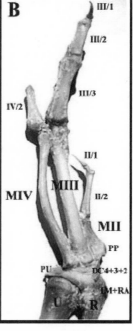

Fig. 5.22. The "flight wing" of ostrich, developmental patterns of cartilage of twenty-day embryo, and adult bony hand (right); note the attenuation of the distal elements. Abbreviations: DC, distal carpal; H, humerus; IM, intermedium; M, metacarpal; PI, pisiform; PP, preaxial process; PU, pseudoulnare; R, radius; RA, radiale; U, ulna; UL, ulnare; 1–3, first to third phalanges; 1, fusion. Courtesy of Martin Kundrát.

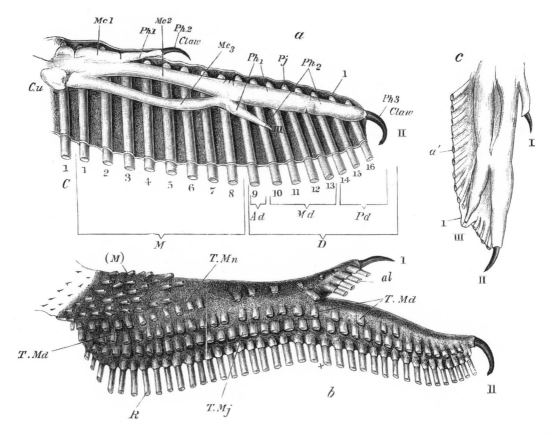

Fig. 5.23. Upper left, manus of ostrich, ventral view, showing attachment of the remiges. Lower left, manus showing dorsal remiges and coverts. Right, digits of embryo, showing remiges and coverts. From F. E. Beddard, *The Structure and Classification of Birds* (New York: Longmans, Green, 1898).

- The wing skeleton is built on the same plan as that of carinates; in *Rhea*, the differentiation of primary and secondary remiges is retained.
- The tail skeleton is reduced, with fusion of terminal vertebrae (pygostyle) in the ostrich (*Struthio*).
- The cerebellum is enlarged as in carinates, projecting forward over the dorsal surface of the midbrain toward the cerebral hemispheres, and possessing the same *arbor vitae* structure as in carinates.
- In *Rhea*, the feathers of the outer digit (II) form a "bastard wing," an adaptation that results in maintenance of a slipstream in flight.
- A vestige of aquintocubitalism, or diastataxy— the absence of the fifth secondary remix from the wing flight feather row (associated with the peculiar method of folding the wing, the ulnar flexure)—is found in the wing of young kiwi (*Apteryx*) and other ratites.
- Strong wing quills are retained in the wing of the cassowary (*Casuarius*).
- The emu (*Dromaius*) embryo exhibits a vestigial first pedal digit, pointed backward—as de Beer put it, "a sign that its ancestors perched on branches of trees. It is unnecessary to press the point further."
- A "flight" carpometacarpus is present, with fusion of several carpal elements (serving for exact positioning of the remiges, or flight feathers, in carinates).
- Other subtle points include the curvatures of the metacarpals II and III and the expansion of the basal phalanx of digit III.[87]

With respect to the cerebellum, the paleo-neuroanatomist Tilly Edinger (1897–1967), of Wellesley College, showed that the "carinate" cerebellum was an adaptation to flight by illustrating its presence in pterosaurs, which possess "a cerebellum thrust forward above the midbrain to adjoin the forebrain as in birds; obviously this is one of the characters distinguishing all pterosauria from other reptiles."[88] As de Beer put it, "There can be no doubt that the parallel development and large-size of the cerebellum in pterosaurs and in birds, by which they both differ from all non-pterosaurian reptiles, are due to the same cause: adaptation to flight."[89] Most interestingly, this same cerebellar structure was present in the urvogel *Archaeopteryx*, thus providing pivotal evidence for its flight capability.

In refutation of Lowe's proposal that the ancestors of ratites (and other flightless birds) never flew, quite similar to that of Huxley, de Beer and others have noted that most of the characters Lowe cited as primitive were in fact the result of paedomorphosis. Some of the more notable neotenous features of the ratites, many of which are seen in the development of rails, include the keel-less sternum, obtuse angle of the scapula and coracoid, retention of skull sutures in ostrich, reduction of wings, and degeneration of the plumage; and de Beer considered that the paleognathous palate was also neotenic.[90] Additionally and importantly, in the ostrich, the young are hatched with a conspicuous yolk sac that must be absorbed, a signal of paedomorphosis. Yolk sac retention and subsequent failure to absorb the yolk is a common problem in newly hatched ostrich chicks and is one of the major causes of mortality in the first two weeks of life.[91] On some South African ostrich farms, "nannies" run the ostrich chicks daily to promote absorption of the yolk sac.

In his seminal article on the evolution of ratites, de Beer summarized his views: "Most unfortunately, the notion spread, a hundred years ago, as a result of T. H. Huxley's work, that Ratites were the most primitive known birds whose ancestors had never flown or even glided, and that therefore Carinates were evolved from them. This is almost a text-book example of how false morphological premises can lead to untenable conclusions: it was contended that as Carinates evolved from Ratites, which had never flown, all the identical similarities between the feathers and arrangement of feathers in *Archaeopteryx* and in Carinate flying birds were the result of independent parallel evolution. This is absurd."[92]

Huxley's and Lowe's Views Still Reign within Paleontology

The paedomorphic origin of flightlessness in terrestrial birds is well documented. As we have seen, much of our knowledge comes from the study of flightless rails, the group of living birds that serves as a laboratory experiment in flightless evolution. Given the overwhelming and preeminent importance of the evolution of flightlessness in Mesozoic as well as Cenozoic birds, which often appear quite dinosaur-like, it is astounding that any modern book on the evolution of birds could omit references to Gavin de Beer or his work. Sadly, however, this is the case with Luis Chiappe's *Glorified Dinosaurs*, published in 2007. For many paleontologists who accept the current orthodoxy on bird origins, the path to flight is quite similar to that envisioned by T. H. Huxley and Percy Lowe. For de Beer, however, the path was clear: "No terrestrial animal became a flying animal directly; it first became an arboreal animal, next a gliding animal, and then, from that stage, a flying animal. This is the pattern of evolutionary advance."[93]

Paedomorphosis-Provided Model

Gavin de Beer's model on the evolution of flightlessness in ratites is applicable to almost all terrestrial flightless birds. As we shall see in the following chapter, the evolution of flightlessness, at all stages, proceeded in the archaic Mesozoic avian radiation just as it did in the Tertiary, with iterative evolution of flightlessness producing neotenous forms that, at least superficially, closely resemble theropod dinosaurs in overall form. Helen James and Storrs Olson nicely explain how the transition to paedomorphic flightlessness can be quite simple and quite rapid: "The genetic mechanism for the evolution of a flightless bird from a flying one is actually quite simple. All birds are flightless

when they are small chicks, and the young of flying birds have the same features that characterize
the adults of flightless birds. . . . Merely by retaining the skeletomuscular structure of infancy into
adulthood—probably by the alteration of a few
regulatory genes—almost any bird species could
become flightless."[94]

Likewise, Bradley Livezey notes that "late
development of the pectoral appendage in birds
makes paedomorphosis the most obvious ontogenetic inference in flightless birds." And, most
instances of avian flightlessness involve an increase in size, since there are no longer weight restrictions associated with flight, and "this in part
reflects peramorphosis of regions other than the
pectoral limb."[95]

Dollo's Law

> The road from Reptiles to Birds is by way of
> Dinosauria to the Ratitae. The bird "phylum"
> was struthious, and wings grew out of rudi-
> mentary forelimbs.
>
> *T. H. Huxley* to *Ernst Haeckel, 1868*

Thomas Huxley's theory of a dinosaurian
origin of birds differed dramatically from today's
orthodox view but coincided in the concept that
birds evolved from earthbound dinosaurs through
a nonflight stage, in the case of Huxley, through
the flightless ratites, which descended directly
from dinosaurs. Such a scenario, then and now,
requires the view that the elongate avian wings
somehow were transformed from already foreshortened forearms. Yet such a transformation is
clearly a problem with respect to the evolutionary
principle of nonreversability known as Dollo's
Law, if valid (described below).

When Gerard Heilmann wrote his influential
book *The Origin of Birds* in 1926, he remarked that
Archaeopteryx showed remarkable similarity to
small theropods such as *Compsognathus*, but by the
principle of Dollo's Law, the absence of clavicles
(furcula) in *Compsognathus* precluded its being ancestral to birds.[96] Heilmann argued that it was improbable that clavicles could reevolve, and for that
and other reasons he favored a common descent for

dinosaurs and birds via the "theocodont," or basal
archosaur, origin hypothesis, which was coupled
with a trees-down origin of flight. The subsequent
discovery of a furcula in dromaeosaurs (and allies)
and other theropods (some less convincing) seemed
to overcome Heilmann's obstacle for many paleontologists.[97] Yet the presence of a well-developed,
indeed *Archaeopteryx*-like furculae in small Triassic basal archosaurs such as *Longisquama* renders
their argument irrelevant, since its early presence
indicates that it is primitive, or plesiomorphic, and
therefore of little phylogenetic importance.

Louis Dollo and His Law

The French-born Belgian paleontologist Louis-
Antoine-Marie-Joseph Dollo (1857–1931) was
known during his lifetime for supervising the
famous excavation in 1878 of numerous complete specimens of the ornithopod *Iguanodon* at a
coal mine near Bernissart, Belgium; many of the
specimens were mounted and are on display at
the Royal Belgian Institute of Natural Sciences in
Brussels. However, Dollo is best known today for
his formulation of Dollo's Law (or Dollo's Principle) in 1893: "An organism is unable to return,
even partially, to a previous stage already realized
in the ranks of its ancestors."[98] According to Dollo's view, a structure that had been lost through the
evolutionary process would not reappear in that
lineage. Richard Dawkins has suggested that the
law is "really just a statement about the statistical
improbability of following exactly the same evolutionary trajectory twice (or, indeed, any particular trajectory), in either direction."[99]

Like most evolutionary and biological "rules"
and "laws" (such as Bergmann's Rule, Allen's
Rule, and Cope's Rule), Dollo's Law or, more
appropriately, Dollo's Principle will surely have
exceptions, particularly at the populational and
species levels, so it is not ironclad, but it does provide a general paradigm in the broad scheme of
vertebrate evolution. For example, it would appear
from what we know of genetics and development
that it is unlikely for blind cave fish, having lost
eyes, to reevolve them into the same biological
structures. Likewise, whales have not evolved any

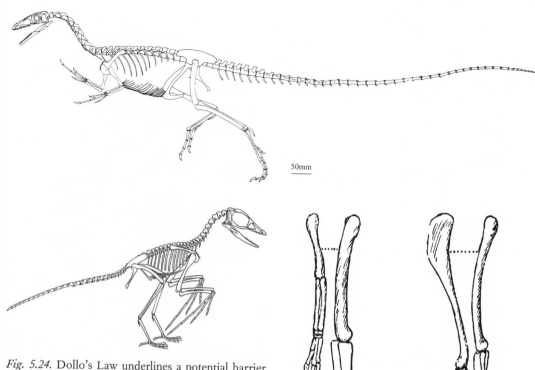

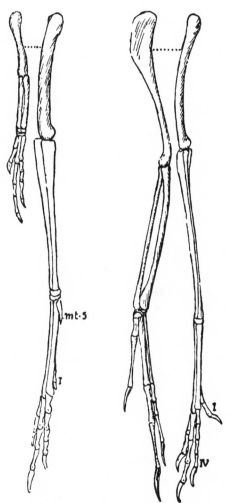

Fig. 5.24. Dollo's Law underlines a potential barrier for a strict theropod origin of birds, unless the fossil record is missing important ghost lineages with elongate forelimbs. Left, skeletons of a Late Jurassic compsognathid coelurosaurian theropod (*Compsognathus*) exemplifying the basal body plan of ancestral theropods, compared to a scansoriopterid (*Scansoriopteryx/Epidendrosaurus*), illustrating the extreme difficulty of reelongating the already greatly foreshortened forelimb of a theropod dinosaur into the elongate forelimbs of birds. Right, comparison of length of forelimbs of *Compsognathus* and *Archaeopteryx*, another illustration of the extreme improbability of reelongating already foreshortened theropod forelimbs. *Compsognathus* from Karin Peyer, "A Reconsideration of *Compsognathus* from the Upper Tithonian Canjuers, Southeastern France," *Journal of Vertebrate Paleontology* 26 (2006): 879–96. Courtesy of Karin Peyer and the Society of Vertebrate Paleontology. © 2006 The Society of Vertebrate Paleontology. Reprinted and distributed with permission of the Society of Vertebrate Paleontology [http://www.vertpaleo.org]. *Scansoriopteryx* from S. A. Czerkas and C. Yuan, "An Arboreal Maniraptoran from Northeast China," *Dinosaur Museum Journal* 1 (2002): 63–95. Copyright Stephen A. Czerkas. Below, adapted and modified from G. Heilmann, *The Origin of Birds* (London: Witherby, 1926).

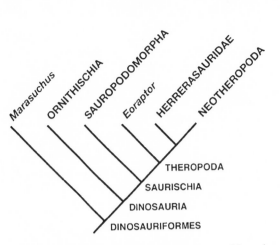

Fig. 5.25. Cladogram of basal Dinosauria. Although there may be some variation in the cladogram from worker to worker, the general scheme remains, so that taxa near the origin of Theropoda (Neotheropoda) all have the same basic body plan (bauplan), with large hind limbs and obligate bipedalism, and with forelimbs reduced to 50 percent or more of the length of the hind limbs. The entire debate centers on whether birds are derived from the far left or far right side of the cladogram—from a common stem basal archosaur origin or directly from highly derived theropods. Modified after P. C. Sereno, "The Phylogenetic Relationships of Early Dinosaurs: A Comparative Report," *Historical Biology* 19, no. 1 (2007): 145–55.

Fig. 5.26. Louis-Antoine-Marie-Joséph Dollo, the French-born, Belgian paleontologist best known for formulating Dollo's Law, best termed a "rule," on the irreversibility of evolution. He is also known for supervising the excavations of the famous multiple *Iguanodon* find in Bernissart, Belgium.

land-based tetrapod lineages from the sea, horses have not evolved a lineage with a pentadactyl foot, and the flightless ratites have not evolved any flying descendants.

Marking the centennial of Dollo's Law, Stephen Jay Gould described the concept less strictly. Using coiling seashells to illustrate a modern version of Dollo's Law, he argued that uncoiled gastropods had become so committed to their morphology that it was extremely unlikely that natural selection could return them to the original coiled state. Gould suggested that "irreversibility" forecloses certain evolutionary pathways once they emerge. As he put it, "Once you adopt the ordinary body plan . . . options are forever closed, and future possibilities must unfold within the limits of inherited design."[100] This is because the loss of a major character complex is generally followed by the loss of genetic structure and associated developmental mechanisms that determine the nature of the character complex.

Certainly, one must view any such biological law with skepticism, and the "law" would almost never apply to microevolution at the populational level, where beak size, for example in Darwin's finches, might vary one way or another depending on ongoing selection pressures and the genetic structure is not lost. Yet the overall concept of Dollo's Law in major evolutionary lineages holds up quite well: once a major structural complex, along with its genetic architecture, has been lost, it would be nearly impossible to expect the roll of the dice to mutate all necessary genes back to their original form.

Biological "laws," however, are not inviolate, and a transgression of Dollo's Law has been discovered in the reevolution of shell coiling in a gastropod after it has lost coiling (an exception to Gould's example), owing to a change in developmental timing, or heterochrony, rather than a change in gene expression.[101] The gastropod case is truly exceptional; these animals develop one type of shell as larvae and another completely new shell as adults, so researchers envisioned a lineage in which the coils were lost in the adult shell but retained in the larvae, so the "coiling" genes were clandestinely retained but eventually lost their larval stage and simply developed uncoiled shells. In other words, by this view only the adult stage was retained, but it retained the genes for coiling. Another exception to Dollo's Law has emerged in stick insects (Phasmidae), a lineage of which reevolved previously lost wings. In this case, the genes for building wings were likely retained as part of a gene and developmental complex for building appendages in general, which is not unexpected in the rigidly metamerically (segmentally) arranged insect (arthropod) body plan, in which genetic structure dictates serially homologous somites, or body segments.[102] One further example is found in the reevolution of sexuality in oribatid mites (Acari), but in general, write the researchers, "degradation of genetic information is sufficiently fast that genes or developmental pathways released from selective pressure will rapidly become nonfunctional."[103] In another study, Charles Marshall, Elizabeth Raff, and Rudolf Raff estimate that "there is a significant probability over evolutionary time scales of 0.5–6 million years for successful reactivation of silenced genes or 'lost' developmental programs." But they also estimate that conversely, reactivation of long-unexpressed genes and dormant developmental pathways is not possible unless function is maintained by other selective constraints.[104]

Evidence for the operation of Dollo's Law at the molecular and organismic levels is both supported and questioned in recent work. Jamie Bridgham, Eric Ortlund, and Joseph Thornton studied the 450-million-year evolution of the vertebrate glucocorticoid receptor protein. This protein originally responded to both cortisol and the hormone aldosterone. Forty million years later, however, in the transition to land, the descendant protein became cortisol-specific: thirty-seven amino acids had changed, with only two necessary to alter the function. The researchers tried in vain to restore the activity by reverting these amino acids. They concluded that "even if selection for the ancestral function were imposed, direct reversal would be extremely unlikely, suggesting an important role for historical contingency in protein evolution." Günter Wagner of Yale called this study "perhaps one of the most important papers in the last 10 to 15 years in evolutionary biology. . . . Not only does the study show how evolution can't go backward, . . . but it also provides a detailed mechanism for why."[105]

Yet examples to the contrary are papers claiming the re-evolution of lost digits in a lizard and the re-evolution of lost mandibular teeth in frogs.[106] In this framework Rachel Collin and Maria Pia Miglietta point out by review that "phylogenetic studies have recently demonstrated cases where seemingly complex features such as digits and wings have been reacquired," pointing to "dwindling evidence for the law-like nature of Dollo's Law" and anticipating "a return to Dollo's original focus on irreversibility of all kinds of changes, not exclusively losses." However, there are divergent views, with Frietson Galis and colleagues asserting that the molecular phylogenetic tree for digit loss is in conflict with evolutionary mechanisms concerning the biogeography of lizards and with morphology-based phylogenies. They argue that recent studies that attempt to refute Dollo's Law, if true, "would change our view of the role of developmental constraints in the evolution of body plans."[107] In addition, many of the studies attempting to refute Dollo's Law are based on either molecular phylogenies or morphological cladistic analyses that are tenuous and subject to change.

Although most evolutionary rules as they apply to reversibility, evolution of body size, surface-volume relationships, color, and so on do have exceptions, such laws do reflect general (but not inviolate) trends in evolution. Cope's Rule, for example—the tendency for organisms in evolving lineages to increase in size over time—has been well established with credible mechanisms that drive the rule at the populational level.[108] Yet in the fossil record many exceptions exist. Classic examples are elephants and hippos, which evolved a smaller size on Mediterranean islands. Human evolution provides additional examples, such as the pygmies of continental Africa and the Pleistocene Indonesian island "hobbit" known as Flores Man. These examples, however, do not involve any dramatic change in overall morphology and general geometric proportions. Importantly, in the case of Dollo's Law, there are among the birds innumerable lineages that have evolved flightlessness, but *there is not a single example of a flightless bird having reevolved flight and thus having reelongated its wings.*

Dollo's Law and Bird Origins

In a review of avian origins, Kevin Padian and Armand de Ricqlès argued, in reference to a body size decrease in association with flight origin: "The closest relatives of *Archaeopteryx* among maniraptorans were generally smaller than average theropods, which is consistent with the hypothesis that protobirds slowed their growth rate, thus miniaturizing their skeletons while *retaining approximately similar geometric proportions* [my emphasis]. This would have made the wings relatively more efficient aerodynamically, because the wing loading, or the ratio of wing area to mass, would have decreased advantageously with reduced size."[109]

A number of workers have argued that the long arms of birds, particularly *Archaeopteryx*, might be derived via heterochrony in theropods (by peramorphosis, in which the descendant resembles the ancestor at an older stage).[110] Others have predicted that "the fossil record will show developmental sequences of theropods in which earlier stages show arms proportionately longer than those of later stages," but in fact, just the opposite is true in modern birds, in which earlier stages invariably show shorter forelimbs than later stages.[111] Yet such proposals remain. For example, tyrannosaurids, giant predators of the last twenty-five million years of the Mesozoic, have been assumed to be heterochronic "peramorphs," the largest having grown from small-bodied forms via developmental acceleration.[112] However, a recent, dazzling, small-bodied basal tyrannosauroid fossil

from the Lower Cretaceous of northeastern China Jehol Biota casts serious doubt on the classical view.[113] *Raptorex*, a 3-meter (10 ft.) tyrannosauroid with a body mass estimated at about 65 kilograms (143 lbs.), or about the size of *Deinonychus*, closely approximates a hypothetical ancestor on the lineage leading to the great *Tyrannosaurus rex*. The transition to "rex" would thus have involved an increase to a body mass about ninety times greater than in the ancestor. *Raptorex* reveals that, at least in the tyrannosaurid lineage, the tiny forelimbs and cursorially adapted hind limbs evolved at relatively small size in the Early Cretaceous, and such a feature therefore "can no longer be explained as a passive allometric consequence of body size increase or the product of extended (peramorphic) growth trajectory."[114]

Still another problem not explicable via peramorphic growth in modern large flightless birds involves the proportions of wing elements. In these birds there is disproportionate reduction of the distal wing elements relative to the proximal ones.[115] In birds that become secondarily flightless, invariably the attenuation of the forelimb is from distal to proximal, so that the hand becomes greatly reduced, followed by the radius-ulna and humerus. In *Archaeopteryx*, the manus is the major elongated complex, and if, for example, a compsognathid did reelongate its forelimbs "retaining approximately similar geometric proportions," the arm would not remotely resemble that of *Archaeopteryx*. And if the long arms could be achieved, what would be the relative proportions of the arm bones? Certainly "compsognathid" arms could not reelongate into *Archaeopteryx* wings. Yet the question is, From whence did the maniraptorans (dromaeosaurids and allies) derive? Thus, regardless of which group either birds and/or maniraptorans are derived, the problem of reelongating wings from already foreshortened forelimbs remains, unless some incredible ghost lineages are still missing from the fossil record.

In the case of dinosaurs, the exact relationships of Triassic dinosaurs are still debated, but all would agree that the basic body plan (bauplan) of the ancestral theropod dinosaur closely resembled that of, say, the familiar *Herrerasaurus*, or even the

Fig. 5.27. Left, silhouette of the primitive Jurassic ceratosaur *Limusaurus*, thought possibly to be near the theropod lineage leading to birds. Note the highly abbreviated forelimbs, illustrating the improbability of reelongating already foreshortened forelimbs. Right, specimen of a Lower Cretaceous specimen of *Sinosauropteryx*, showing the shortened forelimbs characteristic of compsognathids, thought to be near the avian lineage. All of the Chinese "theropod" fossils are now reconstructed with a coating of "dino-fuzz" (protofeathers) that presumably evolved as an insulatory pelt for endothermy. However, there is no credible evidence for such a proposition, and bone histology shows ectothermic growth rings. Left, silhouette rendered following life reconstruction by Portia Sloan. Right, specimen on display at Hong Kong Museum. Reproduced under the terms of the GNU Free Documentation License, Version 1.2, from Wikimedia Commons.

compsognathids.[116] Therefore, now the question should be, Is it biologically possibly to reevolve the already greatly foreshortened forelimbs of such early ancestral theropods as the Late Triassic *Herrerasaurus*, *Coelophysis*, and *Syntarsus* or *Compsognathus?* The answer to this question may be unachievable because the exact nature of developmental plasticity in these groups is unknown. However, an educated guess is: possibly, but not very likely, especially limbs with the same proportions. So, like myriad aspects of the current orthodoxy on bird origins, everything is remotely possible but unlikely following the normal course of evolution as we currently understand it. It is simply unlikely that a species would reevolve elongate arms from already foreshortened forelimbs, and even if this did occur, it would be extremely unlikely that the result would be the typical avian wing.

The other nagging question is, of course, where are the fossils? Although those who advo-cate a dinosaur-bird nexus through common descent from basal archosaurs have been criticized for not having a specific putative ancestor, we are here faced with the same dilemma for a direct theropod origin: ghost lineages with no fossils to substantiate the transition from compsognathid-like forms to proto-maniraptorans, forms that would have to reelongate already foreshortened forelimbs, push the pubis posteriorly as in birds, and delete the rigidly obligately bipedal posture with its attendant bony architecture of the hip region, including a broad sacrum, an open acetabulum, and a supra-acetabular shelf associated with obligate bipedalism.

Any flight origin scenario thus poses a serious problem for an avian ancestry from true theropods (as we currently know them), because all known Late Triassic primitive theropods, which are thought to represent the primitive condition for the group, are obligately bipedal cursors, with massive hind limbs and greatly reduced forelimbs, usually about 40 to 50 percent the length of the hind limbs—about 50 percent in *Juravenator* and *Huaxiagnathus*, 39 percent in *Compsognathus*, and 36 percent in *Sinosauropteryx*.[117] In addition, the individual bone proportions are attenuated distally, so that the humerus is always substantially longer than the ulna (160 percent the length of the ulna in *Huaxiagnathus*), and usually the manus is quite short. This is the case for almost any Late Triassic or Jurassic basal theropod that would approximate the ancestral condition, including *Herrerasaurus*, *Syntarsus*, *Coelophysis*, and *Compsognathus* (including *Sinosauropteryx* and *Sinocalliopteryx*). The Late Jurassic Solnhofen urvogel as well as the Late Jurassic Chinese microraptors, which play the most significant role in modern flight origin scenarios, have forelimbs slightly longer than their hind limbs, not half their length. If birds derived from standard theropods, as mentioned earlier, therefore, elongate forelimbs would have needed to reevolve from already greatly shortened and highly modified appendages. This is true regardless of the group that gave rise to birds, dromaeosaurs or an earlier common ancestor, so that the difficulties that have been enumerated for a theropod origin of birds apply equally to a theropod origin of flying dromaeosaurs and burden the current view of dino-bird relationships with unknown ghost lineages, which have been used to criticize the basal archosaur, common ancestry hypothesis.

There are no reasonable explanations or selective agents as to how such a transformation to a reelongation of the forelimbs, and therefore wings, could possibly occur. John Ostrom was acutely aware of this flaw in his arguments and offered the following explanation. He reasoned that the wing stroke involved in trapping insects was similar to the forearm movement of theropods such as *Deinonychus*, which was similar to a bird's flight stroke.

If vigorous flapping of the feathered forelimbs played a part at any stage in the business of catching prey, the increased surface area of the enlarged contour feathers would undoubtedly have produced some lift during such assaults. From this point, it is a small evolutionary step for selection to improve those features that were important for flapping, leaping attacks on prey—perhaps to "fly

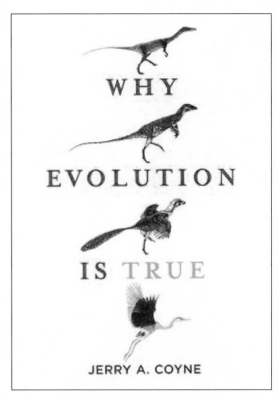

Fig. 5.28. This cover of a recent book by Jerry Coyne illustrates the persistent problem of reelongating already foreshortened forelimbs of dinosaur ancestors in bird evolution. From J. A. Coyne, *Why Evolution Is True* (New York: Penguin, 2009), with permission.

up" after escaping insects: for example, enlargement of the primaries and secondaries and their firm, rigid attachment to the forelimb skeleton; elongation and specialization of the bones of the forelimb and hand . . . enlargement of the pectoral abductor muscles; and stabilization of the shoulder joints by fusion of the clavicle. . . . Thus, selection would tend to improve not only the "flight power" but also the "flight controls"—the associated sensory and motor neural components.[118]

The dino-bird orthodoxy that developed in the 1970s and has continued in modified form to the present was then, and remains, rife with extremely difficult evolutionary pathways. The orthodoxy follows the same pattern of a strict theropod and ground-up flight origins, with forelimbs reelongating under some mysterious selective pressure, in Ostrom's familiar scenario, as insect traps. For all modern birds that we know, however, the situation is just the reverse, with volant birds giving rise iteratively to flightless forms, and in each case with the wings shortening as a result of arrested development, or paedomorphosis, the hand being the first element to diminish. This is clearly a case where Owen was right and Huxley wrong. Below we see Richard Owen correctly "dispensing justice to Huxley and Darwin alike," which should also provide a cautionary note for advocates of today's bird origin orthodoxy. "Science will accept the view of the Dodo as a degenerate Dove rather than as an advanced Dinothere."[119]

The Present Is Not a Key to the Past?

The classic feathered dinosaur *Caudipteryx* that adorned the cover of *Nature* in 1998 provides a classic example of the logic of the current orthodoxy of bird origins that harkens back to the views of Thomas Huxley and Percy Lowe, in which flightless birds are viewed as not having gone through a flight stage but as having evolved directly from earthbound dinosaurs. Flight thus evolved from the ground up. By such an evolutionary scenario, all flight architecture (including the "flight carpometacarpus" and flight feathers) would have evolved under selection pressure for something other than flight, as exaptations, and would later have been preadapted for a flight function. As a consequence, as we have seen, truly bizarre scenarios, such as trapping insects, have been envisioned to explain the evolution of wings. Contrarily, if we view the present as a key to the past, using today's flightless birds as models, then all flightless birds are seen to have passed through a flighted stage, and all of their flight architecture

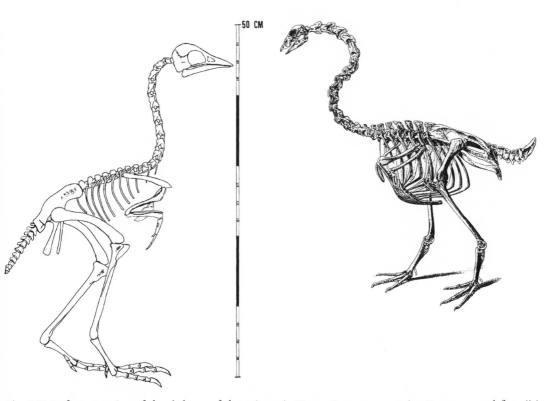

Fig. 5.29. Left, restoration of the skeleton of the enigmatic Upper Cretaceous species *Patagopteryx deferrariisi* (Patagopterygiformes), a flightless bird the size of a chicken, of uncertain affinity. Right, skeleton of New Zealand's extinct South Island flightless goose, *Cnemornis*, which was about a meter (3 ft.) tall and, together with *Patagopteryx*, illustrates a rather typical morphology of flightless birds in general. Note the strong hind limbs and the open fenestra of the pelvis. *Patagopteryx* from H. M. F. Alvarenga and J. F. Bonaparte, "A New Flightless Landbird from the Cretaceous of Patagonia," *Los Angeles County Museum of Natural History, Science Series* 36 (1992): 51–64. Courtesy of Herculano Alvarenga. *Cnemornis* from R. Owen, *Memoirs on the Extinct Wingless Birds of New Zealand*, 2 vols. (London: John van Voorst, 1879).

and flight adaptations originated in the context of selection for aerodynamic design. When one sees highly modified flight adaptations in the Chinese fossils of the Early Cretaceous, even if terrestrial, it is parsimonious to conclude that such animals had a volant ancestry.

To summarize, in birds flightlessness is a pervasive phenomenon, and most cases are the result of paedomorphosis and result in overall similar morphologies, including a trend toward large size, despite disparate ancestries. It was this superficial resemblance of ratites to dinosaurs that misled Thomas Huxley and later Percy Lowe into believing that ratites represented a flightless intermediary stage leading from dinosaurs to modern volant birds. This view was subsequently refuted by Gavin de Beer and others but is again popular today among paleontologists. One of the characteristics of flightless birds that appears iteratively in most all forms is the abbreviated forelimbs, a juvenile feature, and characterized by an attenuation of the forelimb elements from outer to inner (distoproximal). The flightless condition also tends to result in the acquisition of large size, the weight restriction on flight having been lifted. Flightless birds are always, without exception, derived from volant ancestors, and as far as is known, the morphological trends are unidirectional, there being no example of a major evolutionary reversal, or an abrogation of Dollo's Principle. That is, there is no example within any flightless birds lineage of reelongation of already foreshortened forelimbs. Presumably such a trend would have been the same or similar among theropod dinosaurs, and if so, there must exist still unknown ghost lineages if therapods are avian ancestors.

The Chinese Mesozoic Flightless Aviary

Given the incredible penchant for birds to develop paedomorphic flightlessness and the ease with which it can evolve, one of the mysteries of avian evolution has been the near absence of terrestrial flightless birds throughout the Mesozoic. By the end of the Cretaceous there are the giant flightless French bird (probably an enantiornithine) and

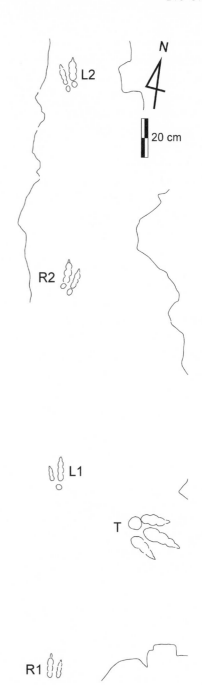

Fig. 5.30. Drawing of tracks of the didactyl deinonychosaur *Dromaeosauripus* (ichnogen.), from the Early Cretaceous southern coast of Korea. Described in J. Y. Kim et al., "New Didactyl Dinosaur Footprints (*Dromaeosauripus hamanensis* ichnogen. et ichnosp. nov.) from the Early Cretaceous Haman Formation, South Coast of Korea," *Palaeogeography, Palaeoclimatology, Palaeoecology* 262 (2008): 72–78. Courtesy of Jeong Yul Kim.

the enigmatic Argentine *Patagopteryx*, but not much more.[120] Many believed that the ratites were Huxley's ancient "waifs and strays," but there is no evidence that any ratite existed before the Cretaceous-Paleogene boundary, and their fossils would have surely been preserved if they had been present.

An intriguing recent discovery involves didactyl fossil footprints from the Lower Cretaceous of South Korea and China that represent large, flightless, presumably sickle-toed dromaeosaurs, which may answer a major question as to why there is an almost complete absence of large, ratitelike flightless birds from the Mesozoic.[121] More and more species are coming to light, including the recently described dromaeosaurid *Balaur bondoc*, a close ally of *Velociraptor*, from the Late Cretaceous of Romania, characterized by a highly retroverted pubis and a pair of hyperextensive pedal claws.[122] Most interestingly, *Balaur* exhibited extensive manual fusion, producing a very avian-like hand with a carpometacarpus, in addition to distal hind-limb fusion. Was *Balaur* an addition to the known flightless avifauna of the Cretaceous, a large, "beefy," flightless predatory bird? Could it be that the flightless dromaeosaurs, of varying sizes and types, underwent a radiation that lasted throughout the Cretaceous and, combined with the herbivorous ornithomimids, were ecological equivalents of the post-Cretaceous carnivorous phorusrhacids and herbivorous ratites, precluding the possibility of any flightless ornithurine evolutionary exploitation of similar ecological roles until after the great extinction event?

In the next chapter, we will see how the Chinese fauna is characterized by numerous flightless forms, most of which are considered to be nonavian, feathered dinosaurs. Were these faunal elements in reality the "hidden birds" of China? Could this be the answer?

6

YALE'S RAPTOR

Toward Consilience,
not Consensus

> Sit down before fact as a little child, be prepared to give up every preconceived notion, follow humbly wherever and to whatever abysses nature leads, or you will learn nothing.
>
> *Thomas H. Huxley to the Reverend Charles Kingsley, 1860*

New World Discovery

Following Thomas Henry Huxley's various pronouncements on the evolutionary relationships of birds and dinosaurs, attention focused on the New World, where two soon-to-be-famous rivals, Edward Drinker Cope (1840–97) and Othniel Charles Marsh (1831–99), were emerging.

New World Figures Take Center Stage

After the publication of Richard Owen's *Archaeopteryx* paper in 1863, Cope, a well-off aspiring paleontologist from Pennsylvania, traveled to Munich to view the Solnhofen fossils at the Bavarian State Collection. There he was shown the specimens by none other than Alfred Opel, who had made the only sketch of Karl Häberlein's *Archaeopteryx*.[1] Afterward Cope traveled to London to study the actual specimen of *Archaeopteryx* at the British Museum, but he had a mild breakdown; it was difficult for him to reconcile the fossils he was seeing with his Quaker creationist background. His father, a wealthy Philadelphia shipping magnate, summoned him home to Pennsylvania, where he soon recovered. It had become apparent to Edward Cope that Darwin's *On the Origin of Species* was much more than a fantasy. He could no longer view the Earth as a six-thousand-year-

old planet; even with his religious beliefs, he had honestly and perhaps unavoidably described *Archaeopteryx* as being "nearer bird than reptile but . . . neither."[2] Despite his lack of formal training, Cope became a professor of zoology at Haverford College (thanks to a donation by his father), where he contributed enormously to the field of vertebrate paleontology.

On his fateful 1863 European fossil tour, Cope had met a fellow American paleontologist, Othniel Charles Marsh. Like Cope, Marsh was on a family-backed tour of paleontological museums, but unlike Cope, he had abandoned all fundamentalist religious beliefs and was well on his way to becoming a full-fledged Darwinian. Following their meeting in Berlin, Cope had proceeded to London, the site of his breakdown, and Marsh traveled first to Bavaria, where he worked the Solnhofen quarries near Pappenheim, then to London to study the *Archaeopteryx*.[3] Like Cope, Marsh had a major sponsor, his excessively rich uncle George Peabody, who funded his education and made a huge donation to Yale University for the construction of a museum, now the Peabody Museum of Natural History. Unsurprisingly, Marsh was appointed Yale's professor of paleontology, initiating a hundred-year tradition of field paleontology that would endure through the tenure of John Os-

Fig. 6.1. Othniel Charles Marsh (left) and Edward Drinker Cope (right). From D. S. Jordon, ed., *Leading American Men of Science* (New York: Henry Holt, 1910).

trom, who was curator emeritus of the Peabody Museum until his death in 2005.

At the end of the Civil War, with Cope and Marsh in similar academic positions and the interior of the American continent ripe for exploration, one of the great sagas in the history of vertebrate paleontology was about to unfold. Known today as the Bone Wars, the intense rivalry between Cope and Marsh for fossil discoveries during America's Gilded Age involved theft, destruction of fossils, bribery, efforts to ruin the credibility of the respective rival, attempts to cut off funding, and personal attacks in scientific publications.[4] Among other revelations, their rivalry was an early demonstration of the lack of objectivity about facts and the truth in the field of paleontology. Perhaps presaging their bitter rivalry, both Cope and Marsh had within a year studied the famous London specimen of *Archaeopteryx* and had reached dramatically divergent conclusions.

Cope jumped to an early lead just before the paleontological exploration of the American West with the discovery in 1868 of the New Jersey dinosaur *Laelaps*. The discovery garnered recognition from none other than Huxley, who graciously acknowledged that both men had independently discovered bipedalism in dinosaurs and that there were some remarkable similarities between the skeletons of birds and dinosaurs. The name *Laelaps*, however, had already been taken, and Marsh renamed it *Dryptosaurus* in 1877. Both Cope and Marsh had an intense hunger for publicity, and while Cope made important scientific discoveries, Marsh sought notoriety through the popular press with tales of Western adventure, encounters with wild animals, tornadoes, and, of course, "Red Indians," establishing the cowboy, mountain man persona of the American vertebrate paleontologist that has endured to this day. Published photographs of Marsh with a pick in one hand and rifle in the other contributed to this image. Marsh also began the tradition of summer student expeditions from Yale and elsewhere. In one photograph from the Yale University expedition of 1870, near Bridger, Wyoming, Marsh stands in the center holding a rifle, flanked by nine students, including George Bird Grinnell, who later wrote that probably none of the lot, except the leader, "had any motive for going other than the hope of adventure with wild animals or wild Indians."[5] A pioneering conservationist, Grinnell was instrumental in founding the Audubon Society. Also in the photograph is Eli Whitney, grandson of the inventor of the cotton gin.

Fig. 6.2. Paleontology at Yale University became famous for its "student expeditions" during the Bone Wars. In this photograph from an 1872 expedition, Yale's Othniel Charles Marsh (center, back) and field assistants are ready to dig fossils in the Wild West. Pictured, left to right, Benjamin Hoppin '72; Thomas H. Russell '72, M.D. '75; Charles D. Hill; James MacNaughton. Courtesy of Peabody Museum of Natural History, Yale University.

The great American "bone rush" was under way, and rich fossil beds lay exposed on the surface, ripe for the picking. Among the most famous deposits were the Upper Cretaceous chalk beds of the western interior seaway known as the Niobrara Formation, where Marsh discovered the large, loonlike, foot-propelled diver *Hesperornis*:

> The first bird discovered in this region was the lower end of the tibia of *Hesperornis*, found by the writer in December, 1870. . . . The extreme cold, and danger from hostile Indians, rendered a careful exploration at that time impossible. . . . In June of the following year, the writer again visited the same region, with a larger party and an escort of United States troops, and was rewarded by the discovery of the skeleton which forms the type of *Hesperornis regalis*. . . . The

results of the trip did not equal our expectations, owing in part to the extreme heat (100° to 120° Fahrenheit, in the shade), which, causing sunstroke and fever, weakened and discouraged guides and explorers alike.[6]

Cutting through the adventuresome hyperbole, Marsh's most significant avian discoveries were two species of toothed birds: the foot-propelled diver *Hesperornis* and the ternlike *Ichthyornis*. In 1875, two years before the discovery of the Berlin *Archaeopteryx*, Marsh published his description of the teeth of *Hespernornis*, which were similar to those of *Archaeopteryx*, essentially peglike, with constrictions between the roots and crowns and with lingual resorption pits. The teeth of *Ichthyornis* were more controversial, but once they were confirmed, Marsh, well aware of the

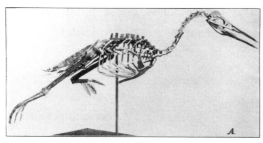

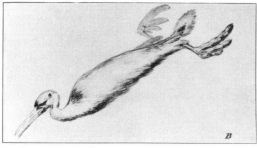

Fig. 6.3. Mounted skeleton and life reconstruction of *Hesperornis regalis*, largest species in the genus, about 1.5 m (5 ft.) long. Note the lobate webbing on the feet, which can be demonstrated from the bones of toes and have a structure similar to that of living grebes. In *Glorified Dinosaurs*, Luis Chiappe incorrectly illustrates this bird with fully webbed feet. From F. A. Lucas, *Animals of the Past* (New York: McClure Phillips, 1901).

controversy surrounding the teeth of *Archaeopteryx* (the existence of which was once rejected by Huxley), wrote a preliminary paper, noting: "The fortunate discovery of these interesting fossils . . . does much to break down the old distinction between Birds and Reptiles, which the *Archaeopteryx* has so materially diminished. . . . The teeth [of *Ichthyornis*] may be regarded as a character inherited from a reptilian ancestry. Their strong resemblance to the teeth of reptiles, in form, structure, and succession, is evidence of this. . . . [*Ichthyornis*] was carnivorous in habit, and doubtless was descended from a long line of rapacious ancestors."[7]

It is interesting here to note that Marsh was careful to stay clear of the dinosaurs and simply point to the reptilian features of his Cretaceous finds. In fact, the teeth of both *Hesperornis* and *Ichthyornis* were quite similar to those of *Archaeopteryx*, but they were equally dissimilar from the dinosaurs then known, both theropods and ornithischians. Huxley was delighted by the discoveries of Marsh, and the two began a correspondence. Huxley still had a deep interest in the evolution of birds, and as we have seen, he was convinced that

Fig. 6.4. Yale University expedition of 1870, near Bridger, Wyoming. At center, O. C. Marsh stands holding a rifle. This was one of the first "student" expeditions, and George Bird Grinnell '70 (third from left) later wrote that probably none of the group, except the leader, "had any motive for going other than the hope of adventure with wild animals or wild Indians." Reclining on the left is Eli Whitney '69, grandson of the inventor of the cotton gin. Courtesy Peabody Museum of Natural History, Yale University.

birds were descended from the dinosaurs through the giant flightless ratites; *Archaeopteryx* was an "intercalary," a sideline. His influence on Marsh is apparent in 1877, when Marsh wrote: "It is now generally admitted by biologists who have made a study of the vertebrates, that Birds have come down to us through the Dinosaurs, and the close affinity of the latter with recent Struthious Birds will hardly be questioned."[8]

But Marsh's primary interest was *Hesperornis*, which he considered a "carnivorous swimming ostrich," and it appeared to satisfy Huxley's view of the evolution of birds. He wrote of Marsh's discoveries: "The discovery of the toothed birds of the cretaceous formation in North America by Professor Marsh completed the series of transitional forms between the birds and reptiles, and removed Mr. Darwin's proposition that 'many animal forms of life have been utterly lost, through which the early progenitors of birds were formerly connected with early progenitors of the other vertebrate classes,' from the region of hypothesis to that of demonstrable fact."[9]

Huxley's fame had risen to stardom, and his friendship with Marsh proved a great nexus. In August 1876, Huxley arrived in the United States for a major lecture tour; he was paid the handsome sum of a thousand dollars for each speaking engagement.[10] The tour brought him to Yale, and Huxley and Marsh finally met, cementing a lasting friendship. At Yale, Huxley was able to study the museum's impressive paleontological collections. Among them were a more or less complete series of horse fossils, from the Early Eocene (fifty-million-year-old) *Hyracotherium* (*Eohippus*), the fox terrier–sized, five-toed ancestral horse, up through the Cenozoic, with diminishing toes all the way to modern horses with their single toe. Huxley was both astounded and delighted to see such a magnificent evolutionary series. Here at Yale was a collection that he considered parallel to none: "I am disposed to think that whether we regard the abundance of material, the number of complete skeletons of the various species, or the extent of geological time covered by the collection, which I had the good fortune to see at New Haven, there is no collection of fossil vertebrates in existence, which can be compared with it."[11]

For Huxley, Marsh's greatest discoveries had been not in the toothed birds but in the history of the horse: "The more I think of it the more clear it is that your great work is the settlement of the pedigree of the horse."[12] The two men bolstered each other's cause and image, and in 1877, Marsh addressed the American Association for the Advancement of Science on the topic of his famous toothed birds as proof of Darwin's ideas:

> To doubt evolution today is to doubt science, and science is only another name for truth. . . . The classes of Birds and Reptiles, as now living, are separated by a gulf so profound that a few years since it was cited as the most important break in the animal series, and one which that doctrine could not bridge over. Since then, as Huxley has clearly shown, this gap has been virtually filled by the discovery of bird-like Reptiles and reptilian Birds. *Compsognathus* and *Archaeopteryx* of the Old World, and *Ichthyornis* and *Hesperornis* of the New, are the stepping stones by which the evolutionist of to-day leads the doubting brother across the shallow remnant of the gulf, once thought impossible.[13]

Marsh gained considerable prestige for Yale in the field of vertebrate paleontology, publishing some three hundred scientific papers and books and describing and naming almost five hundred new species of fossils discovered by him and his collectors. In addition, he was president of the National Academy of Sciences from 1883 to 1895. Following such an illustrious career was a formidable task, but the Yale paleontology tradition was more than adequately carried on by Richard Swann Lull (1867–1957), who accepted a position in 1906 in vertebrate paleontology at Yale and as a curator at the Yale Peabody Museum. Lull served in a curatorial capacity for fifty years, continuing on as curator emeritus after retirement. Unlike Marsh, who did little teaching, Lull taught a popular undergraduate course on organic evolution that eventually morphed into his classic text *Organic Evolution*. Also unlike Marsh, Lull did little to augment the Yale collections, making only three expeditions after coming to Yale, but he directed

Fig. 6.5. Yale's John Ostrom at the *Archaeopteryx* conference, taken at Solnhofen Quarry, October 1984. Photograph by the author.

the move to the present building, designed the museum for educational purposes, and created many scale models and oversaw the mounting of many of the dinosaur skeletons. Most important, perhaps, his dozen or so graduate students included one of the founders of modern vertebrate paleontology, George Gaylord Simpson, a profound scientific thinker whose impact on the field remains immeasurable.

Yale's John Ostrom, Discoverer of Deinonychus

The next significant vertebrate paleontologist at Yale was John Ostrom, who after reading G. G. Simpson's *The Meaning of Evolution,* enrolled at Columbia University to study vertebrate paleontology with Edwin H. (Ned) Colbert. Ostrom joined the Yale faculty in 1961 and had an impressive career, writing a dozen or so books and innumerable scientific papers primarily on dinosaurs and the evolution of birds.

On a yearly field trip to Montana in 1964 Ostrom's academic career took a dramatic turn. He was exploring the Early Cretaceous, 110-million-year-old Cloverly Formation, not far from Billings, known primarily for dinosaur fossils, when he and Grant Meyer chanced upon an incredible fossil claw newly exposed from a cliff face. This predatory claw would forever change the field of vertebrate paleontology.[14] It was large, highly recurved, and quite sharp. Excitement was palpable as the crew painstakingly uncovered the forelimbs and then the rest of the skeleton of a small (slightly over 1 meter [3 ft.] or so in height, and 3.4 meters [11 ft.] in total length), lightly built (about 150 to 175 pounds) theropod that Ostrom knew was completely new to science. The large, lethal,

Fig. 6.6. John Ostrom with a *Deinonychus* skeleton cast. Courtesy of Peabody Museum of Natural History, Yale University.

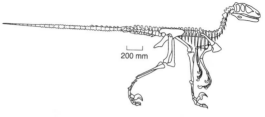

200 mm

Fig. 6.7. Deinonychus (in running pose), the Early Cretaceous dromaeosaurid that was the basis of John Ostrom's revival of the dinosaurian origin of birds and the ground-up origin of avian flight. Modified from J. H. Ostrom, "Osteology of *Deinonychus antirrhoopus,* an Unusual Theropod from the Lower Cretaceous of Montana," *Bulletin of the Peabody Museum of Natural History* 230 (1969): 1–165.

sicklelike claw was attached to the second digit of the foot. Ostrom would name the beast *Deinonychus antirrhopus* (the "counterbalancing terrible claw"). He concluded that *Deinonychus* used these pedal claws to disembowel prey, perhaps leaping, slashing, and killing it with the feet. The discovery of *Deinonychus* permanently changed Ostrom's life as a paleontologist; from then on, he viewed dinosaurs in a new light, as swift, intelligent, and perhaps even hot-blooded, or endothermic. The first "raptor" was popularized by Ostrom's student Robert T. Bakker, who, along with Ostrom, changed the popular view, prevalent since the turn of the century, that dinosaurs were dim-witted, cold-blooded, sluggish reptiles.

John H. Ostrom was the most influential dinosaur paleontologist of the second half of the twentieth century, with the humility of true greatness.[15]

A Friendly Adversary

Ostrom was a very talented vertebrate paleontologist whose primary interest was in dinosaurs. He had a careful and prolific research career, and he was a gentleman. I always enjoyed discussions with John, going back in time to 1973 at Yale University. Although we were adversaries on the question of the origin of birds, we had tremendous mutual personal and professional respect, and we stayed in close contact. It will astound many in the field of paleontology who tend to hate their adversaries and take combative positions that shortly before he died, Ostrom kindly wrote a strong letter of recommendation for me for a distinction. I last saw John in 1999 at a conference on the then newly discovered small, very birdlike theropod *Bambiraptor* held at the Graves Museum of Archaeology and Natural History in Fort Lauderdale, Florida. He was not in good health, but he was cheerful and interested; we had a most cordial meeting, and he told me that I had been a worthy adversary. I always appreciated our friendship and forthright conversations. I had the feeling that although he was quite satisfied that he had solved the problem of the origin of birds, the problem of the origin of avian flight from the ground up was simply not falling into place, and he certainly understood the time problem that had arisen, which

I had previously termed the "temporal paradox." *Deinonychus* was from the Cloverly Formation, approximately 110 million years old; the earliest known bird *Archaeopteryx* was from the Solnhofen limestone, 150 million years old. So, according to Ostrom, here was an animal supposedly close to the ancestry of birds, but some forty million years younger than the urvogel. How could it represent an ancestral bird? Also, if it did, then avian flight must have arisen from the ground up, which required some explaining. Then there arrived the most birdlike dinosaur ever discovered, *Bambiraptor*, from the Late Cretaceous, some seventy million years old, or some eighty million years after *Archaeopteryx*. Obviously, one could not be one's own grandmother!

Ostrom published his work on *Deinonychus* in 1969, in it concluding that his fleet-footed "raptor" dinosaur belonged to a little-known group first described in 1922 as the dromaeosaurs (from *Dromaeosaurus*, "running lizard") but was a distinctive large-brained, agile, and highly active creature with a retractable predatory sickle claw of the second toe.[16] Popularized by the movie *Jurassic Park*, these "raptors" (an unfortunate name because it was already taken by birds of prey) are now familiar to every child in the world.

Deinonychus *and* Archaeopteryx *Share Features*

Following the discovery of *Deinonychus*, Ostrom's brilliant career was at its peak. After years of work on the famous raptor, he decided to return to one of his favorite subjects, pterosaurs. He thus embarked on a European museum tour, and his researches took him to the Teyler Museum in Haarlem, the Netherlands, which housed some Solnhofen pterosaurs from the mid-nineteenth century. On examining a rather scrappy specimen labeled by none other than Hermann von Meyer, who had announced the first *Archaeopteryx* discovery, Ostrom saw that it was named *Pterodactylus crassipes*, yet its features were distinctive from any known pterosaur. The three long fingers each had a sharp claw, and the wrist had a semilunate carpal element similar to that in *Deinonychus*. Eventually he made out the faint impressions of what could be flight feathers. Having earlier studied the Berlin, Maxberg, and London specimens of *Archaeopteryx*, Ostrom knew that he was looking at a fourth specimen of the urvogel. The museum was delighted to know that they were the keepers of a fourth specimen of *Archaeopteryx*, and the curators released it to Yale University for study and publication.[17]

The discovery of the Haarlem specimen was a decisive moment for Ostrom. For him the similarities between the forelimb of *Archaeopteryx* and *Deinonychus* could be no coincidence; he was well aware of the implications of such a discovery for reinvigorating the dinosaurian origin of birds. It had been nearly a half century since Gerard Heilmann in his 1926 *Origin of Birds* had suggested a common ancestry of birds and dinosaurs from arboreal, pseudosuchian thecodonts (basal archosaurs) rather than a direct Huxleyan ancestry of birds from dinosaurs through the flightless ratites.[18]

At the same time, Newcastle University's Alick Walker had discovered evidence, based primarily on the skull, that birds might be descended from primitive crocodylomorphs like the 230-million-year-old *Sphenosuchus*, noting: "The crocodiles as a whole may have descended, perhaps as successive waves, from an unknown stock of late Middle to Upper Triassic reptiles which eventually gave rise to birds."[19] In Walker's view crocodiles and birds were descended from an immediate common ancestor, presumably a thecodontian, and this would remove the theropod dinosaurs from consideration. Although Walker's view never gained widespread support among paleontologists, he made some important points, and his paper, coming at the same time as Ostrom's (Ostrom's a few months following Walker's), helped move the issue of the origin of birds back onto the table for open discussion. To add to the excitement, a specimen of *Velociraptor*, the second discovered dromaeosaur, definitively showed the presence of a furcula (fused clavicles), so for Ostrom, birds had become little more than glorified, feathered, theropod dinosaurs.

Ostrom's new view of avian evolution was the first to spotlight *Archaeopteryx* in almost a century.[20] As we have seen, most paleontologists and evolu-

Fig. 6.8. Life restoration sculptures of *Deinonychus* by Stephen Czerkas of the Dinosaur Museum, Blanding, Utah. Below, *Deinonychus* is reconstructed as a secondarily flightless, feathered avian. Although *Deinonychus* has a hand quite similar to the flight hand of early birds and microraptors, no direct evidence of feathers is present. There is provisional evidence for possible ulnar quill nodes in the allied Late Cretaceous dromaeosaurid *Velociraptor*. Courtesy Stephen A. Czerkas; copyright Stephen A. Czerkas.

tionists, including Huxley and Darwin, had not made much of *Archaeopteryx*, viewing it as an offshoot of the main avian lineage, still retaining such reptilian features as teeth, claws, and a long tail. Not since Marsh, who did view *Archaeopteryx* as a missing link between dinosaurs and birds (but who also viewed *Hesperornis* as a "swimming ostrich"), had such a proposal been made, and this time with new and vigorous evidence from newly discovered fossils.[21] The clincher seemed to be a newly discovered specimen of *Archaeopteryx*, this one a third smaller than the London example, from the Jura Museum in Eichstätt, which was reported in 1973 but had actually been discovered in 1951 and thought to be a juvenile of the Solnhofen theropod *Compsognathus*. This find greatly reinforced Ostrom's view and bolstered his arguments, for if a specimen of the urvogel could have been misidentified as a small theropod dinosaur, it must say something about the similarities of theropods and *Archaeopteryx*.

Although our independent work on the same subject would reach diametrically opposite conclusions, John Ostrom and I approached our trade in virtually the same way. Ostrom was skeptical of cladistics and its tenets. He was a product of the old school of paleontology the likes of Ned Colbert, Al Romer, and G. G. Simpson, who were basically evolutionary systematists, using key characters and overall similarity to reach phylogenetic conclusions. These paleontologists held a keen interest in functional morphology, physiology, stratigraphy, and other aspects of biology and geology that are not normally incorporated into today's scheme of phylogenetic thinking.

Despite his objections to cladistics, Ostrom's dramatic discoveries serendipitously coincided more or less with the introduction and wholesale acceptance of phylogenetic systematics in paleontology. Ostrom preferred the careful analysis of characters and their function to statistical analyses of long lists of coded, often trivial morphological features. Yet when Jacques Gauthier's cladistic analysis of dinosaurs and birds revealed, not surprisingly, that birds are derived theropods, the floodgates were opened, and time and time again, new analyses, using almost the same characters and similar methods, reached the same conclusion.[22] Moreover, many analyses showed that a specific group of dinosaurs, the dromaeosaurs (including Ostrom's *Deinonychus*, and *Velociraptor*), were the most closely related to birds, and this result, again not surprisingly, has also been corroborated many times. Yet Ostrom would not understand the position taken by many current cladists that "there is little to be gained from taking the position that the problems will be solved by finding more fossils" and "for paleontologists faced with controversy, the traditional appeal is for more and better fossils.

In this case [origins], however, methodology—n fossils—will have to be the key to achieving agre ment and converting dissenters."[23] Such a positic is antithetical to the history of Ostrom's dramat revelations and could hardly be given any credibi ity after the last decade of discoveries from Chin

Ostrom failed to see the logic in simply co ing large numbers of simple characters in bina or multistate fashion, and in 1994 he stated h opposition to the obsession among cladists wi ticking morphological characters: "Reasoning such dubious quality demonstrates a fundament flaw in the cladistic methodology. Preoccupatic with compilation of lengthy lists of shared d rived characters at the expense of a well-reason functional analysis of the characters that a shared will result in an erroneous phylogeny e ery time."[24] Ostrom was also concerned by the fa that his *Deinonychus* occurred temporally after t then earliest known bird, *Archaeopteryx*.

Achieving Consilience, not Consensu
Temporal Paradox and the Geologic Record

The "temporal paradox" was initially described I me to designate the apparent problem for a stri dinosaurian origin of birds posed by the stra graphic or time disparity between the occurrence the earliest known bird and its designated ancestr group.[25] *Archaeopteryx*, already a well-develop bird, and the putative dromaeosaurid ancestors birds, thought by John Ostrom to closely resemb the sickle-claw *Deinonychus*, were some forty m lion years apart. The small dromaeosaurid *Bar biraptor*, exhibited in 1999, is some eighty millic years younger than *Archaeopteryx*, but the mo recent discovery of the Early Cretaceous Chine microraptors reduced the time gap to some twent five or so million years. There are, to be sure, old maniraptoran dinosaurs, some approaching t age of *Archaeopteryx*, and even older. Paleontol gists Christopher Brochu and Mark Norell ha analyzed the temporal paradox question by usi several other potential archosaur sister taxa f Aves and have found that these, too, create ten poral paradoxes—that is, long stretches of tin between ancestor and descendant—if they we indeed close to avian ancestry.[26] In 2009, Dong

Fig. 6.9. Illustration modified from a *National Geographic* article on dinosaurs, depicting the current dino-bird orthodoxy, with the putative ancestors geologically much younger than the earliest known bird, *Archaeopteryx*. Paleontologists have attempted to dismiss the confused sequence as irrelevant to the debate, but it is a problem of paramount concern to most workers in the field and especially to the interested public. Adapted and modified from J. Ackerman, "Dinosaurs Take Wing," *National Geographic*, July 1998, 74–99.

Hu and colleagues considered that the discovery of the fully feathered *Anchiornis huxleyi*, considered to be the oldest known troodontid, dating to about 155 million years ago, "refutes the 'temporal paradox.'" The discovery of *Anchiornis*, some five to ten million years older than the iconic urvogel, would appear to eliminate the problem of the temporal paradox for a theropod ancestry as initially conceived, but as noted, it is most likely an archaeopterygid as originally described.[27] Yet another important point emerges that is whatever the ancestor—dinosaur, dinosauromorph, or basal archosaur—there must still remain substantial ghost lineages rather than the smooth, clean lines depicted in most phylogenies, which typically invert descendants and putative ancestors. Also, it is disturbing to see how paleontologists today view fossils of the Late Cretaceous as providing evidence for everything from "insight into the assembly of the modern flight apparatus" to "size evolution preceding avian flight." Today, the geological time scale appears to have little bearing on the interpretation of a fossil's importance in the scheme of evolution. Certainly, there is hardly any smooth transition between any known theropod and the strange, probably avian *Anchiornis*, with its feathered feet and wings and its nontheropod sacral anatomy. What possible ancestral lineage exists for the strange, birdlike caudipterids? Ghost lineages are required all over the place for any hypothesis on bird origins, whether from basal archosaur, dinosauromorph, or advanced theropod, the latter the current consensus view.

One of Huxley's primary concerns regarding the possibility that *Archaeopteryx* might somehow indicate a relationship between birds and dinosaurs was its relatively young geologic age. Both Darwin and Huxley held the view that the succession of life should closely follow the geologic column; in fact, Huxley felt strongly that ancestral birds should be sought in the Triassic or even earlier, in the Late Paleozoic.

Topsy-Turvy Phylogeny from National Geographic

If we examine the typical consensus phylogeny, popularized in articles in *National Geographic*, the small compsognathid *Sinosauropteryx*, thought to possess downlike protofeathers, or dino-fuzz, is normally placed near the base of the cladogram. Next is the Late Cretaceous dromaeosaur *Velociraptor*, followed by the Early Cretaceous flightless *Caudipteryx* (with precise avian arrangement of wing primary feathers), which is followed by its close ally *Protarchaeopteryx*. In sequence is the Late Jurassic *Archaeopteryx*, followed by the Early Cretaceous opposite bird (with an alula) *Eoalulavis* and, finally, a modern crow.[28] The family tree as depicted in the chart accurately reflects the current view, but the sequence is almost the complete reverse of what one would expect in avian evolution from the stratigraphic record, which, except in the case of the popular view of bird origins, is closely correlated with the evolutionary history of vertebrate groups. Adding to the confusion, implicit in this view is the proposition that in the Chinese Early Cretaceous the entire array of avian ancestry was on display, including not only the physical fossil ancestors of birds, some twenty-five to thirty

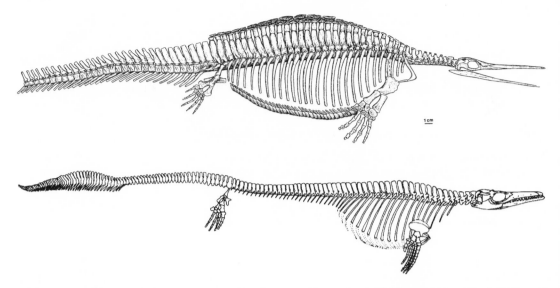

Fig. 6.10. Convergent evolution in marine reptiles: the Middle Triassic *Hupehsuchus* (skeletal reconstruction, above) compared to the Upper Cretaceous *Plotosaurus*, a mosasaur (below). Although the two are not closely related and have no temporal overlap, they would form a monophyletic group if simple cladistic methods were followed. Above, from R. L. Carroll and D. Zhi-Ming, "*Hupehsuchus*, an Enigmatic Aquatic Reptile from the Triassic of China, and the Problem of Establishing Relationships," *Philosophical Transactions of the Royal Society of London B* 331 (1991): 131–53. Courtesy Royal Society of London and Robert Carroll. Below, from D. A. Russell, "Systematics and Morphology of American Mosasaurs (Reptilia, Sauria)," *Peabody Museum of Natural History Bulletin* 23 (1967): 1–237. Courtesy Peabody Museum of Natural History, Yale University.

million years after *Archaeopteryx*, but also all stages of feather evolution, all this when no intermediate feathers of any kind are known in any previously discovered fossil or from any living bird. It would be extraordinary if all stages in the hypothetical evolution of birds coexisted in one small span of geological time.

As Harvard's Al Romer once noted (contrary to the popular practice of phylogenetic systematics), "In discussing fossils, some notion of the geological time scale is necessary."[29] As most stratigraphers and petroleum geologists well know (the latter depend on index fossils to identify age), "Biologic evolutionary history has given us . . . the basis of the unique progressive traditional scale."[30] If time is irrelevant to phylogenetic considerations, it should give one pause to realize that the entire column of Cenozoic stratigraphy, beginning during the Paleocene and continuing through the Middle Pleistocene, has as its basis the North American Land Mammal Ages (NALMA), which was formalized in 1941. NALMA is based on the geographic place-names where individual mammalian species first appeared in particular sections, and the system serves as the source for similar time scales in other continents.[31] It is clear, then, that fossils have traditionally played a huge role not only in phylogenetic considerations but also in deciphering the stratigraphic record. As Columbia's Paul Olsen said in an interview for *American Scientist*: "The study of fossils does yield a unique biological perspective that we can never get from studying only the living world. The crucial time dimension is what makes paleontology worth doing in the first place."[32]

The Recognition of Convergence and Its Attendant Problems

Marine Reptiles Converge: The Case of Hupehsuchus *and Mosasaurs*

In general biology textbooks, the marine reptiles known as ichthyosaurs ("fish reptiles") and modern dolphins are commonly cited as a classic example of convergent evolution. While this convergent pair is easily distinguished and each is placed appropriately in its respective class and order, often the situation is not so simple: massive convergence may conceal true relationships and hide identities. Such an example is provided by the enigmatic Early Triassic, 1-meter-long (3 ft.) aquatic reptile from China named *Hupehsuchus* (and its close cousin *Nanchangosaurus*) and the Cretaceous mosasaurs (aquatic varanid or monitor lizards). These two taxa provide an example in which the simple tabulation of supposed shared, derived characters, or synapomorphies, can produce erroneous results because their bodies have been designed by natural selection for an advanced mode of marine swimming. As Robert Carroll and Zhi-Ming Dong note, "Most of the derived characters appear to be subject to convergence among secondarily aquatic reptiles."[33] *Hupehsuchus* was a marine, diapsid reptile with a girdle and limbs highly adapted for aquatic locomotion.

In trying to assess its phylogenetic status, Carroll and Dong attempted to use typical cladistic methodology, which places *Hupehsuchus* as the sister group of the ichthyosaurs and the nothosaurs, also highly adapted swimmers. They note:

These data might be interpreted as indicating that all early Mesozoic aquatic reptiles shared a common ancestry, distinct from that of the lepidosauromorphs and archosauromorphs. This seems unlikely, however, because ichthyosaurs, nothosaurs, placodonts and thalattosaurs exhibit very different proportions of the trunk and limbs, indicating divergent modes of aquatic locomotion, as well as having very different patterns of skull and vertebral morphology. The problem is to establish whether or not the large number of derived characters shared by these groups are actually homologous. . . . The difficulty of identifying unique derived characters uniting the Hupehsuchia with ichthyosaurs, despite the great number of derived similarities, raises the possibility that skeletal features that are common to secondarily aquatic reptiles might have evolved convergently in each of these groups.[34]

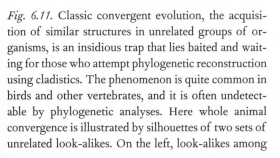

Fig. 6.11. Classic convergent evolution, the acquisition of similar structures in unrelated groups of organisms, is an insidious trap that lies baited and waiting for those who attempt phylogenetic reconstruction using cladistics. The phenomenon is quite common in birds and other vertebrates, and it is often undetectable by phylogenetic analyses. Here whole animal convergence is illustrated by silhouettes of two sets of unrelated look-alikes. On the left, look-alikes among the foot-propelled divers: (A) the Cretaceous toothed bird *Hesperornis* (Hesperornithiformes) and (B) a modern loon (Gaviiformes). On the right, look-alikes among the wing-propelled divers: (C) a Southern Hemisphere diving-petrel (Procellariiformes) and (D) a Northern Hemisphere auk (Charadriiformes). From A. Feduccia, *The Origin and Evolution of Birds* (New Haven: Yale University Press, 1996).

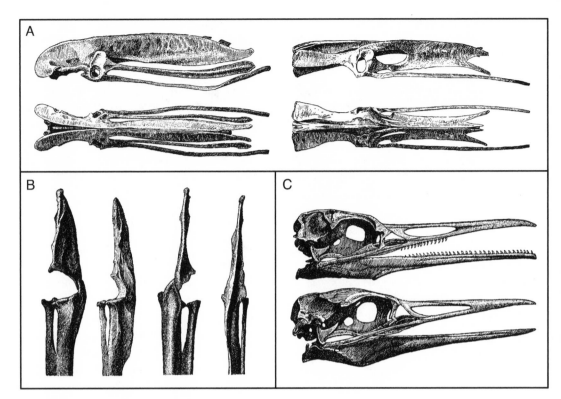

Fig. 6.12. Cladistic analysis has traditionally been incapable of dealing with massive convergence, often combining loons, grebes, and hesperornithiform birds in a single clade. Depicted here is a remarkable example of osteological convergent evolution. (A), pelves in lateral and dorsal views of the toothed Cretaceous *Hesperornis* (left) compared to that of the great crested grebe (*Podiceps cristatus*) (right). (B), cnemial crest of the tibiotarsus, which accommodates the powerful swimming muscles, in two views of *Hesperornis* and the common loon (*Gavia immer*). In *Hesperornis* the cnemial crest is formed embryologically from the patella, in loons, by a projection of the tibia, and in grebes, by contributions from the patella and the tibia, thus proving the similarity through convergence. (C), the skulls of *Hesperornis* (above) compared to the common loon (below). Adapted and modified from G. Heilmann, *The Origin of Birds* (London: Witherby, 1926).

To test their ideas and the efficacy of the cladistic approach in highly convergent organisms further, the authors made comparisons with two additional groups of secondarily aquatic diapsid reptiles, the plesiosaurs and the mosasaurs. These two reptiles are ideal because the specialized body plan of the plesiosaurs is unknown before the Early Jurassic. Mosasaurs, known from the Middle to Late Cretaceous, have a known ancestry from varanid (monitor) lizards, and the living Southeast Asian *Varanus salvator*, a large coastal, seafaring monitor, is a veritable protomosasaur. In the final analysis, plesiosaurs shared twenty-six skeletal synapomorphies with *Hupehsuchus*, and mosasaurs shared twenty-nine. In accordance with the methodology of phylogenetic systematics, convergent or homoplastic characters would be identifiable by the principle of parsimony; that is, if any two groups are united by numerous synapomorphies but fewer characters support an alternative relationship, the smaller set of characters is assumed to be the result of convergence: "Comparison of *Hupehsuchus* and mosasaurs, however, . . . suggests that *most* of the derived characters exhibited by *Hupehsuchus* were convergently acquired by mosasaurs. How could this be recognized if we did not know from other evidence that mosasaurs evolved separately from terrestrial ancestors [varanid lizards] that retained the primitive character state for these features? The principle of parsimony cannot be directly applied in this situation."[35]

The authors point out that most of the skeletal characters associated with an aquatic mode of life are "of such general nature that it is not possible to refute the possibility of their common origin by direct observation." Olivier Rieppel has also discovered that many of the similarities among various marine diapsids are attributable to paedomorphosis: "The same tissues, the same developmental processes, and the same functional explanations may be involved in all groups, but the changes are not strictly homologous because they have occurred independently in each group."[36]

Following this line of inquiry, Carroll and Dong appropriately ask: "Is there any category of characters that is not commonly subject to convergence? Is it ever possible to establish phylogenetic relationships without some specific information regarding the strict homology of the characters in question? The problem of establishing the relationships among the aquatic diapsids suggests that it is not." [37]

The case of marine reptiles sounds unsurprisingly similar to that of the cladistic grouping of highly adapted aquatic birds. Loons have been grouped with grebes and sometimes also with hesperornithiforms despite the mountain of evidence to the contrary.[38] In the case of these three disparate groups of foot-propelled divers, one can demonstrate that they are from separate lineages because the cnemial crest of the tibiotarsus, which houses the powerful swimming muscles, is formed, in *Hesperornis*, embryologically from the patella; in loons, by a projection of the tibia; and in grebes, by contributions from the patella and the tibia. In addition, whereas loons have the anterior toes bound with complete webbing, in grebes, the webbing is lobate, surrounding individual toes, and is accompanied by a unique foot folding mechanism, also present in hesperornithiforms. The case of foot-propelled divers among birds is a classic example of convergence that cladistic methodology is still incapable of dealing with. Clearly, the insidious trap of convergence lies baited and waiting for the unsuspecting victim, who will invariably be

trapped when using naive methodology. *"Cladistics should not become the new 'authority.'"*[39]

Bipedalism

John Ostrom argued that birds and theropods were unique among vertebrates in their possession of an upright bipedal posture, but subsequent discoveries of fossils from disparate groups of tetrapods proved that assumption false and showed

Fig. 6.13. Three silhouettes modified from skeletal reconstructions. Above, *Compsognathus*, a fairly typical Late Jurassic coelurosaurian theropod thought to be close to the ancestry of birds and therefore dromaeosaurs, showing the characteristic drastically reduced forelimbs and massive hind limbs with obligate bipedalism. Middle, the Late Triassic *Effigia*, a crocodylomorph that converged on the theropod body plan. Below, the bipedal Permian bolosaurid reptile *Eudibamus cursoris*, oldest member of the clade Parareptilia. Not to scale. *Compsognathus* from K. Peyer, "A Reconsideration of *Compsognathus* from the Upper Tithonian of Canjuers, Southeastern France," *Journal of Vertebrate Paleontology* 26, no. 4 (2006): 879–896. Courtesy of K. Peyer. *Effigia* from S. J . Nesbitt and M. A. Norell, "Extreme Convergence in the Body Plans of an Early Suchian (Archosauria) and Ornithomimid Dinosaurs (Theropoda)," *Proceedings of the Royal Society B* 273 (2000): 1045–48. *Eudibamus cursoris* from D. S. Berman et al., "Early Permian Bipedal Reptiles," *Science* 290 (2000): 969–72.

that iterative evolution of bipedal, cursorial locomotion, with great reduction of the forelimbs, was
convergently not uncommon. There are numerous examples of living facultatively bipedal lizards (lepidosaurs), exemplified by the Australian
agamid frilled lizard (*Chlamydosaurus*), the Central American iguanid basilisks (*Basiliscus*), and
others. Fossil finds date back to the Triassic South
African Karoo deposits.[40]

In 2000, David Berman and colleagues reported an amazing find, a Lower Permian, 290-
million-year-old reptilian skeleton of a small,
lizardlike bipedal saurian from central Germany,
from an assemblage containing the well-known
North American Permian reptiles *Seymouria* and
Dimetrodon, thus supporting the age.[41] *Eudibamus cursoris*, a 30-centimeter-long (12 in.) swift,
facultatively bipedal bolosaurid parareptile is the
earliest discovered tetrapod to have assumed a bipedal posture using the hind limbs in a parasagittal plane, running on its toes (digitigrade) in rapid
locomotion. Thus, the earliest experiment in an
early form of "theropod locomotion" occurred
some sixty million years before parasagittal archosaurs and the dinosaurs. Moreover, it was not the
only attempt: in the Late Triassic, another group
of reptiles would venture into the same postural
strategy.

This time iterative evolution of bipedalism
occurred within the archosaur assemblage, but
in crocodilians (suchians), not theropods. *Effigia*
("ghost," from the locality) was discovered in un-
prepared blocks of matrix originally excavated in
the 1940s by American Museum of Natural History field parties working the famous Ghost Ranch
Quarry from the Late Triassic Chinle Formation
in New Mexico that produced innumerable skeletons of the small basal theropod *Coelophysis*.[42] The
material was thought until recently to represent a
monospecific assemblage, but as it was restudied,
many new forms emerged, the most interesting of
which was the Triassic equivalent of the Cretaceous ornithomimid dinosaurs, with an edentulous
jaw, massive hind limbs, greatly reduced forelimbs,
and a pubic boot that is one third as long as the pubic shaft. Yet it is clearly a crocodylomorph, with

a crocodile-normal ankle and crocodile-like pes.
Described by Sterling Nesbitt and Mark Norell
of the American Museum of Natural History in
2000, this sensational discovery from the bowels
of the museum's collections showed conclusively
that "some late Triassic suchians may have occupied similar adaptive zones to subsequent clades
of dinosaurs."[43] Without the preserved ankle, just
what might this animal have been diagnosed as
cladistically?

As an aside, the track record of identification
of Triassic archosaurs does not inspire much confidence. Aside from *Effigia*, classic examples include
the Triassic *Procompsognathus* and *Lisboasaurus*.
Jacques Gauthier defined "theropoda . . . to include birds and all saurians that are closer to birds
than they are to sauropodomorphs."[44] In Gauthier's cladistic analysis *Procompsognathus* slots out
at the level of the Ceratosauria; but since that time
Paul Sereno and Rupert Wild have demonstrated
that *Procompsognathus* was incorrectly identified:
"*Procompsognathus triassicus*, long held to be a
primitive theropod, is actually a paleontological chimera composed of the postcranial skeleton
of a . . . ceratosaur and the skull of a basal crocodylomorph."[45] And more recently Fabian Knoll
reviewed the postcranial material of *Procompsognathus* and concluded that "all the material once
named, or assigned to, *Procompsognathus triassicus*
may include three taxa: a coelophysoid theropod
. . . a possible tetanuran theropod . . . and a basal
crocodylomorph."[46] Gauthier's phylogenetic analysis records: "[In] Theropoda, . . . in contrast to
the ancestral condition, the lacrimal forms much
of the skull roof anterior and lateral to the pre-
frontal above the orbit in *Procompsognathus* (Pers.
obs.), Ceratosauria, Carnosauria, Ornithomimi-
dae, [and] Deinonychosauria. . . . The presence of
the apomorphic condition in theropods as diverse
as *Procompsognathus* . . . , *Ceratosaurus* . . . , *Deinonychus* . . . , *Allosaurus* . . . , *Tyrannosaurus* . . . ,
Oviraptor . . . , *Gobipteryx* . . . , and the Ratitae and
Tinami, suggests that the vomers are fused anteriorly in Theropoda generally."[47] Definitive analyses of such poorly studied taxa shed confidence
from phylogenetic analyses that are portrayed as

authoritative. In addition to the problem taxon *Procompsognathus*, there was the Upper Jurassic "missing link" *Lisboasaurus*, considered along with the new taxon *Archaeornithoides* (also based on fragmentary material). *Lisboasaurus* was proposed as the closest nonavian relative of the Avialae but was later shown to be a crocodylomorph.[48] In addition to the ornithomimid-like suchians such as *Effigia*, there were ankylosaur-like suchians, the aetosaurs, and even carnosaur-like suchians in the form of *Postosuchus*, showing that "some of the Late Triassic suchians may have occupied similar adaptive zones to subsequent clades of dinosaurs."[49] Crocs seem to have held their own in the Triassic.

Cursorial Pterosaurs Debut Again

The controversy over pterosaur locomotion goes well back to the first part of the nineteenth century, with early advocates of a batlike posture, such as Austrian paleontologist Othenio Abel in 1925, contrasted with later models of a more birdlike, bipedal posture, advocated by Harry Grovier Seeley, primarily because he erroneously thought that pterosaurs were closely related to birds.[50] The stage was thus set for a controversy that has persisted until just recently but has finally been laid to rest.

As we have seen, overconfidence in the results of phylogenetic systematics can lead to phylogenetic inferences that are misleading, a now classic example being the case for the cursorial model for the origin of flight in birds; but a similar case applies to the origin of flight in pterosaurs, which strikingly parallels the arguments over bird flight. Kevin Padian was a graduate student at Yale University in the late 1970s, and for his thesis he undertook an analysis of locomotion and flight origins in pterosaurs, flying archosaurs (Triassic to Late Cretaceous) that had a leathery, batlike wing membrane for flight. Once thought, but disproved, to be avian predecessors, pterosaurs paralleled birds in many aspects of their evolution through the Mesozoic, losing their teeth and tails, from the Jurassic to the Cretaceous, developing a keeled sternum, fusing bones (synsacrum-fused sacral verte-

brae and notarium-fused thoracic vertebrae), and lightening the skeleton. In 1983 and 1984, Padian published papers on a functional analysis of flying and walking in pterosaurs, and on the origin of pterosaurs.[51] Having been caught up in the cladistic movement, he argued that pterosaurs were the sister group of dinosaurs, and as a consequence of the outcome of the cladogram, he was stuck with a cursorial model for pterosaur locomotion and an origin of flight from the ground up, just as Ostrom was for birds. For Padian, pterosaurs had a fully erect posture; they were bipeds that stood on their toes (digitigrades), much like birds and theropods. Jacques Gauthier, a graduate student at Berkeley in the 1980s, working with Kevin Padian, produced the now classic cladistic analysis of archosaurs and, in the same vein, identified pterosaurs as ornithodirans, the group including the dinosaurs, which fit with Padian's model.

Recently, much of our revolutionary understanding of pterosaurs has come from the discovery of new fossils and trackways by J.-M. Mazin and colleagues and Martin Lockley and his colleagues.[52] Padian's conclusion that pterosaurs were obligate digitigrade bipeds and that, as a consequence, flight in pterosaurs must have originated from the ground up was dealt a fatal blow by both Lockley's trackways and trackways from the primitive pterosaur *Pteraichnus*, discovered by Christopher Bennett.[53] Today hundreds of pterosaur trackways clearly illustrate quadrupedal locomotion, showing that pterosaurs were terrestrial quadrupeds and that much of Padian's cladistically modeled interpretation of pterosaur functional anatomy was erroneous.

The second line of evidence came initially from the discovery of the well-preserved Late Jurassic pterosaur *Sordes pilosus* by David Unwin and Natasha Bakhurina; specimens showed that the main wing membranes attached to the leg as far as the ankle and that a well-developed uropatigium extended between the hind limbs, supported on its rear margin by the fifth toe. The authors suggested that the presence of the uropatagium indicated "a poor terrestrial ability and a 'gravity-assisted' rather than 'ground up' origin

Fig. 6.14. Terrestrial locomotion in pterosaurs. The Triassic basal pterosaur *Dimorphodon*, center, shown errone-ously in a bipedal posture, as envisioned by H. G. Seeley and later by Kevin Padian (based on different reason-ing), compared with the proven posture of pterosaurs as illustrated by *Pteranodon* (above left), *Pterodactylus* (above right), and *Rhamphorhynchus* (below). *Dimorphodon* from H. G. Seeley, *Dragons of the Air* (London: Methuen, 1901); silhouettes after *M. Reichel, 1896–1984: Dessins* (Basel: Geological Institute of Basel Univer-sity, 1985).

of flight for pterosaurs."[54] Decisive anatomical evidence was provided by the discovery of a well-preserved, three-dimensional specimen of the Tri-assic basal pterosaur *Dimorphodon* (Padian's model pterosaur), by paleontologists James Clark and colleagues, which was published as a cover article in *Nature*.[55] This specimen showed unequivocally that this primitive pterosaur was flat-footed and walked on the soles of its feet, a finding quite con-sistent with the fossil footprints. These recent stud-ies not only show that Padian's cursorial model for the origin of pterosaur flight was incorrect but also emphasize the pitfalls of uncritical reliance on cla-distic analysis for biomechanical analyses. Inter-estingly, although Harry Seeley, Thomas Huxley's adversary on the question of the origin of birds, and a Victorian student of pterosaurs, had early on argued against a dinosaur-bird nexus, in his popu-lar 1901 book *Dragons of the Air*, he incorrectly illustrated the basal pterosaur *Dimorphodon* in a bipedal stance, thinking that birds and pterosaurs were allied.[56] At last, the debate on the terrestrial locomotion of pterosaurs appears to be settled.[57]

The pterosaurs were an early but highly successful experiment in vertebrate flight, with extensive leathery wings involving both fore- and hind limbs, and they were competent quadrupedal plantigrades on the ground. Interestingly, cladistic analyses may also be incorrect in recovering pterosaurs as ornithodirans; they may be basal archosaurs or even more primitive diapsids, and the origin of their flight, as in all other vertebrates, was from the trees down.[58] As an aside, a small arboreal pterosaur has now been discovered from the Early Cretaceous of China, showing that even though the vast majority of pterosaurs are known from ancient coastal environments (most likely a preservational phenomenon), some at least inhabited inland forests. "It is very likely that this pterosaur represents a lineage of arboreal creatures that lived and foraged for insects in the gymnosperm forest canopy of Northeast China during the Early Cretaceous.[59]

Despite the overwhelming evidence that pterosaurs basically fit the "old model," a visit to the University of California Museum of Paleontology website reveals, "We can infer that the origin of flight in pterosaurs fits the *'ground up'* evolutionary scenario, supported by the fact that pterosaurs had no evident arboreal adaptations. Some researchers have proposed that the first pterosaurs were bipedal or quadrupedal arboreal gliders, but these hypotheses do not incorporate a robust phylogenetic and functional basis."[60]

Molecular Systematics to the Rescue?
(Except in Paleontology)

The inability of morphological character-based phylogenies to detect obvious convergence was the primary rationale for the interest in molecular systematics, pioneered in birds by Yale University's Charles Sibley, first with starch gel electrophoresis, and then using isoelectric-focusing electrophoresis, immunological comparisons of proteins, and DNA-DNA hybridization. More recently at other laboratories, mitochondrial DNA

and nuclear gene sequencing have been employed. Each time a new technique has been introduced, it has been proclaimed the final solution to avian systematics. Many older avian systematists recall presentations in the late 1960s in which the results of egg-white protein starch gel electrophoresis were hailed as the ultimate solution to systematics, and pleas for major changes in classification were made based on these studies.

Sibley and Ahlquist's DNA hybridization studies had the most significant impact on avian systematics, and their tree, the so-called tapestry, was displayed far and wide. In the final analysis, however, their results were too far-fetched to have credibility and were considered little more than assertions about avian relationships.[61] For example, among their untenable conclusions were that the traditional orders Pelecaniformes (pelicans, boobies, cormorants, and frigatebirds) and the Ciconiiformes (storks, herons, ibises, and flamingos) were each "shown" to be paraphyletic, and individual components were considered interrelated with disparate groups, ranging from penguins, grebes, loons, shearwaters, and albatrosses. Despite the unbelievable DNA hybridization results, and just like the egg-white protein era, there were calls for the total rearrangement of the avian classification. Stephen Jay Gould joined in the euphoria, proclaiming that the problem of avian phylogeny had been solved.[62]

It quickly became apparent that the world of higher-level avian systematics was not to be resolved simply by such an approach, as exemplified by a classic study by Blair Hedges of Pennsylvania State University and Charles Sibley, who performed not only DNA hybridization studies but also DNA sequences of the 12S and a6S rRNA mitochondrial genes.[63] Their results supported the view that the pelicans were most closely related to the strange shoebill, or whale-billed stork, of central African swamps; that the boobies, gannets, cormorants, and anhingas form a clade; that the tropicbirds, long considered the most primitive pelecaniforms, were not closely related to the other taxa; and that the frigatebirds were closely related to the penguins, loons, petrels, shearwaters, and albatrosses (Procellarioidea). The authors claimed that their study provided another example of the incongruence between morphological and genetic traits, noting that the salient defining characters of the pelecaniform birds, the totipalmate foot (all four toes bound by webbing), a gular pouch, a salt gland within the orbit (instead of in a supraorbital groove), and the lack of an incubation pouch (present in other seabirds), are insufficient to conclude monophyly. They concluded that the "morphological characters presumed to be shared-derived are shared-primitive or convergent."[64] Although their results were not widely accepted, if they were true, none of the morphological characters used to define avian fossil clades could be considered reliable, and it would be a refutation of simple character-based phylogenies.

Since that time, more sophisticated global molecular comparisons have been performed using mitochondrial DNA and nuclear gene sequencing (summarized up to 2004 in Cracraft and colleagues' work), but again the results are often so contrary to established conceptions of relationships based on morphology and other lines of evidence that results are difficult to interpret, and perhaps this is exactly what one would expect at this early stage in our understanding of the genome.[65] Most recently a global "phylogenomic" study of the class Aves was made by Shannon Hackett of Chicago's Field Museum of Natural History and seventeen colleagues, in which they examined ~32 kilobases of aligned nuclear DNA sequences from nineteen independent loci for 169 species, representing all major avian groups.[66] Although this study was a major advance in terms of large-scale genomic comparisons, again, at the higher levels, the results are, to say the least, highly controversial, and include findings that: New World vultures (Cathartidae) are allied with Accipitridae; grebes are allied with flamingos; typical pelecaniforms and ciconiiforms form a clade (excluding the tropicbirds (Phaethontidae); tropicbirds, sandgrouse, and pigeons and doves form a clade; core Gruiformes (rails, cranes, and allies), Cuculiformes (cuckoos), and Otitdidae (bustards) form a clade; parrots are the sister group of the Passeriformes; and, perhaps

most heretical, "flighted Tinamiformes [tinamous] arose within the flightless Struthioniformes [ostriches]." It would have been helpful if *Science* had seen fit to tone down the sensationalistic title: "A Phylogenomic Study of Birds Reveals Their Evolutionary History." But, as we saw in chapter 1, the major science journals often appear more prone to tabloid tactics than solid, measured science.

In recent years, there has been an increasing effort to integrate cladistic results based on morphology with molecular data, but again the two data sets are in reality two separate measures of relatedness and are difficult to reconcile. For example, we saw earlier that despite the striking morphological similarity of the early fusiform chordate the fishlike amphioxus to primitive fishes as a putative candidate for approximating a vertebrate "ancestor," the related baglike urochordates show closer molecular affinity to the vertebrates. Cladistic analysis shows the amphioxus to be the sister group of the vertebrate lineage; whole genome comparisons show that the urochordates are most closely allied with the vertebrates despite their lack of similarity to vertebrates in adult structure. Another example is the classic chimpanzee-human comparison, where chimps and people, although morphologically very distinctive, nevertheless share genomes that are 98 percent identical.

Moving to birds, we have seen that the Hawaiian gooselike ducks, the moa-nalos, were more like ratites in their postcranial morphology, were shown to be of anseriform affinity by their gooselike skulls, and in the final analysis were shown to be gooselike duck derivatives. So, although they are flightless geese morphologically, their DNA reveals that they are ducks, perhaps even closely allied with the Pacific black duck of the genus *Anas*.[67]

An even more striking example of parallelism (the convergence of traits arising independently in closely related species) was provided by the radiation of Hawaiian honeyeaters, represented by five endemic species of recently extinct, nectar-feeding songbirds placed in the genera *Moho* and *Chaetoptila*, which exhibit striking similarity to the Australasian honeyeaters (Meliphagidae).[68] Since their discovery by explorer James Cook, the five species of Hawaiian honeyeaters have been logically placed together with the Meliphagidae, because of their similar appearance, including a host of morphological and behavioral traits: long legs, decurved bills, brushy, scrolled tongues, a partial covering of the nostril openings to exclude pollen, and plumage patterns, including a shared trait of yellow feather tufts also occurring in some Australasian species. But by analyzing a combination of nuclear and mitochondrial DNA sequences, a team from the Smithsonian discovered that instead of being allied with the Australasian Meliphagidae, as anticipated, the Hawaiian honeyeaters were instead related to a different group of songbirds from North America and Eurasia, including the waxwings (Bombycillidae), the silky-flycatchers (Ptilogonatidae), and the palmchats (Dulidae). Instead of colonizing by island-hopping from across the Pacific, the Hawaiian honeyeaters came directly from the continents to the north and west. The Hawaiian honeyeaters, now reclassified in the unique family Mohoidae, provide another vivid and spectacular example of how similar foraging strategies can drive convergence and parallelism in avian morphology; without genomic comparisons, their true affinities would have been impossible to ascertain.

Certainly, following the massive cladistic analysis of Bradley Livezey and Richard Zusi in 2006, using 2,954 morphological characters, it would be extraordinarily difficult to justify more studies of this nature.[69] There would be little possibility of any dramatic modification of their tree with the addition of a few more characters, but many of their results, like those of other such studies, are highly suspect, including the alliance of loons and grebes, and are probably the result of groupings of ecomorphs. We can assume, then, that large-scale avian morphological cladistics has been accomplished and is incapable of resolving higher-level relationships with confidence within birds. Noted avian paleontologist Gerald Mayr of the Senckenberg Museum in Frankfurt, Germany, has criticized Livezey and Zusi's approach primarily because the large numbers of simple and homo-

plastic characters may overrule fewer characters of greater phylogenetic significance in the huge data set, and the result is a low ratio of phylogenetic signal to "noise" in the data.[70] Mayr, using a more measured and careful approach, has produced an impressive array of papers on the beautifully preserved Middle Eocene Messel fossils, from near Frankfurt, and has skillfully incorporated the fossils with cladistic and molecular analyses of modern birds.[71] While the results have been intriguing, many are highly controversial and will require further corroboration. Per Ericson and colleagues in 2006 attempted to integrate molecular sequence data with fossil evidence, and many would concur with Scotland and colleagues, who argued that "rigorous and critical anatomical studies of fewer morphological characters, in the context of molecular phylogenies, is a more fruitful approach" and "preferable to compiling larger data matrices of increasingly ambiguous and problematic morphological characters."[72] Yet one must wonder just how such disparate data sets could justifiably be combined into a "total evidence" or "taxonomic congruence" approach, since they are measures of two different aspects of the organism, genotype and phenotype. Why is such an amalgam of these data sets even desirable?

Regrettably, and despite the millions of dollars spent by the National Science Foundation on the new *Tree of Life* project, there is no truly significant breakthrough.[73] Also, the fact that molecular age estimates of ordinal origins for birds and mammals is generally in the range of twofold that of the minimal age as indicated by the fossil record cannot be very encouraging. There is, however, hope that with the new genome science, still in its infancy, major new discoveries, such as the true affinities of urochordates to vertebrates, may be possible, but many will be sorely disappointed that the results do not nicely conform to the morphological evidence. I have suggested previously that the deep avian and mammalian relationships have been difficult (next to impossible) to resolve as a result of the explosive Tertiary radiation following the Cretaceous-Paleogene extinction event sixty-five million years ago.[74] Given that modern orders are all present by fifty million years ago, this can only mean that most avian and mammalian orders arose in a time frame of about ten million years, with very tight clustering of lineages branching off their phyletic nodes. Although there is some hope for the future through the use of whole genome comparisons, the inability of morphology to resolve such issues seems well-established. Given the current state of chaos introduced by various morphological and molecular approaches to avian systematics, for the sake of taxonomic stability, it would seem prudent to stick close to the older classificatory systems that essentially in one form or another followed that of Max Fürbringer, who in 1888 established a foundation of modern avian classification.[75] In the final analysis, as the future unfolds, systematicians may well have to decide whether it is more desirable to produce a phylogenomic or morphological *Tree of Life*.[76]

One message emerges clearly: At the level of species and genus, most methodologies provide reasonable results that are usually fully compatible with molecular analyses. At higher levels from family to order, however, there is hardly any methodology, whether character-based cladistics or molecular comparisons, in which one can have much confidence. Cases of iterative evolution of particular body forms and massive convergence provide a formidable hurdle for any methodology claiming to be able to unlock the long-kept secrets of phylogeny. Still another formidable problem is that in paleontology there is no check from molecular comparisons, making the detection of convergence even more problematic.

Theropods or Hidden Birds?

Most scientists who have ventured into the study of the complex nexus of birds and theropods have been immediately struck with the seemingly intractable problems posed by almost all the sundry interpretations of the putative phylogenies. Recognizing the difficulties and in search of a solution that might bring consilience to the problem, as early as 1900, some quarter century before Gerard Heilmann's *The Origin of Birds*, both American paleontologist Henry Fairfield Osborn in 1900 and South African paleontologist Robert Broom

in 1906 introduced the idea of a common ancestry of birds and dinosaurs, and later in 1913 Broom suggested that birds derived from "groups immediately ancestral to the Theropodous Dinosaurs," in a sense predictive of a pretheropod, arboreal ancestry of birds.[77] Othenio Abel, professor of paleontology at Vienna and later at Göttingen, Germany, suggested that birds and dinosaurs shared a common ancestry: "that the birds and the Theropoda are descended from a common arboreal stem group," the implication being that all dinosaurs were terrestrial only secondarily.[78] This idea was modified by Heilmann, who even suggested that theropods such as *Ornitholestes* and *Struthiomimus* were secondarily terrestrial, having been derived from arboreal ancestors that climbed trees.[79] Through time one sees the confusion creep in. For example, in 1940, C. M. Sternberg identified the jaw of the theropod *Caenognathus* as that of a bird.[80]

The idea was fully explored for the first time in 1988, when Gregory Paul wrote a book entitled *Predatory Dinosaurs of the World*, arguing that a number of theropods were avian derivatives, but the book was so highly speculative, so loaded with bizarre art and unsubstantiated ideas on hot-blooded dinosaurs and associated extreme views, that it never gained momentum.[81] In 1994, George Olshevsky published a somewhat similar view, which he dubbed the "birds came first," or BCF, theory, but proposed that all dinosaurs were derived from birds via early archosaurs.[82] His view was that "on the way to perfecting flight, the lineage leading from reptiles to birds included a number of small, transitional, tree-dwelling animals unlike any other animals we have ever seen."[83] These early publications were followed by a series of books and papers that took this line of reasoning in one form or another, ranging from the view that some nonavian dinosaurs had been misidentified and were really secondarily flightless birds, among them the Late Cretaceous alvarezsaurids, to the view that a number of maniraptoran theropods were secondarily derived from the early avian radiation.[84]

The most influential publication portraying the view that maniraptorans are secondarily derived from birds was Gregory Paul's *Dinosaurs of the Air*, which cut through the time-honored dogma and provided at least a starting point to a possible solution to the seemingly intractable problem of bird origins. Paul presented evidence indicating that many maniraptorans, including dromaeosaurids, troodontids, and oviraptorosaurids, are secondarily flightless birds. In addition, Stephen and Sylvia Czerkas of the Dinosaur Museum in Blanding, Utah, have promoted a similar view in a nicely produced book, largely ignored by the paleontological world, entitled *Feathered Dinosaurs and the Origin of Flight*.[85] Based on recent fossil finds, this view of an avian origin from basal dinosauromorphs rather than theropods was nicely portrayed by Stephen Czerkas's traveling museum exhibit *Feathered Dinosaurs and the Origin of Flight*, organized by the Dinosaur Museum, the Fossil Administration Office of Liaoning, China, and the Liaoning Beipiao China Shihetun Museum of Paleontology. The exhibit was first displayed at the San Diego Museum of Natural History in 2004 and featured lectures by scientists on both sides of the debate, including Larry Martin, Luis Chiappe, and myself. Although the Czerkases' book is not cited frequently in the literature, perhaps because it is considered an in-house publication (like most museum monographs), there is little question that it is of high quality and a substantial contribution to the field.

Microraptors: *Ancestral Dromaeosaurs*

Stephen Czerkas's interest in dinosaurs began as a young child when he began sculpting these mysterious creatures, and this interest carried on into his teens, when he created dinosaurs for movies as well as life-size dinosaur sculptures for museums in Japan, Europe, and the United States. His attention to detail led him to anatomically correct, scientific studies of paleontology and his current research in collaboration with paleontological colleagues from China. For his outstanding artistic and scientific work, Czerkas was awarded an honorary doctorate in 2008.

In 2002, Czerkas and his Chinese coauthors described the Chinese Early Cretaceous *Cryptovolans* (probably congeneric with *Microraptor*) as an ancestral dromaeosaur of "pre-theropod, or non-theropod status."[86] *Cryptovolans* ("hidden flyer") has the typical salient dromaeosaur anatomical distinctions of a stiffened tail (as in ramphorhynchoid pterosaurs), an enlarged second toe claw, and a retroverted, or opisthopubic, pubis (an avian trait). However, given the probability that all microraptors had hind-limb wings attached to a short metatarsus (as evidenced by the recent discovery of the Middle Jurassic *Anchiornis*), with long, recurved pedal claws, they were unlikely to have been efficient ground-dwellers and would have been incapable of using the enlarged claw for predation, as *Deinonychus* likely used its sickle claw. Because microraptors are considered basal within the dromaeosaurs, the sickle claw may have been a climbing adaptation. Unlike true theropods, microraptors also lacked the sacral and femoral anatomy of obligately bipedal theropods, designed for efficient bipedal locomotion. In addition, microraptors had numerous avian features, including an avian hand and distal pubic spoon (hypopubic cup), as opposed to the dinosaurian pubic boot, as well as a reversed hallux for arboreal life. Also unlike theropods, they have a partially closed acetabulum. Most impressively, the hand bones of the wing are virtually identical to those of *Archaeopteryx*, but actually slightly more advanced for the attachment of the primary flight feathers.

The discovery of the so-called four-winged dinosaurs that occured some twenty-five million years or so after *Archaeopteryx*, such as *Microraptor* (*M. zhaoianus* and *M. gui*, including *Cryptovolans*), immediately opened a new dimension to the study of avian evolution and may hold the key to some questions of bird origins. The projection of fully developed flight feathers from the hind limbs along with highly recurved claws can only be interpreted as arboreal adaptations, and precludes their having been capable runners or cursors, as commonly depicted, particularly since the hind-limb remiges came off the tarsometatarsus and because the femora are unusually long and extend somewhat laterally, adapted to splay out in a typical flight profile, as we see in numerous gliding vertebrates, especially mammals.

Adding to the excitement was the discovery of early opposite birds, or enantiornithines, with pennaceous hind-limb flight feathers. At the same time, restudy of the Berlin *Archaeopteryx* showed that the Solnhofen urvogel probably had hind-limb "trousers," most likely a more derived condition than that found in the microraptors, another indication of its mosaic nature.[87] The discovery of the Mid-Jurassic *Anchiornis*, tentatively identified as a troodontid, but more likely an archaeopterygid, with true avian pennaceous flight feathers, seems to have confirmed the hind-limb wing hypothesis.[88] The combined evidence points to the high probability that hind-limb wings were the primitive condition within birds and that William Beebe's 1915 well-known prediction of a missing "tetrapteryx stage" in the evolution of avian flight has been discovered, validating his model.[89] Having studied these amazing Chinese fossils and realizing the difficulties inherent in a ground-up flight origin hypothesis, Zhonghe Zhou and Fucheng Zhang envisioned a "dinosaurian trees-down" hypothesis to account for flight origins in birds, and like a number of subsequent authors, they considered the question of ground-up versus trees-down avian flight origin to be moot at this

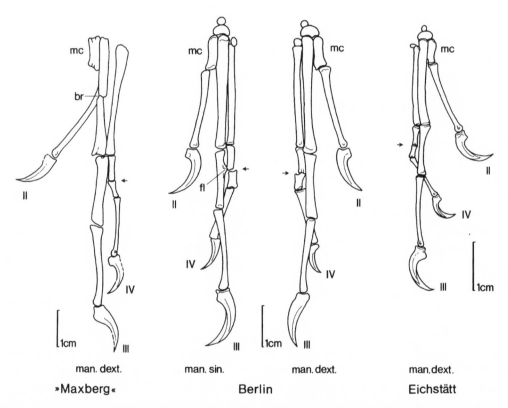

»Maxberg«   Berlin   Eichstätt

Fig. 6.15. Hand skeletons of the Maxberg, Berlin, and Eichstätt *Archaeopteryx* specimens. Note that the toppling or twisting of claws shows that in life the claws were pointed ventrally from below the wing surface, and although such an arrangement would not serve a predatory function, no other orientation would be suitable for trunk climbing. Arrows indicate the joints between the first and second phalanges of digit 3, as numbered here. Abbreviations: br, break of metacar-pal 2 in the Maxberg specimen; fl, flange on the proximal phalanx of left digit 2 in the Berlin specimen; man. dext., right manus; man. sin., left manus; mc, metacarpals. Modified from P. Wellnhofer, "Remarks on the Digit and Pubis Problems of *Archaeopteryx*," in *The Beginnings of Birds*, ed. M. K. Hecht et al. (Eichstätt: Freunde des Jura-Museum, 1985), 113–12. Courtesy of Peter Wellnhofer and Jura-Museum.

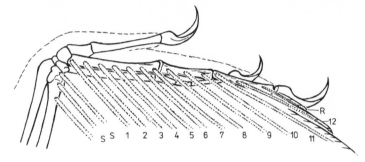

Fig. 6.16. The hand of the microraptor (*Sinornitho-saurus* sp.), above, showing near identical manual morphology to the hand of *Archaeopteryx*, below. (Microraptor hand reversed to correspond to drawing). Above, from Q. Ji et al., "The Distribution of Integumentary Structures in a Feathered Dinosaur," *Nature* 410 (2001): 1084–88. Reprinted by permission from Macmillan Publishers Ltd. Below, from B. Stephan, "Remarks on the reconstruction of the *Archaeopteryx* wing," in *The Beginnings of Birds*, M. K. Hecht et al. (Eichstätt: Freunde des Jura-Museum, 1985), 261–65. Courtesy Jura-Museum, Eichstätt, Germany.

stage of discovery.[90] The obvious question, of course, is why were the four-winged dinosaurs not classified as birds?

Czerkas and colleagues' conclusions that the "microraptor" is a bird, not a theropod dinosaur, and that it "is an ancestral dromaeosaur of pre-theropod, or non-theropod status," have gained considerable momentum.[91] Microraptors lack many of the typical theropod synapomorphies, including the sacral anatomy of obligate bipeds as well as a partially closed hip socket, or acetabulum, and they have numerous avian features, including the diagnostic pubic spoon, a partially reversed hallux, and an avian hand, actually advanced over that seen in the classic urvogel *Archaeopteryx*. As

Zhonghe Zhou noted (regardless of the numbering issue), "If we simply compare the hands of *Archaeopteryx* and some maniraptoran theropods, such as *Microraptor*, they are almost the same in every detail, including the phalangeal formula. If we accept the 'II-III-IV' for modern birds, and assume the same for *Archaeopteryx*, then why not accept the same conclusion for *Microraptor*?"[92]

Stephen Czerkas and his Chinese colleagues emphasized that "both camps have portrayed dromaeosaurs incorrectly as dinosaurs . . . to support their . . . opposing views."[93] For example, Mark Norell, commenting on a small dromaeosaur with feather impressions, noted that nonavian theropods, such as this dromaeosaur, led

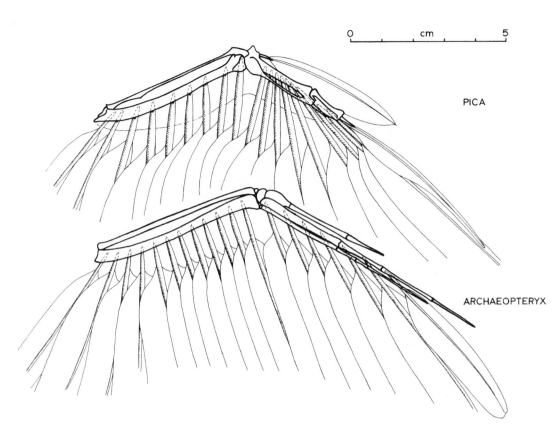

Fig. 6.17. Scale drawings of the skeleton of the forearm of magpie (*Pica pica*) and *Archaeopteryx*, to show the insertion of the primary and secondary flight remiges. The skeletons are drawn in ventral view; the remiges, which insert dorsally on the bones of the hand in *Pica*, and presumably also did in *Archaeopteryx*, are shown in dotted lines where they pass behind the bones. From D. W. Yalden, "Forelimb Function in *Archaeopteryx*," *The Beginnings of Birds*, ed. M. K. Heckt et al. (Eichstätt: Freunde des Jura Museum, 1985), 91–97. Courtesy of Derek Yalden and the Jura-Museum, Eichstätt, Germany.

toward the origins of true birds from the ground up.[94] According to Czerkas and colleagues, "Cladistics has presented a highly misleading interpretation of the evidence by arbitrarily insisting that the ancestral origins of avian flight must have been from an exclusively ground dwelling theropod dinosaur."[95]

Where Does Deinonychus *Fit?*

We saw earlier how the discovery of the human-sized *Deinonychus*, with myriad birdlike features, led John Ostrom to conclude not only that birds were derived from theropod dinosaurs but also that flight must have originated from the ground up. Obviously, if *Deinonychus* were close to the an-cestry of birds, despite its late age, then birds originated flight from a terrestrial creature, and Ostrom struggled to explain the selective agents that could have produced such a tortuous evolutionary event. It is interesting to ponder how the history of the entire dino-bird debate could have turned on a coin, if the discovery of the Chinese arboreal, volant microraptors had preceded that of *Deinonychus*. If so, there would have been little doubt that avian flight had originated from the trees down, and with their very avian appearance, these microraptors would very likely have been identified as avian. If *Deinonychus* and *Velociraptor* had only been discovered later, the probability is that, with their birdlike hands and carpus, they would

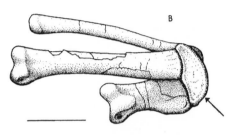

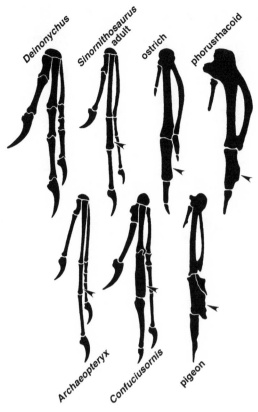

Fig. 6.18. The carpus and metacarpus of the Eichstätt *Archaeopteryx* (A), compared with those of the Early Cretaceous dromaeosaur *Deinonychus antirrhopus* (B). Arrows point to the half-moon-shaped carpal (semilunate carpal element) in each, considered to be a salient feature linking theropods to *Archaeopteryx* and therefore birds. Of course, the assumption was that *Deinonychus* was a theropod dinosaur and not a secondarily flightless derivative of the early avian lineage. If the latter were true, then the digits of *Deinonychus* and all dromaeosaurs would be II, III, and IV, the same as those of birds, and the problems of homology would be resolved without an improbable homeotic frame shift. From J. H. Ostrom, "*Archaeopteryx* and the Origin of Birds," *Biological Journal of the Linnaean Society* 8 (1976): 91–182. Courtesy of The Linnean Society, and John Wiley & Sons, Ltd.

Fig. 6.19. Hands of microraptors, *Deinonychus*, and birds, with a focus on the development of a posterolateral flange on the proximal bone of the central finger (indicated by arrows). Top row, flightless examples; bottom row, flying examples. In both rows hands become increasingly avian progressing to the right. Note that the combination of a well-developed posterolateral flange and a strongly bowed outer metacarpal make the hand of the flightless microraptor *Sinornithosaurus* better suited for supporting primary flight feathers than was the hand of *Archaeopteryx*. From G. S. Paul, *Dinosaurs of the Air: The Evolution and Loss of Flight in Dinosaurs and Birds* (Baltimore: Johns Hopkins University Press, 2002), 160, fig. 9.1. © 2002 Johns Hopkins University Press. Reprinted with permission of The Johns Hopkins University Press.

have readily been described as being secondarily flightless dromaeosaurs, despite their overall dinosaurian appearance. This is exactly what Gregory Paul, Stephen Czerkas, and others have concluded, that the larger dromaeosaurs were indeed secondarily flightless, with Czerkas and colleagues remarking, "The origin of birds stems further back to a common ancestor of pre-theropod sta-

tus that was arboreal. The proto-maniraptoran . . . [scansoriopterids] and . . . [*Microraptor*] are the only known members of such arboreal pretheropods."[96] In contrast, however, Paul views the entire assemblage as being of theropod descent.

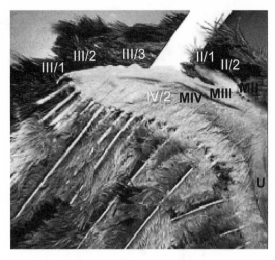

Fig. 6.20. The wing in the adult ostrich (*Struthio camelus*). Note how the avian hand is completely "tied up" by the postpatagium and the insertion of primary feathers, obviating a predatory function, as is commonly illustrated in life reconstructions. Adapted from M. Kundrát, "Primary Chondrification Foci in the Wing Basipodium of *Struthio camelus* with Comments on the Interpretation of Autopodial Elements in Crocodilia and Aves," *Journal of Experimental Zoology (MDE)* 310B (2008): 1–12. Courtesy of Martin Kundrát.

Interestingly, in a review, Zhonghe Zhou, writing on the possibility that "feathered dinosaurs" may be flightless birds, notes, "These ideas have the advantage of explaining why some advanced avian features . . . appeared in some feathered theropods but not in the most basal birds."[97]

Scansoriopterids and Allies

> Homoplasy is the main reason why it is so difficult to pinpoint those non-avian maniraptorans to which birds are most closely related, although it now seems that a curious form called *Epidendrosaurus* shares a few more synapomorphies with *Archaeopteryx* than other taxa do.
>
> *Kevin Padian and Armand de Ricqlès,*
> *"L'Origine et l'évolution des oiseaux: 35 années de progrès,"* Comptes Rendus Palevol 8 (2009, abridged English version)

The Straw Man Syndrome and the Basal Archosaurs

Most scientists voicing opposition to the current dino-bird dogma have pointed to the possibility of an early common descent of birds and dinosaurs from hypothetical basal archosaurs or dinosauromorphs, essentially the Broom/Heilmann hypothesis, which views birds originating from small quadrupedal, jumping and parachuting arboreal archosaurs, and thus flight originating from the trees down. The primary criticisms of those opposing the current dino-bird orthodoxy are that they fail to use the prevailing cladistic methodology and philosophy of the field, and therefore do not propose a "testable" alternative, and that all the proposals are too vague to be adequately tested by their restrictive methodology.[98] Such criticism is typically followed by the pejorative proclamation that any individual who opposes the current orthodoxy and does not use the methodology is not a scientist. Many papers point to individual Triassic archosaurs that have been used to illustrate bird-like characters occurring in archosaurs (or even the prolacertid *Cosesaurus*, with the species name "*aviceps*"), including *Longisquama* and *Scleromochlus*, which have even been termed avimorph thecodonts or similar labels.[99] And as mentioned, the South American Triassic lagosuchids (such as *Marasuchus*, *Lagosuchus*, and *Lagerpeton*) have been thought to be close to the ancestry of dinosaurs and possibly birds.

The term "thecodont," not widely used today, is a collective term used historically to describe Triassic basal archosaurs, but since they do not form a monophyletic assemblage, the term has seen diminished usage. Yet the importance of the fossils is not diminished by our inability to pigeonhole them taxonomically, and paleontologist Robert Carroll of McGill University's Redpath Museum was equally doubtful of theropod systematics, noting that the terms "coelurosaur" and "carnosaur" have been used to describe, respectively, small and large carnivorous theropod dinosaurs. Before the advent of the certitude produced by the new phylogenetic systematics, Carroll wrote that "the 'carnosaur' families may each have evolved separately from different groups that have been classified as coelurosaurs." In addition, he made the important observation that "when attempting to establish relationships of any group with a fossil record, we must emphasize the earliest known members, because they have had the shortest amount of time to evolve new characters since their initial divergence. Hence, they should provide us with the best opportunity to identify the derived features that they share with their closest sister group."[100]

Opponents of the current orthodoxy could equally point to the famous *Deinonychus* and present numerous reasons why it could not be ancestral to birds. Yet what is being emphasized is simply that these Triassic archosaurs are birdlike in many features and should be studied, but they are typically excluded from cladistic analyses for a priori reasons. For example, the Triassic archosaur the rauisuchian *Postosuchus*, were it not for the structure of the ankle, could easily fit in with early dinosaurs in any cladistic analysis, if it were included in the analysis. As Lawrence Witmer has stated, "To be fair, it must be pointed out that they [cladists] rarely include nondinosaurian taxa in the analysis."[101] As if to illustrate this very point, Witmer's statement is in a book in which his article is followed by one by Clark, Norell, and Makovicky entitled "Cladistic Approaches to the Relationships of *Birds to Other Theropod Dinosaurs*."[102] Obviously, with such an approach, any challenge to the current dino-bird dogma will not be put to any substantial and unbiased test. Pointing out deficiencies of any particular basal archosaur as a putative ancestor of birds thus becomes part of a straw man syndrome.

Cladistics and the Origin of Birds

Bearing these deficiencies in mind, Frances James and John Pourtless embarked on a study to test the current orthodoxy with a wider cladistic perspective.[103] Fran James is Pasquale P. Graziadei Professor Emerita at the Department of Biological Science at Florida State University. She has served as president of the American Ornithologists' Union and the American Institute of Biological Sciences, has been awarded medals from the American Ornithologists' Union and the Wilson Ornithological Society for her work in avian ecology, and was elected a fellow of the American Academy of Arts and Sciences in 2001. She became interested in the origin of birds around 2002. At the same time, John Pourtless, student at Florida State University, developed a keen interest in the same problem. Both James and Pourtless were interested in evaluating the consensus hypothesis that birds are theropod dinosaurs using a more thorough and critical approach to cladistics than was evident in the literature supporting the theropod hypothesis. They noted, for example, that there were no published studies that included alternative potential sister-taxa for Aves. James and Pourtless met in 2004 and spent the next five years collaborating on a major review of the literature on the origin of birds, from which they produced two new analyses aimed at assessing in a more comprehensive and conservative phylogenetic framework the strength of support for the theropod hypothesis as evaluated simultaneously with four of its major alternatives (Gregory Paul's "neoflightless" theropod hypothesis; an origin of birds from temporally early, phylogenetically basal archosaurs—here referred to as "thecodonts"; an origin of birds from crocodylomorphs, as postulated by Alick Walker in 1972; and the recent proposal by Evgeny Kurochkin that Aves is diphyletic).[104] In the analysis of a representative matrix from the literature supporting the theropod hypothesis (that of Clark and coworkers from 2002), they found that although there was support for the association of birds and at least the most birdlike maniraptorans (oviraptorosaurs, dromaeosaurs, and troodontids), interrelationships among these taxa were unresolved; a topology was recovered compatible with the proposition that these maniraptoran taxa are birds at all stages of flight and flight loss, and that they belong within Aves at a level of morphological organization more derived than that of *Archaeopteryx*.[105] Furthermore, using Kishino-Hasegawa tests to evaluate the statistical significance of differences in tree length when topological constraints specifying avian status for maniraptorans were enforced, they found that there was no significant difference between a topology in which most maniraptorans were actually birds and a topology in which they were outside of Aves. Thus, the latter hypothesis is not clearly a better explanation of the data than the former. When James and Pourtless simultaneously compared the theropod hypothesis with its four alternatives using a new, conservatively constructed data matrix, they found again that there was support for the avian status of at least some maniraptorans, both from the topologies they recovered and from statistical tests comparing constrained and unconstrained trees, and that a clade of birds and maniraptorans was consistently recovered apart from theropods in a polytomy with other archosaurian lineages, including the Crocodylomorpha. Moreover, using statistical tests, they found that of the four alternative hypotheses they evaluated, the "early" archosaur hypothesis and the crocodylomorph hypothesis were at least as well supported by available data as the theropod hypothesis. Thus, even using cladistic methods, it is clear that the origin of birds is at present unresolved and the strength of support for the theropod hypothesis claimed in the literature is in fact seriously overstated. They concluded that further data and testing is needed to discriminate between the theropod hypothesis and the alternatives supported by their analysis. At the least, their analyses provide further support for the proposition that some maniraptorans were actually birds and belong within Aves.

Summary of Chinese Fossils

At this early stage of discovery considerable speculation remains, but the manner in which most cladograms are rooted and constructed is unlikely to produce any conclusion other than that birds are derived directly from maniraptoran, theropod dinosaurs and that all these basal birds are basically maniraptoran theropods. Yet we have seen the myriad mistakes made by all sides in the debate, and with the recent discoveries of the incredible fossils of the Jehol Biota from China, the entire debate seems to have largely been tossed back to square one. In brief summary, what we see at present—ent from the Chinese ossuary of archosaurs, relevant to the discussion, are:

- Ground-dwelling, turkey-sized, birdlike animals, such as *Caudipteryx* and *Protarchaeopteryx*, with true avian pennaceous feathers, originally described as "feathered dinosaurs" but more likely secondarily flightless birds.
- The four-winged microraptors and allies (*Microraptor*, *Cryptovolans*, and the Middle Jurassic *Anchiornis*), called feathered dromaeosaurs and troodontids, which have modern avian features, an avian hand (more advanced than that of *Archaeopteryx*), an avian foot, and few salient theropod features. They appear to have been fully arboreal, with few adaptations for life on the ground, and their hind-limb wings would be a hindrance in cursorial locomotion.
- The tiny Jurassic scansoriopterids (*Epidendrosaurus*, *Scansoriopteryx*, and *Epidexipteryx*), which may be coeval with temporally or even predate the classic urvogel *Archaeopteryx* and which exhibit features that might be expected in ancestral birds. They have few, if any, salient theropod features.
- Basal coelurosaurian theropods, such as *Sinosauropteryx* and *Sinocalliopteryx*, with filamentous integumentary structures thought by many to be protofeathers but which more likely represent a preserved meshwork of supportive collagen skin fibers.
- A miscellany of other dinosaurs, including the troodontid *Sinovenator*, the oviraptorosaur *Incisivosaurus*, the therizinosaur *Beipiaosaurus* (which, at more than 2 meters [6.5 ft.] in length, is the largest theropod yet discovered from Liaoning), and the ornithischian, primitive ceratopsian *Psittacosaurus*, which also exhibits filamentous skin structures as in *Sinosauropteryx*. In addition, there is *Sinornithosaurus*, described as a feathered dromaeosaur, but which may be of avian status. There is also the earliest known ceratosaur, the Jurassic *Limusaurus*, thought to have a birdlike hand with reduced digits I and V.
- A miscellany of pterosaurs, including primarily pterodactyloids, and a few rhamphorhynchoids, some, such as *Jeholopterus*, exhibiting "hairs"

resembling the hairlike integumentary filaments of the dinosaur *Sinosauropteryx*. There is also the very small, arboreal pterosaur *Nemicolopterus*.

- A large collection of opposite birds, or enantiornithines, which exhibit considerable morphological diversity, produced a major adaptive radiation, and represented the dominant land birds of the Mesozoic.

- A miscellany of surprisingly advanced ornithurine birds, such as the beautifully preserved *Yanornis*, *Yixianornis*, and *Hongshanornis*, illustrating once again the early split between sauriurine and ornithurine birds. We have yet to approximate the date of origin of the ornithurines, which is far more ancient than previously expected.

The Heart of the Problem

> The problem is that we expect too much of morphology in asking it to tell us the genealogy of organisms as well as what they look like.
>
> *Philip Gingerich, "Cladistic Futures," Nature 336 (1988)*

The controversy over the ancestors of birds has been grossly overstated, partly by the adversarial parties themselves, but also by the popular press, which thrives on major scientific disputes. In part the problem is one of semantics, inadequately defining just what is meant by a "dinosaurian origin of birds." As I noted in 1996 in the book that was the precursor to this text: "Are birds derived from dinosaurs? This would depend entirely on what one defines as a dinosaur, or Dinosauromorpha."[106] The various protagonists have always believed birds were nested within the Archosauria, and the two dominant views classically contrast an earlier origin of birds with a common basal archosaur ancestor of theropods and flight evolving from the trees down versus the more popular "dinosaurian" origin of birds, later in time, directly from highly derived theropods, the maniraptorans (or their immediate ancestors), with flight evolving from the ground up. Although the ground-up versus trees-down arguments have been muted by the discovery of arboreal, volant microraptors, such was not

always the case. During the past three decades a ground-up flight origin was considered inextricably linked with a dinosaurian origin of birds, and even now paleontologists still appear to favor a ground-up origin of flight.[107]

Pterosaur Phylogeny Is Just as Troublesome

The agonizing history of deciphering pterosaur relationships bears a striking resemblance to the tortuous bird origins controversy and is still an open question. At various times pterosaurs have been thought to be derived from Permian or Triassic eosuchians (prolacertiforms) or from basal archosaurs and to be a sister group to dinosaurs.[108] Pterosaur flight origins have also been debated as being from the trees down or from the ground up, the latter based on the assumption from cladistics that pterosaurs were the sister group of dinosaurs and therefore obligately bipedal. We know now that pterosaurs were primitively quadrupedal and that flight could have arisen only from the trees down.[109] The debate on pterosaur origins, however, still rages.

Maniraptora and Birds

Substantial evidence indicates that birds and small maniraptoran theropods are closely allied, but the question remains: Are the so-called no-navian maniraptorans ancestral or descendant, or are they all avians, at various stages of flight and flightlessness? Assuming that the thesis presented in this book is correct, or even partially so (that many of these Early Cretaceous feathered, non-avian dinosaurs are early avian descendants), we need to tease out the avian lineage or lineages from those thought to be "true" theropods: coelophysids, compsognathids, ceratosaurids, allosaurids, tyrannosaurids, and others.

We have seen numerous examples in which loss of flight renders various disparate types of birds "dinosaurian" in superficial appearance, and in many cases it is virtually impossible for the current cladistic methodology to deal with secondary loss of flight via paedomorphosis (for example, the gooselike ducks the moa-nalos and *Caudipteryx*). It should give us pause to realize that, given the current cladistic framework, if a modern ratite

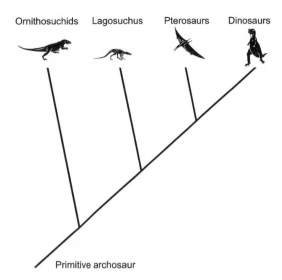

Ornithosuchids   Lagosuchus   Pterosaurs   Dinosaurs

Primitive archosaur

Fig. 6.21. Pterosaurs were once thought to be the sister group of dinosaurs, and therefore bipedal runners, with flight developing from the ground up. Fossils have falsified this interpretation, and the current divergent theories on pterosaur origins remain. Modified from P. Wellnhofer, *The Illustrated Encyclopedia of Pterosaurs* (New York: Crescent Books, 1991).

were discovered in the lacustrine deposits of the Early Cretaceous of China, it would most assuredly be considered a theropod dinosaur, illustrating an early stage in the evolution of flight from the ground up, and most certainly adorned with protofeathers, illustrating all stages of feather evolution. Such a conclusion would be concordant with Thomas Huxley's dinosaurian origin of birds, through the flightless ratites, which was carried on by Percy Lowe and is now, in only slightly modified form, back in vogue.

Yet as noted, most current evidence could easily be interpreted to indicate that Chinese dromaeosaurs are an early remnant of the avian radiation, too specialized to be ancestral, with species at practically all stages of flight and flightlessness, showing that the flightless state in birds is almost as old as birds themselves. Given the tendency of modern birds to become flightless, particularly on islands, the early evolution of flightlessness in birds is not surprising; indeed, it would be surprising had it not occurred. Also not surprising is that

early birds would closely resemble true theropods, just as modern ratites do, albeit superficially, as attested to by Huxley's confused dinosaurian origin of birds hypothesis, which included ornithischian dinosaurs. Yet although birds, particularly the early forms, resemble theropods, their early evolution was from the trees down, as has been the case for all tetrapods that have ventured into the volant realm, and a sophisticated "flight" architecture still resides deep within avian morphology, even though many are secondarily flightless. There is no known example of a flightless bird, living or extinct, that was not derived from a once volant ancestor.

Despite the hyperbole surrounding "feathered dinosaurs," there is to date no credible evidence for the existence of protofeathers, either in fossils or in modern birds, and therefore the best evidence for feather origins still rests with the study of living birds. It is also most likely that the Chinese "dromaeosaurs" are derived from, and not ancestral to, subsequent birds. If one uses the stratigraphic sequence of relevant fossils as a guidepost, one can begin to understand the evolutionary sequence in the early radiation of the class Aves. Major questions can be summarized as: (1) Where do microraptors fit in the evolution of birds? and (2) Are the later dromaeosaurs, such as *Deinonychus*, *Dromaeosaurus*, and *Velociraptor*, as well as oviraptorosaurs and troodontids, and perhaps even therizinosaurs, really secondarily flightless birds? Although they are quite different from the earlier dromaeosaurid microraptors, they are no more distinctive than, say, a clapper rail and the giant Early Tertiary *Diatryma*, or a penguin and a ratite.

Cladistic Approaches Are Sometimes Misleading

Certainly the inability of cladistic methodology to deal with convergence has been illuminated in this debate and is most vividly illustrated by the fact that the vast majority of cladistic hypotheses for recent taxa are continuously falsified by molecular systematics, especially large-scale genomic comparisons. Recent workers on the Early Cretaceous Chinese fossils have been looking at highly derived flightless forms as basal avian ancestors, as dictated by their cladograms. By their arbitrary exclusion of outgroups that might be candidates for ancestry there is no consideration of the possibility that these highly derived microraptors are avian descendants. By viewing these so-called maniraptorans as "derived" one is relieved of the burden of inventing unparsimonious and often bizarre models for everything from a ground-up flight origin to preadaptive scenarios for the evolution of the avian wing, and feathers evolving to insulate hot-blooded dinosaurs, for which there is no evidence. In addition, the problems of inventing a homeotic frame shift to explain digital homology and the temporal paradox fade away. According to the modern dino-bird orthodoxy, virtually all of the complex avian flight architecture takes the form of exaptations. If the sequence is reversed, then all avian flight adaptations evolved directly in an aerodynamic context, with each stage easily explained by natural selection.

Bipedalism was not invented by theropod dinosaurs and carried on by birds; the earliest reptiles to show a trend toward such an anatomical body plan appeared in prearchosaurian lineages, as in the Permian *Eudibamus*. Later in the Triassic, some crocodilians, such as the Late Triassic *Effigia*, also experimented with the same morphology. In addition, a number of predinosaurian lineages, including basal archosaurs, most ornithosuchids (without the ankle preserved), and certainly *Postosuchus*, if thrown into a cladistic analysis instead of being summarily cast out as an outgroup, might well come out at the ceratosaur theropod level in any cladogram.

The inability of cladistic methodology to deal with convergence has been pointed out time and again. What was once "Hennigian cladistics" has now turned into a distinctive phenetic methodology, no more than a distant cousin of the original methodology of Hennig, as nicely summarized by Diana Fisher and Ian Owens: "The phylogenetic approach is a statistical method for analyzing correlations between traits across species."[110] And, like the earlier statistical phenetic approaches of the 1970s, this approach frequently groups ecological morphologies (ecomorphs) instead of clades. If the proposals from the detailed anatomical analyses showing that dromaeosaurs are actually birds, at all stages of flight and flightlessness, turn out to be correct, then the question of bird origins is completely reopened. Also, if true, then both camps in the debate have portrayed dromaeosaurs incorrectly: "Cladistics has presented a highly misleading interpretation of the evidence," and "the origin of birds stems further back to a common ancestor of pre-theropod status."[111] Given their suites of avian characters, the microraptors of China would appear most logically to be birds, regardless of their ancestry. Yet, *whatever microraptors are, so too are birds*.

Confusing Deinonycosaurs

Another question remains: How does one reconcile the question of the relationship of the Early Cretaceous microraptors to Middle and Late Cretaceous deinonychosaurs? Is it possible that such forms as *Deinonychus*, *Dromaeosaurus*, and *Velociraptor*, as well as the troodontids, with their avian-like hands (and possibly therizinosaurs), are actually secondarily flightless birds masquerading as small theropods? These and other important questions will be answered only if we keep an open mind and are not bound to past concepts of relationships. As Storrs Olson noted, "Healthy skepticism is the most powerful tool of science and should be cherished as a welcome anodyne to the complacency of certitude."[112]

Testing the Flight Ancestry Hypothesis

If we hypothesize that some of these Early Cretaceous forms evolved flightlessness soon after the origin of avian flight, we may be able to test the hypothesis through such obvious flight adaptations as the brain, which is adapted for "flight" in parallel fashion in birds and pterosaurs. However, the initial adaptations for flight may have been so limited that features such as a "flight brain" may not have been developed. Patricio Alonso and colleagues, writing on the brain of *Archaeopteryx*, noted that although *Archaeopteryx* exhibits a "flight brain" and is acknowledged to be well on the way toward the modern bird pattern, interestingly, maniraptorans with lesser or no flight ability also "show a trend towards brain enlargement and laterally separated optic lobes."[113]

David Burnham of the University of Kansas has extensively studied *Bambiraptor feinbergi*, a small velociraptorine dromaeosaurid from the Upper Cretaceous Two Medicine Formation of Montana that has been called, before the recent Chinese discoveries, the most birdlike theropod.[114] This beautifully preserved animal was less than a meter long (3 ft.) and weighed some 2 kilograms (4.4 lbs.). *Bambiraptor* conforms to the Dromaeosauridae in having a semilunate carpal bone, a thin, bowed outer metacarpal, a retroverted pubis, a large, retractable pedal ungual on digit II, and a stiffened tail with bony extensions of the prezygapophyses. Unlike the smaller, and volant, *Microraptor* from the Early Cretaceous of China, the bones of *Bambiraptor*'s skull could be individually cast and reassembled. This enabled Burnham to obtain a precise silicone endocast of the brain that was detailed enough to show vascular imprints on the ventral surface of the skull roof, which indicated also that the brain occupied the braincase cavity and therefore illustrated that the cast should be a good indicator of actual brain size. The only other previously described such endocast was of the small, maniraptoran theropod *Troodon*, which also showed a high degree of encephalization.[115] Interestingly, the brain lies mostly behind the orbits, and the optic lobes lie in the midbrain region, in a ventrolateral position, as in birds; and there is some flexure, although not as pronounced as in modern birds. The measured relative brain size of *Bambiraptor* was as large as or larger than that of *Troodon*. The posterior enlargement of the brain is atypical of standard coelurosaurs and even more birdlike.

A well-preserved specimen of a closely related Late Cretaceous maniraptoran, the ovitaptorosaur *Conchoraptor gracilis*, provided an almost complete adult endoneurocranium (actual cavity of brain), which Martin Kundrát analyzed by computed tomography scans to produce a virtual model of the cranial cavity, the most complete maniraptoran endocast to date.[116] The model displays a very avianlike brain, with reduced olfactory bulbs and large cerebral hemispheres contacting the expanded cerebellum, with enlarged cerebellar auricles and with optic lobes displaced latero-ventrally. Although the shortened olfactory tract and overtopping cerebral hemispheres more closely resemble modern birds, most attributes have a less birdlike appearance than corresponding structures in *Archaeopteryx*. Still, the calculation of brain mass relative to body mass indicates that *Conchoraptor* falls within the range of extant birds, in contrast to *Archaeopteryx*, which occupies a marginal range. Kundrát concluded that the brain characteristics, "taken together, suggest that the animal had a keen sense of vision, balance, and coordination."[117] He cautioned that the model does not allow an unambiguous assessment of whether the brain characteristics evolved independently or are an indication of the derivation of oviraptorosaurs from volant ancestors; yet the presence of many of the avian features is difficult to explain in a flightless theropod. In addition, Kundrát notes: "The ventral shift of the optic lobes towards the orbits documents an evolutionary innovation of neurobiotaxic movement of neurons towards the source of optic stimuli and dominance and accuracy of vision of flying theropods. The optic lobes of *Struthio* [ostrich] occupy a position ventral to the posterior part of the cerebral hemispheres, a feature also seen in *Enaliornis* [Early Cretaceous hesperinithiform]."[118] Interestingly, *Struthio*, a recent secondarily flightless bird, has the same cerebral superimposition of the optic lobes as in *Enaliornis*, which is a loonlike ornithurine bird. That the same cerebral geometry should be expressed in *Archaeopteryx*, a flying bird, and in *Conchoraptor* would seem to suggest that "*Conchoraptor* might have evolved from an ancestor with flying capabilities."[119] This hypothesis might also explain why there is such a large suite of volant avian characteristics in oviraptorosaurs, including a pygostyle.[120]

With respect to the small dromaeosaurid *Bambiraptor*, aside from its overall birdlike features, Burnham concluded, "The relatively large brain, overlapping fields of vision, small size, and elongate front limbs might indicate that *Bambiraptor* was arboreal."[121] Certainly the highly recurved manual claws are consistent with the potential

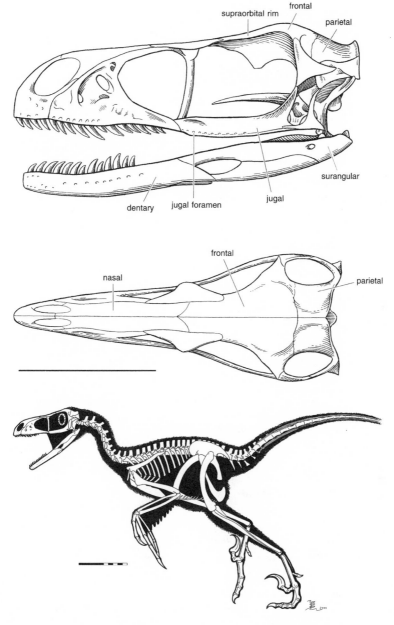

Fig. 6.22. Skeletal reconstructions of *Bambiraptor*. From D. A. Burnham, "New Information on *Bambiraptor feinbergi* (Theropoda: Dromaeosauridae) from the Late Cretaceous of Montana," in *Feathered Dragons: Studies on the Transition from Dinosaurs to Birds*, ed. P. J. Currie et al. (Bloomington: Indiana University Press, 2004), 67–111. Courtesy David A. Burnham. Reprinted with permission.

for arboreal climbing, and such claws, if basal, would be preadapted for a later grasping, seizing, or other predatory function.[122] In a recent study, Phillip Manning and his research group at the University of Manchester have shown, using sophisti-

cated analyses of *Velociraptor* claws, that "dromaeosaurid claws were well-adapted for climbing" with "enhanced climbing abilities."[123]

The final question is, of course, whether this theory of "hidden birds" can be put to a scientific

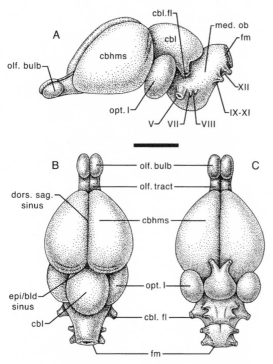

Fig. 6.23. Endocast of the brain of *Bambiraptor*, illustrating avian structure. Shown in lateral, dorsal, and ventral views. From D. A. Burnham, "New Information on *Bambiraptor feinbergi* (Theropoda: Dromaeosauridae) from the Late Cretaceous of Montana," in *Feathered Dragons: Studies on the Transition from Dinosaurs to Birds*, ed. P. J. Currie et al. (Bloomington: Indiana University Press, 2004), 67–111. Courtesy David A. Burnham. Reprinted with permission.

test. Or is the proposal, like much of paleontology, rife with endless speculation? Perhaps the answer lies in the same line of evidence used by Gavin de Beer in determining that the ratites were indeed derived from flying ancestors. The presence of a "flight brain" and a "flight ear" discovered by Alonso and colleagues in *Archaeopteryx* might provide a possible avenue for investigation. As Lawrence Witmer concluded in his commentary on their discovery, "Can it [the flight brain and ear] provide a test of the heretical notion that some of the most bird-like Cretaceous theropods (such as *Velociraptor*) are actually secondarily flightless descendants of early, *Archaeopteryx*-like birds?"[124] Interestingly, these very dromaeosaurs (mani-

raptorans), as noted, have recently been shown to exhibit a trend toward brain enlargement and laterally separated optic lobes. Yet we must ask whether this is a trend or instead the retention of these features from flighted ancestors. Also, the convergence in the elaboration of both visual and equilibrium (vestibular) regions of the brain and associated sense organs in both *Archaeopteryx* and pterosaurs provides critical evidence that "both an aerodynamic wing and a powerful central nervous system are integral to powered flight."[125] Stig Walsh and colleagues, including Witmer, studied the inner ear anatomy as a proxy for deducing auditory capability.[126] They measured the length of the duct of the bony part of the inner ear and showed that the length correlates with average hearing frequencies, concluding that *Archaeopteryx* was a good match for the emu. Such approaches may take on even more importance in the future in assessing the neoflightless hypothesis promoted here, but one could also ask why the sophisticated avian "flight hand" in these early feathered dinosaurs is not considered just as valid evidence for a secondarily flightless status, as it is in the living ratites, which are all considered to have gone through a flight stage.

Flightless Avians

There are other more obvious avians from the Early Chinese Cretaceous, the oviraptorosaurs, such as the flightless *Caudipteryx*, its flight features including an avian hand with an alula, avian skull and teeth, an avian foot, avian gait, and a pygostyle. Instead of being basal, as current paleontological theory advocates, the oviraptorosaurs, along with microraptors and possibly later dromaeosaurs (which may well have been feathered), are best interpreted as derived from the basal avian radiation.[127] If this is proven to be true, cladistic methodology will have consistently reversed the true evolutionary sequence of dromaeosaurs and birds. Microraptors represent a remnant of the early avian radiation, exemplifying all stages of flight and flightlessness, and *Caudipteryx* and oviraptorosaurs are secondarily flightless birds and, in that sense, are derived, not basal. The association of the Troodontidae and the Dromaeosauridae is well established, as indicated by close alliance with the Avialae and as designated by the group name Paraves. There are simply no reasonable morphological definitions of these taxa; they are placed within clades based on the outcome of tenuous cladistic analyses, and many results may hang by a spider's thread.

One thing is clear; the basal members of most groups have long, pennaceous feathers on their lower legs and feet, suggesting that the common ancestor also had feathered feet.[128] Often misinterpreted as agile runners (in attempts to accommodate the cladogram), these basal paravians, with their elongate foot feathers, could not have been anything other than arboreal gliders, and those that were not show characteristics indicative of their having been secondarily flightless at various stages. Yet is there any putative ultimate ancestor among the known earlier theropods? Even though the recent discoveries from China have virtually closed the gap between *Archaeopteryx* and more modern birds through the discoveries of a variety of intermediate "basal" birds, and perhaps taken us back a step with *Anchiornis*, the huge gulf between these avians and true theropods remains impassable, requiring among other things a reelongation of already foreshortened forelimbs to form a wing and a complete renovation of the body type (bauplan) from bipedal runner to sprawling climber. Are we any nearer to discovering animals close to the ultimate avian ancestor than we are to discovering the true ancestry of pterosaurs? Given the iterative trends toward a birdlike morphology, or "ornithization," during the Mesozoic, in diverse groups ranging from such basal archosaurs as the ornithosuchids to such dinosaurs as the ornithomimids and alvarezsaurids, we now need to tease out those "hidden birds" masquerading as theropods.

Many students of dinosaur evolution have been ensnared by the insidious trap of convergent evolution. As Larry Martin has written, "The cladograms were correct in embedding some putative dinosaurs within birds, but were incorrect in their relationship to the dinosaur radiation as a whole. . . . The common ancestor of such a grouping must have looked like a bird and lacked most salient dinosaurian features."[129] If the new view presented here proves correct, it would also cast off almost all of the persistent and seemingly intractable problems that have historically been associated with the most recent version of a theropod origin of birds, namely: flight from the ground up, feathers evolving for endothermic dinosaurs, and morphological mismatches. The other, and perhaps most critical, problem with the current orthodoxy is that it necessitates that all the sophisticated avian aerodynamic flight architecture, including the avian perching hallux, evolved in a terrestrial, nonflight context, which is almost non-Darwinian.

In this new view, *Archaeopteryx* remains the classic basal bird, although with a fairly advanced flight architecture, including an avian brain, a flight hand, and perfectly formed avian pennaceous feathers. In addition to solving most of the intractable problems related to early avian evolution, this view of derived maniraptorans answers the nagging question of why virtually no flightless birds are known from the Mesozoic until the Late Cretaceous. They were present throughout the Cretaceous, as *hidden birds*, and some of their larger, putative secondarily flightless derivatives, such as the ratite-sized deinonychosaurs, are now known from the Chinese Early Cretaceous through the discovery of didactyl tracks.[130]

The Definition of Aves

We can provisionally compose an interim morphological definition of birds based on the salient features of Aves. *Birds are mesotarsal bipedal archosaurs with pennaceous feathers and a tridactyl avian "flight" hand* (digits II, III, and IV herein). Other important features include: a bowed posterior (outer) metacarpal (IV) for feather support (slight in most urvogels); the longest middle digit (III) with a posterior flange along the proximal phalanx expanded distally or flattened; flight feathers anchored primarily to the anterior metacarpal (III), to the basal and terminal phalanges of the middle digit (primaries), and to the dorsum of the ulna (secondaries); reduced, semilunate carpals (the

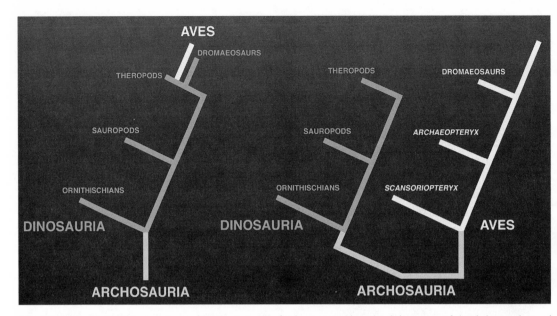

Fig. 6.24. Divergent interpretations of bird origins. Left, the current dino-bird dogma, with birds being derived relatively late in time directly from highly derived theropods and flight originating from the ground up. Right, the major competing common descent model, showing scansoriopterids as an approximation of the basal avian and flight originating from the trees down. Reprinted with permission. Copyright Stephen A. Czerkas.

carpal trochlea of modern birds); a foot with re-versed digit I (hallux); a pretibial bone fused with either the astragalus or the calcaneum; and a retro-verted pubis with a pubic spoon (accommodating suprapubic muscles).

It will only be through considering the totality of evidence from various branches of science that we can ever hope to solve the mystery of bird ori-gins and the origin of avian flight. The frequently used phrase "birds are living dinosaurs" does little more than dampen research, because if it were true, then any fossil specimen with feathers in the Meso-zoic could automatically be considered a feathered dinosaur. With the recent spectacular discovery of birdlike fossil footprints with a clearly preserved hallux from the Late Triassic or slightly younger, Zhonghe Zhou correctly notes, "It is probably too early to declare that it is time to abandon debate on the theropod origin of birds. . . . Abandoning debate may succeed in concealing problems rather than finding solutions to important scientific ques-tions."[131] Like the pterosaurs, the journey from the earliest primitive avian forms to the ultimate an-cestor of birds remains a mystery. The problem of avian origins is far from being resolved.

Epilogue

Students entering any field of science would be well-advised to read Martin Schwartz's essay in the *Journal of Cell Science* entitled "The Importance of Stupidity in Scientific Research." As Schwartz states: "One of the beautiful things about science is that it allows us to bumble along, getting it wrong time after time, and feel perfectly fine as long as we learn something each time. . . . The more comfort-able we become with being stupid, the deeper we will wade into the unknown and the more likely we are to make big discoveries."[132] Regrettably, to-day's field of phylogenetic systematics, a mélange of ideology and methodology, does not permit the luxury of being stupid. Students of paleontology slot their newly discovered fossils into prepro-grammed cladistic formulae, using human-coded skeletal features and excluding outgroups a priori. Each time, they get a precise answer. There is no

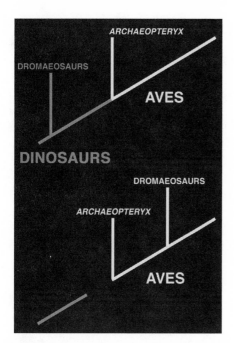

Fig. 6.25. Above, interpretation based on cladistic analyses in which theropod dinosaurs are the direct avian ancestors and dromaeosaurs represent no-navian dinosaurs, along with its corollary, that flight developed from the ground up. Below, interpretation in which dromaeosaurs are shown as being derived from the early avian radiation, some of which lost flight ability; but avian flight originated from the trees down, taking advantage of the cheap energy provided by gravity. The lower line may represent not actual dinosaurs but a separate arboreal avian lineage, perhaps derived from basal archosaurs. Such a view would approximate the hypothesis of a common descent of birds and dinosaurs. Reproduced with permission. Copyright Stephen A. Czerkas.

Fig. 6.26. Hypothetical proavis according to Hans Steiner. This model is based on an arboreal basal archosaur (thecodont) ancestor, a trees-down model for flight origin, and the classical scale-to-feather model. Missing from Steiner's interpretation are the hindlimb wings, which would produce a model like William Beebe's tetrapteryx stage and seen in microraptors. There is also no substantial propatagium, which was present in *Archaeopteryx* and may have been present in the proavis. The general concept, however, does not diverge strongly from a general anatomical pattern of that discovered in the Jurassic scansoriopterids and *Anchiornis*, which are very tenuously, and for no solid reasons, classified as theropods. From H. Steiner, "Das Problem der Diastataxie des Vogelflugels," *Jenaische Zeitschrift für Naturwissenschaften Gesellschaft* 83 (1918): 279–300.

bumbling along, no agony, and hardly any elation. The answer is the truth until someone playing the same "board game" by the same set of rigid rules somehow gets a different answer, and if you don't play by the rules, you are not a scientist.

As this interesting saga of deciphering bird origins continues, it is obvious that most workers on both sides of the debate have been at least in part wrong, though often for the right reasons. The widespread notion among paleontologists that "the change in consensus on the origin of birds was more the result of a radical change in methods and

philosophy than the discovery of new fossils" is an erroneous belief based on a faith-based reliance on the methodology of phylogenetic systematics, the results of which often produce featherweight phylogenies that are routinely challenged or falsified by molecular, genomic systematics.[133] Indeed, the various versions of the theropod origin of birds, from Thomas Huxley to John Ostrom, were based not on methodology or a rigid "scientific method" but on the discovery of new fossils, and the recent revolution in vertebrate paleontology was likewise the exclusive result of the massive, spectacular discoveries from the Chinese Early Cretaceous and Jurassic. Methodology has not led the field out of the wilderness; rather, the unsung heroes in this odyssey are those industrious peasant farmers of Liaoning Province who uncovered the critical evidence. For future advances in the field a new breed of scientists must break away from binding and stultifying methodological straitjackets and view the world of origins anew, seeking new discoveries. Before such daunting problems the scientist must work toward future approximations and not dwell on the last analysis.

APPENDIX 1

A SLIVER OF URVOGEL BONE: IMPLICATIONS FOR ENDOTHERMY

· ·

Hot-Blooded Dinosaurs Emerge

It all began with a paper in *Nature* in 1974 by Robert Bakker and Peter Galton entitled "Dinosaur Monophyly and a New Class of Vertebrates," which first proposed that all dinosaurs were hot-blooded (endothermic) with high exercise metabolism, at the same activity level maintained by living birds and mammals.[1] Consequently, they proposed the erection of a new Class Dinosauria, which would include as subclasses the Saurischia, Aves, and Ornithischia. The arguments for dinosaurian endothermy, mostly flawed, are discussed in great detail elsewhere but are still debated.[2] As a brief synopsis, although the argument for endothermic dinosaurs was introduced by Yale's John Ostrom as early as 1969, based primarily on upright (erect) dinosaurian posture, the torch for dinosaur endothermy was primarily carried and pushed to the extreme by Robert Bakker beginning in the 1970s and going forward.[3] Debate went back and forth, from those viewing dinosaurs as large active Mesozoic ectotherms, perhaps homeothermic, to those who saw them as true endotherms.[4]

Although a rigorous debate continues, a more reasoned argument has emerged, that dinosaur giants could have been quite active during the excessively warm climates of the Mesozoic, becoming facultative or inertial homeotherms (gigantotherms). This argument is based on simple extrapolations from surface-to-volume relationships, an idea dating back to G. S. Wieland in 1942.[5] The overall conclusion from studies based on size-related "inertial" homeothermy is that in the warm climates of the Mesozoic, large ectothermic dinosaurs would have been reasonably stable thermally, in essence homeotherms—that is, having a stable body temperature throughout the day. By this passive and energetically inexpensive thermal mechanism, dinosaurs would have avoided the burden of having to consume huge amounts of food to generate endogenously produced body heat. In addition, there is simply no imaginable selective advantage for endothermy in large reptiles during the excessively warm period of the dominance of dinosaurian behemoths, and there are numerous reasons for endogenously produced endothermy to be maladaptive during such an era, although such a view would not encompass smaller theropods such as *Compsognathus*.

Ostrom's view evolved over time into thinking that endothermy was more or less restricted to theropods, based on his discovery in the late 1960s of the highly active, predatory *Deinonychus*, and he openly opposed any view of endothermy in the large sauropods.[6] In an article in *National Geographic*, one sees his train of thought, which basically posited that small endothermic theropods were clothed in an insulatory pelt of feathers or protofeathers and that birds arose as "feathered dinosaurs": "*Archaeopteryx* supports two theories: warm-bloodedness in dinosaurs and dinosaurian ancestry of birds. The theropods . . . and especially the smaller kinds like *Compsognathus, Deinonychus*, and the struthiomimids, I suspect may have been true endotherms. And add to that the evidence that *Archaeopteryx* and other birds evolved from a small theropod dinosaur."[7] For Ostrom *Archaeopteryx* thus provided evidence on feather origins and for endothermic dinosaurs, an early version of the current orthodoxy on bird origins, and we see this view expressed as early as 1974 in Bakker and Galton's paper.[8]

Thus, the prevailing view that emerged from the 1970s, and has continued as the orthodoxy in the field, is that dinosaurs evolved endothermy, became insulated with protofeathers, and eventually gave rise to birds, which inherited their endothermy from their dinosaurian precursors. This view has characterized and guided interpretation of the recently discovered Chinese Mesozoic fossils, and the discovery of the first "feathered dinosaur," *Sinosauropteryx*, although later shown to be a mistake, added fuel to an already out-of-control blaze of enthusiasm for endothermic dinosaurs as avian precursors.[9] Providing a glimpse of the flavor of the argument, Philip Currie noted: "The warm-blooded proponents gained support in 1996 by the discovery in China of a pair of smaller "feathered" theropods [*Caudipteryx* and *Protarchaeopteryx*]. . . . The discovery of these specimens has given additional support to the hypothesis that theropod dinosaurs were the direct ancestors of birds, and that some theropods were endothermic."[10] As we have seen, however, the problem is that the two "feathered dinosaurs" were in all likelihood secondarily flightless birds, a type of Mesozoic kiwi.

In contrast to this prevailing view on endothermy in early birds and their putative dinosaurian antecedents, comparative physiologist John Ruben has long argued, based on data from the muscle physiology of extant reptiles, that the urvogel *Archaeopteryx* was a flying ectotherm.[11] Based on Ruben's estimates, in *Archaeopteryx* the total flight muscle would have been about 9 percent of an approximately 200-gram (7 oz.) body mass (in modern birds it averages 25 percent, ranging from 11.5 to 43.9 percent). Ectothermic muscle, approximately twice as "capable" as modern avian flight muscle, would have given the urvogel the necessary equipment for powered flight. Although debate continues, there is little question that *Archaeopteryx* and early birds, with their trunk-climbing abilities, would have encountered little problem generating the necessary power for flight. Paul Sereno and Chenggang Rao followed this assessment, stating that "*Archaeopteryx* may have been a 'flying ectotherm' capable of flight over short distances, powered by lightweight anaerobic flight muscles and an ectothermic physiology as suggested by Ruben."[12] Obviously, if *Archaeopteryx* were ectothermic and assumed to be derived from small theropods, then it would be difficult to argue that birds had evolved from endothermic theropods.

Resolving the Conflict

Two conflicting views of early avian physiology have thus been proposed: an endothermic version based on the orthodox paleontological view that birds are living "feathered dinosaurs," derived from insulated, endothermic dinosaurs; and the view that early birds were ectothermic and that feathers were not present in the antecedents of birds but arose as a salient feature of the clade Aves.[13] When these views were first expressed, there seemed little hope that the conflict would ever be resolved with any degree of certainty, but as the recent Mesozoic Chinese discoveries emerged, several important findings were made. First, it became apparent that there were two dichotomous clades of Mesozoic birds: the primitive, predominant land birds of the Age of Reptiles, the opposite birds, or enantiornithines (Subclass Sauriuriae); and the more advanced ornithurines, with a sophisticated flight apparatus, the group that would ultimately give rise to the post-Cretaceous modern avian radiation. Although it cannot be established with certainty, the striking resemblance of the skull of *Archaeopteryx* to that of some enantiornithean, sauriurine birds has led to the conclusion that the urvogel may indeed be a member of that clade, but the lack of synapomorphies, or derived characters, linking *Archaeopteryx* and enantiornithines makes the proposition only a tentative hypothesis.[14]

Bone Microstructure to the Rescue

In recent years some paleontologists have attempted to assess dinosaur life-history parameters, particularly growth rates, through paleohistology.[15] Bone histology, especially Haversian bone and fibro-lamellar bone, has historically been unable to resolve whether animals were ectothermic or endothermic. As John Ruben has written, "Alligator growth rates are virtually indistinguishable from estimated growth rates for the bipedal theropod *Troodon*," and "the fibro-lamellar bone is absent in many small, rapidly growing endotherms, and its presence in a labyrinthodont amphibian and in some clearly ectothermic cotylosaurs, pelycosaurs, and dicynodont therapsids is particularly puzzling."[16] In addition, the presence of varied forms of endothermy in some sea turtles, sharks, billfish, and tuna and a brooding python casts doubts on the correlation of gait and metabolic physiology.

Part of the problem stems from biothermal semantics; there is no clear-cut division between ectothermy and endothermy, owing to the multifarious adaptive biothermal strategies found in tetrapods. Vertebrate physiology simply does not easily dichotomize into a binary character, "ectothermy or endothermy," there being every conceivable intermediate condition. The situation is made more complex by the numerous living "advanced" neornithine birds (considered an endothermic clade) that are marginally endothermic, notably forms like West Indian todies (Todidae) and African wood-hoopoes (Pheniculidae), which are not absolute and comprehensive homeotherms, and the well-known turkey vulture (*Cathartes aura*), which normally lowers its body temperature about at night 6°C (11°F) until it reaches 34°C (93°F)—in other words, it becomes mildly hypothermic. Some small species with excessive surface exposure, such as hummingbirds (Trochilidae), are capable of entering a state of torpor, or profound hypothermia, in which they are unresponsive to most stimuli; their oxygen consumption can drop by some 75 percent when the body temperature drops by 10°C (18°F). By far the most dramatic example of torpor is exemplified by the common poorwills (*Phalaenoptilus nuttallii*: Caprimulgiformes), which hibernate at a body temperature of 6°C (11°F) for up to two to three months during the winter; these birds normally require some seven hours to warm up.[17]

We must therefore exercise extreme caution when drawing conclusions on the physiology of primitive Mesozoic birds, which may have ranged from near complete ectothermy or quasi-ectothermy through a gradient approaching that of the level of the poorwill all the way to what is seen in most modern birds. We will never know for sure. However, the discovery of growth rings, or lines of arrested growth (LAGs), typical of living ectotherms (including crocodiles and dinosaurs) in Mesozoic enantiornithine birds led researchers to conclude that these birds were possibly ectothermic or quasi-ectothermic and to suggest that "these birds differed physiologically from their living relatives and were not fully homeothermic."[18]

Cyclical growth rings (tree ring–like growth lines) in cortical bone are reflective of a slowing or halting of deposition in response to seasonal changes or periodicity that result in compact bone being interrupted by dense lines of arrested growth. In general, living ectotherms display LAGs, their growth halting seasonally, and living endotherms do not. Thus, the discovery that growth rings were present in the long bones of Mesozoic enantiornithine birds and the Late Cretaceous flightless Argentine bird *Patagopteryx* was the cause of considerable excitement, especially given the fact that no such growth rings had been observed in any living ornithurine birds or in the Cretaceous ornithurines *Hesperornis* and *Ichthyornis*.[19] However, as we saw back in the 1970s, Haversian bone, thought to be a litmus test for endothermy, proved to be widespread in vertebrates and thus anything but a guidepost for metabolic parameters. So, as with Haversian bone, tremendous variation exists in the deposition of LAGs, and although such growth rings are not typically present in ornithurines, they have been found in moa, the extant parrot *Amazona*, and the giant Eocene ground bird *Diatryma*.[20] In moa, for example, it is estimated that in some instances it took almost a decade to attain skeletal maturity.[21]

A more alarming cautionary discovery relates to a provocative study of bone histology of the Plio-Pleistocene Balearic Island cave goat, *Myotragus*, which survived for some 5.2 million years on Majorca before being extirpated following the arrival of humans about three thousand years ago. The intriguing aspect of this small goat (smaller than a medium-sized dog) is its long-term persistence on an energy-poor Mediterranean island, an environment normally conducive to proliferation of reptiles. Indeed, it is in such environments that reptiles frequently displace mammals because of their relatively slow and flexible growth rates; their low metabolic rates permit them to operate effectively with low energy flow.[22] In an attempt to discover more about this strange mammal, Meike Köhler and Salvador Moyà-Solà examined the bone histology of *Myotragus*. They were amazed to discover that its bone exhibited lines of arrested development, a trait normally found only in ectothermic reptiles. Unlike any other known mammal, *Myotragus* grew at slow and flexible rates, similar to living crocodilians and other ectothermic reptiles, attaining maturity at approximately twelve years, extremely late for a bovid. They suggested that such developmental and physiological plasticity were crucial to the survival of *Myotragus* and perhaps other large mammals on the resource-limited Mediterranean Islands, and that similar life-history traits will be discovered in other large insular mammals, such as the dwarf elephants, hippos, and deer.

The studies of bone histology of particularly the Balearic goat and New Zealand moa sound a stark cautionary note concerning interpretation of bone histology of extinct organisms. Given the paucity of data from living reptiles, and especially extant birds, any comprehensive conclusions on metabolic physiology deduced from fossil bone histology would appear premature. As a single

example, one would like to know the differences in bone histology among the cranes, which have a large range of growth patterns. Sandhill cranes (*Grus canadensis*) at four to five months attained 91 percent of adult weight, whereas crowned-cranes (*Balearica regulorum*) of that age had attained less than 60 percent of adult weight.[23]

Archaeopteryx *Bone*

Although bone from enantiornithine birds has been well known for some time, the missing piece of the puzzle was the urvogel *Archaeopteryx*. At last, in 2009, Gregory Erickson of Florida State University and colleagues were able use bone histology to examine a sliver of bone from the Munich specimen of *Archaeopteryx*, as well as bones from the basal birds *Jeholornis* and *Sapeornis* and the small dromaeosaurid *Mahakala*. Their prediction, based on their view that *Archaeopteryx* was a "feathered dinosaur," was that the long bones in basal birds "formed using rapidly growing, well-vascularized woven tissue typical of non-avialan dinosaurs."[24] Yet, on examination, they discovered the opposite, that long bones of *Archaeopteryx* and other basal birds (and *Mahakala*) examined are composed of nearly avascular parallel-fibered bone, which is among the slowest growing and is common and characteristic of ectothermic reptiles. If the *Archaeopteryx* specimens (or most) represent a single species, the near doubling in size between the Eichstätt and Solnhofen specimens would indicate a typical indeterminate, ectothermic growth pattern, as seen now from the new paleohistological findings. Although Erickson and colleagues clearly noted that "these findings dispute the hypothesis that non-avialan dinosaur growth and physiology were inherited in totality by the first birds." they somehow concluded that "the unexpected histology of *Archaeopteryx* . . . is actually consistent with retention of the phylogenetically earlier paravian dinosaur condition when size is considered" and that "the first birds were simply feathered dinosaurs with respect to growth and energetic physiology."[25]

Their actual histological finding, which they found difficult to accept as simple data, was "that the long bones of *Archaeopteryx* are composed of slow growing, reptilian grade, parallel-fibered bone," which they found "surprising." As they noted, "Well-vascularized woven bone was expected. The finding of the same peculiar matrix and vascular pattern in the slightly more derived bird *Jeholornis* indicates that the *Archaeopteryx* histology was not aberrant, but is typical of the basal avialan condition." Yet they could reach only one conclusion, dictated by phylogenetics, that "*Archaeopteryx* was simply a feathered and presumably volant dinosaur."[26] Indeed, only because of the need in modern phylogenetic systematics to "accommodate the cladogram" could one possibly reach any conclusion other than that these birds were quite reptilian and most likely quasi-ectothermic and that feathers did not evolve as an insulatory mechanism in their precursors. As we have seen, the entire geometry of the traditional cladogram of dinosaur-bird relationships may be incorrect; layering assumptions on the currently accepted phylogeny may therefore be extremely misleading.[27] One might also consider the alternative to one of their primary questions based on a traditional theropod ancestry of birds—that is, "how birds became miniaturized." Did, in fact, birds actually become miniaturized, or were their antecedents roughly similar in size (or even smaller) to the famous urvogel? Let the evidence speak for itself!

There are numerous problems with the osteohistological approach, exemplified by a study by Matthias Starck and Anusuya Chinsamy in which they showed, using Japanese quail (*Coturnix japonica*), that bone deposition rates vary considerably in response to variation in environmental conditions. They concluded that "bone deposition rates measured in extant birds cannot simply be extrapolated to their fossil relatives."[28] They further suggested that the techniques and assumptions used by paleohistologists in aging and tissue-formation characterizations have been "premature and inaccurate," and a number of workers have expressed reservations and urged caution in making sweeping generalizations on extrapolations of metabolic status from bone histology. However, in the paper by Erickson and colleagues there is never any consideration of the far more parsimonious hypoth-

esis to explain the nature of the urvogel's primitive bone histology—that is, that *Archaeopteryx* and basal birds are part of a disparate clade of avialans that includes most of the groups now considered maniraptoran theropods, which in the Mesozoic exhibit all stages of flight and flightlessness.[29]

Two major points emerge: (1) deduction of physiological parameters from bone histology is an interesting but precarious approach that requires extraordinary caution, and much more work on modern tetrapods must be accomplished before adequate comparisons can be accomplished; but (2), whatever the case, the presence of treelike growth rings in long bones is more reasonably explained as an indication of ectothermy or quasi-ectothermy than any other interpretation. One can picture a feathered but quasi-ectothermic *Archaeopteryx*, living in the warm, tropical setting of the forested edge of Late Jurassic Solnhofen lagoons, thermoregulating using ambient heat, somewhat similar to the greater roadrunner (*Geococcyx californianus*), but with considerably less thermoregulatory ability. In the roadrunner, body temperature on cold nights may decline by about 4°C (7°F), but after sunrise the bird will bask in solar radiation to warm ectothermally to its normal body temperature.[30] *Archaeopteryx* was an active climber and flier but was almost certainly not endothermic; neither were its putative dinosaurian ancestors.

John Ruben, Terry Jones, and Nick Geist concluded from their studies of reproductive paleophysiology and earlier work by Jaap Hillenius on nasal turbinates, that although many dinosaurs and early birds were likely to have been homeothermic, their lack of nasal respiratory turbinates indicates that they were more likely to have maintained ectothermic metabolic rates, like modern reptiles, during periods of rest and routine activity.[31] The turbinate bones are thin, scroll-like bones or cartilages in the nasal cavity of all reptiles, birds, and mammals, consisting of two distinctive sets of elements, the respiratory turbinates (lined with moist respiratory epithelia) and the olfactory turbinates, which are located out of the main path of respired air and are lined with olfactory (sensory) epithelia that contain the receptors for the sense of smell. Olfactory turbinates occur ubiquitously in all known reptiles, birds, and mammals and are not associated with endothermy. However, there is a fundamental association of respiratory turbinates with endothermy. They occur in all extant terrestrial birds and mammals, and they have no analogues or homologues among living reptiles and amphibians. In mammals and birds, endothermy is clearly linked with high levels of oxygen consumption and elevated rates of lung ventilation, exceeding reptilian rates by about twenty times. The respiratory turbinates have the effect of creating an intermittent countercurrent exchange of respiratory heat and water between respired air and the moist epithelial linings, resulting in "excess" water vapor condensing on the turbinate surfaces, where it is reclaimed. Thus, substantial water and heat are conserved rather than lost to the ambient environment. Ruben and colleagues concluded that "in the absence of respiratory turbinates, continuously high rates of oxidative metabolism and endothermy might well be unsustainable insofar as respiratory water and heat loss rates would frequently exceed tolerable levels."[32] Not surprisingly, there is no evidence of respiratory turbinates in the urvogel *Archaeopteryx*, enantiornithine birds, or any known dinosaurs, including maniraptorans.

Hypotheses of dinosaurian endothermy go way back and have traditionally relied on correlations of metabolic rate with weakly supported criteria, including everything from predator-prey ratios, trackways, posture, bone histology, putative but unproven protofeathers, and growth rates. But all correlations are equivocal, seeming to reflect an approach that is designed to "prove" endothermy in dinosaurs more by a verificationist method than by testing hypotheses. Many of the latest attempts to establish endothermy in dinosaurs have relied on growth rates, but there is at present not sufficient information on living reptiles to draw substantive conclusions, and it is significant that growth rates in some alligators are virtually indistinguishable from the estimates of growth rates for the dromaeosaurid *Troodon*.[33]

In one other recent attempt to test whether dinosaurs might have been endothermic, Herman

Pontzer and colleagues applied two new methods for estimating metabolic rates to fourteen extinct species of dinosaurs, ranging from small birdlike forms to the gigantic *Tyrannosaurus*.[34] The models are designed to predict metabolic rate on the basis of anatomy and the energetic cost of walking or running. Unfortunately, many of their assumptions are questionable, because the net cost of locomotion is roughly similar in all vertebrates, with the net cost decreasing as the body mass increases, as comparative physiologist Knut Schmit-Neilson noted long ago.[35]

Aside from all other biomechanical considerations, it is difficult to imagine any physiological scenario in which it would be selectively advantageous to be a giant reptilian endotherm in the excessively warm, monotonous tropical climates of the Cretaceous Period, when dinosaurs flourished. Indeed, it is reptilian ectotherms that trend toward gigantism in today's tropics. Why endothermy, with its dramatic increase in energetic cost and concomitant food burden, when ectothermic reptiles thrive in tropical settings? What about endothermy in brooding pythons, sea turtles, billfish, and some sharks? Metabolic parameters involve extremely complex biological phenomena that are often not resolved by simplistic correlation science. However, apparently contradicting the conclusions by Erickson and colleagues, Pontzer and colleagues concluded that, of the fourteen species examined only the smallest, *Archaeopteryx* (estimated at 0.25 kilograms), had estimated locomotor metabolic rates that fell within or near the range seen in modern ectotherms.[36]

Paleontologists in particular appear to have a fixation on proving that dinosaurs and their putative descendents, the early birds, were endothermic, feathered dinosaurs, and there has been a rash of papers on this topic during the past four decades; yet the evidence is simply not there, and confusion abounds. In the final analysis, despite all the confusion and evidence to the contrary, endothermic dinosaurs have crept into today's textbooks, and in 2009, reporting on the paper by Erickson and colleagues, the *New York Times* proclaimed: "The 'early bird' *Archaeopteryx* may not be a bird, after all."[37]

APPENDIX 2

THE PERSISTING PROBLEM OF AVIAN DIGITAL HOMOLOGY

··

Early in the debate (1984) it was noted that if one key *synapomophy* [referring to the digits] were falsified, it would reduce all the others to the status of parallelism or homoplasy.

M. E. Howgate, "Back to the Trees for Archaeopteryx *in Bavaria," 1985*

Community of embryonic structure reveals community of descent.

Charles Darwin, On the Origin of Species, *1859*

In cladistic analyses all characters are treated equally, and so-called key or trump characters, which involve character complexes of varying complexity with developmental interconnectivity, are of no greater weight than any other. Because of equal weighting of characters in phylogenetics, one can legitimately ask with justification why almost every major paper or book on the topic of avian evolution by traditional cladists invariably gives disproportionate attention to these very key characters—the problem of digital homology and others—that are supposed to be of no greater weight than any other. The answer is simple. Overturning only one key synapomorphy in the bird-theropod nexus has the capability of reducing the entire suite of putative synapomorphies to convergent characters, generally co-correlated, and often related to the adaptations for a particular mode of life, in the case of dinosaurs to the rigid structure of a bipedal, mesotarsal vertebrate.

Among the important morphological problems for the current theropod hypothesis for the origin of birds, however, has remained the problem of digital homology. Richard Owen in 1836 identified the three remaining bird digits as II, III, IV, whereas William Kitchen Parker in 1888 numbered them I, II, III.[1] The debate has continued to the present, with paleontologists preferring an avian manus composed of digits I, II, and III, given that the manus of neotetanurine theropods seems to be reduced to digits I, II, and III. In contrast, developmental biologists have uncovered overwhelming evidence based on embryonic connectivity for a manus composed of digits II, III, and IV.

Perhaps the most significant character complex thought to link birds to theropod dinosaurs is the possession by both groups of a hand reduced from the basic pentadactyl, or five-fingered, hand to a tridactyl hand, with three remaining digits in the adult. Obviously, if the tridactyl manus of theropods and the tridactyl manus of birds are composed of different digits, the changes in manual morphology cannot be homologous across taxa, and making an evolutionary switch from one to the other would seemingly involve insurmountable difficulties. Digital homology determines the homology of numerous carpal elements, including the semilunate carpal that has been so prominent in arguments for a theropod origin of birds since John Ostrom's discovery of *Deinonychus* in 1969. Thus, if the manual digits of birds and neotetanurine theropods are not homologous, multiple statements of homology used to link the two are falsified. Yet the hand of dromaeosaurs exhibits extreme similarity to that of early birds, including the presence of a semilunate carpal element.

Many have cautioned that most of the characters linking birds and theropod dinosaurs are plesiomorphic (primitive) and that convergence rather than homology may explain other similar characters. Because anatomical complexity and developmental interconnectivity of the manus are highly constrained and complex, phylogenetic hypotheses must be concordant with these data in order to be valid.[2] Thus, paleontologist James Clark optimistically noted that the "only substantive problem with the theropod-bird hypothesis remains the discrepancy between the homology of the digits of the manus as indicated by the fossils and the development of extant birds."[3] Although the basic tree topology of the currently accepted theropod hypothesis is recovered whether hand characters are used or deleted, concern over the digits of the hand as a trump character is precisely the reason so much research has been focused on this problem.[4] For example, Luis Chiappe in *Glorified Dinosaurs* spent five pages on the problem of digital homology, but there is little space given to the two hundred or so other characters used in the modern cladistic analyses of birds and dinosaurs.[5]

The digits of tridactyl theropods are best identified as digits I, II, and III, based on indisputable fossil evidence showing progressive reduction and subsequent loss of digits IV and V during theropod evolution, as revealed by the well-preserved hand of the Late Triassic basal theropod (possibly a basal saurischian) *Herrerasaurus* and the more advanced, basal theropod *Dilophosaurus*.[6] The pattern of digital reduction in theropods is unusual in vertebrates, and results in a functionally grasping, raking hand. The definitive trend in all dinosaurs is for combined symmetrical reduction of the medial and lateral pedal digits (bilateral digital reduction, or BDR), typical of tetrapods that reduce their digits, but asymmetrical reduction of the lateral manual digits (lateral digital reduction, or LDR). This single character complex presents an impressive synapomorphy for the clade. Thus, the postaxial or lateral digital reduction of digits IV and V may be the most salient synapomorphy of Dinosauria. A comparison of digital reduction in a broad array of tetrapods supports a common pattern, with digits I and V, the outer mesial and lateral digits, respectively, being the first lost in a lineage. This pattern is termed Morse's Law of Digital Reduction; thus the retention in theropod hands of digits I, II, and III violates the pattern of reduction consistently found in tetrapods. Max Hecht and Bessie Hecht summarized: "In order to relate birds to the saurischian or theropod clade,

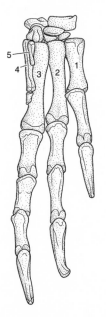

Fig. A1. The left hand of the late Triassic, basal dinosaur (possibly theropod) *Herrerasaurus*, typical of many late Triassic theropods, illustrating a pentadactyl hand but showing digits 4 (IV) and 5 (V) greatly reduced, illustrating the dinosaur manual morphology with a tridactyl hand composed of digits I, II, III, a key synapomorphy for Dinosauria. Note that the longest finger in primitive dinosaurs is not the middle finger, as in *Archaeopteryx* (and birds), *Deinonychus*, and most tetrapods including man, but finger 3, which is aberrant. Note also the primitive nature of the carpal elements, with nothing remotely resembling a semilunate carpal or any birdlike features. From A. Feduccia, "1,2,3 = 2,3,4: Accommodating the Cladogram," *Proceedings of the National Academy of Sciences* 96 (1999): 4740–742. Copyright 1999 National Academy of Science, U.S.A.; modified after P. C. Sereno, *Journal of Vertebrate Paleontology* 13 (1994): 425–450.

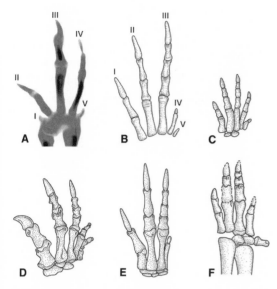

Fig. A2. The pentadactyl hand of a 14(–15)-day-old ostrich (A), showing the anlagen for digits I, "thumb" (left), and V (right), compared to the hands of various dinosaurs, illustrating the reduction of digits IV and V (disappearance of V in [E]). (B) *Herrerasaurus* (putative basal theropod, Late Triassic); (C) *Lesothosaurus* (basal ornithischian, Late Triassic); (D) *Plateosaurus* (basal sauropodomorph, Late Triassic); (E) *Syntarsus/Coelophysis* (theropods, Late Triassic); (F) *Hypsilophodon* (ornithopod, Early Cretaceous). Drawn to same scale, modified from a variety of drawings. From A. Feduccia, "Birds Are Dinosaurs: Simple Answer to a Complex Problem," *Auk* 119 (2002): 1187–201; modified from drawings primarily in D. B. Weishampel et al., eds., *The Dinosauria*, 2nd ed. (Berkeley: University of California Press, 2004); ostrich embryo from A. Feduccia and J. Nowicki, "The Hand of Birds Revealed by Early Ostrich Embryos," *Naturwissenschaften* 89 (2002): 391–93.

it is necessary to deny the reductive amniote sequence which has been inferred from many studies on the development of the manus of modern birds."[7]

Neil Shubin of the University of Chicago supported the identification of the digits of the avian hand as I, II, and III rather than II, III, and IV, basing his conclusion largely on "phylogenetic analysis that relies on the total evidence"—that is, on the truth of the theropod hypothesis for the origin of birds. Yet he cautioned that without "fossils

of theropods, and the phylogenetic interpretations that they imply, the II-III-IV interpretation of the homologies of the avian digits would, perhaps, be a more likely interpretation."[8] It would appear, therefore, that the more fundamental problem is largely philosophical and methodological: "Developmental biologists use conservation of embryonic patterning to establish homology, while . . . paleontologists use the methodology of phylogenetic systematics to define homology a posteriori from cladistic analysis of multiple synapomorphies."[9] Paleontologist Luis Chiappe, recognizing this dilemma, correctly noted that "the debate . . . highlights the methodological chasm between researchers endorsing the cladistic notion of homology, which is based on how congruent an interpretation is with similar interpretations about other structures, and those using a different concept of homology, based on the similarity of developmental pathways."[10]

Given this background, in 1995, I and developmental biologist Ann Burke, then at the University of North Carolina (now at Wesleyan University, Connecticut), who had done considerable earlier work on the development of the vertebrate hand at Harvard during her graduate student years, embarked on a study of developmental connectivity in the avian hand in an attempt to determine the identity of the remaining avian digits. To gain a firmer understanding of the homologies of the avian digits, we examined and compared the fore- and hind-limb developmental patterns of an archosaur (alligator), a primitive anapsid reptile (turtle), and the chick and some other bird species, including the ostrich, a cormorant, and a duck.[11] Our work owed a great debt to earlier workers such as Nils Holmgren and others, and the previous work of Richard Hinchliffe of the University of Wales, Gerd Müller of the University of Vienna, and Pere Alberch, whose studies of comparative embryology advanced the field to the level of serious consideration of homology of digits.[12]

The manus in all the forms examined begins development as a series of condensations. As development proceeds, a postaxial condensation extends distally toward the precursors of distal carpal IV and metacarpal IV; this distal extension is

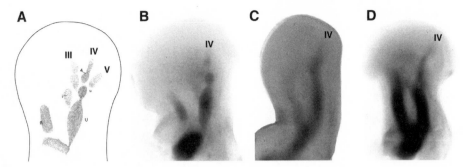

A   **B**   **C**   **D**

Fig. A3. The developing primary axis. (A) Diagrammatic illustration of developing turtle limb bud showing general topography (precondensations of: R, radius; U, ulna; u, ulnare; I, intermedium; 4, carpal four; and digits III–V, 3–5). The primary axis therefore is highly conserved developmentally and invariably identifies digit IV, 4. (B–D) Developing limb buds in a turtle, bird, and alligator, respectively, illustrating the conserved developmental pattern of the primary axis. From A. Feduccia, "1,2,3 = 2,3,4: Accommodating the Cladogram," *Proceedings of the National Academy of Sciences* 96 (1999): 4740–42. Copyright 1999 National Academy of Sciences, U.S.A. (A) modified from A. C. Burke and P. Alberch, "The Development and Homology of the Chelonian Carpus and Tarsus," *Journal of Morphology* 186 (1985): 119–31; (C) and (D), photographs from specimens prepared by A. C. Burke.

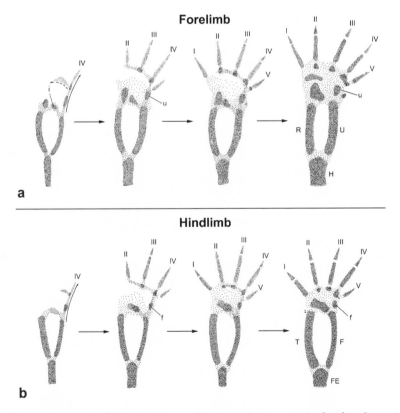

Forelimb

a

Hindlimb

b

Fig. A4. Schematic representation of the sequences and connectivity patterns in the chondrogenic development of (a) the forelimb and (b) the hind limb in *Alligator*. The primary axis and the digital arch in the forelimb are indicated by arrows. Modified after G. B. Müller and P. Alberch, "Ontogeny of the Limb Skeleton in *Alligator mississippiensis*: Developmental Invariance and Change in the Evolution of Archosaur Limbs," *Journal of Morphology* 203 (1990): 151–64. Courtesy of Gerd Müller.

termed the "primary axis," a conserved embryonic landmark in the developing hands of all amniotes. Digit IV forms as the distal development of this primary axis and is therefore easily identified by its developmental connectivity. Immediately posterior to the primary axis and digit IV, digit V develops. Digits III, II, and I develop in a posterior to anterior arch, the digital arch. This constrained pattern is evident in all the forms we examined. Moreover, developmental patterns in the manus in these taxa are mirrored in the pes, as one can see from the figures of chick and ostrich development. Summarizing, in all amniotes stereotyped early developmental patterns in the fore- and hind limbs are conserved, they are characterized by a Y-shaped condensation, and the sequence of digital formation is posterior to anterior, with digit IV the first formed as a developmental organizer. The

Y-shaped condensation we observed represents the embryonic humerus branching distally into the radius and ulna as development continues. During this process, the ulna (postaxial element) dominates, extends, and branches to give rise to the intermedium and ulnare. Turtles and alligators are pentadactyl, and there is consequently no difficulty in identification of the specific digits and their embryonic connectivity. In birds, the first digit to form is in the identical position and has the same embryonic connections as the first digit to form in the turtle and alligator. Accepting the principle of connectivity, we thus identified this digit in the manus of birds as digit IV, in both the fore- and hind limb. As we noted, there are some lizards (for example, the African *Chamaesaura*) in which digit IV develops as the organizer transitorily but regresses so that the adult hand is composed of digits

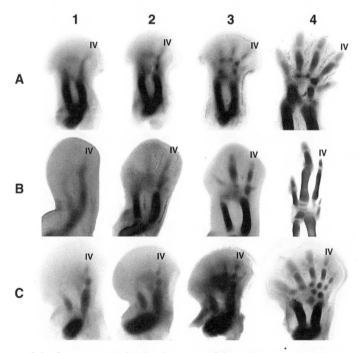

Fig. A5. Dorsal views of the four stages in the development of the right manus in (A) a crocodilian (*Alligator*); (B) a chicken (*Gallus*); and (C) a turtle (*Chelydra*). The stages in column 1 represent the early appearance of the primary axis of the developing hand. Column 2 shows the early digital arch. This pattern has been described for numerous species by many classical morphologists and more recently in the chicken, turtles, alligators, lizards, and mice. Columns 3 and 4 show subsequent development of the digits. Note the transient appearance of digit 5 in *Gallus*. Digit 4 (IV) is labeled in each specimen as a landmark. Images are not to scale. From A. C. Burke and A. Feduccia, "Developmental Patterns and Identification of Homologies in the Avian Hand," *Science* 278 (1997): 666–69. Reprinted with permission from AAAS.

II and III only. However, we found no evidence of this in birds. To the contrary, this observation suggests how the theropod hand might have developed with a typical amniote "primary axis" but aberrantly retained digits I, II, III in the adult.

Historically, the primary argument advanced by paleontologists in defense of the identification of the digits of the bird wing as I, II, and III, as in theropods, aside from the cladogram showing birds as dinosaurs, is the phalangeal formula. In the plesiomorphic pentadactyl manus of archosaurs, the phalangeal formula for digits I, II, and III is the same as in both the surviving three digits of the forelimbs of theropods and those in *Archaeopteryx*. Arguments of this nature fail to point out that the phalangeal formula (P) of P2-3-4-x-x (x = absence), for digits I, II, and III in theropods and for the digits of the hand in *Archaeopteryx*, is that found in basal archosaurs; the character is plesiomorphic and therefore not phylogenetically informative. Basal archosaurs begin with an ancestral formula, P2-3-4–5-3, which is reduced to P2-3-4–3-2 in basal theropods, to P2-3-4–1-x, and ultimately to P2-3-4-x-x, with even more reduction in tyrannosaurids. Phalangeal formulae are in any case quite labile; for example, pedal phalangeal formulae in *Archaeopteryx* differ across specimens, and in the theropods, phalangeal counts can vary even within species.[13] Also, as noted in the text, in no known ceratosaur do the three digits exhibit a typical theropod phalangeal formula of P2,3,4, demonstrating that the assumed conservatism of this formula for theropods is not absolute.[14] This lability is also demonstrated by the relatively rapid reduction of wing phalangeal formulae to P2,3,2 in the Lower Cretaceous primitive toothed bird *Sapeornis*, as well as in the flightless bird (sometimes termed a feathered theropod) *Caudipteryx*. This reduction continues in modern birds, with P2,2,1 in *Gallus* (sometimes reduced to P1,2,1) and P2,3,1 in *Struthio*. The phalangeal formula for the South American hoatzin (*Opisthocomus*) is P2,3,1 in the juvenile but P1,2,1 in the adult. In Cretaceous enantiornithine birds, phalangeal formulae are varied: P2,3,3 (*Eoalulavis*), P2,3,1 (*Concornis*), and P2,3,2 (*Protopteryx*). However, contrary to the case of *Archaeopteryx*, in which the phalangeal formula of 2,3,4 is used to show its putative affinity with theropods, in the case of *Caudipteryx*, the avian phalangeal formula of 2,3,2 is *not* used to argue for avian affinity, because it was predetermined by phylogenetic analysis that it was a feathered dinosaur instead of a "Mesozoic kiwi."

If, as argued, the digits of the hand of birds were II, III, and IV, then, to account for the phalangeal formula in the manus of *Archaeopteryx*, it is necessary to postulate a symmetrical reduction of one distal phalanx per digit. Such a transformation is basically a one-step developmental event, as can be demonstrated in experiments on limb development in the lizard *Lacerta viridis*. In this lizard, it was found, at a specific level of reduced cell division in the limb buds, that each of the three middle digits lost its terminal phalanx.[15] Thus, in the foot a normal phalangeal formula of 2,3,4,5,4 became a formula of 0,2,3,4,3, with a similar reduction pattern for the forelimb (P2,3,4,5,4 becoming P0,2[or 1],3,4,3). Use of the basal reptile phalangeal formula would identify the digits in such experimental limb buds in *Lacerta* as I, II, and III, whereas in fact we know them from other evidence to be II, III, and IV. Thus—returning to the question of the evolution of birds—loss of a terminal phalange from each of the digits II, III, and IV would produce the same phalangeal formula as in the three digits of *Archaeopteryx*. The *Lacerta* experiments provide an instructive model, showing that such transformation is feasible and not simply theoretical. Such a reduction of distal phalanges simultaneously and symmetrically in all digits was accomplished experimentally by blockage of bone morphogenetic protein 4 signaling, which mediates apoptosis in the avian limb bud.[16]

Compelling evidence in support of the embryological view that the wing of modern birds appears to retain digits II, III, and IV is found by comparison of the development of the fore- and hind limbs of modern birds.[17] Such an identification is based on classical criteria for establishment of homology, including timing (digit IV, the first to form), position (digit IV, postaxially positioned

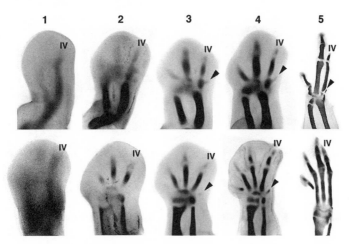

Fig. A6. Comparison of right manus (top row) and pes (bottom row) development in the chicken, dorsal views. Note the transient appearance of digit V in the manus and pes (arrowheads). These comparisons—of manus with four developing digits and pes with all five—leave little doubt that the avian hand contains digits II, III, and IV. Digit IV is labeled in each specimen as a landmark. Images are not to scale. From A. C. Burke and A. Feduccia, "Developmental Patterns and Identification of Homologies in the Avian Hand," *Science* 278 (1997): 666–69. Reprinted with permission from AAAS.

as in the leg bud), and connections (ulnare at base of digit IV). The embryological data are clearly a potential problem for the current orthodoxy of a linear dinosaur-bird nexus. Rather than accept this as a falsification of the theropod hypothesis, some have chosen instead to devise highly innovative attempts to accommodate the cladogram by an explanation of a major shift in digital evolution within theropods.

In a provocative 1999 paper, Günter Wagner and Jacques Gauthier of Yale University integrated paleontological and embryological data by proposing a "frame shift" hypothesis.[18] They accepted the embryological evidence that the embryonic condensations that give rise to the digits of the manus of birds are II, III, and IV. However, they separated the identification of digital condensations (for example, condensations C2, C3, and C4) from the later specification of its definitive phenotypic morphology (digits II, III, and IV, and so on). According to their view, repositioning of expression domains of genes controlling digit identity within the developing limb bud could result in a "frame shift" or "homeotic shift" whereby digits I, II, and III of theropods have become

shifted during the evolution of birds to the digit blastemas or condensations of birds, C2, C3, and C4. Thus, avian digit condensations are correctly identified as C2, C3, C4, but developmental mechanisms that specify morphological identity have shifted so that condensation C2 generates theropod digit I morphologically; and likewise in sequence, C3 generates digit II morphology, and C4, digit III morphology. If correct, birds would still be nested within the same clade as maniraptoran theropods, and the seemingly intractable dilemma between embryological and paleontological data would appear to be somewhat resolved.

A number of recent molecular studies have claimed to support both the classic II, III, IV avian hand and the Wagner frame shift hypothesis. In 2004, Alexander Vargas and John Fallon argued that in both mouse forelimb and hind-limb buds and chick hind-limb buds, prospective digit I is characterized by Hoxd13 expression alone while digits II–V have Hoxd13 as well as Hoxd12 expression.[19] Because the anterior wing bud digit in birds has only Hoxd13 expression, they argued that it is digit I on the basis of molecular homology. Frietson Galis of the University of Leiden and Richard

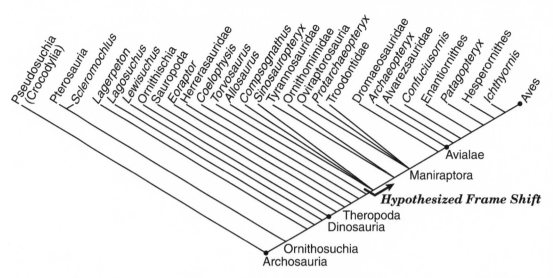

Fig. A7. Consensus cladogram depicting the phylogenetic relationships among archosaurs, dinosaurs, and birds, showing the proposed magical frame shift whereby the embryonic condensation for digit II (CII) is forced into the developmental trajectory of digit I (DI), CII into DII, and CIV into DIII. Such a phenomenon, if possible, would reconcile the embryological data demonstrating that the avian hand is embryologically represented by manual condensation for CII, CIII, and CIV, and would "accommodate the cladogram" for an avian origin from advanced theropod dinosaurs with manual digits I, II, and III. Yet there is little reason to assume that such a unique homeotic shift would have occurred, and there is no reasonable selective reason for such a dramatic transformation, "somewhere between *Allosaurus* and birds." From G. P. Wagner and J. A. Gauthier, "1,2,3 = 2,3,4: A Solution to the Problem of the Homology of the Digits in the Avian Hand," *Proceedings of the National Academy of Sciences* 96 (1999): 5111–16. Copyright 1999 National Academy of Sciences, USA.

Hinchliffe countered convincingly that the molecular evidence cited by Vargas and Fallon is insufficient for their conclusion, because the mutants cited (talpid and Hoxd deletion mutants) show only a weak correlation of Hoxd12/13 expression with digit identity.[20] Then in 2005, Monique Welton and colleagues found a condensation-specific Sox9 molecular domain in a digit I position in the chick wing bud, concluding, "We have found molecular evidence of a digit I domain in the chicken wing that is specified by early patterning mechanisms, but fails to undergo terminal differentiation."[21] Most recently, in 2011 a Japanese team lead by Koji Tamura used transplantation and cell-labeling experiments to conclude that the posteriormost wing digit does not correspond to digit IV of the hind limb and that because the progenitor of the last digit segregates early from the zone of polarizing activity, it lies in the domain of digit III specification.[22] Yet, despite the intriguing new putative molecular evidence for both a II, III, IV and a I, II, III avian wing, these findings simply add to our lack of a comprehensive understanding of the molecular mechanisms underlying digit identity, and the application of findings of a typical pentadactyl amniote limb to the highly modified and condensed avian hand may not be appropriate at this primitive level of our understanding. In addition, a nagging problem is that the Tamura team discovered that by day 3.5 of development a shift occurred, causing progenitor region cells for digit IV to move forward and grow into digit III. A similar shift occurred for the digits that become I and II.

The more fundamental problem lies in establishing that specific molecular domains represent signatures for particular digits and that these specific domains can be shifted from one area to another in an evolutionary transformation. We know that the molecule Sonic hedgehog (Shh) regulates

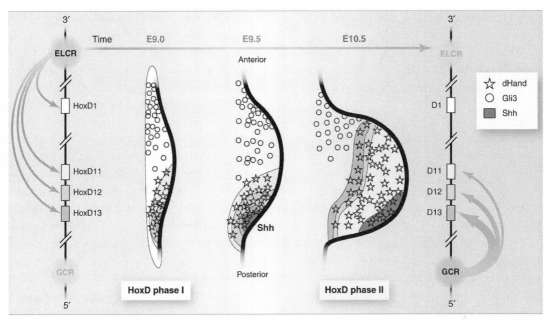

Fig. A8. Patterning of the budding limb. Biphasic regulation of the 5′HoxD gene cluster during limb bud outgrowth. Left, the early overlapping expression of domains of HoxD genes. Right, Shh expression domain subsequently extends and is displaced distally. Note that the second phase initiates morphogenesis of the most distal limb structures, the digits. Shh and Gli3 are dispensable for generating limb skeletal development but are required for specifying digit identity; yet their activation is during phase 2, which is still limb bud stage. From J. Deschamps, "Hox Genes in the Limb: A Play in Two Acts," *Science* 304 (2004): 1610–11. Reprinted with permission from AAAS.

both the number and type of digits in the manus and pes. In addition, Shh determines digit identity by forming a gradient across the limb, stimulating the formation of different digit morphologies in response to different Shh concentrations. High concentrations for a short duration lead to an anterior digit, whereas continued exposure leads to equivalent cells forming posterior digit types.[23] Assessing unknown digit identities by experimentally manipulating Shh gradients in a dramatically modified and transformed crown group of birds (extant and common ancestors) is extremely difficult. Gene networks of the limb bud are complex, and molecular developmental biologists are far from agreed on whether digit identity can be assigned by simple combinations of Hox gene expression. Also, the action of Shh occurs quite early in development, but pentadactyl avian hands have been discovered much later in development, as we shall see in ostrich embryos.[24] As late as 2008 developmental biologist Matthew Towers and colleagues identified the avian digits II, III, IV in a study of chick wing digit patterning.[25]

Moreover, one should consider that without the cladogram, no one would ever have considered such proposals: the frame shift hypothesis was a clever proposal to adjust the developmental program to accommodate the cladogram. In other words, it is an ad hoc auxiliary hypothesis designed to narrow the range of falsifying observation statements of the theropod hypothesis orthodoxy. Other objections have been mounted, significant among which is the observation that a homeotic frame shift in digital identity, as proposed, between *Allosaurus* and birds, "without any further anatomical changes does not appear to lead to an adaptive advantage."[26] Indeed, there is no conceivable adaptive advantage: "there is as yet no adaptive significance that would overcome the evolutionary constraint."[27] Although theoretically possible, there are too many unknowns to accept the frame shift as a convincing theory, and it is biologically implausible in the course of actual evolution. Such a proposal involves numerous new assumptions and factors that are introduced without any substantiating evidence, and the proposal is largely theoretical. The primary objections are summarized here:

- Deleterious mutations in tightly constrained developmental processes can have negative pleiotropic effects.
- Fore- and hind limbs have the same highly conserved developmental pattern; yet the frame shift would have occurred in the forelimb only, and not in the serially homologous pedal digits.
- Despite other theoretical scenarios, no homeotic frame shift is demonstrable for digits in manus or pes, or across any other broad embryonic field, between any two known amniote groups (although one such event, albeit unconvincingly, is proposed for an aberrant Australian skink),[28] and if true could not be used as a general mechanism for amniotes.
- Such a frame shift must explain not only digit identity but equivalent integrated changes in the carpus, which it does not.
- There would be no selective adaptive advantage to a digital frame shift in evolution somewhere between *Allosaurus* and birds. Such a development shift would be unique among amniotes as a shift across a broad embryonic field between two higher groups and would imply neutral mechanisms.
- József Zákány and colleagues showed that patterning of the budding limb involves biphasic regulation of the 5'HoxD gene cluster during limb outgrowth. However, the second phase initiates morphogenesis of the most distal limb structures, the actual digits. Shh and Gli3 are dispensable for generating limb skeletal development but are required for specifying digit identity; yet their activation is during phase 2, which is still limb bud stage.[29]

Still another issue is that a frame shift hypothesis assumes that neotetanurine theropods would have had an embryonic C4 condensation. Since digit IV in neotetanurine theropods is reduced or absent, C4 would also be reduced or absent. If C4 was absent, the identity of digit III clearly could not be imposed upon it. Condensation 4 would have to be of sufficient size to generate a long metacarpal plus four phalangeal elements for a posterior digit of *Archaeopteryx*; even in modern birds, there is a substantial digit IV, especially in the embryo. Therefore, a frame shift hypothesis must explain not only the change in the identity of the digits but also the necessary integrated changes in the carpus. The difficulty, of course, is that the distal semilunate carpal (a key synapomorphy of dromaeosaurs and *Archaeopteryx*) would have to be shifted posteriorly so that it retained its position at the base of theropod metacarpals I and II, by the frame shift hypothesis, formed from bird C2 and C3. Such a shift would involve identity changes and modification of the distal carpal element of the embryonic limb, and even greater morphological reconstruction of the limb than had been assumed by the frame shift hypothesis. In sum, a shift that would affect four digit condensations simultaneously, as well as the entire carpus, is unknown in amniotes.

It has been argued that the detailed morphology of the carpus and phalangeal elements of *Archaeopteryx* and *Deinonychus* and other dromaeosaurid theropods indicates that the elements in question must be homologous. By this view, the three wing digits of *Archaeopteryx* are identified as I, II, and III, but this assumes that dromaeosaurs are indeed dinosaurs and not derived flightless birds. I made this same mistake in 1996 when I wrote *The Origin and Evolution of Birds*, in assuming that *Deinonychus* was a typical theropod dinosaur, and therefore it would follow that the manual and carpal elements must have been different from those of birds. With the additional evidence coming in from Chinese fossils during the past decade, it is now clear that dromaeosaurs have a very close affinity with Aves, and their manual and carpal elements are homologous with those of *Archaeopteryx*. As Luis Chiappe noted in 2007: "The remarkable anatomical resemblance between the hands of *Archaeopteryx* and the dromaeosaurid *Deinonychus* supports the homology of the fingers of these animals. . . . There is no

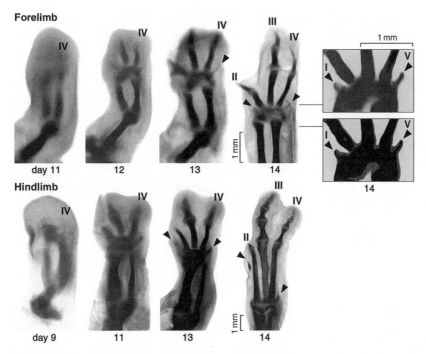

Fig. A9. Comparison of right-hand views of manus (top row) and pes (bottom row) development in *Struthio*, dorsal views, days 8–14 (or 15). Arrows point to condensations for digits I and V in manus and metatarsals I and V in pes; days given below each specimen. High-contrast photographs of enlargement of area of condensation for DI and DV, far right. Images are not to scale. From A. Feduccia and J. Nowicki, "The Hand of Birds Revealed by Early Ostrich Embryos," *Naturwissenschaften* 89 (2002): 391–93.

reason not to assume the same correspondence to the pentadactyl hand of primitive tetrapods for both *Archaeopteryx* and *Deinonychus*."[30] In addition, new evidence from the wing bones of *Velociraptor*, showing possible quill knobs, if correct, would indicate an avian wing, adding support to the view that these Cretaceous, earthbound "raptors" are in reality secondarily flightless birds.[31] To assume otherwise would invoke convergence, a widely held view before the discoveries in China of the Early Cretaceous microraptors, which are extremely birdlike basal dromaeosaurs. One could go as far to say that whatever microraptors are, so too are birds. It has even been proposed that birds evolved from dinosaurs that had a hand with digits II, III, and IV, despite the convincing paleontological evidence to the contrary. Despite this interesting new proposal, embryological connectivity still favors a II-III-IV identity for the bird wing and I-II-III for the basic theropod hand. As we shall see, the actual digital morphology as well as the carpus of microraptors are remarkably similar to that of *Archaeopteryx*, and this has prompted the proposal that there may be two separate lineages, a true "theropod" lineage with a manus composed of digits I, II, and III (*Herrerasaurus, Coelophysis, Compsognathus, Allosaurus, Tyrannosaurus*, and so on), and the true "avian" lineage (*Archaeopteryx*, dromaeosaurs, including microraptors, and secondarily flightless forms), with a manus composed of digits II, III, and IV. As Zhonghe Zhou noted, "If we simply compare the hands of *Archaeopteryx* and some maniraptoran theropods, such as *Microraptor*, they are almost the same in every detail, including the phalangeal formula, so in that sense the "identity" of the hand and digits of these two groups is without question. If we accept the 'II-III-IV' for modern birds, and assume the same for *Archaeopteryx*, then why not accept the same conclusion for *Microraptor*?"[32]

Digit I Discovered: Ratites Give Up the Ghost of the Ancestral Hand

> A problem with the interpretation of the bird hand [as II-III-IV] is that the anlagen of all five digits never form in any known bird.
> Kevin Padian, "The False Issues of Bird Origins: An Historiographic Perspective" (2001)

The most decisive embryological evidence for a pentadactyl avian hand would be the presence of an atavistic anterior digit I in modern birds. Although numerous developmental papers had supported the identification of the digits of the hand of birds as II, III, and IV, the one thing missing was an attempt to locate digit I (the homologue of the thumb), a problem noted by a number of paleontologists who number the manual digits of birds the same as those of theropods: I, II, III. In 2001, I and Julie Nowicki, then a graduate study of developmental biology, embarked on a study of ostrich embryology in an attempt to discover the primitive configuration of the manual digits of birds, an unsolved problem dating back to German anatomist Johann Friedrich Meckel in 1821.[33] The rationale behind our studies was that paleognaths are probably the most primitive living birds and, in addition, the ratites exhibit dramatic signs of paedomorphosis. Therefore, following in the framework of Karl Ernst von Baer on the principle of the conservatism of early embryos and the observation that embryos of higher forms in early stages of development resemble the embryos of ancestral forms, it seemed likely that if the primitive hand of birds were to be discovered, the ostrich was perhaps the best place to look. Like the appearance of a tail in the embryo of a human at week eight and its subsequent disappearance by week ten, it was our prediction that we might discover the first digit of birds in the early embryo before it, too, disappeared.

We obtained and incubated ostrich eggs in the laboratory and examined embryos at one-day intervals to develop a series of specimens representing early development of the appendicular skeleton. A key problem with earlier studies is that they concentrated on later stages of development, and

one of our more interesting discoveries was that within the forty-two-day incubation period for ostrich, most of the major skeletal features develop between days 8 and 15. Thus, by day 20 the ostrich embryo represents a small adult with respect to most skeletal features. Most studies have relied on the aberrant chicken and have concentrated on the second half of the ontogeny, usually near hatching, so that they were dealing with essentially an adult. We discovered that the development of the ostrich manus (and pes for comparison) follows almost exactly the typical pattern of amniote development and connectivity, with a primary axis developing and identifying digit IV in both manus and pes. As in the chicken, a postaxial nubbin develops as the condensation for digit V, but to our astonishment and delight, an identical nubbin could also be observed at day 14 on the preaxial side; it was in fact the preaxial nubbin representing the condensation for the putative anlage for digit I. The grounds for this interpretation, which was the only obvious one, were: (1) its location in the exact anatomical position where putative digit I/metacarpal I should appear; (2) its morphological similarity to the postaxial condensation representing the anlage for digit V; (3) its extension distally beyond the base of the other developing manual metacarpals and digits; and (4) the fact that the larger, centrally located digits would be asymmetrically skewed postaxially if they represented digits I, II, III.

Digit I in the ostrich was confirmed more recently by similar studies by Martin Kundrát of the Slovak Academy of Sciences. In addition, Monique Welten and colleagues in 2005, working at the molecular level, reported the appearance of a prechondrogenic molecular domain (*Sox9*, specific for condensation and expressed briefly in anterior mesenchyme in a position appropriate for digit I), in the chick wing bud.[34] *Sox9* expression in this digit I position is not followed by cartilage matrix synthesis, and thus the wing is briefly pentadactyl, confirming the embryologists' identification of digits II, III, IV in a carinate bird.[35]

Recent findings from embryology thus consistently interpret the digits of the avian hand as II, III, and IV and show that the primitive avian hand was pentadactyl, as quoted below:

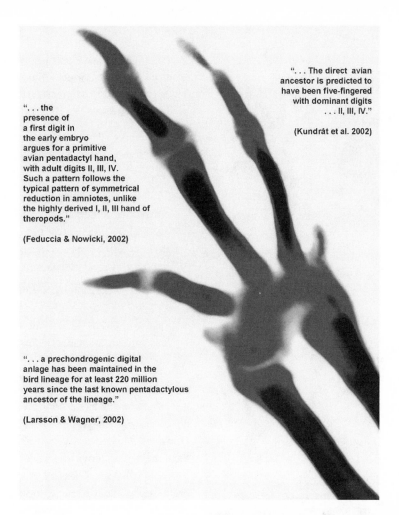

"...the presence of a first digit in the early embryo argues for a primitive avian pentadactyl hand, with adult digits II, III, IV. Such a pattern follows the typical pattern of symmetrical reduction in amniotes, unlike the highly derived I, II, III hand of theropods."

(Feduccia & Nowicki, 2002)

"...The direct avian ancestor is predicted to have been five-fingered with dominant digits ...II, III, IV."

(Kundrát et al. 2002)

"...a prechondrogenic digital anlage has been maintained in the bird lineage for at least 220 million years since the last known pentadactylous ancestor of the lineage."

(Larsson & Wagner, 2002)

Fig. A10. The pentadactyl hand of birds was revealed for the first time in 2002 in a fourteen-day-old ostrich embryo. That same year, two other studies employing different, indirect approaches reached the same conclusion. Since then, the first digit of the ostrich has been confirmed by Martin Kundrát. Right, the author in 2001 after having discovered the first digit of the hand of birds in the ostrich embryo. From A. Feduccia and J. Nowicki, "The Hand of Birds Revealed by Early Ostrich Embryos," *Naturwissenschaften* 89 (2002): 391–93. Photograph by Brian Nalley.

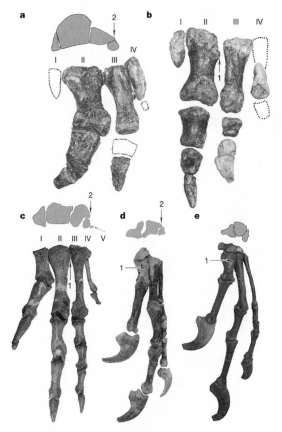

Fig. A11. Theropod manual morphologies as represented by several nonavian theropods. (a, b) Ceratosaur *Limusaurus* (IVPP V 15923 and 15924); (c) basal theropod *Dilophosaurus* (UCMP 37302); (d) tyrannosauroid *Guanlong* (IVPP V14531); and (e) dromaeosaurid *Deinonychus* (YPM 5206). 1, dorsolateral process; 2, metacarpal IV located ventral to metacarpal III. Note that the three metacarpals of *Guanlong* and *Deinonychus* display many similarities to metacarpals II–IV of *Limusaurus* and *Dilophosaurus*. Interestingly, many metacarpal features, such as the contacts among the three metacarpals and the morphology of the lateral metacarpal, were previously considered to be tetanuran synapomorphies, but in fact they can be better interpreted as retained unchanged from the condition in nontetanuran theropods if the three metacarpals of tetanurans are identified as II-III-IV. From X. Xu et al., "A Jurassic Ceratosaur from China Helps Clarify Avian Digital Homologies," *Nature* 459 (2009): 940–44. Copyright 2009. Reprinted by permission from Macmillan Publishers Ltd.

"The presence of a first digit in the early embryo argues for a primitive avian pentadactyl hand, with adult digits II, III, IV . . . unlike the highly derived I, II, III hand of theropods."

"The direct avian ancestor is predicted to have been five-fingered with dominant digits . . . II, III, IV."

"The existence of five discrete metacarpal condensations in the 16-day embryo of *Struthio* argues for unique linear patterning process for each, and these are . . . digits 2,3,4 originating from metacarpal condensations 2,3,4."

"A full pentadactyl prechondrogenic digital anlage has been maintained in the bird lineage for at least 220 million years since the last known pentadactylous ancestor of the lineage."[36]

Interestingly, Hans Larsson and Günter Wagner note that this age is congruent with both a dinosaurian and a nondinosaurian origin of birds.[37] However, if the ancestor of birds were a dinosaur, then the only known theropods of that age are forms like the Late Triassic *Herrerasaurus* (possible pretheropod), and *Coelophysis*, which are already committed to a highly derived pattern of postaxial reduction, the former clearly preserving vestigial digits (metacarpals) IV and V. In order for there to be a pentadactyl ground plan, one would have to invoke a basal archosaur or early dinosauromorph ancestor that had an undifferentiated, pentadactyl hand, a form such as the Triassic dinosauromorph *Marasuchus* (lagosuchid), considered by many to be close to the ancestry of dinosaurs.

In summary, the embryological approach is based on classical approaches to homology and reliance on the principles of developmental position and anatomical connectivity. Developmental biologists typically use conservation of embryonic patterning to establish homology, whereas most paleontologists use the methodology of phylogenetic systematics to define homology a posteriori from cladistic analyses. Until substantial evidence to the contrary becomes available, it is more plausible to consider the avian hand II-III-IV, in contrast to the highly derived grasping, raking theropod hand

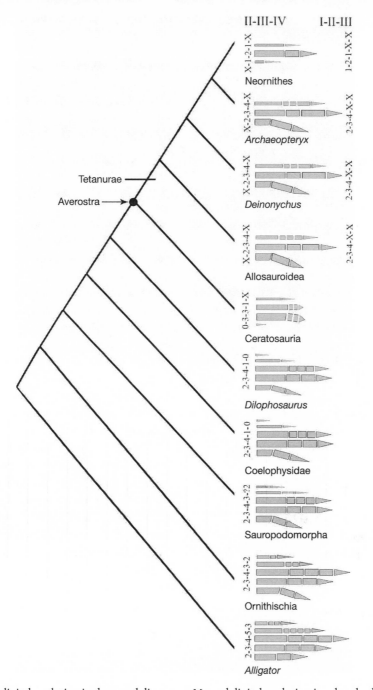

Fig. A12. Manual digital evolution in theropod dinosaurs. Manual digital evolution involves both BDR (bilateral digital reduction) and LDR (lateral digital reduction) in theropod dinosaurs. The shift to BDR in ceratosaurs is coincident with features indicating a reduction in the grasping function of the manus. In ceratosaurs, the manus is small, the manual phalanges are abbreviated, and the claws are nonraptorial. This supports the hypothesis that a grasping function constrained the hand to LDR in nontetanuran theropods. If BDR applies to the more inclusive Averostra, as the II-III-IV hypothesis suggests, early stages of tetanuran evolution must have involved loss of the already highly reduced metacarpal I, reduction in the length of metacarpal II, and the reappearance of additional phalanges on metacarpal IV. Both I-II-III and II-III-IV hypotheses can claim support from morphological data, but the II-III-IV hypothesis is more parsimonious when developmental data from extant birds are considered. From X. Xu et al., "A Jurassic Ceratosaur from China Helps Clarify Avian Digital Homologies," *Nature* 459 (2009): 940–44. Copyright 2009. Reprinted by permission from Macmillan Publishers Ltd.

with digits I-II-III. As Gerhard Müller and Pere Alberch concluded from their study in the 1990s, "The developmental evidence overwhelmingly supports the 2-3-4 theory of the wing skeleton in birds."[38] In 2008, pioneering developmental biologist Richard Hinchliffe concluded: "Evidence for 'frame shift' is speculative and insufficiently convincing to support reinterpretation of the wing digits as 1-2-3. Evidence that wing digits have been correctly identified as 2-3-4 continues to provoke doubts about the 'dinosaur-bird' theory. The presence of well-defined feathers in the bird-like hands of early Cretaceous dromaeosaurs known as microraptor and their descendents, earth-bound didactyl forms such as *Velociraptor* and *Deinonychus* may be due to their being hidden birds, with secondarily flightless descendants, thus suggesting that birds may derive from a lineage separate from that of theropods."[39] Martin Kundrát concluded: "Relationships between chondrogenic foci and ossified patterns of the wing autopodium in the ostrich support the argument that digit identities in the three-fingered hand of *Archaeopteryx* are 2-3-4."[40]

The recently discovered Jurassic ceratosaur *Limusaurus*, if properly interpreted, has provided a provocative new twist to the problem of avian digital homology, but the conclusions concerning the fossil itself and its meaning await further discovery.[41] The alternative, frame shift hypothesis

continues to be investigated and has been widely accepted by paleontologists because it conforms to the cladogram—as well as some recent genetic studies—showing that birds are directly derived from highly advanced theropod dinosaurs.[42] However, even though the evidence for avian digital homology is conflicting, if the cladogram did not exist, it is highly doubtful that such a proposal would have been investigated. A final important consideration is this: it should give one pause to realize that if normally undetectable homeotic shifts are common in vertebrate history, adult phenotypic morphology (vis-à-vis fossils) would be rendered irrelevant to the study of phylogenetic systematics.

Finally, although there has been a recent concerted effort to provide convincing evidence for a I, II, III avian hand as part of the overall theory of a dinosaur-bird nexus via a homeotic frame shift, not only is the evidence premature but, if one considers the avian clade to include the birdlike dromaeosaurs, then the hand is morphologically quite distinctive from that of typical theropods. Regardless, *Science* magazine loves a sensationalistic story, and despite numerous developmental and genetic analyses supporting both sides of the digital homology debate, one of their writers came up with a headline more suited to tabloid journalism à la *National Enquirer*: "Dinos Gave Birds the Finger."[43]

NOTES

Introduction

1. "*T. rex* Discoveries," *USA Today*, 29 December 2009, 8D; R. E. Schmid, "From *T. rex* to Chicken: The Dino-Bird Connection," *Discovery News* (AP, 24 April 2008); H. Gee, "Birds Are Dinosaurs: The Debate Is Over," *Nature Science* Update (1998), www.nature.com/nsu/980702–8.html; A. Milner, "Book Reviewed: *Glorified Dinosaurs: The Origin and Early Evolution of Birds* by Luis Chiappe," *Nature* 447 (2007): 147–48.

2. K. R. Popper, *The Myth of the Framework: In Defense of Science and Rationality*, ed. M. A. Notturno (London: Routledge, 1994), 16.

3. K. Padian, "The Continuing Debate over Avian Origins," review of *The Origin and Evolution of Birds*, by A. Feduccia, *American Scientist* 85 (1997): 178–80; P. Dodson, "Origin of Birds: The Final Solution?" *American Zoologist* 40 (2000): 504–12.

4. J. Avise, *On Evolution* (Baltimore: Johns Hopkins University Press, 2007), ch. 11, "Cladists in Wonderland," 103–16; D. Joravsky, *The Lysenko Affair* (Chicago: University of Chicago Press, 1986).

5. F. C. James and J. A. Pourtless IV, "Cladistics and the Origin of Birds: A Review and Two New Analyses," *Ornithological Monographs* 66 (2009): 1–79; A. G. Kluge, "Philosophical Conjectures and Their Refutation," *Systematic Biology* 50 (2001): 322–30.

6. Dinosaur Mailing List [dinosaur@usc.edu; sponsored by the University of Southern California]; M. Mortimer, "'Tis Time to Get Medieval on Alan Feduccia," Dinosaur Mailing List, 3 January 2003, 22:36:31–0880.

7. B. C. Livezey and R. L. Zusi, "Higher-Order Phylogeny of Modern Birds (Theropoda, Aves: Neornithes) Based on Comparative Anatomy: I.\ Methods and Characters," *Bulletin of the Carnegie Museum of Natural History* 37 (2006): 1–544; G. Mayr, "Avian Higher-Level Phylogeny: Well-Supported Clades and What We Can Learn from a Phylogenetic Analysis of 2954 Morphological Characters," *Journal of Zoological Systematics and Evolutionary Research* 46 (2007): 63–72.

8. D. Haussler, "Genome 10K: A Proposal to Obtain Whole-Genome Sequence for 10,000 Vertebrate Species," *Journal of Heredity* 100 (2009): 659–74; E. C. Hayden, "10,000 Genomes to Come: Vertebrates in Line for Massive Sequencing Project," *Nature* 462 (2009): 21.

9. F. R. Shapiro, ed., *The Yale Book of Quotations* (New Haven: Yale University Press, 2006), 670; D. Normile, "Research Kicks into Gear after a Long, Uphill Struggle," *Science* 291 (2001): 237–38; D. Walton, "Oldest Known Bird Found in China," *Globe and Mail*, 15 June 2006; M. A. Norell and J. A. Clarke, "Fossil Fills and Critical Gap in Avian Evolution," *Nature* 409 (2001): 181–84; A. H. Turner et a.,

"A Basal Drom aeosaurid and Size Evolution Preceding Avian Flight," *Science* 317 (2007): 1378–81; Q. Ji et al., "Two Feathered Dinosaurs from Northeastern China," *Nature* 393 (1998): 753–61.

10. A. Sibley, "Feathered Dinosaurs and the Disneyfication of Palaeontology," Creation Science Movement, 16 September 2005, https://www .csm.org.uk/news.php?viewmessage=34&PHPS ESSID=ff69d2422904538d52b4938269938e5d.

11. D. R. Prothero, *Evolution: What the Fossils Say and Why It Matters* (New York: Columbia University Press, 2007).

12. M. Crichton, "Aliens Cause Global Warming," speech delivered at California Institute of Technology, Pasadena, CA, 17 January 2003.

Chapter 1: Romancing the Dinosaurs

Epigraph: K. Thomson, "Dinosaurs, the Media and Andy Warhol," *American Scientist* 90, no. 3 (2002): 222.

1. J. Gauthier, "Saurischian Monophyly and the Origin of Birds," in *The Origin of Birds and the Evolution of Flight*, ed. K. Padian, *Memoirs of the California Academy of Sciences* 8 (1986): 1–55.

2. A. Feduccia, *The Origin and Evolution of Birds* (New Haven: Yale University Press, 1996).

3. E. Mayr, review of *The Origin and Evolution of Birds*, by A. Feduccia, *American Zoologist* 37 (1997): 210–11; W. J. Bock, review of *The Origin and Evolution of Birds*, by A. Feduccia, *Auk* 114 (1997): 531–34; J. Ruben, review of *The Origin and Evolution of Birds*, by A. Feduccia, *Bioscience* 47 (1997): 392–94.

4. D. Zalewski, "Bones of Contention," *Lingua Franca* 6 (1996): 22–24.

5. M. A. Norell and L. M. Chiappe, "Flight from Reason," review of *The Origin and Evolution of Birds*, by A. Feduccia, *Nature* 384 (1996): 230.

6. P. Dodson, "Origin of Birds: The Final Solution?" *American Zoologist* 40 (2000): 504–12.

7. H. Gee, "Birds Are Dinosaurs: The Debate Is Over," *Nature Science* Update (1998), www .nature.com/nsu/980702–8.html.

8. R. M. Prum, "Why Ornithologists Should Care about the Theropod Origin of Birds," *Auk* 119 (2002): 1–17.

9. Christopher A. Brochu, "Progress and Future Directions in Archosaur Phylogenetics," *Journal of Paleontology* 75 (2001): 1185–201.

10. R. M. Prum, "Are Current Critiques of the Theropod Origin of Birds Science? Rebuttal to Feduccia (2002)," *Auk* 120 (2003): 550–61.

11. F. C. James and J. A. Pourtless IV, "Cladistics and the Origin of Birds: A Review and Two New Analyses," *Ornithological Monographs* 66 (2009): 1–78.

12. P. Gingerich, "Cladistic Futures," *Nature* 336 (1988): 628.

13. R. A. Jenner, "The Scientific Status of Metazoan Cladistics: Why Current Research Practice Must Change," *Zoologica Scripta* 33 (2004): 293–310.

14. R. Dalton, "Chasing the Dragons," *Nature* 406 (2000): 930–32.

15. J. Wang, "Scientists Flock to Explore China's 'Site of the Century,'" *Science* 279 (1998): 1626–27.

16. Dalton, "Chasing the Dragons."

17. R. Stone, "Altering the Past: China's Faked Fossils Problem," *Science* 330 (2010): 174–41.

18. C. P. Sloan, "Feathers for *T. rex?*" *National Geographic*, November 1999, 98–107.

19. S. L. Olson, "Open Letter to Peter Raven, Chairman, Committee on Research and Exploration, National Geographic Society," 1999.

20. S. A. Czerkas and X. Xu, "A New Toothed Bird from China," *Dinosaur Museum Journal* 1 (2002): 43–61; Z. Zhou, J. A. Clarke, and F. Zhang, "*Archaeoraptor*'s Better Half," *Nature* 420 (2002): 285.

21. Olson, "Open Letter."

22. C. Holden, "Florida Meeting Shows Perils, Promise of Dealing for Dinos," *Science* 288 (2000): 238–39.

23. D. A. Burnham, "Paleoenvironment, Paleoecology, and Evolution of Maniraptoran 'Dinosaurs'" (PhD diss., University of Kansas, Lawrence, 2007).

24. C. Holden, "Dinos and Turkeys: Connected by DNA?" *Science* 288 (2000): 238.

25. J. Wells, *Icons of Evolution* (Washington, DC: Regnery, 2000), 131, 132–133.

26. Ibid.

27. S. L. Olson, review of *New Perspectives on the Origin and Early Evolution of Birds: Proceedings of the International Symposium in Honor of John H. Ostrom*, by J. Gauthier and L. F. Gall, *Auk* 119 (2002): 1202–4.

28. P. E. Fisher et al., "Cardiovascular Evidence for an Intermediate or Higher Metabolic Rate in an

Ornithischian Dinosaur," *Science* 288 (2000): 503–5.

29. V. Morell, "Revealing a Dinosaur's Heart of Stone," *Science* 288 (2000): 416–17.

30. T. Radford, "Dinosaur Heart a Shock to Evolution," *Guardian*, 21 April 2000.

31. T. Friend, "Debate over Dino 'Heart' Gets Blood Pumping," *USA Today* (Science), 3 October 2000.

32. E. Stokstad, "Doubts Raised about Dinosaur Heart," *Science* 291 (2001): 811.

33. T. Rowe, E. F. McBride, and P. C. Sereno, "Dinosaur with a Heart of Stone," *Science* 291 (2001): 783.

34. Friend, "Debate over Dino 'Heart.'"

35. Center for the Exploration of the Dinosaurian World, North Carolina State University and the North Carolina Museum of Natural History, www.dinoheart.org.

36. T. P. Cleland et al., "Histological, Chemical, and Morphological Reexamination of the 'Heart' of a Small Cretaceous *Thescelosaurus*," *Naturwissenschaften* 98 (2011): 203–11.

37. R. E. Barrick and W. J. Showers, "Thermophysiology of *Tyrannosaurus rex*: Evidence from Oxygen Isotopes," *Science* 265 (1994): 222–24.

38. Y. Kolodny et al., "Dinosaur Bones: Fossils or Pseudomorphs? The Pitfalls of Physiology Reconstruction from Apatitic Fossils," *Palaeogeography, Palaeoclimatology, Palaeoecology* 126 (1996): 161–71.

39. J. Ruben, "The Evolution of Endothermy in Mammals and Birds," *Annual Review of Physiology* 57 (1995): 69–95.

40. A. Bernard et al., "Regulation of Body Temperature by Some Mesozoic Marine Reptiles, *Science* 328 (2010): 1379–82.

41. J. O. Farlow, "Thermal Energetics and Thermal Biology," in Weishampel et al., *Dinosauria*, 43–55.

42. M. H. Schweitzer et al., "Soft-Tissue Vessels and Cellular Preservation in *Tyrannosaurus rex*," *Science* 307 (2005): 1952–55; M. H. Schweitzer, J. L. Wittmeyer, and J. R. Horner, "Soft Tissue and Cellular Preservation in Vertebrate Skeletal Elements from the Cretaceous to the Present," *Proceedings of the Royal Society B* 274 (2007): 183–97; M. H. Schweitzer et al., "Analyses of Soft Tissue from *Tyrannosaurus rex* Suggest the

Presence of Protein," *Science* 316 (2007): 277–80; J. M. Asara et al., "Protein Sequences from *Mastodon* and *Tyrannosaurus rex* Revealed by Mass Spectrometry," *Science* 316 (2007): 280–85; C. L. Organ et al., "Molecular Phylogenetics of Mastodon and *Tyrannosaurus rex*," *Science* 320 (2008): 499.

43. Asara et al., "Protein Sequences from *Mastodon* and *Tyrannosaurus rex*."

44. E. Ratliff, "Origin of Species: How a *T. rex* Femur Sparked a Scientific Smackdown," *Wired*, 22 June 2009.

45. B. Yeoman, "Schweitzer's Dangerous Discovery," *Discover*, April 2006, 37–41, 77.

46. Ibid.

47. T. G. Kaye, G. Gaugler, and Z. Sawlowicz, "Dinosaurian Soft Tissues Interpreted as Bacterial Biofilms," *PLoS ONE* 3, no. 7 (2008): e2808, DOI: 10.1371/journal.pone.0002808.

48. J. Hecht, "*T. rex* 'Tissue' May Just Be Bacterial Scum," *New Scientist* News service (NewScientist.com), 2008.

49. Kaye, Gaugler, and Sawlowicz, "Dinosaurian Soft Tissues."

50. E. Willerslev et al., "Ancient Biomolecules from Deep Ice Cores Reveal a Forested Southern Greenland," *Science* 317 (2007): 111–14.

51. R. Weiss, "*T. rex* Closer to Gizzards than Lizards," *Washington Post*, 25 April 2008, A02.

52. P. A. Pevzner, S. Kim, and J. Ng, "Comment on 'Protein Sequences from Mastodon and *Tyrannosaurus rex* Revealed by Mass Spectrometry,'" *Science* 321 (2008): 1040b; M. Buckley et al., "Comment on 'Protein Sequences from Mastodon and *Tyrannosaurus rex* Revealed by Mass Spectrometry,'" *Science* 319 (2008): 33c.

53. M. Fitzgibbon and M. McIntosh, "Protein Sequences from Mastodon and *Tyrannosaurus rex* Revealed by Mass Spectrometry," http://pubs.acs.org, 2009.

54. Pevzner, Kim, and Ng, "Comment on 'Protein Sequences from *Mastodon* and *Tyrannosaurus rex*.'"

55. S. Salzburg, GenoEvoPseudo, http://genome.fieldofscience.com, 22 August 2008.

56. M. H. Schweitzer et al., "Biomolecular Characterization and Protein Sequences of the Campanian Hadrosaur *B. canadensis*," *Science* 324 (2009): 626–31.

57. M. Bern, B. S. Phinney, and D. Goldberg, "Reanalysis of *Tyrannosaurus rex* Mass Spectra," *Journal of Proteome Research* 8 (2009): 1–5.

58. D. O. Carter, D. Yellowlees, and M. Tibbett, "Cadaver Decomposition in Terrestrial Ecosystems," *Naturwissenschaften* 94 (2007): 12–24.

59. Yeoman, "Schweitzer's Dangerous Discovery."

60. J. Hitt, "Paleontology: New Discoveries Hint There's a Lot More in Fossil Bones than We Thought," *Discover*, October 2005.

61. S. R. Woodward, N. J. Weyand, and M. Bunnell, "DNA Sequence from Cretaceous Period Bone Fragment," *Science* 266 (1994): 1229–32.

62. J. N. Wilford, "A Scientist Says He Has Isolated Dinosaur DNA," *New York Times*, 18 November 1984.

63. M. A. Schweitzer, "Blood from Stone," *Scientific American*, December 2010, 62–69.

64. J. Horner and J. Gorman, *How to Build a Dinosaur: Extinction Doesn't Have to Be Forever* (Boston: Dutton, 2009).

65. E. J. Koller and C. Fisher, "Tooth Induction in Chick Epithelium: Expression of Quiescent Genes for Enamel Synthesis," *Science* 207 (1980): 993–95; T. A. Mitsiadis et al., "Development of Teeth in Chick Embryos after Mouse Neural Crest Transplantations," *Proceedings of the National Academy of Sciences* 100 (2003): 6541–45; M. P. Harris et al., "The Development of Archosaurian First-Generation Teeth in a Chicken Mutant," *Current Biology* 16 (2006): 371–77.

66. M. Henderson and A. Sage, "Birds with Teeth Turn the Clock Back 70m Years," *Times* (London), 4 June 2003.

67. Yeoman, "Schweitzer's Dangerous Discovery."

68. K. Thomson, "Dinosaurs, the Media and Andy Warhol," *American Scientist* 90 (2002): 222.

69. Gee, "Birds Are Dinosaurs"; L. M. Witmer, "Flying Feathers," review of *The Origin and Evolution of Birds*, by A. Feduccia, *Science* 276 (1997): 1209–10; L. M. Chiappe, *Glorified Dinosaurs: The Origin and Early Evolution of Birds* (New York: John Wiley and Sons, 2007); L. M. Chiappe and G. J. Dyke, "The Mesozoic Radiation of Birds," *Annual Review of Ecology and Systematics* 33 (2002): 91–124; M. A. Norell and X. Xu, "Feathered Dinosaurs," *Annual Review of Earth and Planetary Sciences* 33 (2005): 277–99.

70. A. Feduccia, *The Age of Birds* (Cambridge, MA: Harvard University Press, 1980).

71. Feduccia, *Origin and Evolution of Birds*.

72. S. L. Olson, review of *The Beginnings of Birds: Proceedings of the International Archaeopteryx Conference*, ed. M. K. Hecht et al., *American Scientist* 75, no. 1 (1987): 74–75.

73. M. K. Hecht et al., eds., *The Beginnings of Birds: Proceedings of the International Archaeopteryx Conference* (Eichstätt: Freunde des Jura-Museums, 1985).

74. A. Feduccia, T. Lingham-Soliar, and J. R. Hinchliffe, "Do Feathered Dinosaurs Exist? Testing the Hypothesis on Neontological and Paleontological Evidence," *Journal of Morphology* 266 (2005): 125–66.

75. M. E. Howgate, "Back to the Trees for *Archaeopteryx* in Bavaria," *Nature* 313 (1985): 435–36.

76. A. Feduccia, "A Colorful Mesozoic Menagerie," review of *Feathered Dinosaurs: The Origin of Birds*, by J. Long, *Trends in Ecology and Evolution* 24 (2009): 416–17.

77. Feduccia, Lingham-Soliar, and Hinchliffe, "Do Feathered Dinosaurs Exist?"; T. Lingham-Soliar, "Evolution of Birds: Ichthyosaur Integumental Fibers Conform to Dromaeosaur ProtoFeathers," *Naturwissenschaften* 90 (2003): 428–32; T. Lingham-Soliar, "The Dinosaurian Origin of Feathers: Perspectives from Dolphin (Cetacea) Collagen Fibers," *Naturwissenschaften* 90 (2003): 563–67; T. Lingham-Soliar, A. Feduccia, and X. Wang, "A New Chinese Specimen Indicates That 'ProtoFeathers' in the Early Cretaceous Theropod Dinosaur *Sinosauropteryx* Are Degraded Collagen Fibres," *Proceedings of the Royal Society B* 274 (2007): 1823–29 (published same year as *Glorified Dinosaurs*, but included here for the reader).

78. W. Bock, "The Arboreal Theory for the Origin of Birds," in Hecht et al., *Beginnings of Birds*, 199–207; W. Bock, "The Arboreal Origin of Avian Flight," *Memoirs of the California Academy of Sciences* 8 (1986): 57–72; W. Bock and P. Bühler, "Origin of Birds: Feathers, Flight and Homoiothermy," *Archaeopteryx* 13 (1995): 5–13.

79. U. M. Norberg, "Flying, Gliding, and Soaring," in *Functional Vertebrate Morphology*, ed. M. Hildebrand et al. (Cambridge, MA: Harvard University Press, 1985), 129–58, 391–93; U. M.

Norberg, *Vertebrate Flight* (Berlin: Springer, 1990); U. M. Norberg, "How a Long Tail and Changes in Mass and Wing-Shape Affect the Cost for Flight in Animals," *Functional Ecology* 9 (1995): 48–54.

80. Q. Ji et al., "Two Feathered Dinosaurs from Northeastern China," *Nature* 393 (1998): 753–61; J. Lü et al., "Oviraptorosaurs Compared to Birds," in *Proceedings of the Fifth Symposium of the Society of Avian Paleontology and Evolution*, ed. Z. Zhou and F. Zhang (Beijing: Science Press, 2002), 175–89; T. Maryańska, H. Osmólska, and M. Wolsan, "Avian Status for Oviraptorosauria," *Acta Paleontologica Polonica* 47 (2002): 97–116; S. J . Czerkas, ed., *Feathered Dinosaurs and the Origin of Flight* (Blanding, UT: Dinosaur Museum, 2002).

81. Lü et al., "Oviraptorosaurs Compared to Birds."

82. Czerkas, *Feathered Dinosaurs and the Origin of Flight*; S. J. Czerkas and C. Yuan, "An Arboreal Maniraptoran from Northeast China," *Dinosaur Museum Journal* 1 (2002): 63–95; S. J. Czerkas et al., "Flying Dromaeosaurs," *Dinosaur Museum Journal* 1 (2002): 97–126.

83. G. S. Paul, *Dinosaurs of the Air: The Evolution and Loss of Flight in Dinosaurs and Birds* (Baltimore: Johns Hopkins University Press, 2002); G. Olshevsky, "A Revision of the Parainfraclass Archosauria Cope, 1869, Excluding the Advanced Crocodylia," *Mesozoic Meanderings* 2 (1994): 1–196; Feduccia, Lingham-Soliar, and Hinchliffe, "Do Feathered Dinosaurs Exist?"

84. R. H. Sawyer and L. W. Knapp, "Avian Skin Development and the Evolutionary Origin of Feathers," *Journal of Experimental Zoology* 298B (2003): 57–72; R. H. Sawyer et al., "Origin of Feathers: Feather β-Keratins Are Expressed in Discrete Cell Populations of Embryonic Scutate Scales," *Journal of Experimental Zoology* 295B (2003): 12–24; R. H. Sawyer et al., "Evolutionary Origin of the Feather Epidermis," *Developmental Dynamics* 232 (2004): 256–67; L. L. Alibardi, W. Knapp, and R. H. Sawyer, "Beta-Keratin Localization in Developing Alligator Scales and Feathers in Relation to the Development and Evolution of Feathers," *Journal of Submicroscopy, Cytology and Pathology* 38, nos. 2–3 (2006): 175–92 (published too late for *Glorified Dinosaurs* but included here for the reader).

85. Chiappe, *Glorified Dinosaurs*, 139; J. H. Ostrom, "*Archaeopteryx* and the Origin of Birds," *Biological Journal of the Linnaean Society* 8 (1976): 91–182.

86. D. W. Yalden, "Forelimb Function in *Archaeopteryx*," in Hecht et al., *Beginnings of Birds*, 91–97; D. W. Yalden, "Climbing *Archaeopteryx*," *Archaeopteryx* 15 (1997): 107–8.

87. A. Feduccia, "Evidence from Claw Geometry Indicating Arboreal Habits for *Archaeopteryx*," *Science* 259 (1993): 790–93.

88. J. A. Hopson, "Ecomorphology of Avian and Nonavian Theropod Phalangeal Proportions: Implications for the Arboreal versus Terrestrial Origin of Bird Flight," in *New Perspectives on the Origin and Early Evolution of Birds*, ed. J. A. Gauthier and L. F. Gall (New Haven: Peabody Museum of Natural History, Yale University, 2001), 211–35.

89. Z. Zhou and J. O. Farlow, "Flight Capability and Habits of *Confuciusornis*," in Gauthier and Gall, *New Perspectives on the Origin and Early Evolution of Birds*, 237–54.

90. K. Padian, "Dinosauria," in *Encyclopedia of Dinosaurs*, ed. P. J. Currie and K. Padian (San Diego, CA: Academic Press, 1997), 175–79.

91. S. Chatterjee, *The Rise of Birds: 225 Million Years of Evolution* (Baltimore: Johns Hopkins University Press, 1997).

92. M. J. Benton, *Vertebrate Paleontology*, 3rd ed. (Oxford: Blackwell, 2005).

93. L. M. Chiappe, "Climbing *Archaeopteryx*? A Response to Yalden," *Archaeopteryx* 15 (1997): 109–12.

94. K. Padian and L. M. Chiappe, "The Origin of Birds and Their Flight," *Scientific American*, February 1998, 38–47.

95. Chiappe, *Glorified Dinosaurs*.

96. M. J. Benton, "Origin and Relationships of Dinosauria," and M. C. Langer, "Basal Saurischia," in *The Dinosauria*, ed. D. B. Weishampel, P. Dodson, and H. Osmólska, 2nd ed. (Berkeley: University of California Press, 2004), 7–19, 25–46.

97. A. Feduccia, *The Origin and Evolution of Birds*, 2nd ed. (New Haven: Yale University Press, 1999).

98. James and Pourtless, "Cladistics and the Origin of Birds"; L. D. Martin, "A Basal Archosau-

rian Origin of Birds," *Acta Zoologica Sinica* 50 (2004): 978–90.

99. G. Heilmann, *The Origin of Birds* (London: Witherby, 1926).

100. J. N. Choiniere et al., "A Basal Alvarezsauroid Theropod from the Early Late Jurassic of Xinjiang, China," *Science* 327 (2010): 571–74.

101. James and Pourtless, "Cladistics and the Origin of Birds."

102. Gauthier, "Saurischian Monophyly."

103. J. H. Ostrom, "The Origin of Birds," *Annual Review of Earth and Planetary Sciences* 3 (1975): 55–77.

104. J. A. Gauthier and K. Padian, "Phylogenetic, Functional, and Aerodynamic Analyses of the Origin of Birds and Their Flight," in Hecht et al., *Beginnings of Birds*, 185–97.

105. Feduccia, Lingham-Soliar, and Hinchliffe, "Do Feathered Dinosaurs Exist?"; Paul, *Dinosaurs of the Air*.

106. Benton, *Vertebrate Paleontology*; Benton, "Origins and Relationships of Dinosauria"; M. J. Benton, "*Scleromochlus taylori* and the Origin of Dinosaurs and Pterosaurs," *Philosophical Transactions of the Royal Society of London*, ser. B, 354 (1999): 1423–46.

107. S. Chatterjee, "*Postosuchus*, a New Thecodontian Reptile from the Triassic of Texas and the Origin of Tyrannosaurs," *Philosophical Transactions of the Royal Society of London*, ser. B, 309 (1985): 395–460.

108. Benton, "Origins and Relationships of Dinosauria."

109. James and Pourtless, "Cladistics and the Origin of Birds."

110. Benton, "*Scleromochlus taylori*."

111. E. N. Kurochkin, "Basal Diversification of Aves," in *Evolution of the Biosphere and Biodiversity: To 70 Years of A. Yu. Roʒanov* [in Russian], ed. S. V. Rozhnov (Moscow: Toravirtzestvo Scientific Editions KMK, 2006), 219–32; E. N. Kurochkin, "Parallel Evolution of Theropod Dinosaurs and Birds" [in Russian], *Zoologicheskii Zhurnal* 85 (2006): 283–97; English version, *Entomological Review* 86 (2006): S45–58.

112. Paul, *Dinosaurs of the Air*.

113. Feduccia, Lingham-Soliar, and Hinchliffe, "Do Feathered Dinosaurs Exist?"; Czerkas, *Feathered Dinosaurs and the Origin of Flight*; Feduccia, *Origin and Evolution of Birds*.

114. F. D. Peat, *From Certainty to Uncertainty: The Story of Science and Ideas in the Twentieth Century* (Washington, DC: National Academies Press, 2002).

115. Chatterjee, "*Postosuchus*."

116. L. Huxley, *Life and Letters of Thomas Henry Huxley*, 2 vols. (London: Macmillan, 1990); www.dinoheart.org.

Chapter 2: What Did Evolution's High Priest Say?

Epigraph: C. Darwin to T. H. Huxley, 8 August 1860, in *Life and Letters of Charles Darwin*, vol. 2, ed. Francis Darwin (New York: D. Appleton, 1911), 123–24.

1. T. H. Huxley to E. Haeckel, 1874, in L. Huxley, *Life and Letters of Thomas Henry Huxley*, vol. 1 (London: Macmillan, 1900), 425.

2. S. Conway Morris, *The Crucible of Creation: The Burgess Shale of the Rise of Animals* (Oxford: Oxford University Press, 1998).

3. L. Z. Holland and N. D. Holland, "Evolution of Neural Crest and Placodes: Amphioxus as a Model for the Ancestral Vertebrate?" *Journal of Anatomy* 199 (2001): 85–98.

4. H. Gee, "The Amphioxus Unleashed," *Nature* 453 (2008): 999–1000; N. H. Putnam et al., "The Amphioxus Genome and the Evolution of the Chordate Karyotype," *Nature* 453 (2008): 1064–72.

5. Gee, "Amphioxus Unleashed."

6. T. V. Venkatesh et al., "Sequence and Developmental Expression of Amphioxus *AmphiNk2*–1: Insights into the Evolutionary Origin of the Vertebrate Thyroid Gland and Forebrain," *Development, Genes and Evolution* 209 (1999): 254–59.

7. Huxley to Haeckel, 1874.

8. A. Wagner, "Neue Beiträge zur Kenntnis der urweltlichen Fauna des lithographischen Schiefers; V. *Compsognathus longipes* Wagner," *Abhandlungen der Bayerischen Akademie der Wissenschaften* 9 (1861): 30–38; K. W. Barthel, N. H. M. Swinburne, and S. C. Morris, *Solnhofen: A Study in Meso ʒ oic Paleontology* (Cambridge: Cambridge University Press, 1990).

9. A. Feduccia, *The Origin and Evolution of Birds* (New Haven: Yale University Press, 1996).

10. C. Darwin, *On the Origin of Species by Means of Natural Selection; or, The Preservation of*

Favoured Races in the Struggle for Life (London: John Murray, 1859).

11. Feduccia, *Origin and Evolution of Birds*; J. H. Ostrom, "Introduction to *Archaeopteryx*," in *The Beginnings of Birds: Proceedings of the International Archaeopteryx Conference*, ed. M. K. Hecht et al. (Eichstätt: Freunde des Jura-Museums, 1985), 9–20.

12. A. Desmond, *Archetypes and Ancestors: Palaeontology in Victorian London, 1850–1875* (Chicago: University of Chicago Press, 1982), 127.

13. I. Sidgwick, "A Grandmother's Tales," *Macmillan's Magazine*, October 1898, 433–34.

14. R. Owen, "Darwin on the Origin of Species," *Edinburgh Review* 3 (1860): 487–532.

15. Desmond, *Archetypes and Ancestors*.

16. R. Owen, *On the Anatomy of Vertebrates*, vol. 2, *Birds and Mammals* (London: Longmans, Green, 1866), 12; B. K. Hall and R. Amundson, eds., *On the Nature of Limbs: A Discourse* [1849], by Richard Owen, with essays by R. Amundson et al. (Chicago: University of Chicago Press, 2007).

17. Hall and Amundson, *On the Nature of Limbs*.

18. R. Owen, "On the *Archaeopteryx* of von Meyer, with a Description of the Fossil Remains of a Long-Tailed Species from the Lithographic Stone of Solnhofen," *Philosophical Transactions of the Royal Society of London* 153 (1863): 33–47.

19. T. H. Huxley, "On the Animals Which Are Most Nearly Intermediate between Birds and Reptiles," *Annals and Magazine of Natural History, London* 2 (1868): 66–75.

20. F. Darwin, *The Life and Letters of Charles Darwin*, 2 vols. (London: John Murray, 1887).

21. C. R. Darwin, *On the Origin of Species by Means of Natural Selection; or, The Preservation of Favoured Races in the Struggle for Life*, 6th ed. (London: John Murray, 1872).

22. A. Desmond, *Huxley: From Devil's Disciple to Evolution's High Priest* (Reading, MA: Addison-Wesley, 1997), 355.

23. Huxley, "On the Animals Which Are Most Nearly Intermediate"; P. Chambers, *Bones of Contention: The Archaeopteryx Scandals* (London: John Murray, 2002), 23; T. H. Huxley, "Further Evidence on the Affinity between the Dinosaurian Reptiles and Birds," *Proceedings of the Geological Society of London* 26 (1870): 12–31, 28. L. Dollo, "Troisième note sur les Dinosauriens

de Bernissart," *Bulletin du Musée royal d'Histoire Naturelle de Belgique* 2 (1883): 85–126, 87.

24. T. H. Huxley, "On the Animals Which Are Most Nearly Intermediate," 75; Desmond, *Huxley*; M. Foster and E. R. Lankester, eds., *The Scientific Memoirs of Thomas Henry Huxley*, 5 vols. (London: Macmillan, 1898–1902).

25. M. A. Di Gregorio, "The Dinosaur Connection: A Reinterpretation of T. H. Huxley's Evolutionary View," *Journal of the History of Biology* 15 (1982): 397–418.

26. L. M. Witmer, "Perspectives on Avian Origins," in *Origins of the Higher Groups of Tetrapods*, ed. H.-P. Schultz and L. Trueb (Ithaca, NY: Cornell University Press, 1991), 427–66, 431; Huxley, "On the Animals Which Are Most Nearly Intermediate."

27. T. H. Huxley, "On the Classification of Birds and on the Taxonomic Value of the Modifications of Certain of the Cranial Bones Observable in That Class," *Proceedings of the Zoological Society of London* (1867): 415–72, 419.

28. Owen, *On the Anatomy of Vertebrates*; Desmond, *Archetypes and Ancestors*, 145.

29. G. de Beer, "The Evolution of Ratites," *Bulletin of the British Museum (Natural History)* 4 (1956): 59–70.

30. Desmond, *Huxley* (Huxley's "Rattlesnake" Diary).

31. Hall and Amundson, *On the Nature of Limbs*; S. J. Gould, *The Panda's Thumb* (New York: W. W. Norton, 1980), ch. 16, "Flaws in a Victorian Veil."

32. P. Chambers, *Bones of Contention: The Archaeopteryx Scandals* (London: John Murray, 2002); Desmond, *Huxley*, 358.

33. C. Darwin, *The Descent of Man and Selection in Relation to Sex* (London: John Murray, 1871).

34. E. Hitchcock, *Ornithichnology: Description of the Foot Marks of Birds (Ornithichnites) on New Red Sandstone of the Connecticut Valley, Especially Its Fossil Footmarks* (Boston: William White, 1836); D. R. Dean, "Hitchcock's Dinosaur Tracks," *American Quarterly* 21 (1969): 639–44.

35. R. N. Melchor et al., "Bird-Like Fossil Footprints from the Late Triassic," *Nature* 417 (2002): 936–38; J. F. Genise et al., "Application of Neoichnological Studies to Behavioral and Taphonomic Interpretation of Fossil Bird-Like Tracks from Lacustrine Settings: The Late Triassic–Early

Jurassic?" *Palaeogeography, Palaeoclimatology, Palaeoecology* 272 (2009): 143–61.

36. L. D. Martin, "Mesozoic Birds and the Origin of Birds," in *Origins of the Higher Groups of Tetrapods: Controversy and Consensus*, ed. H. P. Schultz and L. Trueb (Ithaca, NY: Cornell University Press, 1991), 485–540; L. D. Martin, "The Enantiornithines: Terrestrial Birds of the Cretaceous in Avian Evolution," *Courier Forschungsinstitut Senckenberg* 181 (1995): 23–36.

37. Desmond, *Archaetypes and Ancestors*, 128.

38. R. Owen, "On the Bone of an Unknown Struthious Bird from New Zealand," *Proceedings of the Zoological Society of London* (1839): 169–70.

39. R. Owen, *Memoirs on the Extinct Wingless Birds of New Zealand* (London: John van Voors, 1879).

40. Desmond, *Huxley*, 358.

41. Huxley, "On the Animals Which Are Most Nearly Intermediate."

42. Ibid.

43. Ibid.; Chambers, *Bones of Contention*.

44. Desmond, *Huxley*; Chambers, *Bones of Contention*.

45. Huxley, "On the Animals Which Are Most Nearly Intermediate"; Chambers, *Bones of Contention*, 233.

46. H. G. Seeley, "On the Classification of the Fossil Animals Commonly Called Dinosauria," *Proceedings of the Royal Society of London* 43 (1887): 221–28.

47. H. Seeley, *Dragons of the Air: An Account of Extinct Flying Reptiles* (London: Methuen, 1901).

48. T. H. Huxley, "Further Evidence of the Affinity Between the Dinosaurian Reptiles and Birds," *Proceedings of the Geological Society of London* 26 (1870): 12–31, 31; Witmer, "Perspectives on Avian Origins," 435–36; T. H. Huxley, "On the Classification of the Dinosauria, with Observations on the Dinosauria of the Trias.," *Proceedings of the Geological Society of London* 26 (1870): 32–38.

49. B. Mudge, "Are Birds Derived from Dinosaurs?" *Kansas City Review of Science* 3 (1879): 224–26; L. Dollo, "Première note sur les Dinosauriens de Bernissart," *Bulletin du Musée royal d'Histoire Naturelle de Belgique* 1 (1882): 161–80; L. Dollo, "Note sur le présence chez les oiseaux du 'troisième trochanter' des Dinosauriens et sur la fonction de celui-ci," *Bulletin de Musée royal d'Histoire Naturelle de Belgique* 2 (1883): 13–20;

W. Dames, "Über *Archaeopteryx*," *Paläontologische Abhandlungen* 2, no. 3 (1884): 119–98; W. K. Parker, "Remarks on the Skeleton of *Archaeopteryx* and on the Relations of the Bird to the Reptile," *Geological Magazine* 1 (1864): 55–57; M. Fürbringer, *Untersuchungen zur Morphologie und Systematik der Vogel, zugleich ein Beitrag zur Anatomie der Stütz- und Bewegungsorgane*, 2 vols. (Amsterdam: T. J. Van Holkema, 1888); H. F. Osborn, "Reconsideration of the Evidence for a Common Dinosaur-Avian Stem in the Permian," *American Naturalist* 34 (1900): 777–99; R. Broom, "On the South African Pseudosuchian *Euparkeria* and Allied Genera," *Proceedings of the Zoological Society of London* (1913): 619–33; G. Heilmann, *The Origin of Birds* (London: Witherby, 1926); G. G. Simpson, "Fossil Penguins," *Bulletin of the American Museum of Natural History* 87 (1946): 1–99; de Beer, "Evolution of Rarities"; A. S. Romer, *Vertebrate Paleontology*, 3rd ed. (Chicago: University of Chicago Press, 1966).

50. Simpson, "Fossil Penguins," 94.

51. T. H. Huxley to Charles Darwin, 1869, in Huxley, *Life and Letters of Thomas Henry Huxley*, 1:300.

52. Darwin, *Origin of Species*.

53. R. W. Storer, "Evolution of the Diving Birds," *Proceedings of the Twelfth International Ornithological Congress* (1960): 694–707.

54. J. Cracraft, "Avian Evolution, Gondwana Biogeography and the Cretaceous-Tertiary Mass Extinction Event," *Proceedings of the Royal Society B* 268 (2001): 459–69; G. Mayr, "A New Basal Galliform Bird from the Middle Eocene of Messel (Hessen, Germany)," *Senckenbergiana lethaea* 80 (2000): 45–57; G. J. Dyke and B. E. Gulas, "The Fossil Galliform Bird *Paraortygoides* from the Eocene of the United Kingdom," *American Museum Novitates* 3360 (2002): 1–14; A. Feduccia, "'Big Bang' for Tertiary Birds?" *Trends in Ecology and Evolution* 18, no. 4 (2003): 172–76.

55. P. Shipman, *Taking Wing: Archaeopteryx and the Evolution of Bird Flight* (New York: Simon and Schuster, 1988); D. R. Prothero, *Evolution: What the Fossils Say and Why It Matters* (New York: Columbia University Press, 2007).

56. T. H. Huxley, "Further Evidence of the Affinity between Dinosaurian Reptiles and Birds," *Quarterly Journal of the Geological Society of London* 26 (1870): 12–31; Prothero, *Evolution*.

57. Huxley, "On the Animals Which Are Most Nearly Intermediate," 75; T. H. Huxley, "On the Evidence as to the Origin of Existing Vertebrate Animals," in M. Foster and E. R. Lankester, *The Scientific Memoires of Thomas Henry Huxley, 1898–1902*, 4 vols. (London: Macmillan, 1876), 4:164–87; Desmond, *Archetypes and Ancestors*, 129.

58. Huxley to Haeckel, 1874, in Huxley, *Life and Letters of Thomas Henry Huxley*, 1:303; Desmond, *Huxley*, 360.

59. T. H. Huxley, "Remarks upon *Archaeopteryx lithographica*," *Proceedings of the Royal Society of London* 16 (1868): 243–48.

60. Prothero, *Evolution*.

61. Darwin, *Descent of Man*.

62. Desmond, *Huxley*.

63. Desmond, *Huxley*; Chambers, *Bones of Contention*.

64. M. H. Schweitzer et al., "Biomolecular Characterization and Protein Sequences of the Campanian Hadrosaur *B. canadensis*," *Science* 324 (2009): 626–61.

65. De Beer, "Evolution of Ratites"; Simpson, "Fossil Penguins."

66. Huxley, "On the Animals Which Are Most Nearly Intermediate."

67. L. M. Witmer, "Perspectives on Avian Origins," in *Origins of the Higher Groups of Tetrapods*, ed. H.-P. Schultze and L. Trueb (Ithaca, NY: Cornell University Press, 1991), 427–66.

68. Mudge, "Are Birds Derived from Dinosaurs?"

69. M. J. Benton, *Vertebrate Palaeontology*, 3rd ed. (Oxford: Blackwell Science, 2005).

70. G. S. Paul, *Dinosaurs of the Air: The Evolution and Loss of Flight in Dinosaurs and Birds* (Baltimore: Johns Hopkins University Press, 2002).

71. Heilmann, *Origin of Birds*.

72. Ibid.

73. H. N. Bryant and A. P. Russell, "The Occurrence of Clavicles within Dinosauria: Implications for the Homology of the Avian Furcula and the Utility of Negative Evidence," *Journal of Vertebrate Paleontology* 13 (1993): 171–84; R. S. Tykoski et al., "A Furcula in the Coelophysid Theropod *Syntarsus*," *Journal of Vertebrate Paleontology* 22, no. 3 (2002): 728–33; T. Jones et al., "Non-Avian Feathers in a Late Triassic Archosaur," *Science* 291 (2000): 1899–902; A. M. Yates and E. C. Vasconcelos, "Furcula-Like Clavicles in the

Prosauropod Dinosaur *Massospondylus*," *Journal of Vertebrate Paleontology* 25 (2005): 466–68.

74. S. Chatterjee, "Cranial Anatomy and Relationships of a New Triassic Bird from Texas," *Philosophical Transactions of the Royal Society of London*, ser. B, 309 (1991): 395–460.

75. J. H. Ostrom, "The Bird in the Bush," *Nature* 353 (1991): 212.

76. S. Chatterjee, *The Rise of Birds: 225 Million Years of Evolution* (Baltimore: Johns Hopkins University Press, 1997); W. J. Bock, review of *The Rise of Birds*, by S. Chatterjee, *Auk* 115 (1998): 808–9.

77. Melchor, de Valais, and Genise, "Bird-Like Fossil Footprints from the Late Triassic"; S. de Valais and R. N. Melchor, "Ichnotaxonomy of Bird-Like Footprints: An Example from the Late Triassic–Early Jurassic of Northwest Argentina," *Journal of Vertebrate Paleontology* 28 (2008): 145–59.

78. J. F. Genise et al., "Application of Neoichnological Studies to Behavioural and Taphonomic Interpretation of Fossil Bird-Like Tracks from Lacustrine Settings: The Late Triassic–Early Jurassic? Santo Domingo Formation, Argentina," *Palaeogeography, Palaeoclimatology, Palaeoecology* 272 (2009): 143–61.

Chapter 3: The Iconic Urvogel

Epigraph: T. H. Huxley, "On the Animals Which Are Most Nearly Intermediate between the Birds and Reptiles," *Annals and Magazine of Natural History, London* 2 (1868): 66–75.

1. J. H. Ostrom, "Introduction to *Archaeopteryx*," in *The Beginnings of Birds: Proceedings of the International Archaeopteryx Conference*, ed. M. K. Hecht et al. (Eichstätt: Freunde des Jura-Museums, 1985), 9–20; P. Chambers, *Bones of Contention: The Archaeopteryx Scandals* (London: John Murray, 2002).

2. J. A. Wagner, "Über ein neues, angeblich mit Vogelfedern versehenes Reptil aus dem Solenhofener lithographischen Schiefer," *Sitzungsberichte der Bayerischen Akademie der Wissenschaften* 9 (1861): 146–54.

3. Ibid.

4. J. Evans, "On Portions of a Cranium and of a Jaw, in the Slab Containing the Fossil Remains of the *Archaeopteryx*," *Natural History Review*, n.s., 5 (1865): 415–21.

5. A. Desmond, *Archetypes and Ancestors: Palae-ontology in Victorian London, 1850–1875* (Chicago: University of Chicago Press, 1984); L. M. Witmer, "Perspectives on Avian Origins," in *Origins of the Higher Groups of Tetrapods*, ed. H.-P. Schultze and L. Trueb (Ithaca, NY: Cornell University Press, 1991), 427–66.

6. R. Owen, "On the *Archaeopteryx* of von Meyer, with a Description of the Fossil Remains of a Long-Tailed Species, from the Lithographic Stone of Solenhofen," *Philosophical Transactions of the Royal Society of London* 153 (1863): 33–47.

7. W. Dames, "Über Brustbein, Schulter- und Beckengurtel der *Archaeopteryx*," *Sitzungsberichte der Preussischen Akademie der Wissenschaften* 2 (1897): 818–34.

8. B. Petronievics, "Über die Berliner *Archaeornis*," *Geologica Balkanskoga poluostrova* 8 (1925): 37–84; B. Petronievics, "Nouvelles recherches sur l'ostéologie des Archaeornithes," *Annales Paléontologica* 16 (1927): 39–55.

9. J. H. Ostrom, "The Yale *Archaeopteryx*: The One That Flew the Coop," in Hecht et al., *Beginnings of Birds*, 359–69.

10. J. H. Ostrom, "*Archaeopteryx*: Notice of a 'New' Specimen," *Science* 170 (1970): 537–38.

11. P. Wellnhofer, "Das fünfte Skelettexemplar von *Archaeopteryx*," *Paleontographica*, ser. A, 147 (1974): 169–216.

12. P. Wellnhofer, "A New Specimen of *Archaeopteryx* from the Solnhofen Limestone," *Los Angeles County Museum of Natural History, Science Series* 36 (1992): 3–23; A. Elzanowski, "Archaeopterygidae (Upper Jurassic of Germany)," in *Mesozoic Birds: Above the Heads of Dinosaurs*, ed. L. M. Chiappe and L. Witmer (Berkeley: University of California Press, 2002), 129–59.

13. P. Wellnhofer, "Das siebte Exemplar von *Archaeopteryx* aus den Solnhofener Schichten," *Archaeopteryx* 11 (1993): 1–48.

14. Ibid.

15. P. Wellnhofer and M. Röper, "Das neunte *Archaeopteryx*-Exemplar von Solnhofen," *Archaeopteryx* 23 (2005): 3–21.

16. G. Mayr, B. Pohl, and D. S. Peters, "A Well-Preserved *Archaeopteryx* Specimen with Theropod Features," *Science* 310 (2005): 1483–86; G. Mayr et al., "The Tenth Skeletal Specimen of *Archaeopteryx*," *Zoological Journal of the Linnaean Society* 149 (2007): 97–116.

17. S. C. Bennett, "Ontogeny and *Archaeopteryx*," *Journal of Vertebrate Paleontology* 28 (2008): 535–42.

18. M. A. Houck, J. A. Gauthier, and R. E. Strauss, "Allometric Scaling in the Earliest Fossil Bird, *Archaeopteryx lithographica*," *Science* 247 (1990): 195–98; P. Senter and J. H. Robins, "Taxonomic Status of the Specimens of *Archaeopteryx*," *Journal of Vertebrate Paleontology* 23 (2003): 961–65; P. Christiansen, "Allometry in Phylogeny and *Archaeopteryx*," *Journal of Vertebrate Paleontology* 26 (2006): 480–86.

19. A. Feduccia, L. D. Martin, and S. Tarsitano, "*Archaeopteryx* 2007: *Quo Vadis?*" *Auk* 124 (2007): 373–80.

20. Ibid.; W. J. Bock, "The Arboreal Origin of Avian Flight," *Memoirs of the California Academy of Sciences* 8 (1986): 57–72.

21. Desmond, *Archetypes and Ancestors*, 128.

22. G. de Beer, *Archaeopteryx lithographica*: *A Study Based on the British Museum Specimen* (London: British Museum of Natural History, 1954).

23. G. Heilmann, *The Origin of Birds* (London: Witherby, 1926); R. T. Bakker, "Dinosaur Renaissance," *Scientific American* 232, no. 4 (1975): 58–78.

24. R. T. Bakker, *The Dinosaur Heresies: New Theories Unlocking the Mystery of the Dinosaurs and Their Extinction* (New York: William Morrow, 1986).

25. T. Monmaney, "The Dinosaur Heretic," *New Yorker*, 31 May 1993, 41–52.

26. L. M. Witmer, "The Debate on Avian Ancestry: Phylogeny, Function, and Fossils," in Chiappe and Witmer, *Mesozoic Birds*, 3–30.

27. A. Feduccia and H. B. Tordoff, "Feathers of *Archaeopteryx*: Asymmetric Vanes Indicate Aerodynamic Function," *Science* 203 (1979): 1021–22.

28. R. Å. Norberg, "Function of Vane Asymmetry and Shaft Curvature in Bird Flight Feathers: Inferences on Flight Ability of *Archaeopteryx*," and S. Rietschel, "Feathers and Wings of *Archaeopteryx*, and the Question of Her Flight Ability," in Hecht et al., *Beginnings of Birds*, 303–18, 251–60.

29. U. M. Norberg, *Vertebrate Flight* (Berlin: Springer, 1990).

30. D. B. O. Savile, "Adaptive Radiation of the Avian Wing," *Evolution* 11 (1957): 212–24.

31. O. Heinroth, "Die Flügel von *Archaeopteryx*," *Journal für Ornithologie* 71 (1923): 277–83.

32. A. Feduccia, "Evidence from Claw Geometry Indicating Arboreal Habits of *Archaeopteryx*," *Science* 259 (1993): 790–93.

33. N. Longrich, "Structure and Function of Hindlimb Feathers in *Archaeopteryx lithographica*," *Paleobiology* 32 (2006): 417–31.

34. P. Christiansen and N. Bonde, "Body Plumage in *Archaeopteryx*: A Review, and New Evidence from the Berlin Specimen," *Comptes Rendus Palevol* 3 (2004): 99–118.

35. Longrich, "Structure and Function."

36. H. Tischlinger and D. M. Unwin, "UV-Untersuchungen des Berliner Exemplars von *Archaeopteryx lithographica* H. v. Meyer 1861 und der isolierten Archaeopteryx-Feder," *Archaeopteryx* 22 (2004): 17–50; H. Tischlinger, "Ultraviolet Light of Fossils from the Upper Jurassic Plattenkalks of Southern Frankonia," *Zitteliana*, ser. B, 26 (2005): 26; H. Tischlinger, "Der Eichstätter Archaeopteryx im Langwelligen UV-Licht," *Archaeopteryx* 20 (2002): 21–38; H. Tischlinger, "Neue Information zum Berliner Exemplar von *Archaeopteryx lithographica* H. v. Meyer 1861," *Archaeopteryx* 23 (2005): 33–50.

37. C. W. Beebe, "A Tetrapteryx Stage in the Ancestry of Birds," *Zoologica* 2 (1915): 39–52.

38. R. O. Prum, "Paleontology: Dinosaurs Take to the Air," *Nature* 421 (2003): 323–24.

39. K. N. Whetstone, "Braincase of Mesozoic Birds. I. New Preparation of the 'London' Specimen," *Journal of Vertebrate Paleontology* 2 (1983): 439–52.

40. A. D. Walker, "The Braincase of *Archaeopteryx*," in Hecht et al., *Beginnings of Birds*, 123–24.

41. P. Bühler, "On the Morphology of the Skull of *Archaeopteryx*," in Hecht et al., *Beginnings of Birds*, 135–40.

42. B. M. Haubitz et al., "Computed Tomography of *Archaeopteryx*," *Paleobiology* 14 (1988): 206–13.

43. A. Elzanowski and P. Wellnhofer, "Cranial Morphology of *Archaeopteryx*: Evidence from the Seventh Specimen," *Journal für Ornithologie* 135 (1994): 331; A. Elzanowski and P. Wellnhofer, "The Skull and the Origin of Birds," *Archaeopteryx* 13 (1994): 41–46; A. Elzanowski and P. Wellnhofer, "Cranial Morphology of *Archaeopteryx*: Evidence from the Seventh Specimen," *Journal of Vertebrate Paleontology* 16 (1996): 81–94.

44. A. Elzanowski, "Archaeopterygidae (Upper Jurassic of Germany)," in Chiappe and Witmer, *Mesozoic Birds*, 129–59.

45. Mayr, Pohl, and Peters, "Well-Preserved Archaeopteryx."

46. D. W. Yalden, "Forelimb Function in *Archaeopteryx*," in Hecht et al., *Beginnings of Birds*, 91–97.

47. Feduccia, "Evidence from Claw Geometry."

48. D. W. Yalden, "Climbing *Archaeopteryx*," *Archaeopteryx* 15 (1997): 107–8.

49. Ibid.; A. V. L. Pike and D. P. Maitland, "Scaling of Bird Claws," *Journal of the Zoological Society of London* 262 (2004): 73–81.

50. J. H. Ostrom, "*Archaeopteryx* and the Origin of Birds," *Biological Journal of the Linnaean Society* 8 (1976): 91–182; L. M. Chiappe, *Glorified Dinosaurs: The Origin and Early Evolution of Birds* (New York: John Wiley and Sons, 2007), 139; C. L. Glen and M. B. Bennett, "Foraging Modes of Mesozoic Birds and Non-Avian Theropods," *Current Biology* 17 (2007): R911–12.

51. P. L. Manning et al., "Dinosaur Killer Claws or Climbing Crampons?" *Biology Letters* 2 (2006): 110–12.

52. P. L. Manning et al., "Biomechanics of Dromaeosaurid Dinosaur Claws: Application of X-ray Microtomography, Nanoindentation, and Finite Element Analysis," *Anatomical Record* 292 (2009): 1397–405; D. Burnham et al., "Tree-Climbing—a Fundamental Avian Adaptation," *Journal of Systematic Palaeontology* (2011): 103–7.

53. L. D. Martin, and J.-D. Lim, "Soft Body Impressions of the Body of *Archaeopteryx*," *Current Science* 89 (2005): 1089–90.

54. Mayr, Pohl, and Peters, "Well-Preserved Archaeopteryx."

55. K. M. Middleton, "The Morphological Basis of Hallucal Orientation in Extant Birds," *Journal of Morphology* 250 (2001): 51–60.

56. A. Feduccia, *The Origin and Evolution of Birds*, 2nd ed. (New Haven: Yale University Press, 1999).

57. Bock, "Arboreal Origin of Avian Flight."

58. P. Wellnhofer, "*Archaeopteryx*: Zur Lebensweise Solnhofener Urvögel," *Fossilien* 5 (1995): 296–307.

59. A. Elzanowski, "The Life Style of *Archaeopteryx* (Aves)," Seventh International Symposium on Mesozoic Terrestrial Ecosystems, *Special Publication* 7 (Buenos Aires, 2001), 91–99.

60. P. J. Griffiths, "The Claws and Digits of *Archaeopteryx lithographica*," *Geobios* 16 (1993): 101–6.

61. Wellnhofer, "*Archaeopteryx*: Zur Lebensweise Solnhofner Urvögel."

62. F. E. Zeuner, *Fossil Orthoptera Ensifera* (London: British Museum of Natural History, 1939).

63. D. A. Burnham, "Paleoenvironment, Paleoecology, and Evolution of Maniraptoran 'Dinosaurs'" (PhD diss., University of Kansas, Lawrence, 2007).

64. Griffiths, "Claws and Digits of *Archaeopteryx lithographica*."

65. S. L. Olson, review of *The Beginnings of Birds: Proceedings of the International Archaeopteryx Conference*, ed. M. K. Hecht et al., *American Scientist* 75 (1987): 74–75.

66. J. A. Hopson, "Ecomorphology of Avian and Nonavian Theropod Phalangeal Proportions: Implications for the Arboreal versus Terrestrial Origin of Bird Flight," in *New Perspectives on the Origin and Early Evolution of Birds*, ed. J. Gauthier and L. F. Gall (New Haven: Peabody Museum of Natural History, Yale University, 2001), 212–35.

67. Z. Zhou and J. O. Farlow, "Flight Capability and Habits of *Confuciusornis*," in Gauthier and Gall, *New Perspectives on the Origin and Early Evolution of Birds*, 237–54.

68. M. Reichel, *Manfred Reichel, 1896–1984: Dessins* (Basel: Geological Institute of Basel University, 1984).

69. P. Wellnhofer, *Archaeopteryx: Der Urvögel von Solnhofen* (Munich: Verlag Dr. Friedrich Pfeil, 2008); P. Wellnhofer, *Archaeopteryx: Icon of Evolution*, trans. F. Haase (Munich: Verlag Dr. Friedrich Pfeil, 2009).

70. L. D. Martin and J. D. Stewart, "Implantation and Replacement of Bird Teeth," *Smithsonian Contributions to Paleobiology* 89 (1999): 295–300; A. Feduccia, "Birds Are Dinosaurs: Simple Answer to a Complex Problem," *Auk* 119 (2002): 1187–201.

71. A. D. Buscalioni et al., "The Upper Jurassic Maniraptoran Theropod *Lisboasaurus estesi* (Guimarota, Portugal) Reinterpreted as a Crocodilomorph," *Journal of Vertebrate Paleontology* 16 (1996): 358–62.

72. P. J. Currie, "Bird-Like Characteristics of the Jaws and Teeth of Troodontid Theropods (Dinosauria: Saurischia)," *Journal of Vertebrate Paleontology* 7 (1987): 72–81; M. A. Norell and S. H. Hwang, "A Troodontid Dinosaur from Ukhaa Tolgod (Late Cretaceous Mongolia)," *American Museum Novitates* 3446 (2004): 1–9.

73. P. D. Alonso et al., "The Avian Nature of the Brain and Inner Ear of *Archaeopteryx*," *Nature* 430 (2004): 666–69; L. M. Witmer, "Inside the Oldest Bird Brain," *Nature* 430 (2004): 619–20.

74. L. M. Witmer et al., "Neuroanatomy of Flying Reptiles and Implications for Flight, Posture and Behaviour," *Nature* 425 (2003): 950–53; D. M. Unwin, "Smart-Winged Pterosaurs," *Nature* 425 (2003): 910–11.

75. S. A. Walsh et al., "Inner Ear Anatomy Is a Proxy for Deducing Auditory Capability and Behavior in Reptiles and Birds," *Proceedings of the Royal Society B* 276 (2009): 1355–60.

76. Feduccia, *Origin and Evolution of Birds*.

77. Ostrom, "*Archaeopteryx* and the Origin of Birds"; Feduccia, *Origin and Evolution of Birds*.

78. G. de Beer, "*Archaeopteryx* and Evolution," *Advancement of Science* 11 (1954): 160–70; G. L. Stebbins, "Mosaic Evolution: An Integrating Principle for the Modern Synthesis," *Experientia* 39 (1983): 823–34; R. A. Barton and P. H. Harvey, "Mosaic Evolution of Brain Structure in Mammals," *Nature* 405 (2000): 1055–58; T. C. Rae, "Mosaic Evolution in the Origin of the Hominoidea," *Folia Primatologica* 70 (1999): 125–35.

79. See on the topic Bock, "Arboreal Origin of Avian Flight"; Heilmann, *Origin of Birds*; Feduccia, *Origin and Evolution of Birds*; J. H. Ostrom, "Bird Flight: How Did It Begin?" *American Scientist* 67 (1979): 46–56; J. H. Ostrom, "The Cursorial Origin of Avian Flight," *Memoirs of the California Academy of Sciences* 8 (1986): 73–81; O. C. Marsh, "Odontornithes: A Monograph on the Extinct Toothed Birds of North America," *Report of the U.S. Geological Exploration of the Fortieth Parallel* (Washington, DC: US Government Printing Office, 1880); and J. H. Ostrom, "On the Origin of Birds and of Avian Flight," in *Major Features of Vertebrate Evolution*, ed. D. P. Prothero and R. M. Schoch (Knoxville: University of Tennessee Press, 1994), 160–77. Quotation from Marsh, "Odontornithes."

80. Ostrom, "Cursorial Origin of Avian Flight."

81. Ostrom, "Origin of Birds and Avian Flight."

82. C. J. Pennycuick, "Mechanical Constraints on the Evolution of Flight," *Memoirs of the California Academy of Sciences* 8 (1986): 83–98.

83. S. W. Williston, "Are Birds Derived from Dinosaurs?" *Kansas City Review of Science* 3 (1879): 457–60.

84. F. Nopsca, "Ideas on the Origin of Flight," *Proceedings of the Zoological Society of London* (1907): 223–36; F. Nopsca, "On the Origin of Flight in Birds," *Proceedings of the Zoological Society of London* (1923): 463–77. Quotation from Nopsca, "Ideas on the Origin of Flight."

85. Nopsca, "Ideas on the Origin of Flight."

86. Ostrom, "Cursorial Origin of Avian Flight."

87. P. Burgers and L. M. Chiappe, "The Wing of *Archaeopteryx* as a Primary Thrust Generator," *Nature* 399 (1999): 60–62.

88. S. Chatterjee and R. J. Templin, "The Flight of *Archaeopteryx*," *Naturwissenschaften* 90 (2003): 27–32.

89. L. M. Chiappe, "Climbing *Archaeopteryx?* A Response to Yalden," *Archaeopteryx* 15 (1997): 109–12.

90. Chiappe, *Glorified Dinosaurs*.

91. K. Padian, "Stages in the Origin of Bird Flight: Beyond the Arboreal—Cursorial Dichotomy," in Gauthier and Gall, *New Perspectives on the Origin and Early Evolution of Birds*, 255–72.

92. K. Padian and L. M. Chiappe, "The Origin of Birds and Their Flight," *Scientific American*, February 1998, 38–47.

93. J. H. Ostrom, "*Archaeopteryx:* Notice of a 'New' Specimen"; Ostrom, "*Archaeopteryx* and the Origin of Flight," *Quarterly Review of Biology* 49 (1974): 27–47; R. T. Bakker, "Dinosaur Renaissance," *Scientific American* 232, no. 4 (1975): 58–78; Ostrom, "New Ideas About Dinosaurs," *National Geographic* 154 (1978): 152–85; Ostrom, "The Meaning of *Archaeopteryx*," in Hecht et al., *Beginnings of Birds*, 161–76; Ostrom, "Cursorial Origin of Avian Flight"; Ostrom, "On the Origin of Birds and of Avian Flight."

94. S. M. Gatesy and K. P. Dial, "Locomotor Modules and the Evolution of Avian Flight," *Evolution* 50 (1996): 331–40.

95. Norberg, *Vertebrate Flight*.

96. R. L. Nudds and G. J. Dyke, "Narrow Primary Feather Rachises in *Confuciusornis* and *Archaeopteryx* Suggest Poor Flight Ability," *Science* 328 (2010): 887–89.

97. X. Zheng et al., "Comment on 'Narrow Primary Feather Rachises in *Confuciusornis* and *Archaeopteryx* Suggest Poor Flight Ability,'" *Science* 330 (2010): 320-c; G. S. Paul, "Comment on 'Narrow Primary Feather Rachises in *Confuciusornis* and *Archaeopteryx* Suggest Poor Flight Ability,'" *Science* 330 (2010): 320-b; R. L. Nudds and G. J. Dyke, "Response to Comments on 'Narrow Primary Feather Rachises in *Confuciusornis* and *Archaeopteryx* Suggest Poor Flight Ability,'" *Science* 330 (2010): 320-d.

98. S. Perkins, "Earliest Birds Didn't Make a Flap," *Science News*, 5 June 2010, 12; S. Perkins, "Bird Brain? Cranial Scan of Fossil Hints at Flight Capability," *Science News*, 7 August 2004, 86.

99. L. Dingus and T. Rowe, *The Mistaken Extinction: Dinosaur Evolution and the Origin of Birds* (New York: W. H. Freeman, 1998).

100. W. J. Bock, review of *The Mistaken Extinction: Dinosaur Evolution and the Origin of Birds*, by L. Dingus and T. Rowe, *Auk* 116 (1999): 566–68.

101. Bock, "Arboreal Origin of Avian Flight"; W. J. Bock, "The Role of Adaptive Mechanisms in the Origin of the Higher Levels of Organization," *Systematic Zoology* 14 (1965): 272–87.

102. K. D. Dial, "Wing-Assisted Incline Running and the Evolution of Flight," *Science* 299 (2003): 402–4; K. P. Dial, B. E. Jackson, and P. Sefre, "A Fundamental Avian Wing-Stroke Provides a New Perspective on the Evolution of Flight," *Nature* 451 (2008): 985–89.

103. J. Ruben, "Reptilian Physiology and the Flight Capacity of *Archaeopteryx*," *Evolution* 45 (1991): 1–17.

104. R. Hertel and K. E. Campbell, Jr., "The Antitrochanter of Birds: Form and Function in Balance," *Auk* 124 (2007): 78–805.

105. B. Handwerk, "Wing Angle May Be Key to Bird Flight Origins," *National Geographic News*, 23 January 2008.

106. Dial, Jackson, and Sefre, "Fundamental Avian Wingstroke," emphasis added.

107. P. Senter, "Scapular Orientation in Theropods and Basal Birds, and the Origin of Flapping Flight," *Acta Palaeontologica Polonica* 51 (2006): 305–13.

108. D. R. Prothero, *Evolution: What the Fossils Say and Why It Matters* (New York: Columbia University Press, 2007).

109. R. Kipling, *Just So Stories* (New York: Doubleday, 1902).

110. Prothero, *Evolution*.

111. Bock, "Arboreal Origin of Avian Flight"; Feduccia, *Origin and Evolution of Birds*.

112. Bock, "Arboreal Origin of Avian Flight."

Chapter 4: Mesozoic Chinese Aviary Takes Form

1. A. Feduccia, "Mesozoic Aviary Takes Form," *Proceedings of the National Academy of Sciences* 103 (2006): 5–6.

2. A. Feduccia, "Explosive Evolution in Tertiary Birds and Mammals," *Science* 267 (1995): 637–38; P. Houde and S. L. Olson, "Paleognathous Carinate Birds from the Early Tertiary of North America," *Science* 214 (1981): 1236–37.

3. G. Mayr, *Paleogene Fossil Birds* (Berlin: Springer, 2009).

4. C. Walker, "New Subclass of Birds from the Cretaceous of South America," *Nature* 292 (1981): 51–53.

5. E. N. Kurochkin, "A True Carinate Bird from Lower Cretaceous Deposits in Mongolia and Other Evidence of Early Cretaceous Birds in Asia," *Cretaceous Research* 6 (1985): 271–78; E. N. Kurochkin, "A New Order of Birds from the Lower Cretaceous of Mongolia" [in Russian], *Doklady Academii Nauk SSSR* 262 (1982): 452–55.

6. J. L. Sanz and J. F. Bonaparte, "A New Order of Birds (Class Aves) from the Lower Cretaceous of Spain," *Los Angeles County Museum of Natural History, Science Series* 36 (1992): 39–50.

7. P. C. Sereno and C. Rao, "Early Evolution of Avian Flight and Perching: New Evidence from the Lower Cretaceous of China," *Science* 255 (1992): 845–48; P. C. Sereno, C. Rao, and J. Ki, "*Sinornis santensis* (Aves: Enantiornithes) from the Early Cretaceous of Northeastern China," in *Mesozoic Birds: Above the Heads of Dinosaurs*, ed. L. M. Chiappe and L. M. Witmer (Berkeley: University of California Press, 2002), 184–208.

8. J. L. Sanz et al., "A New Lower Cretaceous Bird from Spain: Implications for the Evolution of Flight," *Nature* 382 (1996): 442–45; J. L. Sanz et al., "A Nestling Bird from the Early Cretaceous of Spain: Implications for Avian Skull and Neck Evolution," *Science* 276 (1997): 1543–46; J. L. Sanz et al., "The Birds from the Lower Cretaceous of Las Hoyas (Province of Cuenca, Spain)," in Chiappe and Witmer, *Mesozoic Birds*, 209–29.

9. Z. Zhou, "Discovery of New Cretaceous Birds in China," *Courier Forschungsinstitut Senckenberg* 181 (1995): 9–23.

10. M.-M. Chang, ed., *The Jehol Biota: The Emergence of Feathered Dinosaurs, Beaked Birds, and Flowering Plants* (Shanghai: Shanghai Scientific and Technical Publishers, 2003).

11. F. Zhang, Z. Zhou, and L.-H. Hou, "Birds," in Chang, *Jehol Biota*, 170–92; Z. Zhou and L.-H. Hou, "The Discovery and Study of Mesozoic Birds in China," in Chiappe and Witmer, *Mesozoic Birds*, 160–83; Z. Zhou and F. Zhang, "Mesozoic Birds of China: An Introduction of Review," *Acta Zoologica Sinica* 50 (2004): 913–20; Z. Zhou, "The Origin and Early Evolution of Birds: Discoveries, Disputes, and Perspectives from Fossil Evidence," *Naturwissenschaften* 91 (2004): 455–71; L. M. Chiappe, *Glorified Dinosaurs: The Origin and Early Evolution of Birds* (New York: John Wiley and Sons, 2007).

12. Zhou, Barrett, and Hilton, "Exceptionally Preserved Lower Cretaceous Ecosystem"; Z. Zhou, "Evolutionary Radiation of the Jehol Biota: Chronological and Ecological Perspectives," *Geological Journal* (2006): 377–93.

13. Z. Guo, J. Liu, and X. Wang, "Effect of Mesozoic Volcanic Eruptions in the Western Liaoning Province, China on Palaeoclimate and Palaeoenvironment," *Science in China*, ser. D, 46 (2003): 1261–72; P. M. Barrett and J. M. Hilton, "The Jehol Biota (Lower Cretaceous, China): New Discoveries and Future Prospects," *Integrative Zoology* 1 (2006): 15–17.

14. L.-H. Hou and Z. Lui, "A New Fossil Bird from Lower Cretaceous of Gansu and Early Evolution of Birds," *Scientia Sinica* 27 (1984): 1296–302.

15. H.-L. You et al., "A Nearly Modern Amphibious Bird from the Early Cretaceous of Northwestern China," *Science* 312 (2006): 1640–43; J. D. Harris et al., "A Second Enantiornithean (Aves: Ornithothoraces) Wing from the Early Cretaceous Xiagou Formation near Changma, Gansu Province, People's Republic of China," *Canadian Journal of Earth Science* 43 (2006): 547–54; P. Dodson, "Imperfect Fossils," *American Paleontologist* 15, no. 2 (2007): 26–30.

16. Zhang, Zhou, and Hou, "Birds"; A. Feduccia, *The Origin and Evolution of Birds* (New Haven: Yale University Press, 1996).

17. Sereno, Rao, and Ki, "*Sinornis santensis*."

18. L. D. Martin and Z. Zhou, "The Skull of *Cathayornis*," *Nature* 389 (1997): 556.

19. L. D. Martin, "The Enantiornithines: Terrestrial Birds of the Cretaceous in Avian Evolution," *Courier Forschungsinstitut Senckenberg* 181 (1995): 23–36.

20. Chiappe, *Glorified Dinosaurs*; Martin, "Enantiornithines"; L. M. Chiappe and C. A. Walker, "Skeletal Morphology and Systematics of the Cretaceous Euenantiornithes (Ornithothoraces: Enantiornithes)," in Chiappe and Witmer, *Mesozoic Birds*, 240–67.

21. E. M. Morschhauser et al., "Anatomy of the Early Cretaceous Birds *Rapaxavis pani*, a New Species from Liaoning Province, China," *Journal of Vertebrate Paleontology* 29 (2009): 545–54.

22. Chiappe and Walker, "Skeletal Morphology."

23. C. A. Walker, E. Buffetaut, and G. J. Dyke, "Large Euenantiornithine Birds from the Cretaceous of Southern France, North America and Argentina," *Geological Magazine* 144 (2007): 977–86; L. M. Chiappe et al., "A New Enantiornithine Bird from the Late Cretaceous of the Gobi Desert," *Journal of Systematic Paleontology* 5, no. 2 (2007): 193–208.

24. G. J. Dyke and R. L. Nudds, "The Fossil Record and Limb Disparity of Enantiornithines, the Dominant Flying Birds of the Cretaceous," *Lethaia* 42 (2009): 248–54.

25. L.-H. Hou et al., "A Beaked Bird from the Jurassic of China," *Nature* 377 (1995): 616–18; L.-H. Hou et al., "*Confuciusornis sanctus*, a New Late Jurassic Sauriurine Bird from China" [in Chinese], *Chinese Science Bulletin* 40 (1995): 726–29.

26. Zhang, Zhou, and Hou, "Birds."

27. Zhou and Zhang, "Mesozoic Birds of China."

28. L. M. Chiappe et al., "Life History of a Basal Bird: Morphometrics of the Early Cretaceous *Confuciusornis*," *Biology Letters* 4 (2008): 719–23.

29. Q. Ji, L. M. Chiappe, and S. Ji, "A New Late Mesozoic Confuciusornithid from China," *Journal of Vertebrate Paleontology* 19 (1999): 1–7; F. Zhang, Z. Zhou, and M. Benton, "A Primitive Confuciusornithid Bird from China and Its Implications for Early Avian Flight," *Science in China*, ser. D, 51 (2008): 625–39.

30. Z. Zhang et al., "Diversification in an Early Cretaceous Avian Genus: Evidence from a New Species of *Confuciusornis* from China," *Journal of Ornithology* 150 (2009): 783–90.

31. K. Padian and L. M. Chiappe, "The Origin of Birds and Their Flight," *Scientific American* 278 (1998): 38–47.

32. S. L. Olson, review of "Anatomy and Systematics of the Confuciusornithidae (Theropoda: Aves) from the Late Mesozoic of China," *Auk* 117 (2000): 836–39.

33. L. M. Chiappe et al., "Anatomy and Systematics of the Confuciusornithidae (Theropoda: Aves) from the Late Mesozoic of China," *Bulletin of the American Museum of Natural History* 242 (1999): 1–89.

34. Olson, review of "Anatomy and Systematics."

35. L. D. Martin et al., "*Confuciusornis sanctus* Compared to *Archaeopteryx lithographica*," *Naturwissenschaften* 85 (1998): 286–89; Z. Zhou and J. O. Farlow, "Flight Capability and Habits of *Confuciusornis*," in *New Perspectives on the Origin and Early Evolution of Birds*, ed. J. Gauthier and L. F. Gall (New Haven: Peabody Museum of Natural History, Yale University, 2001), 237–54.

36. A. Chinsamy-Turan, *The Microstructure of Dinosaur Bone* (Baltimore: Johns Hopkins University Press, 2005).

37. F. Zhang et al., "Some Microstructure Difference among *Confuciusornis*, Alligator and a Small Theropod Dinosaur, and Its Implications," *Palaeoworld* 11 (1999): 296–309.

38. A. J. de Ricqlès et al., "Osteohistology of *Confuciusornis sanctus* (Theropod: Aves)," *Journal of Vertebrate Paleontology* 23, no. 2 (2003): 373–86.

39. Feduccia, *Origin and Evolution of Birds*.

40. Z. Zhou and F. Zhang, "Discovery of an Ornithurine Bird and Its Implications for Early Cretaceous Avian Radiation," *Proceedings of the National Academy of Sciences* 102 (2005): 18998–9002.

41. Zhang, "Diversification in an Early Cretaceous Avian Genus"; Padian and Chiappe, "Origin of Birds and Their Flight"; Olson, review of "Anatomy and Systematics"; Chiappe et al., "Anatomy and Systematics"; Martin et al., "*Confuciusornis sanctus*"; Zhou and Farlow, "Flight Capability"; Chinsamy-Turan, *Microstructure of Dinosaur Bone*; Zhang et al., "Some Microstructure Difference"; de Ricqlès et al., "Osteohistology of *Confuciusornis sanctus*"; Zhou and Zhang, "Discovery of an Ornithurine Bird"; Z. Zhou and F. Zhang, "A Long-Tailed, Seed-Eating Bird from the Early Cretaceous of China,"

Nature 418 (2002): 405–9; Z. Zhou and F. Zhang, "Largest Bird from the Early Cretaceous and Its Implications for the Earliest Avian Ecological Diversification," *Naturwissenschaften* 89 (2002): 34–38; Z. Zhou and F. Zhang, "Anatomy of the Primitive Bird *Sapeornis chaoyangensis* from the Early Cretaceous of Liaoning, China," *Canadian Journal of Earth Sciences* 40 (2003): 731–47; Z. Zhou and F. Zhang, "*Jeholornis* Compared to *Archaeopteryx*, with a New Understanding of the Earliest Avian Evolution," *Naturwissenschaften* 90 (2003): 220–25; C. Gao et al., "A New Basal Lineage of Early Cretaceous Birds from China and Its Implications on the Evolution of the Avian Tail," *Palaeontology* 51 (2008): 775–91; Z. Zhou, F. Zhang, and Z. Li, "A New Lower Cretaceous Bird from China and Tooth Reduction in Early Avian Evolution," *Proceedings of the Royal Society B* 277 (2010): 161–64.

42. Zhou and Zhang, "*Jeholornis* Compared to *Archaeopteryx*."

43. Zhou and Zhang, "Long-Tailed, Seed-Eating Bird."

44. Zhou and Zhang, "Anatomy of the Primitive Bird."

45. Gao, "New Basal Lineage."

46. Zhou, Zhang, and Li, "New Lower Cretaceous Bird."

47. F. Zhang and Z. Zhou, "A Primitive Enantiornithine Bird and the Origin of Feathers," *Science* 290 (2000): 1955–59.

48. F. Zhang et al., "Early Diversification of Birds: Evidence from a New Opposite Bird," *Chinese Science Bulletin* 46 (2001): 945–49; Zhou and Zhang, "Mesozoic Birds of China."

49. L.-H. Hou et al., "New Early Cretaceous Fossil from China Documents a Novel Trophic Specialization for Mesozoic Birds," *Naturwissenschaften* 91 (2004): 22–25.

50. J. K. O'Connor et al., "Phylogenetic Support for a Specialized Clade of Cretaceous Enantiornithine Birds with Information from a New Species," *Journal of Vertebrate Paleontology* 29 (2009): 188–204.

51. Z. Zhang et al., "The First Mesozoic Heterodactyl Bird from China," *Acta Geologica Sinica* 80 (2006): 631–35 (English ed.); M. G. Lockley et al., "Earliest Zygodactyl Bird Feet: Evidence from Early Cretaceous Roadrunner-Like Tracks," *Naturwissenschaften* 94 (2007): 657–65.

52. Z. Zhou, J. Clarke, and F. Zhang, "Insight into Diversity, Body Size and Morphological Evolution from the Largest Early Cretaceous Enantiornithine Bird," *Journal of Anatomy* 212 (2008): 565–77.

53. Z. Zhou, and F. Zhang, "Two New Ornithurine Birds from the Early Cretaceous of Western Liaoning, China," *Chinese Science Bulletin* 46 (2001): 945–50; J. A. Clarke, Z. Zhou, and F. Zhang, "Insight into the Evolution of Avian Flight from a New Clade of Early Cretaceous Ornithurines from China and the Morphology of *Yixianornis grabaui*," *Journal of Anatomy* 208 (2006): 287–308; Z. Zhou, F. Zhang, and A. Li, "A New Basal Ornithurine Bird (*Jianchangornis microdonta* Gen. et Sp. Nov.) from the Lower Cretaceous of China," *Vertebrata PalAsiatica* 47 (2009): 299–310; Z. Zhou and L. D. Martin, "Distribution of the Predentary in Mesozoic Ornithurine Birds," *Journal of Systematic Palaeontology* 9 (2011): 25–31.

54. Clarke, Zhou, and Zhang, "Insight into the Evolution of Avian Flight"; J. A. Clarke and M. A. Norell, "The Morphology and Phylogenetic Position of *Apsaravis ukhaana* from the Late Cretaceous of Mongolia," *American Museum Novitates* 3387 (2002): 1–46.

55. Z. Zhou and F. Zhang, "Discovery of an Ornithurine Bird and Its Implication for Early Cretaceous Avian Radiation," *Proceedings of the National Academy of Sciences* 102 (2005): 18998–9002.

56. Ibid.

57. A. Elzanowski, "Skulls of *Gobipteryx* (Aves) from the Upper Cretaceous of Mongolia," *Palaeontologica Polonica* 37 (1977): 153–65.

58. Z. Zhou and F. Zhang, "A Beaked Ornithurine Bird (Aves, Ornithurae) from the Lower Cretaceous of China," *Zoologica Scripta* 35 (2006): 363–73.

59. M. W. Browne, "Feathery Fossil Hints Dinosaur-Bird Link," *New York Times*, 18 October 1996, 1, 10.

60. A. Gibbons, "New Feathered Fossil Brings Dinosaurs and Birds Closer Together," *Science* 274 (1996): 720–21; V. Morell, "The Origin of Birds: The Dinosaur Debate," *Audubon* 99, no. 2 (1997): 36–45.

61. Gibbons, "New Feathered Fossil"; V. Morell, "Warm-Blooded Dino Debate Blows Hot and Cold," *Science* 265 (1994): 188.

62. Morell, "Origin of Birds."

63. Morell, "Warm-Blooded Dino"; V. Morell, "A Cold, Hard Look at Dinosaurs," *Discover* 17, no. 2 (1996): 98–108.

64. R. Monastersky, "Paleontologists Deplume Feathery Dinosaur," *Science News* 151 (1997): 271.

65. Q. Ji and S.-A. Ji, "On the Discovery of the Earliest Known Bird Fossil in China and the Origin of Birds," *Chinese Geology* 233 (1996): 30–33.

66. P. Chen, Z. Dong, and S. Zhen, "An Exceptionally Well-Preserved Theropod Dinosaur from the Yixian Formation of China," *Nature* 391 (1998): 147–52; P. J. Currie and P.-J. Chen, "Anatomy of *Sinosauropteryx prima* from Liaoning, Northeastern China," *Canadian Journal of Earth Science* 38 (2001): 1705–27.

67. Chen, Dong, and Chen, "Anatomy of *Sinosauropteryx prima*."

68. J. Ackerman, "Dinosaurs Take Wing," *National Geographic*, July 1998, 74–99.

69. X. Xu, "Dinosaurs," in Chang, *Jehol Biota*, 141.

70. S. Xu, Z. Tang, and X. Wang, "A Therizinosaurid Dinosaur with Integumentary Structures from China," *Nature* 399 (1999): 350–54; S. Ji et al., "A New Giant Compsognathid Dinosaur with Long Filamentous Integuments from the Lower Cretaceous of Northeastern China," *Acta Geologica Sinica* 81, no. 1 (2007): 8–15; A. Feduccia, T. Lingham-Soliar, and J. R. Hinchliffe, "Do Feathered Dinosaurs Exist? Testing the Hypothesis on Neontological and Paleontological Evidence," *Journal of Morphology* 266 (2005): 125–66; Zhou and Zhang, "Mesozoic Birds of China."

71. Ji et al., "New Giant Compsognathid Dinosaur."

72. L. D. Martin and S. A. Czerkas, "The Fossil Record of Feather Evolution in the Mesozoic," *American Zoologist* 40 (2000): 687–94.

73. D. G. Homberger and K. N. de Silva, "Functional Microanatomy of the Feather-Bearing Avian Integument: Implications for the Evolution of Birds and Avian Flight," *American Zoologist* 40 (2000): 553–74.

74. J. H. Ostrom, "The Osteology of *Compsognathus longipes*," *Zitteliana* 4 (1978): 73–118.

75. U. B. Göhlich and L. M. Chiappe, "A New Carnivorous Dinosaur from the Late Jurassic Solnhofen Archipelago," *Nature* 440 (2006): 329–32.

76. Ibid.

77. X. Xu, "Scales, Feathers and Dinosaurs," *Nature* 440 (2006): 287–88.

78. X. Xu et al., "Basal Tyrannosauroids from China and Evidence for ProtoFeathers in Tyrannosauroids," *Nature* 431 (2004): 680–84.

79. E. Stokstad, "*T. rex* Clan Evolved Head First," *Nature* 306 (2004): 211.

80. K. Peyer, "A Reconsideration of *Compsognathus* from the Upper Tithonian of Canjuers, Southeastern France," *Journal of Vertebrate Paleontology* 26, no. 4 (2006): 879–96.

81. Feduccia, Lingham-Soliar, and Hinchliffe, "Do Feathered Dinosaurs Exist?"

82. A. W. A. Kellner, "Fossilized Theropod Soft Tissue," *Nature* 379 (1996): 32.

83. C. dal Sasso and M. Signore, "Exceptional Soft-Tissue Preservation in a Theropod Dinosaur from Italy," *Nature* 392 (1998): 383–87.

84. D. B. O. Savile, "Gliding and Flight in the Vertebrates," *American Zoologist* 2 (1962): 161–66.

85. S. J. Gould and E. Vrba, "Exaptation: A Missing Term in the Science of Form," *Paleobiology* 8 (1982): 4–15.

86. R. O. Prum and A. H. Brush, "The Evolutionary Origin and Diversification of Feathers," *Quarterly Review of Biology* 77 (2002): 261–95.

87. Feduccia, Lingham-Soliar, and Hinchliffe, "Do Feathered Dinosaurs Exist?"

88. P. F. A. Maderson, "On How an Archosaurian Scale Might Have Given Rise to an Avian Feather," *American Naturalist* 176 (1972): 424–28; P. F. A. Maderson and L. Alibardi, "The Development of the Sauropsid Integument: A Contribution to the Problem of the Origin and Evolution of Feathers," *American Zoologist* 40 (2000): 513–29; P. F. A. Maderson et al., "Towards a Comprehensive Model of Feather Regeneration," *Journal of Morphology* 270 (2009): 1166–208; R. O. Prum, "The Development and Evolutionary Origin of Feathers," *Journal of Experimental Zoology Part B* 285 (1999): 291–306; R. O. Prum and A. H. Brush, "Which Came First, the Feather or the Bird?" *Scientific American*, March 2004, 72–81.

89. D. Dhouailly and M. H. Hardy, "Retinoic Acid Causes the Development of Feathers in the Scale-Forming Integument of the Chick Embryo," *Wilhelm Roux's Archives* 185 (1978): 195–201; D. Dhouailly, "Early Events in Retinoic Acid-Induced Ptilopody in the Chick Embryo," *Roux's*

Archives of Developmental Biology 192 (1983): 21–27; R. Cadi et al., "Use of Retinoic Acid for the Analysis of Dermal-Epidermal Interactions in the Tarsometatarsal Skin of the Chick Embryo," *Developmental Biology* 100 (1983): 489–95; C. M. Chuong, "The Making of a Feather: Homeoproteins, Retinoids and Adhesion Molecules," *Bioessays* 15 (1993): 513–21; H. Zhou and L. Niswander, "Requirement for BMP Signaling in Interdigital Apoptosis and Scale Formation," *Science* 272 (1996): 738–41; C. M. Chuong and R. Widelitz, "Feather Morphogenesis: A Model of the Formation of Epithelial Appendages," in *Molecular Basis of Epithelial Appendage Morphogenesis*, ed. C. M. Chuong (Austin, TX: Landes Bioscience, 1999), 57–73; Zhang and Zhou, "Primitive Enantiornithine Bird"; R. H. Sawyer et al., "Evolutionary Origin of the Feather Epidermis," *Developmental Dynamics* 232 (2005): 256–67.

90. A. Feduccia, "Birds Are Dinosaurs: Simple Answer to a Complex Question," *Auk* (1972): 1187–201; N. Holmgren, "Studies on the Phylogeny of Birds," *Acta Zoologica* 36 (1955): 243–328; A. Feduccia, pers. obs.

91. L. Alibadi et al., "Evolution of Hard Proteins in the Sauropsid Integument in Relation to the Cornification of Skin Derivatives in Amniotes," *Journal of Anatomy* 214 (2009): 560–86; Maderson and Alibardi, "Development of the Sauropsid Integument."

92. R. H. Sawyer et al., "Origin of Feathers: Feather Beta (β) Keratins Are Expressed in Discrete Epidermal Cell Populations of Embryonic Scutate Scales," *Journal of Experimental Zoology Part B* 295 (2003): 12–24; R. H. Sawyer et al., "Evolutionary Origin of the Feather Epidermis," *Developmental Dynamics* 232 (2004): 256–67.

93. T. Lingham-Soliar, "Evolution of Birds: Ichthyosaur Integumental Fibers Conform to Dromaeosaur Protofeathers," *Naturwissenschaften* 90 (2003): 428–32.

94. T. Lingham-Soliar, "The Dinosaurian Origin of Feathers: Perspectives from Dolphin (Cetacea) Collagen Fibers," *Naturwissenschaften* 90 (2003): 563–67; X. Xu, Z. Zhou, and R. O. Prum, "Branched Integumental Structures in *Sinornithosaurus* and the Origin of Birds," *Nature* 410 (2001): 200–204.

95. Feduccia, Lingham-Soliar, and Hinchliffe, "Do Feathered Dinosaurs Exist?"

96. Lingham-Soliar, Feduccia, and Wang, "New Chinese Specimen."

97. A. W. A. Kellner et al., "The soft tissue of *Jeholopterus* (Pterosauria, Anurognathidae, Batrachognathinae) and the wing structure of the pterosaur wing membrane," *Proceedings of the Royal Society* B 277 (2010): 321–29.

98. Prum and Brush, "Which Came First?"

99. Ibid.

100. Lingham-Soliar, "Dinosaurian Origin of Feathers"; L. V. Kukhareva and R. K. Ileragimov, *The Young Mammoth from Magadan* [in Russian] (Leningrad: Nauka, 1981).

101. Prum and Brush, "Which Came First?"

102. P. J. Currie and P.-J. Chen, "Anatomy of *Sinosauropteryx prima* from Liaoning, Northeastern China," *Canadian Journal of Earth Science* 38 (2001): 1705–27.

103. John Ruben, pers. comm., 2009.

104. X. Xu et al., "A New Feather Type in a Nonavian Theropod and the Early Evolution of Feathers," *Proceedings of the National Academy of Sciences* 106 (2009): 832–34.

105. Ibid.

106. C. L. Organ and J. Adams, "The Histology of Ossified Tendon in Dinosaurs," *Journal of Vertebrate Paleontology* 25 (2005): 602–13; C. L. Organ, "Biomechanics of Ossified Tendons in Ornithopod Dinosaurs," *Paleobiology* 32 (2006): 652–65; R. T. Holmes and C. L. Organ, "An Ossified Tendon Trellis in *Chasmosaurus* (Ornithischia: Ceratopsidae)," *Journal of Paleontology* 81 (2007): 411–14.

107. T. Lingham-Soliar, "Dinosaur Protofeathers: Pushing Back the Origin of Feathers into the Middle Triassic," *Journal of Ornithology*, 51 (2010): 193–200; Chen, Dong, and Zhen, "Well-Preserved Theropod Dinosaur"; X. T. Zheng et al., "An Early Cretaceous Heterodontosaurid Dinosaur with Filamentous Integumentary Structures," *Nature* 458 (2009): 333–36.

108. Zheng, You, Xu, and Dong, "Early Cretaceous Heterodontosaurid."

109. Y. Wu, P. Xi, and J. Y. Qu, "Depth-Resolved Fluorescence Spectroscopy Reveals Layered Structure of Tissue," *Optical Express* 13 (2004): 382–88.

110. G. Mayr et al., "Bristle-Like Integumentary Structures at the Tail of the Horned Dinosaur *Psittacosaurus*," *Naturwissenschaften* 89 (2002): 361–65.

111. Lingham-Soliar, "Dinosaur Protofeathers."

112. Lingham-Soliar, Feduccia, and Wang, "New Chinese Specimen"; K. Sanderson, "Bald Dino Casts Doubt on Feather Theory," *NatureNews*, 23 March 2007.

113. J. Vinther et al., "The Color of Fossil Feathers," *Biology Letters* 4 (2008): 522–25; J. Vinther et al., "Structural Coloration in a Fossil Feather," *Biology Letters* 6 (2010): 128–31; P. G. Davis and D. E. G. Briggs, "Fossilization of Feathers," *Geology* 23 (1995): 783–86.

114. C. Wasmeier et al., "Melanosomes at a Glance," *Journal of Cell Science* 121 (2008): 3995–99; F. Zhang et al., "Fossilized Melanosomes and the Colour of Cretaceous Dinosaurs and Birds," *Nature* 463 (2010): 1075–78.

115. F. Zhang et al., "Fossilized Melanosomes."

116. Ibid.; Feduccia, *Origin and Evolution of Birds*, 378.

117. T. Lingham-Soliar, "The Evolution of the Feather: *Sinosauropteryx*, a Colourful Tail," *Journal of Ornithology* DOI: 10.1007/s10336-010-0620-y.

118. K. W. Barthel, N. H. M. Swinburne, and S. C. Morris, *Solnhofen: A Study in Mosozoic Paleontology* (Cambridge: Cambridge University Press, 1990).

119. R. S. Sansom et al., "Non-Random Decay of Chordate Characters Causes Bias in Fossil Interpretation," *Nature* 463 (2010): 797–800; D. E. G. Briggs, "Decay Distorts Ancestry," *Nature* 463 (2010): 741–42.

120. A. G. Sharov, "An Unusual Reptile from the Lower Triassic of Fergana" [in Russian], *Paleontologiceskij Zurnal* 1 (1970): 127–30.

121. H. Haubold and E. Buffetaut, "Une nouvelle interprétation de *Longisquama insignis*, reptile énigmatique du Trias supérieur d'Asie centrale," *Comptes Rendus de l'Academie des Sciences*, ser. 2A, 305 (1987): 65–70.

122. T. D. Jones et al., "Nonavian Feathers in a Late Triassic Archosaur," *Science* 288 (2000): 2202–5.

123. Sharov, "Unusual Reptile."

124. S. Voigt et al., "Feather-Like Development of Triassic Diapsid Skin Appendages," *Naturwissenschaften* 96 (2009): 81–86.

125. Z. Zhou and F. Zhang, "Origin of Feathers—Perspectives from Fossil Evidence," *Science Progress* 84 (2002): 87–104.

126. S. Voigt et al., "Feather-Like Development."

127. L. D. Martin, "A Basal Archosaur Origin of Birds," *Acta Zoological Sinica* 50 (2004): 978–90; L. D. Martin, "Origin of Avian Flight: A New Perspective," *Oryctos* 7 (2008): 45–54.

128. E. Stokstad, "Feathers, or Flight of Fancy?" *Science* 288 (2000): 2124–25.

129. Martin, "Basal Archosaur Origin."

130. A. G. Sharov, "New Flying Reptiles from the Mesozoic of Kazakhstan and Kirghizia [in Russian]," *Trudy of the Paleontological Institute Moscow* 130 (1971): 104–13.

131. G. J. Dyke, R. L. Nudds, and J. M. V. Raynor, "Flight of *Sharovipteryx*: The World's First Delta-Winged Glider," *Journal of Evolutionary Biology* 19 (2006): 1040–43.

132. N. C. Fraser et al., "A New Gliding Tetrapod (Diapsida: Archosauromorpha) from the Upper Triassic (Carnian) of Virginia," *Journal of Vertebrate Paleontology* 27 (2007): 261–65.

133. A. R. Jusufi et al., "Active Tails Enhance Arboreal Acrobatics in Geckos," *Proceedings of the National Academy of Sciences* 105 (2008): 4215–19.

134. T. Laman, "Wild Gliders," *National Geographic*, October 2000, 68–85.

135. M. Kessler, letter to the editor, *National Geographic*, February 2001.

136. U. M. Norberg, *Vertebrate Flight* (Berlin: Springer, 1990).

137. Zhang and Zhou, "Primitive Enantiornithine Bird."

138. Ibid.

139. Zhou and Zhang, "Origin of Feathers."

140. S.-T. Zheng, Z.-H. Zhang, and L.-H. Hou, "A New Enantiornithine Bird with Four Long Rectrices from the Early Cretaceous of Northern Hebei," *China Acta Geologica Sinica* 81 (2007): 703–8.

141. Ibid.

142. R. Zhang et al., "A Juvenile Coelurosaurian Theropod from China Indicates Arboreal Habits," *Naturwissenschaften* 89 (2002): 394–98.

143. S. J. Czerkas and C. Yuan, "An Arboreal Maniraptoran from Northeast China," *Dinosaur Museum Journal* 1 (2002): 63–95.

144. J. D. Harris, "'Published Works' in the Electronic Age: Recommended Amendments to Articles 8 and 9 of the Code," *Bulletin of Zoological Nomenclature* 61, no. 3 (2004): 138–148.

145. J. Gauthier, "Saurischian Monophyly and the Origin of Birds," in *The Origin of Birds and the Evolution of Flight,* ed. K. Padian, *Memoirs of the California Academy of Sciences* 8 (1986): 1–55.

146. S. Czerkas, pers. comm., in press.

147. Czerkas and Yuan, "Arboreal Maniraptoran."

148. Zhang, Zhou, Xu, and Wang, "Juvenile Coelurosaurian Theropod."

149. G. Heilmann, *The Origin of Birds* (London: Witherby, 1926).

150. F. Zhang et al., "A Bizarre Maniraptoran from China with Elongated Ribbon-Like Feathers," *Nature* 455 (2008): 1105–8.

151. X. Wang et al., "Stratigraphy and Age of the Daohugou Bed in Ningcheng, Inner Mongolia," *Chinese Science Bulletin* 50 (2005): 2369–76; K. Gao and D. Ren, "Radiometric Dating of Ignimbrite from Inner Mongolia Provides No Indication of a Post-Middle Jurassic Age for the Daohugou Beds," *Acta Geologica Sinica* (English ed.), 80 (2006): 42–45.

152. Zhang et al., "Bizarre Maniraptoran from China."

153. R. Barsbold et al., "New Oviraptorosaur (Dinosauria, Theropoda) from Mongolia: The First Dinosaur with a Pygostyle," *Acta Palaeontologica Polonica* 45 (2000): 97–106; R. Barsbold et al., "A Pygostyle in a Non-Avian Theropod," *Nature* 403 (2000): 135.

154. T. He, X.-L. Wang, and Z. Zhou, "A New Genus and Species of Caudipterid Dinosaur from the Lower Cretaceous Jiufotang Formation of Western Liaoning, China," *Vertebrata PalAsiatica* 7 (2008): 178–89.

155. Zhang et al., "Bizarre Maniraptoran from China."

156. P. J. Mackovicky and M. A. Norell, "Troodontidae," in *The Dinosauria,* ed. D. B. Weishampel, P. Dodson, and H. Osmólska, 2nd ed. (Berkeley: University of California Press, 2004), 184–195; D. J. Varricchio, "Troodontidae," in *Encyclopedia of Dinosaurs* (San Diego, CA: Academic, 1997), 749–54; D. A. Russell, "A New Specimen of *Stenonychosaurus* from the Oldman Formation (Cretaceous of Alberta)," *Canadian Journal of Earth Sciences* 6 (1969): 595–612; P. J. Currie, "Cranial Anatomy of *Stenonychosaurus inequalis* (Saurischia, Theropoda) and Its Bearing on the Origin of Birds," *Canadian Journal of*

Earth Sciences 22 (1985): 1643–58; P. J. Currie, "Bird-Like Characteristics of the Jaws and Teeth of Troodontid Theropods (Dinosauria, Saurischia)," *Journal of Vertebrate Paleontology* 7 (1987): 72–81; D. A. Russell and Z. M. Dong, "A Nearly Complete Skeleton of a Troodontid Dinosaur from the Early Cretaceous of the Ordos Basin, Inner Mongolia, China," *Canadian Journal of Earth Sciences* 30 (1993): 2163–73; P. J. Currie and Z. Dong, "New Information on Cretaceous Troodontids (Dinosauria, Theropoda) from the People's Republic of China," *Canadian Journal of Earth Sciences* 38 (2001): 1753–66.

157. X. Xu et al., "A Basal Troodontid from the Early Cretaceous of China," *Nature* 415 (2002): 780–84.

158. X. Xu and M. A. Norell, "A New Troodontid Dinosaur from China with Avian-Like Sleeping Posture," *Nature* 431 (2004): 838–41.

159. D. J. Varricchio et al., "Avian Paternal Care Had Dinosaur Origin," *Science* 322 (2008): 1826–28; R. O. Prum, "Who's Your Daddy?" *Science* 322 (2009): 1799–800.

160. J. L. Kavanau, "Secondary Flightless Birds or Cretaceous Non-Avian Theropods?" *Medical Hypotheses* 74 (2010): 275–76.

161. Q. Ji et al., "First Avialan Bird from China (*Jinfengopteryx elegans* gen. et sp. nov.)," *Geological Bulletin of China* 24, no. 3 (2005): 197–205; S. Ji and W. Ji, "*Jinfengopteryx* Compared to *Archaeopteryx,* with Comments on the Mosaic Evolution of Long-Tailed Avialan Birds," *Acta Geologica Sinica* (English ed.), 81 (2007): 337–43; Chiappe, *Glorified Dinosaurs*; X. Xu and M. A. Norell, "Non-Avian Dinosaur Fossil from the Lower Cretaceous Jehol Group of Western Liaoning, China," *Geological Journal* 41 (2006): 419–37; A. H. Turner et al., "A Basal Dromaeosaurid and Size Evolution Preceding Avian Flight," *Science* 317 (2007): 1378–81.

162. Turner et al., "Basal Dromaeosaurid."

163. X. Xu and F. Zhang, "A New Maniraptoran Dinosaur from China with Long Feathers on the Metatarsus," *Naturwissenschaften* 92, no. 4 (2005): 173–177.

164. P. J. Currie, "Bird-Like Characteristics of the Jaws and Teeth of Troodontid Theropods (Dinosauria: Saurischia)," *Journal of Vertebrate Paleontology* 7 (1987): 72–81; M. A. Norell and

S. H. Hwang, "A Troodontid Dinosaur from Ukhaa Tolgod (Late Cretaceous Mongolia)," *American Museum Novitates* 3446 (2004): 1–9.

165. X. Xu et al., "A New Feathered Maniraptoran Dinosaur That Fills a Morphological Gap in Avian Origins," *Chinese Science Bulletin* 54 (2009): 430–35; D. Hu et al., "A Pre-*Archaeopteryx* Troodontid Theropod from China with Long Feathers on the Metatarsus," *Nature* 461 (2009): 640–43.

166. B. C. Livesey, "Morphological Corollaries and Ecological Implications of Flightlessness in the Kakapo (Psittaciformes: *Strigops habroptilus*," *Journal of Morphology* 213 (1992): 105–45; D. V. Merton, "Kakapo," in *Reader's Digest Complete Book of New Zealand Birds*, ed. C. J. R. Robertson (Sydney: Reader's Digest Service, 1985), 242–43.

167. Hu et al., "Pre-*Archaeopteryx* Troodontid."

168. Q. Li et al., "Plumage Color Patterns of an Extinct Dinosaur," *Science* DOI: 10.1126/science.1186290.

169. F. Hertel and K. E. Campbell, Jr., "The Antitrochanter of Birds: Form and Function in Balance," *Auk* 124 (2007): 789–805.

170. A. D. Walker, "Evolution of the Pelvis in Birds and Dinosaurs," in *Problems in Vertebrate Evolution*, ed. S. M. Andrews et al., *Linnean Society Symposium Series*, no. 4 (1977): 319–58.

171. L. D. Martin and S. A. Czerkas, "The Fossil Record of Feather Evolution in the Mesozoic," *American Zoologist* 40 (2000): 687–94.

172. M. T. Carrano, "Locomotion in Non-Avian Dinosaurs: Integrating Data from Hindlimb Kinematics, In Vivo Strains, and Bone Morphology," *Paleobiology* 24 (1998): 450–69.

173. Hertel and Campbell, "Antitrochanter of Birds."

174. R. E. Brown and A. C. Cogley, "Contributions of the Propatagium to Avian Flight," *Journal of Experimental Zoology* 276 (1996): 112–24.

175. L. D. Martin and J.-D. Lim, "Soft Body Impression of the Hand of *Archaeopteryx*," *Current Science* 89 (2005): 1089–90.

176. Heilmann, *Origin of Birds*.

177. O. W. M. Rauhut, "The Interrelationships and Evolution of Basal Theropod Dinosaurs," *Special Papers in Palaeontology* 69 (2003): 1–213.

178. X. Xu et al., "A Jurassic Certosaur from China Helps Clarify Avian Digital Homologies," *Nature* 459 (2009): 940–44.

179. M. H. Kaplan, "Dinosaur's Digits Show How Birds Got Wings," *Nature News*, 17 June 2009.

180. Ibid.

181. X. Xu et al., "Jurassic Ceratosaur."

182. Ibid.

183. A. Feduccia, "Birds Are Dinosaurs: Simple Answer to a Complex Problem," *Auk* 119 (2002): 1187–201.

184. Feduccia, Lingham-Soliar, and Hinchliffe, "Do Feathered Dinosaurs Exist?"; Feduccia, "Birds Are Dinosaurs."

185. X. Xu, "Dinosaurs," in Chang, *Jehol Biota*, 109–27.

186. Q. Ji et al., "Two Feathered Dinosaurs from Northeastern China," *Nature* 393 (1998): 753–61.

187. H. Gee, "Birds Are Dinosaurs: The Debate Is Over," *Nature Science Update* (1998), www.nature.com/nsu/980702-8.html.

188. Ji Q. et al., "Two Feathered Dinosaurs."

189. H. Gee, "Birds Are Dinosaurs."

190. L. M. Witmer, "The Debate on Avian Ancestry: Phylogeny, Function and Fossils," in Chiappe and Witmer, *Mesozoic Birds*, 3–30.

191. T. D. Jones et al., "Cursoriality in Bipedal Archosaurs," *Nature* 406 (2000): 716–18; G. J. Dyke and M. A. Norell, "*Caudipteryx* as a Non-Avian Theropod Rather Than a Flightless Bird," *Acta Palaeontologica Polonica* 50 (2005): 101–16.

192. Z. Elżanowski, "A Comparison of the Jaw Skeleton in Theropods and Birds, with a Description of the Palate in the Oviraptoridae," *Smithsonian Contributions to Paleobiology* 89 (1999): 311–23; Gauthier, "Saurischian Monophyly."

193. T. Maryańska, H. Osmólska, H. M. Wolsan, "Avialan Status for Oviraptorosauria," *Acta Palaeontoliga Polonica* 47 (2002): 97–116; J. Lü et al., "Oviraptorosaurs Compared to Birds," in *Proceedings of the 5th Symposium of the Society of Avian Paleontology and Evolution*, ed. Z. Zhou and R. Zhang (Beijing: Science Press, 2002), 175–89.

194. L.-H. Hou et al., *Mesozoic Birds from Western Liaoning in China* (Liaoning: Liaoning Science and Technology Publishing, 2002); R. O. Prum, "Moulting Tail Feathers in a Juvenile Oviraptorisaur," *Nature* 468 (2010): E1, DOI: 10.1038/nature09480.

195. X. Xu et al., "An Unusual Oviraptorosaurian Dinosaur from China," *Nature* 419 (2001): 291–93.

196. Chang, ed., *Jehol Biota*.
197. P. Currie, "Dromaeosauridae" and "Feathered Dinosaurs," in *Encyclopedia of Dinosaurs*, ed. P. Currie and K. Padian (San Diego, CA: Academic Press, 1997), 194–95, 241.
198. J. Ruben, "The Evolution of Endothermy in Mammals and Birds: From Physiology to Fossils," *Annual Review of Physiology* 57 (1995): 69–95; J. Ruben, "The Evolution of Endothermy in Mammals, Birds and Their Ancestors," *Society of Experimental Biology, Science Series* 59 (1996): 347–76; J. Ruben and W. J. Hillenius, "Were Dinosaurs 'Cold-' or 'Warm-Blooded'?: 'Cold-blooded,'" *Natural History*, May 2005, 50–51; D. E. Quick and J. A. Ruben, "Cardio-Pulmonary Anatomy in Theropod Dinosaurs: Implications from Extant Archosaurs," *Journal of Morphology* 270 (2009): 1232–46.
199. M. J. Benton, "Dinosaurs," *Current Biology* 19 (2009): R318–23.
200. Zhou and Zhang, "Mesozoic Birds of China."
201. P. J. Currie et al., *Feathered Dragons: Studies on the Transition from Dinosaurs to Birds* (Bloomington: Indiana University Press, 2004).
202. M. A. Norell and X. Xu, "Feathered Dinosaurs," *Annual Review of Earth and Planetary Sciences* 33 (2005): 277–99.
203. D. R. Prothero, *Evolution: What the Fossils Say and Why It Matters* (New York: Columbia University Press, 2007).
204. W. D. Allmon, P. H. Kelley, and R. M. Ross, *Stephen Jay Gould: Reflections on His View of Life* (Oxford: Oxford University Press, 2009).
205. G. de Beer, "The Evolution of Ratites," *Bulletin of the British Museum (Natural History)* 4 (1956): 59–70.
206. X. Xu, Z.-L.Tang, and X.-L.Wang, "A Therizinosauroid Dinosaur with Integumentary Structures from China," *Nature* 399 (1999): 350–54.
207. Xu et al., "New Feather Type in a Nonavian Theropod."
208. X.-T. Zheng, et al. "An Early Cretaceous heterodontosaurid Dinosaur with Filamentous Integumentary Structures," *Nature* 458 (2009): 333–36.
209. T. Lingham-Soliar, "Dinosaur Protofeathers: Pushing Back the Origin of Feathers into the Middle Triassic?" *Journal of Ornithology* 151 (2010): 193–200; G. Mayr, "Response to Lingham-Soliar: Dinosaur Protofeathers . . . ,"

Journal of Ornithology 151 (2010): 523–24; T. Lingham-Soliar, "Response to Comments by G. Mayr . . . ," *Journal of Ornithology* 151 (2010): 519–21.
210. X. Xu, X. Wang, and X. Wu, "A Dromaeosaurid Dinosaur with a Filamentous Integument from the Yixian Formation of China," *Nature* 401 (1999): 262–66; Q. Ji et al., "The Distribution of Integumentary Structures in a Feathered Dinosaur," *Nature* 410 (2001): 1084–87.
211. M. A. Norell, "The Proof Is in the Plumage," *Natural History* 10 (2001): 58–63.
212. Feduccia, Lingham-Soliar, and Hinchliffe, "Do Feathered Dinosaurs Exist?"
213. Prum and Brush, "Which Came First?"
214. Ibid.
215. E. Gong et al., "The Birdlike Raptor *Sinornithosaurus* Was Venomous," *Proceedings of the National Academy of Sciences* 107 (2010): 766–68.
216. X. Xu, Z. Zhou, and X. Wang, "The Smallest Known Non-Avian Theropod Dinosaur," *Nature* 408 (2000): 705–8.
217. X. Xu et al., "Four-Winged Dinosaurs from China," *Nature* 421 (2003): 335–40; R. O. Prum, "Paleontology: Dinosaurs Take to the Air," *Nature* 421 (1003): 323–24; E. Stokstad, "Four-Winged Dinos Create a Flutter," *Science* 299 (2003): 491.
218. S. A. Czerkas et al., "Flying Dromaeosaurs," in *Feathered Dinosaurs and the Origin of Flight*, ed. Czerkas (Blanding, UT: Dinosaur Museum, 2002), 97–126; P. Senter et al., "Systematics and Evolution of Dromaeosauridae (Dinosauria, Theropoda)," *Bulletin of the Gunma Museum of Natural History* 8 (2004): 1–20.
219. S. H. Hwang et al., "New Specimens of *Microraptor zhaoianus* (Theropoda: Dromaeosauridae) from Northeastern China," *American Museum Novitates* 3381 (2002): 1–44.
220. Zhou and Zhang, "Mesozoic Birds of China."
221. C. W. A. Beebe, "Tetrapteryx Stage in the Ancestry of Birds," *Zoologica* 2 (1915): 38–52.
222. N. Longrich, "Structure and Function of Hindlimb Feathers in *Achaeopteryx lithographica*," *Paleobiology* 32, no. 3 (2006): 417–31; F. Zhang and Z. Zhou, "Leg Feathers in an Early Cretaceous Bird," *Nature* 431 (2004): 925; X. Xu and F. Zhang, "A New Maniraptoran Dinosaur from China with Long Feathers on the Metatarsus," *Naturwissenschaften* 92 (2005): 173–77.

223. Xu et al., "A New Maniraptoran."

224. Xu et al., "Four-winged Dinosaurs."

225. D. W. E. Hone et al., "The Extent of the Pre-served Feathers on the Four-winged Dinosaur *Microraptor gui* under Ultraviolet Light," *PLoS ONE* 5 (2010): e9223.

226. N. Longrich, "Structure and Function of Hindlimb Feathers"; H. Tischlinger and D. M. Unwin, "UV-Untersuchungen des Berliner Exemplars von *Archaeopteryx lithographica* H. v. Meyer 1861 und der Isolierten Archaeopteryx-Feder [Ultra-violet Light Investigation of the Berlin Example of *Archaeopteryx lighographica* H. v. Meyer 1861 . . .]," *Archaeopteryx* 22 (2004): 17–50; H. Tischlinger, "Ultraviolet Light of Fossils from the Upper Jurassic Plattenkalks of Southern Frankonia," *Zitteliana* B 26 (2005): 26; H. Tischlinger, "Der Eichstätter Archaeopteryx im Langwelligen UV-Licht," [The Eichstätt Specimen of *Archaeopteryx* under Longwave Ul-traviolet Light], *Archaeopteryx* 20 (2002): 21–38; H. Tischlinger, "Neue Information Zum Berliner Exemplar von *Archaeopteryx lithographica* H. v. Meyer 1861," *Archaeopteryx* 23 (2005): 33–50.

227. S. Chatterjee and R. J. Templin, "Biplane Wing Planform and Flight Performance of the Feath-ered Dinosaur *Microraptor gui*," *Proceedings of the National Academy of Sciences* 104 (2007): 1576–80.

228. M. Davis (producer), "The Four-Winged Dinosaur," *NOVA*, 2008, http://www.pbs.org/wgbh/nova/microraptor/.

229. D. E. Alexander, *Nature's Flyers: Birds, Insects, and the Biomechanics of Flight* (Baltimore: Johns Hopkins University Press, 2002).

230. M. Davis (producer), video extra, part 5, for "The Four-Winged Dinosaur," *NOVA*, http://www.pbs.org/wgbh/nova/microraptor/extras.html.

231. Ibid., part 2.

232. Davis, "Four-Winged Dinosaur."

233. Ibid.

234. D. E. Alexander et al., "Model Tests of Gliding with Different Hindwing Configurations in the Four-winged Dromaeosaurid *Microraptor gui*," *Proceedings of the National Academy of Sciences* 107 (2010): 2972–76; J. Ruben, "Paleobiology and the Origins of Avian Flight," *Proceedings of the National Academy of Sciences* 107 (2010): 2733–34.

235. Norell, *Unearthing the Dragon.*

236. Ji et al., "Distribution of Integumentary Structures."

237. Chiappe, *Glorified Dinosaurs.*

238. B. C. Livezey and R. L. Zusi, "Higher-Order Phylogeny of Modern Birds (Theropoda, Aves: Neornithes) Based on Comparative Anatomy," *Bulletin of the Carnegie Museum of Natural History*, no. 37 (2007).

239. G. Mayr, "Avian Higher-Level Phylogeny: Well-Supported Clades and What We Can Learn from a Phylogenetic Analysis of 2954 Morphological Characters," *Journal of Systematic Evolutionary Research* 46, no. 1 (2007): 63–72.

240. Chang, *Jehol Biota.*

241. P. Dodson, "Origin of Birds: The Final Solu-tion," *American Zoologist* 40, no. 4 (2000): 504–12.

242. X. Xu et al., "An *Archaeopteryx*-like Theropod from China and the Origin of Avialae," *Nature* 475 (2011): 465–70.

Chapter 5: Big Chicks

Epigraph: W. Wordsworth, "The Rainbow," in *The Complete Poetical Works of William Wordsworth* (London: Macmillan, 1888).

1. C. Linnaeus, *Philosophia Botanica* (Stockholm, 1751).

2. T. H. Huxley, review of *On the Origin of Species . . .* , by C. Darwin, *Westminster Re-view* (April 1860).

3. C. Lyell, *The Principles of Geology*, 3 vols. (London: Murray, 1830–33).

4. W. Whewell, review of *Principles of Geology*, vol. 2, by Charles Lyell, *Quarterly Review* 47 (1832): 103–23.

5. T. H. Huxley to C. Lyell, 1874, in L. Huxley, *Life and Letters of Thomas Henry Huxley*, vol. 1 (London: Macmillan, 1900), 173.

6. Huxley to C. Darwin, ibid., 176.

7. B. K. Hall and W. M. Olson, *Key Words and Concepts in Evolutionary Developmental Biology* (Cambridge, MA: Harvard University Press, 2003); S. J. Gould, *Ontogeny and Phylogeny* (Cambridge, MA: Harvard University Press, 1977).

8. P. Alberch et al., "Size and Shape in Ontogeny and Phylogeny," *Paleobiology* 5 (1979): 296–317; E. J. McNamara, ed., *Evolutionary Change and Heterochrony* (New York: John Wiley and Sons, 1995).

9. M. L. McKinney and K. J. McNamara, *Heterochrony: The Evolution of Ontogeny* (New York: Plenum, 1991).

10. G. de Beer, "The Evolution of Ratites," *Bulletin of the British Museum (Natural History)* 4 (1956): 59–70.

11. B. C. Livezey, "Heterochrony and the Evolution of Avian Flightlessness," in McNamara, *Evolutionary Change and Heterochrony*, 169–93.

12. Gould, *Ontogeny and Phylogeny*.

13. Ibid.

14. L. Bolk, "On the Problem of Anthropogenesis," *Proceedings of the Section of Sciences, Koninklijke Akademie van Wetenschappen te Amsterdam* 29 (1926): 465–75.

15. K. Miller, "What Does It Mean to Be One of Us?" (photographs by Lennart Nilsson), *Life*, November 1996, 38–56; Wordsworth, "The Rainbow."

16. W. Garstang, "The Theory of Recapitulation: A Critical Re-Statement of the Biogenetic Law," *Zoological Journal of the Linnaen Society* 35 (1922): 81–101.

17. L. N. Trut, "Early Canid Domestication: The Farm-Fox Experiment," *American Scientist* 87 (1999): 160–69; L. N. Trut, I. Z. Plyusnina, and I. N. Oskina, "An Experiment on Fox Domestication and Debatable Issues of Evolution of the Dog," *Russian Journal of Genetics* 40 (2004): 644–55.

18. J. F. Gudersnatsch, "Feeding Experiments on Tadpoles: The Influence of Specific Organs Given as Food on Growth and Differentiation," *Archive Entwicklungsmech* 35 (1912): 457–83; A. Dawson, "Neoteny and the Thyroid in Ratites," *Reviews of Reproduction* 1 (1996): 78–81.

19. Dawson, "Neoteny and the Thyroid in Ratites."

20. A. Dawson et al., "Ratite-Like Neoteny Induced by Neonatal Thyroidectomy of European Starlings, *Sturnus vulgaris*," *Journal of Zoology, London* 232 (1994): 633–39.

21. Garstang, "Theory of Recapitulation."

22. De Beer, "Evolution of Ratites"; Livezey, "Heterochrony and the Evolution of Avian Flightlessness"; A. Feduccia, *The Origin and Evolution of Birds* (New Haven: Yale University Press, 1996).

23. G. G. Simpson, "Fossil Penguins," *Bulletin of the American Museum of Natural History* 87 (1946): 1–99.

24. L. M. Witmer and K. D. Rose, "Biomechanics of the Jaw Apparatus of the Gigantic Eocene Bird *Diatryma*: Implications for Diet and Mode of Life," *Paleobiology* 17 (1991): 95–120.

25. S. J. Gould, "Play It Again, Life," *Natural History*, February 1986, 23–26.

26. A. S. Romer, *Vertebrate Paleontology*, 3rd ed. (Chicago: University of Chicago Press, 1966).

27. L. G. Marshall, "The Terror Birds of South America," *Scientific American* 270, no. 2 (1994): 90–95; M. F. Alvarenga and E. Höfling, "Systematic Revision of the Phorusrhacidae (Aves: Ralliformes)," *Papéis Avulsos de Zoologia* 43, no. 4 (2003): 55–91.

28. L. M. Chiappe and S. Bertelli, "Skull Morphology of Giant Terror Birds," *Nature* 443 (2006): 929.

29. R. M. Chandler, "The Wing of *Titanis walleri* (Aves: Phorusrhacidae) from the Late Blancan of Florida," *Bulletin Florida Museum Natural History, Biological Sciences Series* 36, no. 6 (1994): 175–80; B. McFadden et al., "Revised Age of the Late Neogene Terror Bird (*Titanis*) in North America during the Great American Interchange," *Geology* 35, no. 2 (2007): 123–26.

30. G. Mayr, "Old World Phorusrhacids (Aves, Phorusrhacidae): A New Look at *Strigogyps* ('*Aenigmavis*') *sapea* (Peters 1987)," *PaleoBios* 25, no. 1 (2005): 11–16.

31. D. A. Roff, "The Evolution of Flightlessness: Is History Important?" *Evolutionary Ecology* 8 (1994): 639–57; B. C. Livezey, "Evolution of Flightlessness in Rails (Gruiformes: Rallidae): Phylogenetic, Ecomorphological, and Ontogenetic Perspective," *American Ornithologists' Union Monograph* 59 (2003).

32. P. V. Rich, "The Dromornithidae," *Bureau of Mineral Resources Bulletin* 184 (Canberra: Australian Government Publishing, 1979); P. F. Murray and D. Megirian, "The Skull of Dromornithid Birds: Anatomical Evidence for Their Relationship to Anseriformes (Dromornithidae, Anseriformes)," *Records of the South Australian Museum* 31 (1998): 51–97.

33. C. Darwin, *On the Origin of Species* (London: John Murray, 1859).

34. A. R. Wallace, *Island Life; or, The Phenomena and Causes of Insular Faunas and Floras, Including a Revision and Attempted Solution of the Problem of Geological Climates* (London: Macmillan, 1880).

35. R. Owen, *On the Anatomy of Vertebrates*, vol. 2: *Birds and Mammals* (London: Longmans, Green, 1866); A. Desmond, *Archetypes and Ancestors: Palaeontology in Victorian London, 1850–1875* (Chicago: University of Chicago Press, 1982).

36. M. Fürbringer, *Untersuchungen zur Morphologie und Systematik der Vögel, zugleich ein Beitrag zur Anatomie der Stütz- und Bewegungsorgane*, 2 vols. (Amsterdam: T. J. Van Holkema, 1888); R. Broom, "On the Early Development of the Appendicular Skeleton of the Ostrich, with Remarks on the Origin of Birds," *Transactions of the South African Philosophical Society* 16 (1906): 355–68; J. E. Duerden, "Methods of Degeneration in the Ostrich," *Journal of Genetics* 9 (1920): 131–93; E. Stresemann, "Sauropsida: Aves," in *Handbuch der Zoologie*, vol. 7(2) (Berlin: De Gruyter, 1934); W. K. Gregory, "Remarks on the Origins of the Ratites and Penguins," *Proceedings of the Linnean Society of New York* 45–46 (1935): 1–18.

37. De Beer, "Evolution of Ratites."

38. H. E. Strickland and A. G. Melville, *The Dodo and Its Kindred* (London: Reeve, Benham and Reeve, 1848).

39. Livezey, "Evolution of Flightlessness"; S. L. Olson, "Evolution of the Rails of the South Atlantic Islands (Aves: Rallidae)," *Smithsonian Contributions to Zoology* 152 (1973): 1–53.

40. D. W. Steadman, "Two New Species of Rails (Aves: Rallidae) from Mangaia, Southern Cook Islands," *Pacific Studies* 40 (1986): 27–43; D. W. Steadman, "Prehistoric Extinctions of Pacific Island Birds: Biodiversity Meets Zooarchaeology," *Science* 267 (1995): 1123–31.

41. Olson, "Evolution of the Rails."

42. B. K. McNab, "Energy Conservation and the Evolution of Flightlessness in Birds," *American Naturalist* 144 (1994): 628–42.

43. J. Diamond, "Flightlessness and Fear of Flying in Island Species," *Nature* 293 (1981): 507–8.

44. J. Cracraft, "Phylogeny and Evolution of Ratite Birds," *Ibis* 115 (1974): 494–521; J. Cracraft, "Cladistic Analysis and Vicariance Biogeography," *American Scientist* 71 (1983): 273–81; J. Cracraft, "Avian Evolution, Gondwana Biogeography and the Cretaceous-Tertiary Mass Extinction Event," *Proceedings of the Royal Society B* 268 (2001): 459–69; J. Cracraft, "Gondwana Genesis," *Natural History* 110 (2001): 64–73.

45. C. G. Sibley and J. E. Ahlquist, *Phylogeny and Classification of Birds: A Study in Molecular Evolution* (New Haven: Yale University Press, 1990).

46. P. Houde and S. L. Olson, "Paleognathous Carinate Birds from the Early Tertiary of North America," *Science* 214 (1981): 1236–37.

47. P. Houde, *Paleognathous Birds from the Early Tertiary of the Northern Hemisphere* (Cambridge, MA: Nuttall Ornithological Club, 1988).

48. Ibid.

49. A. Cooper et al., "Complete Mitochondrial Genome Sequences of Two Extinct Moas Clarify Ratite Evolution," *Nature* 409 (2001): 704–7; A. Cooper et al., "Independent Origins of New Zealand Moas and Kiwis," *Proceedings of the National Academy of Sciences* 89 (1992): 8741–44.

50. R. Dawkins, *The Ancestor's Tale* (Boston: Houghton Mifflin, 2004).

51. J. Harshman et al., "Phylogenomic Evidence for Multiple Losses of Flight in Ratite Birds," *Proceedings of the National Academy of Sciences* 36 (2008): 13462–67; M. J. Phillips et al., "Tinamous and Moa Flock Together: Miochondrial Genome Sequence Analysis Reveals Independent Losses of Flight among Ratites," *Systematic Biology* 59 (2010): 90–107.

52. S. L. Olson and H. F. James, "Descriptions of Thirty-Two New Species of Birds from the Hawaiian Islands; Part I: Non-Passeriformes," *Ornithological Monographs* 45 (1991): 1–88.

53. S. L. Olson and A. Wetmore, "Preliminary Diagnoses of Extraordinary New Genera of Birds from Pleistocene Deposits in the Hawaiian Islands," *Proceedings of the Biological Society of Washington* 89 (1976): 247–58.

54. Olson and James, "Descriptions of Thirty-Two New Species."

55. Ibid.

56. E. E. Paxinos et al., "mtDNA from Fossils Reveals a Radiation of Hawaiian Geese Recently Derived from the Canada Goose (*Branta canadensis*)," *Proceedings of the National Academy of Sciences* 99 (2002): 1399–404.

57. M. D. Sorenson et al., "Relationships of the Extinct Moa-Nalos, Flightless Hawaiian Waterfowl, Based on Ancient DNA," *Proceedings of the Royal Society B* 266 (1999): 2187–93.

58. Paxinos et al., "mtDNA from Fossils Reveals a Radiation."

59. A. Feduccia, "'Big Bang' for Tertiary Birds?" *Trends in Ecology and Evolution* 18 (2003): 172–76.

60. Huxley quoted in Desmond, *Archetypes and Ancestors*.

61. Olson, "Evolution of the Rails of the South Atlantic Islands."

62. Olson and James, "Descriptions of Thirty-Two New Species."

63. B. Silkas, S. L. Olson, and R. C. Fleischer, "Rapid, Independent Evolution of Flightlessness in Four Species of Pacific Island Rails (Rallidae): An Analysis Based on Mitochondrial Sequence Data," *Journal of Avian Biology* 33 (2002): 5–14.

64. Paxinos et al., "mtDNA from Fossils Reveals a Radiation."

65. T. J. Horder, "Gavin Rylands de Beer: How Embryology Foreshadowed the Dilemmas of the Genome," *Nature Reviews Genetics* 7 (2006): 892–98.

66. Ibid.

67. De Beer, *Embryos and Ancestors*.

68. Gould, *Ontogeny and Phylogeny*.

69. G. de Beer, *Archaeopteryx lithographica: A Study Based on the British Museum Specimen* (London: British Museum of Natural History, 1954).

70. De Beer, "Evolution of Ratites."

71. R. W. Richards, *The Tragic Sense of Life: Ernst Haeckel and the Struggle over Evolutionary Thought* (Chicago: University of Chicago Press, 2008).

72. M. K. Richardson, "Haeckel's Embryos Continued," *Science* 281 (1998): 1289; M. K. Richardson and G. Keuck, "Haeckel's ABC of Evolution and Development," *Biological Reviews* 77 (2002): 495–528.

73. E. Mayr, "Recapitulation Reinterpreted: The Somatic Program," *Quarterly Review of Biology* 69, no. 2 (1994): 223–32; J. F. Meckel, *System der Vergleichenden Anatomie*, 7 vols. (Halle: Rengersche Buchhandlung, 1821).

74. Gould, *Ontogeny and Phylogeny*; Mayr, "Recapitulation Reinterpreted."

75. T. H. Huxley, "On the Classification of Birds and on the Taxonomic Value of the Modifications of Certain of the Cranial Bones Observable in That Class," *Proceedings of the Zoological Society of London* (1867): 415–72.

76. W. P. Pycraft, "On the Morphology and Phylogeny of the Palaeognathae (Ratitae and Crypuri) and Neognathae (Carinatae)," *Transactions of the Zoological Society of London* 15 (1900): 149–290.

77. E. Buffetaut et al., "A Large French Cretaceous Bird," *Nature* 377 (1995): 110.

78. P. R. Lowe, "Studies and Observations Bearing on the Phylogeny of the Ostrich and Its Allies," *Proceedings of the Zoological Society of London* 98 (1928): 185–247; P. R. Lowe, "On the Primitive Characters of the Penguins, and Their Bearing on the Phylogeny of Birds," *Proceedings of the Zoological Society of London* 102 (1933): 483–541; P. R. Lowe, "On the Relationships of the Struthiones to the Dinosaurs and the Rest of the Avian Class, with Special Reference to the Position of *Archaeopteryx*," *Ibis* 77 (1935): 398–432; P. R. Lowe, "Some Additional Anatomical Factors Bearing on the Phylogeny of the Struthiones," *Proceedings of the Zoological Society of London* 112 (1942): 1–20; P. R. Lowe, "Some Additional Remarks on the Phylogeny of the Struthiones," *Ibis* 86 (1944): 37–42; P. R. Lowe, "An Analysis of the Characters of *Archaeopteryx* and *Archaeornis*: Were They Reptiles or Birds?" *Ibis* 86 (1944): 517–43.

79. Lowe, "Some Additional Anatomical Factors."

80. Lowe, "Studies and Observations."

81. Ibid.

82. Lowe, "On the Primitive Characters of the Penguins."

83. Lowe, "Some Additional Anatomical Factors."

84. P. R. Lowe, "A Description of *Atlantisia rogersi*, the Diminutive and Flightless Rail of Inaccessible Island (Southern Atlantic), with Some Notes on Flightless Rails," *Ibis* 70 (1928): 99–131.

85. S. J. Czerkas, ed., *Feathered Dinosaurs*.

86. Simpson, "Fossil Penguins."

87. De Beer, "Evolution of Ratites"; de Beer, "Evolution of Flying and Flightless Birds."

88. T. Edinger, "The Brain of Pterodactylus," *American Journal of Science* 239 (1941): 665–82.

89. De Beer, "Evolution of Ratites."

90. Olson, "Evolution of the Rails."

91. B. M. Dzoma and G. M. Dorrestein, "Yolk Sac Retention in the Ostrich (*Struthio camelus*): Histopathologic, Anatomic, and Physiologic Considerations," *Journal of Avian Medicine and Surgery* 15, no. 2 (2001): 81–89.

92. De Beer, "Evolution of Ratites."

93. De Beer, "Evolution of Flying and Flightless Birds."

94. H. F. James and S. L. Olson, "Flightless Birds," *Natural History* 92 (1983): 30–40.

95. Livezey, "Heterochrony and the Evolution of Avian Flightlessness."

96. G. Heilmann, *The Origin of Birds* (London: Witherby, 1926).

97. K. Padian, "The False Issues of Bird Origins: An Historiographic Perspective," in *New Perspectives on the Origin and Early Evolution of Birds*, ed. J. Gauthier and L. F. Gall (New Haven: Peabody Museum of Natural History, Yale University, 2001), 485–99; L. M. Chiappe, *Glorified Dinosaurs: The Origin and Early Evolution of Birds* (New York: John Wiley and Sons, 2007).

98. L. Dollo, "Les lois de l'évolution," *Bulletin Belge de Géologie* 7 (1893): 164–67.

99. S. J. Gould, *Eight Little Piggies: Reflections in Natural History* (New York: W. W. Norton, 1993); R. Dawkins, *The Blind Watchmaker* (1996; repr., New York: W. W. Norton, 1996).

100. Gould, *Eight Little Piggies*.

101. R. Collin and R. Cipriani, "Dollo's Law and the Re-Evolution of Shell Coiling," *Proceedings of the Royal Society B* 270 (2003): 2551–55.

102. M. F. Whiting, S. Bardler, and T. Maxwell, "Loss and Recovery of Wings in Stick Insects," *Nature* 421 (2003): 264–67.

103. K. Domes et al., "Reevolution of Sexuality Breaks Dollo's Law," *Proceedings of the National Academy of Sciences* 104 (2007): 7139–44.

104. C. R. Marshall, E. C. Raff, and R. A. Raff, "Dollo's Law and the Death and Resurrection of Genes," *Proceedings of the National Academy of Sciences* 91 (1994): 12283–87.

105. J. T. Bridgham, E. Ortlund, and J. Thornton, "An Epistatic Ratchet Constrains the Direction of Glucocorticoid Receptor Evolution," *Nature* 461 (2009): 515–19; M. Torrice, "Dylan to Darwin: Don't Look Back," *Science Now*, 23 September 2009.

106. T. Kohlsdorf and G. P. Wagner, "Evidence for the Reversibility of Digit Loss: A Phylogenetic Study of Limb Evolution in the Genus *Bachia* (Gymnophthalmidae: Squamata)," *Evolution* 60 (2006): 1896–1912; J. J. Weins, "Re-Evolution of Lost Mandibular Teeth in Frogs after More Than

200 Million Years, and Re-Evaluating Dollo's Law," *Evolution* 65 (2011): 1283–96.

107. R. Collin and M. P. Miglietta, "Reversing Opinions on Dollo's Law," *Trends in Ecology and Evolution* 23 (2008): 602–9; F. Galis et al., "Dollo's Law and the Irreversibility of Digit Loss in Bachia," *Evolution* 64 (2010): 2466–76; T. Kohlsdorf et al., "Data and Data-Interpretation in the Study of Limb Evolution: A Reply to Galis et al., on the Re-Evolution of Digits in the Lizard Genus *Bachia*," *Evolution* 64 (2010): 2477–85.

108. J. G. Kingsolver and D. W. Pfennig, "Individual-Level Selection as a Cause of Cope's Rule of Phyletic Size Increase," *Evolution* 58 (2004): 1608–12; D. W. E. Hone and M. J. Benton, "The Evolution of Large Size: How Does Cope's Rule Work?" *Trends in Ecology and Evolution* 20, no. 1 (2005): 4–6.

109. K. Padian and A. de Ricqlès, "L'Origine et l'évolution des oiseaux: 35 années de progrès," *Comptes Rendus Palevol* 8 (2009): 257–80.

110. J. Long and K. McNamara, "Heterochrony in Dinosaur Evolution," in *Evolutionary Change and Heterochrony*, ed. K. McNamara (New York: John Wiley and Sons, 1995), 151–68.

111. A. Vargas, "Evolution of Arm Size in Theropod Dinosaurs: A Developmental Hypothesis," *Noticiario Mensual, Museo Nacional de Historia Natural* (Santiago, Chile) 338 (1999): 16–19.

112. X. Xu et al., "Basal Tyrannosauroids from China and Evidence for Protofeathers in Tyrannosauroids," *Nature* 431 (2004): 680–84; G. M. Erickson et al., "Gigantism and Comparative Life-History Parameters of Tyrannosaurid Dinosaurs," *Nature* 433 (2004): 772–75.

113. P. C. Sereno et al., "Tyrannosaurid Skeletal Design First Evolved at Small Body Size," *Sciencexpress*, 17 September 2009, doi:10.1126/science.1177428.

114. Ibid.

115. Livezey, "Heterochrony and the Evolution of Avian Flightlessness"; W. Montagna, "A Re-Investigation of the Development of the Wing of the Fowl," *Journal of Morphology* 76 (1945): 87–113.

116. P. Senter, "A New Look at the Phylogeny of Coelurosauria (Dinosauria: Theropoda)," *Journal of Systematic Paleontology* 5 (2004): 429–63; M. C. Langler and M. J. Benton, "Early

Dinosaurs: A Phylogenetic Study," *Journal of Systematic Paleontology* 4 (2006): 309–58; P. C. Sereno, "The Phylogenetic Relationships of Early Dinosaurs: A Comparative Report," *Historical Biology* 19 (2007): 145–55.

117. S. Ji et al., "A New Giant Compsognathid Dinosaur with Long Filamentous Integuments from the Lower Cretaceous of Northeastern China," *Acta Geologica Sinica* 81, no. 1 (2007): 8–15.

118. J. H. Ostrom, "*Archaeopteryx* and the Origin of Flight," *Quarterly Review of Biology* 49 (1974): 27–47.

119. Owen quoted in Desmond, *Archetypes and Ancestors*.

120. Buffetaut, "Large French Cretaceous Bird"; Chiappe, *Glorified Dinosaurs*.

121. S. Zhen et al., "Dinosaur and Bird Footprints from the Lower Cretaceous of Emei County, Sichuan, China," *Memoirs of the Beijing Natural History Museum* 54 (1994): 106–20. J. Y. Kim et al., "New Didactyl Dinosaur Footprints (*Dromaeosauripus hamanensis* ichnogen. et ichnosp. nov.) from the Early Cretaceous Haman Formation, South Coast of Korea," *Palaeogeography* 262 (2008): 72–78; L.-D. Xing et al., "The Earliest Known Deinonychosaur Tracks from the Jurassic-Cretaceous Boundary in Hebei Province, China," *Acta Palaeontologica Sinica* 48 (2009): 662–71.

122. Z. Csiki et al., "An Aberrant Island-Dwelling Theropod Dinosaur for the Late Cretaceous of Romania," *Proceedings of the National Academy of Sciences* 107 (2010): 15357–61; H.-D. Sues, "An Unusual Dinosaur from the Late Cretaceous of Romania and the Island Rule," *Proceedings of the National Academy of Sciences* 107 (2010): 15310–11.

Chapter 6: Yale's Raptor

Epigraph: T. H. Huxley to Rev. C. Kinglsey, 1860, in L. Huxley, *Life and Letters of Thomas Henry Huxley*, vol. 1 (London: Macmillan, 1900), 217.

1. J. H. Ostrom, "The Yale *Archaeopteryx*: The One That Flew the Coop," in *The Beginnings of Birds: Proceedings of the International Archaeopteryx Conference*, ed. M. K. Hecht et al. (Eichstätt: Freunde des Jura-Museums, 1985), 359–67.

2. P. Chambers, *Bones of Contention: The Archaeopteryx Scandals* (London: John Murray, 2002).

3. Ibid.

4. K. Thomson, *The Legacy of the Mastodon: The Golden Age of Fossils in America* (New Haven: Yale University Press, 2008); D. R. Wallace, *The Bonehunter's Revenge: Dinosaurs, Greed, and the Greatest Scientific Feud of the Gilded Age* (New York: Houghton Mifflin, 1999).

5. D. S. Jordon, ed., *Leading American Men of Science* (New York: Henry Holt, 1910).

6. O. C. Marsh, *Odontornithes: A Monograph on the Extinct Toothed Birds of North America*, Report of the U. S. Geological Exploration of the Fortieth Parallel, no. 7 (Washington, DC, 1880), 2.

7. O. C. Marsh, "Notice of a New and Remarkable Fossil Bird," *American Journal of Science*, 3rd ser., 4 (1872): 344.

8. O. C. Marsh, "Introduction and Succession of Vertebrate Life in America," *American Journal of Science*, 3rd ser., 14 (1877): 337–78.

9. A. Desmond, *Huxley: From Devil's Disciple to Evolution's High Priest* (Reading, MA: Addison-Wesley, 1997); Lecture to the Royal Society, 19 March 1875, in *The Scientific Memoirs of Thomas Henry Huxley*, vol. 4, ed. M. Foster and E. R. Lankester (London: Macmillan, 1902).

10. Desmond, *Huxley*, 471.

11. Huxley to C. King, 1876, in Huxley, *Life and Letters of Thomas Henry Huxley*, 464.

12. Huxley to O. C. Marsh, 1876, in ibid., 463.

13. Marsh, "Introduction and Succession of Vertebrate Life," 352.

14. J. H. Ostrom, "Terrible Claw," *Discovery* 5 (1969): 1–9.

15. R. T. Bakker, "Anatomical and Ecological Evidence of Endothermy in Dinosaurs," *Nature* 238 (1972): 81–85; R. T. Bakker, "Dinosaur Renaissance," *Scientific American* 232, no. 4 (1975): 58–78; R. T. Bakker, *The Dinosaur Heresies: New Theories Unlocking the Mystery of the Dinosaurs and Their Extinction* (New York: William Morrow, 1986); A. Desmond, *The Hot-Blooded Dinosaurs* (New York: Dial, 1976).

16. J. H. Ostrom, "Osteology of *Deinonychus antirrhopus*, an Unusual Theropod from the Lower Cretaceous of Montana," *Bulletin of the Peabody Museum of Natural History* 30 (1969): 1–165; E. H. Colbert and D. A. Russell, "The Small Dinosaur *Dromaeosaurus*," *American Museum Novitates* 2380 (1969): 1–49.

17. J. H. Ostrom, "*Archaeopteryx*: Notice of a 'New' Specimen," *Science* 170 (1970): 537–38.

18. G. Heilmann, *The Origin of Birds* (London: Witherby, 1926).

19. A. D. Walker, "New Light on the Origin of Birds and Crocodiles," *Nature* 237 (1972): 257–63.

20. J. H. Ostrom, "The Ancestry of Birds," *Nature* 242 (1973): 136; J. H. Ostrom, "The Origin of Birds," *Annual Review of Earth and Planetary Sciences* 3 (1975): 55–77; J. H. Ostrom, "*Archaeopteryx* and the Origin of Birds," *Biological Journal of the Linnean Society* 8 (1976): 91–182.

21. Desmond, *Huxley*, 472.

22. J. A. Gauthier, "Saurischian Monophyly and the Origin of Birds," in *The Origin of Birds and the Evolution of Flight*, ed. K. Padian, *Memoirs of the California Academy of Sciences* 8 (1986): 1–55.

23. Ibid.; L. M. Witmer, "The Debate on Avian Ancestry: Phylogeny, Function, and Fossils," in *Mesozoic Birds: Above the Heads of Dinosaurs*, ed. L. M. Chiappe and L. Witmer (Berkeley: University of California Press, 2002), 3–30.

24. J. H. Ostrom, "On the Origin of Birds and of Avian Flight," in *Major Features of Vertebrate Evolution*, ed. D. R. Prothero and R. M. Schoch (Knoxville: University of Tennessee Press, 1994), 160–77.

25. A. Feduccia, *The Origin and Evolution of Birds* (New Haven: Yale University Press, 1996).

26. C. A. Brochu and M. A. Norell, "Time and Trees: A Quantitative Assessment of Temporal Congruence in the Bird Origins Debate," in *New Perspectives on the Origin and Early Evolution of Birds*, ed. J. Gauthier and L. F. Gall (New Haven: Peabody Museum of Natural History, Yale University, 2001), 511–35.

27. D. Hu et al., "A Pre-*Archaeopteryx* Troodontid Theropod from China with Long Feathers on the Metatarsus," *Nature* 461 (2009): 640–43.

28. J. Ackerman, "Dinosaurs Take Wing," *National Geographic*, July 1998, 74–99.

29. A. S. Romer, *Vertebrate Paleontology*, 3rd ed. (Chicago: University of Chicago Press, 1966).

30. W. B. Harland et al., *A Geologic Time Scale* (Cambridge: Cambridge University Press, 1990). F. M. Gradstein, J. G. Ogg, and A. G. Smith, eds., *A Geologic Time Scale* (Cambridge: Cambridge University Press, 2004).

31. M. O. Woodburne, "A Prospectus of the North American Mammal Ages," in *Cenozoic Mammals of North America*, ed. M. O. Woodburne (Berkeley: University of California Press, 1987), 285–90.

32. E. Wailoo, Interview with Paul Olsen, *American Scientist* 76 (1988): 276–81.

33. R. L. Carroll and Z.-M. Dong, "*Hupehsuchus*, an Enigmatic Aquatic Reptile from the Triassic of China, and the Problem of Establishing Relationships," *Philosophical Transactions of the Royal Society of London*, ser. B, 331 (1991): 131–53.

34. Ibid.

35. Ibid.

36. Ibid.; O. Rieppel, "*Helveticosaurus zollingeri* Peyer (Reptilia, Diapsida) Skeletal Paedomorphosis, Functional Anatomy and Systematic Affinities," *Palaeontographica*, ser. A, 208 (1989): 123–52.

37. Carroll and Dong, "*Hupehsuchus*," emphasis in original.

38. Feduccia, *Origin and Evolution of Birds*.

39. D. M. Irwin, "Dead Branches on the Tree of Life," review of *In Search of Deep Time: Beyond the Fossil Record to a New History of Life*, by H. Gee, *Nature* 403 (2000): 480–81, emphasis added.

40. R. L. Carroll and P. Thompson, "A Bipedal Lizardlike Reptile from the Karoo," *Journal of Paleontology* 56 (1982): 1–10.

41. D. S. Berman et al., "Early Permian Bipedal Reptiles," *Science* 290 (2000): 969–72.

42. S. J. Nesbitt and M. A. Norell, "Extreme Convergence in the Body Plans of an Early Suchian (Archosauria) and Ornithomimid Dinosaurs (Theropoda)," *Proceedings of the Royal Society B* 273 (2000): 1045–48.

43. Ibid.

44. Gauthier, "Saurischian Monophyly," 18.

45. P. C. Sereno and R. Wild, "*Procompsognathus*: Theropod, 'Thecodont' or Both?" *Journal of Vertebrate Paleontology* 12 (1992): 435–58.

46. F. Knoll, "On the *Procompsognathus* Postcranium (Late Triassic, Germany)," *Geobios* 41 (2008): 779–86.

47. Gauthier, "Saurischian Monophyly," 18.

48. A. R. Milner and S. E. Evans, "The Upper Jurassic Diapsid *Lisboasaurus estesi*: A Maniraptoran Theropod," *Palaeontology* 34 (1991): 503–13; A. Elzanowski and P. Wellnhofer, "A New Link between Theropods and Birds from the Cretaceous of Mongolia," *Nature* 359 (1992): 821–23; A. Elzanowski and P. Wellnhofer, "Skull of *Archaeornithoides* from the Upper Cretaceous of Mongolia," *American Journal of Science* 293

(1993): 235–52; A. D. Buscalioni et al., "The Upper Jurassic Maniraptoran Theropod *Lisboasaurus estesi* (Guimarota, Portugal) Reinterpreted as a Crocodilomorph," *Journal of Vertebrate Paleontology* 16 (1996): 358–62.

49. Nesbitt and Norell, "Extreme Convergence in the Body Plans."

50. O. Abel, *Geschichte und Methode der Rekonstruktion vorzeitlicher Wirbeltiere* (Jena: G. Fischer, 1925); H. G. Seeley, *Dragons of the Air: An Account of Extinct Flying Reptiles* (New York: D. Appleton, 1901).

51. K. Padian, "A Functional Analysis of Flying and Walking in Pterosaurs," *Paleobiology* 9 (1983): 218–39; K. Padian, "Osteology and Functional Morphology of *Dimorphodon macronyx* (Buckland) (Pterosauria: Rhamphorhynchoidea) Based on New Material in the Yale Peabody Museum," *Postilla* 189 (1983): 1–44; K. Padian, "The Origin of Pterosaurs," in *Third Symposium on Mesozoic Terrestrial Ecology* (Tübingen: Attempto, 1984), 163–68.

52. J.-M. Mazin et al., "Des pistes de pterosaurs dans le Tithonien de Crayssac (Quercy, France)," *Comptes Rendus de l'Academie des Sciences*, ser. 2a, 321 (1995): 417–24; M. G. Lockley et al., "The Fossil Trackway *Pteraichnus* Is Pterosaurian, not Crocodilian: Implications for the Global Distribution of Pterosaur Tracks," *Ichnos* 4 (1995): 7–20.

53. S. C. Bennett, "Terrestrial Locomotion of Pterosaurs: Reconstruction Based on *Pteraichnus* Trackways," *Journal of Vertebrate Paleontology* 17 (1997): 104–13.

54. D. M. Unwin and N. N. Bakhurina, "*Sordes pilosus* and the Nature of the Pterosaur Flight Apparatus," *Nature* 371 (1994): 62–64. .

55. J. M. Clark et al., "Foot Posture in a Primitive Pterosaur," *Nature* 391 (1998): 886–89.

56. Seeley, *Dragons of the Air*.

57. P. Wellnhofer, "A Short History of Pterosaur Research," *Zitteliana* B28 (2008): 7–19.

58. D. M. Unwin, "Pterosaurs: Back to the Traditional Model?" *Trends in Ecology and Evolution* 14 (1999): 263–68; D. M. Unwin and D. M. Henderson, "On the Trail of the Totally Integrated Pterosaur," *Trends in Ecology and Evolution*, 17, no. 2 (2002): 58–59; D. M. Unwin, *The Pterosaurs: From Deep Time* (New York: Pi, 2006); R. Wild,

"Flugsaurier aus der Obertrias von Italien," *Naturwissenschaften* 71 (1984): 1–11.

59. X. Wang et al., "Discovery of a Rare Arboreal Forest-Dwelling Flying Reptile (Pterosauria, Pterodactyloidea) from China," *Proceedings of the National Academy of Sciences* 105 (2008): 1983–87.

60. University of California Museum of Paleontology, Berkeley, website, 2008, http://www.ucmp.berkeley.edu/vertebrates/flight/pter.html.

61. C. G. Sibley and J. E. Ahlquist, *Phylogeny and Classification of Birds: A Study in Molecular Evolution* (New Haven: Yale University Press, 1990).

62. S. J. Gould, "A Clock of Evolution," *Natural History* 94, no. 4 (1985): 12–25.

63. S. B. Hedges and C. G. Sibley, "Molecules vs. Morphology in Avian Evolution: The Case of the 'Pelecaniform' Birds," *Proceedings of the National Academy of Sciences* 91 (1994): 9861–65.

64. Ibid.

65. J. Cracraft et al., "Phylogenetic Relationships among Modern Birds (Neornithes): Toward an Avian Tree of Life," in *Assembling the Tree of Life*, ed. J. Cracraft and M. J. Donoghue (Oxford: Oxford University Press, 2004), 468–89.

66. S. J. Hackett et al., "A Phylogenomic Study of Birds Reveals Their Evolutionary History," *Science* 320 (2008): 1763–68.

67. M. D. Sorenson et al., "Relationships of the Extinct Moa-Nalos, Flightless Hawaiian Waterfowl, Based on Ancient DNA," *Proceedings of the Royal Society B* 266 (1999): 2187–93.

68. R. C. Fleischer, H. F. James, and S. L. Olson, "Convergent Evolution of Hawaiian and Australo-Pacific Honeyeaters from Distant Songbird Ancestors," *Current Biology* 18 (2008): 1927–31; I. J. Lovette, "Convergent Evolution: Raising a Family from the Dead," *Current Biology* 18 (2008): R1132–34.

69. B. C. Livezey and R. L. Zusi, "Higher-Order Phylogeny of Modern Birds (Theropoda, Aves: Neornithes) Based on Comparative Anatomy; I. Methods and Characters," *Bulletin of the Carnegie Museum of Natural History* 37 (2006): 1–544; B. C. Livezey and R. L. Zusi, "Higher-Order Phylogeny of Modern Birds (Theropoda, Aves: Neornithes) Based on Comparative Anatomy; II. Analysis and Discussion," *Zoological Journal of the Linnaean Society* 149 (2007): 1–95.

70. G. Mayr, "Avian Higher-Level Phylogeny: Well-Supported Clades and What We Can Learn from a Phylogenetic Analysis of 2954 Morphological Characters," *Journal of Zoological Systematics and Evolutionary Research* 46 (2007): 63–72.

71. H. F. James, "Paleogene Fossils and the Radiation of Modern Birds," *Auk* 122 (2005): 1049–54; G. Mayr, "The Contribution of Fossils to the Reconstruction of the High-Level Phylogeny of Birds," *Species, Phylogeny and Evolution* 1 (2007): 59–64.

72. P. G. P. Ericson et al., "Diversification of Neoaves: Integration of Molecular Sequence Data and Fossils," *Biological Letters* 2 (2006): 543–47; R. W. Scotland, R. G. Olmstead, and J. R. Bennett, "Phylogeny Reconstruction: The Role of Morphology," *Systematic Biology* 52 (2003): 539–48.

73. Cracraft et al., "Phylogenetic Relationships among Modern Birds."

74. A. Feduccia, "'Big Bang' for Tertiary Birds," *Trends in Ecology and Evolution* 18 (2003): 172–76; A. Feduccia, "Explosive Evolution in Tertiary Birds and Mammals," *Science* 267 (1995): 637–38.

75. M. Fürbringer, *Untersuchungen zur Morphologie und Systematik der Vogel, zugleich ein Beitrag zur Anatomie der Stütz- und Bewegungsorgane*, 2 vols. (Amsterdam: T. J. Van Holkema, 1888).

76. D. Haussler, "Genome 10K: A Proposal to Obtain Whole-Genome Sequence for 10,000 Vertebrate Species," *Journal of Heredity* 100 (2009): 659–74; E. C. Hayden, "10,000 Genomes to Come: Vertebrates in Line for Massive Sequencing Project," *Nature* 462 (2009): 21.

77. H. F. Osborn, "Reconstruction of the Evidence for a Common Dinosaur-Avian Stem in the Permian," *American Naturalist* 34 (1900); R. Broom, "On the Early Development of the Appendicular Skeleton of the Ostrich, with Remarks on the Origin of Birds," *Transactions of the South African Philosophical Society* 16 (1906): 355–68; R. Broom, "On the South African Pseudosuchian *Euparkeria* and Allied Genera," *Proceedings of the Zoological Society of London* (1913): 619–33.

78. O. Abel, "Die Vorfahren der Vögel und ihre Lebensweise," *Verhandlungen, Zoologisch-Botanische Gesellschaft in Wien* 61 (1911): 144–91.

79. Heilmann, *Origin of Birds*.

80. C. M. Sternberg, "A Toothless Bird from the Cretaceous of Alberta," *Journal of Paleontology* 14 (1940): 81–85.

81. G. S. Paul, *Predatory Dinosaurs of the World: A Complete Illustrated Guide* (New York: Simon and Schuster, 1988).

82. G. Olshevsky, "A Revision of the Parainfraclass Archosauria Cope, 1869, Excluding the Advanced Crocodylia," *Mesozoic Meanderings* 2 (1994): 1–196; T. Maryańska, H. Osmólska, and H. M. Wolsan, "Avialan Status for Oviraptorosauria," *Acta Palaeontoligica Polonica* 47 (2002): 97–116.

83. Olshevsky, "Revision of the Parainfraclass Archosauria Cope."

84. Marayańska, Osmólska, and Wolson, "Avialan Status for Oviraptorosauria"; K. Padian "Basal Avialae," in *The Dinosauria*, ed. D. Weishampel, P. Dodson, and H. Osmólska, 2nd ed. (Berkeley: University of California Press, 2004), 210–31; G. S. Paul, *Dinosaurs of the Air: The Evolution and Loss of Flight in Dinosaurs and Birds* (Baltimore: Johns Hopkins University Press, 2002); S. J. Czerkas, ed., *Feathered Dinosaurs and the Origin of Flight*, The Dinosaur Museum Journal, vol. 1 (Blanding, UT: Dinosaur Museum, 2002); S. A. Czerkas et al., "Flying Dromaeosaurs," *Dinosaur Museum Journal* 1 (2002): 97–126; A. Feduccia, "Birds Are Dinosaurs: Simple Answer to a Complex Question," *Auk* 119 (2002): 1187–201; L. D. Martin, "A Basal Archosaurian Origin of Birds," *Acta Zoologica Sinica* 50 (2004): 978–90; A. Feduccia, T. Lingham-Soliar, and J. R. Hinchliffe, "Do Feathered Dinosaurs Exist? Testing the Hypothesis on Neontological and Paleontological Evidence," *Journal of Morphology* 266 (2005): 125–66.

85. Czerkas et al., "Flying Dromaeosaurs."

86. Ibid.

87. H. Tischlinger and D. M. Unwin, "Neue Informationen zum Berliner Exemplar von *Archaeopteryx lithographica* H. v. Meyer 1861 und der Isolierten *Archaeopteryx*-Feder," *Archaeopteryx* 22 (2004): 17–50; N. Longrich, "*Archaeopteryx*: Two Wings or Four?" *Paleobiology* 32 (2006): 417–31.

88. D. Hu et al., "A Pre-Archaeopteryx Troodontid Theropod from China with Long Feathers on the Metatarsus," *Nature* 461 (2009): 640–43.

89. C. W. Beebe, *A Tetrapteryx Stage in the Ancestry of Birds* (New York: Zoologica, 1915).

90. Z. Zhou and F. Zhang, "Mesozoic Birds of China: An Introduction of Review," *Acta Zoologica Sinica* 50 (2004): 913–20.

91. Czerkas, Zhang, Li, and Li, "Flying Dromaeosaurs."

92. Z. Zhou, "The Origin and Early Evolution of Birds: Discoveries, Disputes, and Perspectives from Fossil Evidence," *Naturwissenschaften* 91 (2004): 455–71.

93. Czerkas, Zhang, Li, and Li, "Flying Dromaeosaurs."

94. M. A. Norell, "The Proof Is in the Plumage," *Natural History* 110, no. 6 (2001): 58–63.

95. Czerkas, Zhang, Li, and Li, "Flying Dromaeosaurs."

96. Ibid.

97. Zhou, "Origin and Early Evolution of Birds."

98. R. O. Prum, "Are Current Critiques of the Theropod Origin of Birds Science? Rebuttal to Feduccia (2002)," *Auk* 120 (2003): 550–61.

99. A. R. Milner, "*Cosesaurus*—the Last Proavian?" *Nature* 315 (1985): 544; J. Welman, "*Euparkeria* and the Origin of Birds," *South African Journal of Science* 91 (1995): 533–37; Ostrom, "On the Origin of Birds"; L. D. Martin, "Origin of Avian Flight: A New Perspective," *Oryctos* 7 (2008): 45–54; M. J. Benton, "*Scleromochlus taylori* and the Origin of Dinosaurs and Pterosaurs," *Philosophical Transactions of the Royal Society of London*, ser. B, 354 (1999): 1423–46; S. Tarsitano and M. K. Hecht, "A Reconsideration of the Reptilian Relationships of *Archaeopteryx*," *Zoological Journal of the Linnean Society of London* 69 (1980): 257–63.

100. R. L. Carroll, *Vetebrate Paleontology and Evolution* (New York: W. H. Freeman, 1988).

101. Witmer, "Debate on Avian Ancestry."

102. J. M. Clark, M. A. Norell, and P. J. Makovicky, "Cladistic Approaches to the Relationships of Birds to Other Theropod Dinosaurs," in Chiappe and Witmer, *Mesozoic Birds*, 31–61, emphasis added.

103. F. C. James and J. A. Pourtless, "Cladistics and the Origin of Birds: A Review and Two New Analyses," *Ornithological Monographs* 66 (2009).

104. Paul, *Dinosaurs of the Air*; A. D. Walker, "New Light on the Origin of Birds and Crocodiles," *Nature* 237 (1972): 257–63; E. N. Kurochkin, "Parallel Evolution of Theropod Dinosaurs and Birds," *Zoologicheskii Zhurnal* 85 (2006): 283–97 [English version, *Entomological Review* 86, S45–58].

105. Clark, Norell, and Makovicky, "Cladistic Approaches to the Relationships of Birds to Other Theropod Dinosaurs."

106. Feduccia, *Origin and Evolution of Birds*.

107. Norell, "Proof Is in the Plumage"; L. M. Chiappe, *Glorified Dinosaurs: The Origin and Early Evolution of Birds* (New York: John Wiley and Sons, 2007).

108. P. Wellnhofer, *The Illustrated Encyclopedia of Pterosaurs* (New York: Crescent Books, 1991); S. C. Bennett, "The Phylogenetic Position of the Pterosauria within the Archosauromorpha," *Zoological Journal of the Linnaean Society* 118 (1996): 261–308; D. M. Unwin, *Pterosaurs from Deep Time* (New York: Pi, 2006); D. W. E. Hone and M. J. Benton, "An Evaluation of the Phylogenetic Relationships of the Pterosaurs among Archosauromorph Reptiles," *Journal of Systematic Paleontology* 5 (2007): 465–69.

109. Unwin, *Pterosaurs from Deep Time*.

110. D. O. Fisher and I. P. F. Owens, "The Comparative Method in Conservation Biology," *Trends in Ecology and Evolution* 19 (2004): 391–98.

111. Czerkas et al., "Flying Dromaeosaurs."

112. S. L. Olson, review of *New Perspectives on the Origin and Early Evolution of Birds, Proceedings of the International Symposium in Honor of John H. Ostrom*, ed. J. Gauthier and L. F. Gall, *Auk* 119, no. 4 (2001): 1202–5.

113. P. D. Alonso et al., "The Avian Nature of the Brain and Inner Ear of *Archaeopteryx*," *Nature* 430 (2004): 666–69; J. Franzosa, "Evolution of the Brain in Theropoda (Dinosauria)" (PhD diss., University of Texas, Austin, 2004).

114. D. A. Burhnam et al., "Remarkable New Bird-like Dinosaur (Theropoda: Maniraptora) from the Upper Cretaceous of Montana," *University of Kansas Paleontological Institute, Paleontological Contributions*, n.s., 13 (2000): 1–14; D. A. Burnham, "New Information on *Bambiraptor feinbergi* (Theropoda: Dromaeosauridae) from the Late Cretaceous of Montana," in *Feathered Dragons: Studies on the Transition from Dinosaurs to Birds*, ed. P. J. Currie et al. (Bloomington: Indiana University Press, 2004), 67–111; D. A. Burnham, "Paleoenvironment, Paleoecology, and Evolution of Maniraptoran 'Dinosaurs'" (PhD diss., University of Kansas, Lawrence, 2007).

115. P. J. Currie, "Cranial Anatomy of *Stenony-chosaurus inequalis* (Saurischia, Theropoda) and Its Bearing on the Origin of Birds," *Canadian Journal of Earth Sciences* 22 (1985): 1643–58.

116. M. Kundrát, "Avian-Like Atributes of a Virtual Brain Model of the Oviraptorid Theropod *Conchoraptor gracilis*," *Naturwissenschaften* 94 (2009): 499–504.

117. Ibid.

118. Ibid.

119. Ibid.

120. R. Barsbold et al., "New Oviraptorosaur (Dinosauria, Theropoda) from Mongolia: The First Dinosaur with a Pygostyle," *Acta Palaeontologica Polonica* 45 (2000): 97–106; R. Barsbold et al., "A Pygostyle in a Non-Avian Theropod," *Nature* 403 (2000): 135.

121. Burnham, *Paleoenvironment, Paleoecology, and Evolution*.

122. P. Senter, "Comparison of Forelimb Function between *Deinonychus* and *Bambiraptor*," *Journal of Vertebrate Paleontology* 26 (2006): 897–906; A. Gishlick, "The Function of the Manus and Forelimb of *Deinonychus antirrhopus* and Its Importance for the Origin of Flight," in Gauthier and Gall, *New Perspectives on the Origin and Early Evolution of Birds*, 302–18.

123. P. L. Manning et al., "Biomechanics of Dromaeosaurid Dinosaur Claws: Application of X-ray Microtomography, Nanoidentation, and Finite Element Analysis," *Anatomical Record* 292 (2009): 1397–405.

124. L. M. Witmer, "Inside the Oldest Bird Brain," *Nature* 430 (2004): 619–20.

125. Ibid.

126. S. A. Walsh et al., "Inner Ear Anatomy Is a Proxy for Deducing Auditory Capability and Behavior in Reptiles and Birds," *Proceedings of the Royal Society B* 276 (2009): 1355–60.

127. A. H. Turner, P. J. Makovicky, and M. A. Norell, "Feather Quill Knobs in the Dinosaur *Velociraptor*," *Science* 317 (2007): 1721.

128. L. Witmer, "Feathered Dinosaurs in a Tangle," *Nature* 461 (2009): 601–2.

129. Martin, "Basal Archosaurian Origin of Birds."

130. R. Li et al., "Behavioral and Faunal Implications of Early Cretaceous Deinonychosaur Trackways from China," *Naturwissenschaften*

95 (2007): 185–91; L.-D. Xing et al., "The Earliest Known Deinonychosaur Tracks from the Jurassic-Cretaceous Boundary in Heibei Province, China," *Acta Palaeontologica Sinica* 48 (2009): 662–71.

131. Zhou, "Origin and Early Evolution of Birds."

132. M. A. Schwartz, "The Importance of Being Stupid in Scientific Research," *Journal of Cell Science* 121 (2008): 1771.

133. Padian and Ricqlès, "L'Origine et l'évolution des oiseaux."

Appendix 1: A Sliver of Urvogel Bone

1. R. T. Bakker and P. M. Galton, "Dinosaur Monophyly and a New Class of Vertebrates," *Nature* 248 (1974): 168–72.

2. A. Feduccia, *The Origin and Evolution of Birds* (New Haven: Yale University Press, 1996); V. Morell, "A Cold, Hard Look at Dinosaurs," *Discover* 17 (1996): 98–102; J. O. Farlow, P. Dodson, and A. Chinsamy, "Dinosaur Biology," *Annual Review of Ecology and Systematics* 26 (1995): 445–71; D. E. Quick and J. A. Ruben, "Cardio-Pulmonary Anatomy in Theropod Dinosaurs: Implications from Extant Archosaurs," *Journal of Morphology* 270 (2009): 1232–46; J. Ruben, "The Evolution of Endothermy in Mammals, Birds and Their Ancestors," *Society of Experimental Biology, Science Series* 59 (1996): 347–76; A. Chinsamy and W. J. Hillenius, "Physiology of Non-Avian Dinosaurs," in *The Dinosauria*, 2nd ed., ed. D. G. Weishampel, P. Dodson, and H. Osmólska (Berkeley: University of Calfornia Press, 2004), 643–59.

3. J. H. Ostrom, "Terrestrial Vertebrates as Indicators of Mesozoic Climates," *Proceedings of the North American Paleontological Convention*, pt. D (1969): 347–76; R. T. Bakker, "Dinosaur Renaissance," *Scientific American* 232, no. 4 (1975): 58–78; R. T. Bakker, *The Dinosaur Heresies: New Theories Unlocking the Mystery of the Dinosaurs and Their Extinction* (New York: William Morrow, 1986).

4. A. Feduccia, "Dinosaurs as Reptiles," *Evolution* 27 (1973): 166–69.

5. N. Hotton III, "An Alternative to Dinosaur Endothermy: The Happy Wanderers," in *A Cold Look at the Warm-Blooded Dinosaurs*, ed. R. D. A. Thompson and E. C. Olson, American Asso-

ciation for the Advancement of Science Selected Symposium, 28 (Boulder, CO: Westview, 1980), 311–50.

6. J. H. Ostrom, "Romancing the Dinosaurs," *Sciences*, May–June 1987, 56–63.

7. J. H. Ostrom, "New Ideas about Dinosaurs," *National Geographic* 154 (1978): 152–85.

8. K. Padian and L. M. Chiappe, "The Origin of Birds and Their Flight," *Scientific American* 278, no. 2 (1998): 38–47; Bakker and Galton, "Dinosaur Monophyly."

9. T. Lingham-Soliar, A. Feduccia, and X. Wang, "A New Chinese Specimen Indicates That 'Protofeathers' in the Early Cretaceous Theropod Dinosaur *Sinosauropteryx* Are Degraded Collagen Fibers," *Proceedings of the Royal Society B* 274 (2007): 1823–29.

10. P. Currie in *Encyclopedia of Dinosaurs*, ed. P. J. Currie and K. Padian (San Diego, CA: Academic Press, 1977), 194–95, 241.

11. J. Ruben, "Reptilian Physiology and the Flight Capacity of *Archaeopteryx*," *Evolution* 45 (1991): 1–17.

12. P. C. Sereno and C. Rao, "Early Evolution of Avian Flight and Perching: New Evidence from the Lower Cretaceous of China," *Science* 255 (1992): 845–48.

13. Feduccia, *Origin and Evolution of Birds*.

14. L. D. Martin and Z. Zhou, "*Archaeopteryx*-Like Skull in Enantiornithine Bird," *Nature* 389 (1997): 556.

15. K. Padian et al., "Dinosaurian Growth Rates and Bird Origins," *Nature* 412 (2001): 405–12; G. M. Erickson, K. Curry-Rogers, and S. Yerby, "Dinosaur Growth Patterns and Rapid Avian Growth Rates," *Nature* 412 (2001): 429–33; G. M. Erickson, "Assessing Dinosaur Growth Patterns: A Microscopic Revolution," *Trends in Ecology and Evolution* 20 (2005): 677–84; G. M. Erickson et al., "Was Dinosaurian Physiology Inherited by Birds? Reconciling Slow Growth in *Archaeopteryx*," *PLoS ONE* 4, no. 10 (2009): e7390; M. J. Benton, "Dinosaurs," *Current Biology* 19, no. 8 (2009): R318–23.

16. J. Ruben, "The Evolution of Endothermy in Mammals and Birds: From Physiology to Fossils," *Annual Review of Physiology* 57 (1996): 69–95.

17. Feduccia, *Origin and Evolution of Birds*.

18. A. Chinsamy, *The Microstructure of Dinosaur Bone: Deciphering Biology with Fine-Scale Techniques* (Baltimore: Johns Hopkins University Press, 2005); A. Chinsamy, L. Chiappe, and P. Dodson, "Growth Rings in Mesozoic Avian Bones: Physiological Implications for Basal Birds," *Nature* 368 (1994): 196–97; A. Chinsamy, "Bone Microstructure in Early Birds," in *Mesozoic Birds: Above the Heads of Dinosaurs*, ed. L. M. Chiappe and L. M. Witmer (Berkeley: University of California Press, 2002), 421–31.

19. A. Chinsamy, L. Chiappe, and P. Dodson, "Mesozoic Avian Bone Microstructure: Physiological Implications," *Paleobiology* 21 (1995): 561–74.

20. S. T. Turvey, O. R. Green, and R. N. Holdaway, "Cortical Growth Marks Reveal Extended Juvenile Development in New Zealand Moa," *Nature* 435 (2005): 940–43.

21. Ibid.

22. M. Köhler and S. Moyà-Solà, "Physiological and Life History Strategies of a Fossil Large Mammal in a Resource-Limited Environment," *Proceedings of the National Academy of Sciences* 106 (2009): 20354–58.

23. R. S. Miller and J. P. Hatfield, "Age Ratios of Sandhill Cranes," *Journal of Wildlife Management* 38 (1974): 234–42; D. E. Pomeroy, "Growth and Plumage Changes of the Grey-Crowned Crane *Balearica regulorum gibbericeps*," *Bulletin of the British Ornithologists' Club* 100, no. 4 (1980): 219–22.

24. Erickson et al., "Was Dinosaurian Physiology Inherited by Birds?"

25. Ibid.

26. Ibid.

27. F. C. James and J. A. Pourtless, Jr., "Cladistics and the Origin of Birds: A Review and Two New Analyses," *Ornithological Monographs* 66 (2009); A. Feduccia, T. Lingham-Soliar, and J. R. Hinchliffe, "Do Feathered Dinosaurs Exist? Testing the Hypothesis on Neontological and Paleontological Evidence," *Journal of Morphology* 266 (2005): 125–66.

28. J. M. Starck and A. Chinsamy, "Bone Microstructure and Developmental Plasticity in Birds and Other Dinosaurs," *Journal of Morphology* 254 (2002): 232–46.

29. Feduccia, Lingham-Soliar, and Hinchliffe, "Do Feathered Dinosaurs Exist?"

30. Ruben, "Evolution of Endothermy."

31. J. A. Ruben, T. D. Jones, and N. R. Geist, "Respiratory and Reproductive Paleophysiol-

ogy of Dinosaurs and Early Birds," *Physiological and Biochemical Zoology* 76 (2003): 141–64; W. J. Hillenius, "The Evolution of Nasal Turbinates and Mammalian Endothermy," *Paleobiology* 18 (1992): 17–29; W. J. Hillenius, "Turbinates in Therapsids: Evidence for Late Permian Origins of Mammalian Endothermy," *Evolution* 48 (1994): 207–29.

32. Ruben, Jones, and Geist, "Respiratory and Reproductive Paleophysiology of Dinosaurs."

33. J. A. Ruben, "The Evolution of Endothermy in Mammals and Birds," *Annual Rerview of Physiology* 57 (1995): 69–95.

34. H. Pontzer, V. Allen, and J. R. Hutchinson, "Biomechanics of Running Indicates Endothermy in Bipedal Dinosaurs," *PLoS ONE* 4, no. 11 (2009): e7783.

35. Schmidt-Nilsen, *Animal Physiology*, 5th ed. (Cambridge: Cambridge University Press, 1997).

36. Pontzer, Allen, and Hutchinson, "Biomechanics of Running."

37. J. N. Wilford, "The 'Early Bird' *Archaeopteryx* May Not Be a Bird After All," *New York Times*, 8 October 2009.

Appendix 2: The Persisting Problem of Avian Digital Homology

General note: Much of this section is derived from A. Feduccia, T. Lingham-Soliar, and J. R. Hinchliffe, "Do Feathered Dinosaurs Exist? Testing the Hypothesis on Neotological and Paleontological Evidence," *Journal of Morphology* 266, no. 2 (2005): 125–66, which discusses the problem of digital homology in greater detail.

1. R. Owen, "Aves," in *Todd's Cyclopedia in Anatomy and Physiology* 1 (1836): 265–358; W. K. Parker, "On the Structure and Development of the Wing in the Common Fowl," *Proceedings of the Royal Society of London* 42 (1888): 52–58.

2. Feduccia, Lingham-Soliar, and Hinchliffe, "Do Feathered Dinosaurs Exist?"

3. J. M. Clark, review of *Origins of the Higher Groups of Vertebrates: Controversy and Consensus, Journal of Vertebrate Paleontology* 12 (1992): 532–36.

4. G. P. Wagner and J. A. Gauthier, "1,2,3 = 2,3,4: A Solution to the Problem of the Homology of the Digits in the Avian Hand," *Proceedings of the National Academy of Sciences* 96 (1999): 5111–16.

5. L. M. Chiappe, *Glorified Dinosaurs: The Origin and Early Evolution of Birds* (New York: John Wiley and Sons, 2007).

6. M. C. Langer, "Basal Saurischia," in *The Dinosauria*, ed. D. B. Weishampel, P. Dodson, and H. Osmólska, 2nd ed. (Berkeley: University of California Press, 2004), 25–46; P. C. Sereno, "Shoulder Girdle and Forelimb of *Herrerasaurus," Journal of Vertebrate Paleontology* 13 (1993): 425–50.

7. M. K. Hecht and B. M. Hecht, "Conflicting Developmental and Paleontological Data: The Case of the Bird Manus," *Acta Palaeontologica Polonica* 38 (1994): 329–38.

8. N. H. Shubin, "History, Ontogeny, and Evolution of the Archetype," in *Homology: The Hierarchical Basis of Comparative Biology*, ed. B. K. Hall (San Diego: Academic Press, 1994), 249–69.

9. A. C. Burke and A. Feduccia, "Developmental Patterns and Identification of Homologies in the Avian Hand," *Science* 278 (1997): 666–69.

10. Chiappe, *Glorified Dinosaurs*.

11. Burke and Feduccia, "Developmental Patterns and Identification"; J. R. Hinchliffe, "The Forward March of the Bird-Dinosaurs Halted?" *Science* 278 (1997): 596–97.

12. J. R. Hinchliffe, "'One, Two, Three' or 'Two, Three, Four': An Embryologist's View of the Homologies of the Digits and Carpus of Modern Birds," in *The Beginnings of Birds: Proceedings of the International Archaeopteryx Conference*, ed. M. K. Hecht et al. (Eichstätt: Freunde des Jura-Museums, 1985), 141–48; G. B. Müller and P. Alberch, "Ontogeny of the Limb Skeleton in *Alligator mississippiensis*: Developmental Invariance and Change in the Evolution of Archosaur Limbs," *Journal of Morphology* 203 (1990): 151–64; A. C. Burke and P. Alberch, "The Development and Homology of the Chelonian Carpus and Tarsus," *Journal of Morphology* 186 (1985): 119–31.

13. P. Wellnhofer, "A New Specimen of Archaeopteryx from the Solnhofen Limestone," *Los Angeles County Museum of Natural History, Science Series* 36 (1992): 3–23; R. D. Dahn and J. F. Fallon, "Interdigital Regulation of Digit Identity and Homeotic Transformation by Modulated BMP Signaling," *Science* 289 (2000): 438–41; G. Drossopoulou et al., "A Model for Anteroposterior Patterning of the Vertebrate Limb Based on Sequential Long- and Short-Range Shh

Signalling and Bmp Signalling," *Development* 127 (2000): 1337–48.

14. X. Xu et al., "A Jurassic Ceratosaur from China Helps Clarify Avian Digital Homologies," *Nature* 459 (2009): 940–44.

15. A. Raynaud, "Developmental Mechanisms Involved in the Embryonic Reduction of Limbs in Reptiles," *International Journal of Developmental Biology* 34 (1980): 233–43; A. Raynaud and M. Clergue-Gazeau, "Identification des doigts réduits ou manquants dans les pattes des embryons de Lézard vert (*Lacerta viridis*) traités par la cytosine-arabinofuranoside: Comparaison avec les réductions digitales naturelles des espèces de reptiles serpentiformes," *Archives of Biology* (Brussels) 97 (1986): 79–299.

16. H. Zhou and L. Niswander, "Requirement for BMP Signaling in Interdigital Apoptosis and Scale Formation," *Science* 272 (1996): 738–41.

17. Burke and Feduccia, "Developmental Patterns and Identification"; Hinchliffe, "Forward March of the Bird-Dinosaurs Halted?"

18. Wagner and Gauthier, "1,2,3 = 2,3,4"; A. Feduccia, "1,2,3 = 2,3,4: Accommodating the Cladogram," *Proceedings of the National Academy of Sciences* 96 (1999): 4740–42.

19. A. O. Vargas and J. Fallon, "Birds Have Dinosaur Wings: The Molecular Evidence," *Journal of Experimental Zoology (MDE)* 304B (2004): 85–89.

20. F. Galis, M. Kundrát, and J. A. Metz, "Hox Genes, Digit Identities and the Theropod/Bird Transition," *Journal of Experimental Zoology (MDE)* 304B (2005): 198–205; J. R. Hinchliffe, "Bird Wing Digits and Their Homologies: Reassessment of Developmental Evidence for a 2,3,4 Identity," *Oryctos* 7 (2008): 7–12.

21. M. C. M. Welton et al., "Gene Expression and Digit Homology in the Chicken Embryo," *Evolution and Development* 7 (2005): 18–28.

22. K. Tamura et al., "Embryological Evidence Identifies Wing Digits in Birds as Digits 1, 2, and 3," *Science* 331 (2011): 753–57.

23. C. J. Tabin and A. P. McMahon, "Grasping Limb Patterning," *Science* 321 (2008): 350–352; Y. Yang et al., "Relationship between Dose, Distance and Time in Sonic Hedgehog-Mediated Regulation of Anteroposterior Polarity in the Chick Limb," *Development* 124 (1997): 4393–404.

24. A. Feduccia and J. Nowicki, "The Hand of Birds Revealed by Early Ostrich Embryos," *Naturwissenschaften* 89 (2002): 391–93.

25. J. R. Hinchliffe, in Feduccia, Lingham-Soliar, and Hinchliffe, "Do Feathered Dinosaurs Exist?"; M. Towers et al., Integration of Growth and Specification in Chick Wing Digit-Patterning," *Nature* 452 (2008): 882–86.

26. F. Galis, M. Kundrát, and J. A. Metz, "Why Five Fingers? Evolutionary Constraints on Digit Numbers," *Trends in Ecology and Evolution* 16 (2001): 637–46; A. Feduccia, "Birds Are Dinosaurs: Simple Answer to a Complex Problem," *Auk* 119 (2002): 1187–201. Quotation from Galis, Kundrát, and Metz, "Why Five Fingers?"

27. Galis, Kundrát, and Metz, "Why Five Fingers?"

28. M. D. Shapiro, "Developmental Morphology of Limb Reduction in *Hemiergis* (Squamata: Scincidae): Chondrogenesis, Osteogenesis, and Heterochrony," *Journal of Morphology* 254 (2002): 211–31.

29. J. Zákány, J. Kmita, and M. Duboule, "A Dual Role for Hox Genes in Limb Anterior-Posterior Asymmetry," *Science* 304 (2004): 1669–72; J. Deschamps, "Hox Genes in the Limb: A Play in Two Acts," *Science* 304 (2004): 1610–11.

30. Chiappe, *Glorified Dinosaurs*.

31. A. Turner, P. J. Makovicky, and M. A. Norell, "Feather Quill Knobs in the Dinosaur *Velociraptor*," *Science* 317 (2007): 1721; Feduccia, Lingham-Soliar, and Hinchliffe, "Do Feathered Dinosaurs Exist?"; Feduccia, "Birds Are Dinosaurs"; G. S. Paul, *Dinosaurs of the Air: The Evolution and Loss of Flight in Dinosaurs and Birds* (Baltimore: Johns Hopkins University Press, 2002).

32. Z. Zhou, "The Origin and Early Evolution of Birds: Discoveries, Disputes, and Perspectives from Fossil Evidence," *Naturwissenschaften* 91 (2004): 455–71.

33. A. Feduccia and J. Nowicki, "The Hand of Birds Revealed by Early Ostrich Embryos," *Naturwissenschaften* 89 (2002): 391–93.

34. M. Kundrát et al., "Pentadactyl Pattern of the Avian Autopodium and Pyramid Reduction Hypothesis," *Journal of Experimental Zoology* 294 (2002): 152–59; M. C. M. Welten et al., "Gene

Expression and Digit Homology in the Chicken Embryo Wing," *Evolution and Development* 7 (2005): 18–28.

35. M. Kundrát, "Primary Chondrification Foci in the Wing Basipodium of *Struthio camelus* with Comments on the Interpretation of Autopodial Elements in Crocodilia and Aves," *Journal of Experimental Zoology (MDE)* 310B (2008): 1–12; Welten et al., "Gene Expression and Digit Homology in the Chicken Embryo Wing."

36. Quotations, in order, from: Feduccia and Nowicki, "Hand of Birds Revealed"; Kundrát et al., "Pendactyl Pattern of the Avian Autopodium"; Kundrát, "Primary Chondrification Foci"; H. C. E. Larsson and G. P. Wagner, "Pentadactyl Ground State of the Avian Wing," *Journal of Experimental Zoology* 294 (2002): 146–51.

37. Larsson and Wagner, "Pentadactyl Ground State of the Avian Wing."

38. Müller and Alberch, "Ontogeny of the Limb Skeleton in *Alligator mississippiensis*."

39. R. Hinchliffe, "Bird Wing Digits and Their Homologies: Reassessment of Developmental Evidence for 2, 3, 4," *Oryctos* 7 (2008): 7–12.

40. Kundrát et al., "Pentadactyl Pattern of the Avian Autopodium."

41. Xu et al., "A Jurassic Ceratosaur from China Helps Clarify Avian Digital Homologies"; A. O. Vargas, G. P. Wagner, and J. A. Gauthier, "*Limusaurus* and Bird Digit Identity," *Nature Precedings*, http://precedings.nature.com/documents/3828/version/1/files/npre 20093828–1.pdf.

42. G. P. Wagner, "The Developmental Evolution of Avian Digital Homology: An Update," *Theory in Biosciences* 124 (2005): 165–83; A. O. Vargas and G. P. Wagner, "Frame-Shifts of Digit Identity in Bird Evolution and Cyclopamine-Treated Wings," *Evolution and Development* 11, no. 2 (2009): 163–69; K. Tamura et al., "Embryological Evidence Identifies Wing Digits in Birds as Digits 1, 2, and 3," *Science* 331 (2011): 753–57.

43. E. Pennisi, "Dinos Gave Birds the Finger," *Science Now*, 10 February 2011.

GENERAL REFERENCES

Anderson, J. S., and H.-D. Sues, eds. *Major Transitions in Vertebrate Evolution*. Bloomington: Indiana University Press, 2007.

Bakker, R. T. *The Dinosaur Heresies: New Theories Unlocking the Mystery of the Dinosaurs and Their Extinction*. New York: William Morrow, 1986.

———. "Dinosaur Renaissance." *Scientific American* 232, no. 4 (1975): 58–78.

Barthel, K. W., N. H. M. Swinburne, and S. C. Morris. *Solnhofen: A Study in Mesozoic Paleontology*. Cambridge: Cambridge University Press, 1990.

de Beer, G. *Archaeopteryx lithographica: A Study Based upon the British Museum Specimen*. London: British Museum of Natural History, 1954.

———. "The Evolution of Ratites." *Bulletin of the British Museum (Natural History)* 4 (1956): 59–70.

Benton, M. J. *Vertebrate Paleontology*. 3rd ed. Oxford: Blackwell, 2005.

Carrano, M. T., et al. *Amniote Paleobiology: Perspectives on the Evolution of Mammals, Birds, and Reptiles*. Chicago: University of Chicago Press, 2006.

Carroll, R. L. *Vertebrate Paleontology and Evolution*. New York: W. H. Freeman, 1988.

Chambers, P. *Bones of Contention: The Archaeopteryx Scandals*. London: John Murray, 2002.

Chang, M.-M., ed. *The Jehol Biota: The Emergence of Feathered Dinosaurs, Beaked Birds, and Flowering Plants*. Shanghai: Shanghai Scientific and Technical Publishers, 2003.

Chatterjee, S. *The Rise of Birds: 225 Million Years of Evolution*. Baltimore: Johns Hopkins University Press, 1997.

Chiappe, L. M. *Glorified Dinosaurs: The Origin and Early Evolution of Birds*. New York: John Wiley and Sons, 2007.

Chiappe, L. M., and G. Dyke. "The Beginnings of Birds: Recent Discoveries, Ongoing Arguments, and New Directions." In *Major Transitions in Vertebrate Evolution*, ed. J. S. Anderson and H.-D. Sues, 303–36. Bloomington: Indiana University Press, 2007.

Chiappe, L. M., and G. Dyke. "The Mesozoic Radiation of Birds." *Annual Review of Ecology and Systematics* 33 (2002): 91–124.

Chiappe, L. M., and L. Witmer, eds. *Mesozoic Birds: Above the Heads of Dinosaurs*. Berkeley: University of California Press, 2002.

Currie, P. J., and K. Padian, eds. *Encyclopedia of Dinosaurs*. San Diego, CA: Academic, 1997.

Desmond, A. *Archetypes and Ancestors: Palaeontology in Victorian London, 1850–1875*. Chicago: University of Chicago Press, 1982.

———. *The Hot-Blooded Dinosaurs*. New York: Dial, 1976.

———. *Huxley: From Devil's Disciple to Evolution's High Priest*. Reading, MA: Addison-Wesley, 1997.

Desmond, A., and J. Moore. *Darwin*. New York: Warner Books, 1992.

Dingus, L., and T. Rowe. *The Mistaken Extinction: Dinosaur Evolution and the Origin of Birds*. New York: W. H. Freeman, 1997.

Farlow, J. O., and M. K. Brett-Surman, eds. *The Complete Dinosaur*. Bloomington: Indiana University Press, 1997.

Fastovsky, D. E., and D. B. Weishampel. *The Evolution and Extinction of the Dinosaurs*. 2nd ed. Cambridge: Cambridge University Press, 2005.

Feduccia, A. *The Age of Birds*. Cambridge, MA: Harvard University Press, 1980.

———. *The Origin and Evolution of Birds*. New Haven: Yale University Press, 1996.

———. *The Origin and Evolution of Birds*. 2nd ed. New Haven: Yale University Press, 1999.

Gauthier, J. "Saurischian Monophyly and the Origin of Birds." *Memoirs of the California Academy of Sciences* 8 (1986): 1–55.

Hecht, M. K., et al., eds. *The Beginnings of Birds: Proceedings of the International Archaeopteryx Conference*. Eichstätt: Freunde des Jura-Museums, 1985.

Heilmann, G. *The Origin of Birds*. London: Witherby, 1926.

Long, J., and P. Schouten. *Feathered Dinosaurs: The Origin of Birds*. Oxford: Oxford University Press, 2010.

Lucas, A. M., and P. R. Stettenheim. *Avian Anatomy: Integument*. 2 vols. Agricultural Handbook no. 362. Washington, DC: US Government Printing Office, 1972.

Lyons, S. L. *Thomas Henry Huxley: The Evolution of a Scientist*. Amherst, NY: Prometheus Books, 1999.

Norberg, U. M. *Vertebrate Flight*. Berlin: Springer, 1990.

Norell, M. A., E. S. Gaffney, and L. Dingus. *Discovering Dinosaurs in the American Museum of Natural History*. New York: Alfred A. Knopf, 1995.

Ostrom, J. H. "*Archaeopteryx* and the Origin of Birds." *Biological Journal of the Linnaean Society* 8 (1976): 91–182.

———. "*Archaeopteryx* and the Origin of Flight." *Quarterly Review of Biology* 49 (1974): 27–47.

———. "Bird Flight: How Did It Begin?" *American Scientist* 67 (1979): 46–56.

———. "The Origin of Birds." *Annual Review of Earth and Planetary Sciences* 3 (1975): 55–77.

Padian, K., and L. M. Chiappe. "The Origin and Early Evolution of Birds." *Biological Reviews of the Cambridge Philosophical Society* 73 (1998): 1–42.

Parsons, K. M. *Drawing Out Leviathan: Dinosaurs and the Science Wars*. Bloomington: Indiana University Press, 2001.

———. *The Great Dinosaur Controversy: A Guide to the Debates*. Santa Barbara, CA: ABC-CLIO, 2004.

Paul, G. S. *Dinosaurs of the Air: The Evolution and Loss of Flight in Dinosaurs and Birds*. Baltimore: Johns Hopkins University Press, 2002.

———. *Predatory Dinosaurs of the World: A Complete Illustrated Guide*. New York: Simon and Schuster, 1988.

———. *The Princeton Field Guide to the Dinosaurs*. Princeton, NJ: Princeton University Press, 2010.

Romer, A. S. *Vertebrate Paleontology*. 3rd ed. Chicago: University of Chicago Press, 1966.

Rong, J.-Y., et al., eds. *Originations, Radiations and Biodiversity Changes*. Beijing: Science Press, 2006.

Ruse, M., and J. Travis. *Evolution: The First Four Billion Years*. Cambridge, MA: Harvard University Press, 2009.

Sereno, P. "The Evolution of the Dinosaurs." *Science* 284 (1999): 2137–47.

Shipman, P. *Taking Wing: Archaeopteryx and the Evolution of Bird Flight*. New York: Simon and Schuster, 1998.

Unwin, D. M. *The Pterosaurs: From Deep Time*. New York: Pi, 2005.

Weishampel, D. B., Dodson, P., and Osmólska, H., eds. *The Dinosauria*. 2nd ed. Berkeley: University of California Press, 2004.

Wellnhofer, P. *The Illustrated Encyclopedia of Pterosaurs*. New York: Crescent Books, 1991.

———. *Archaeopteryx: Icon of Evolution*. Trans. F. Haase. Munich: Verlag Dr. Friedrich Pfeil, 2009.

INDEX

· ·

Page numbers in *italics* indicate illustrations and captions. Page numbers followed by *t* indicate tables.